GROUNDWATER AND SOIL REMEDIATION
PROCESS DESIGN AND COST ESTIMATING OF PROVEN TECHNOLOGIES

Marve Hyman
R. Ryan Dupont

American Society of Civil Engineers
1801 Alexander Bell Drive
Reston, Virginia 20191-4400

Abstract: This book provides the theory and application of proven soil and groundwater remediation techniques with an emphasis on the integration of remediation technologies into a process design scheme. Chemical engineering techniques are applied to civil/environmental engineering and hydrogeologic surface and subsurface contamination problems so that feasible remediation solutions can be identified from competing alternatives. The steps in systematic process design and process control are described in detail for a variety of soil and groundwater remediation techniques. Applications of traditional cost estimating methods as well as those developed especially for remediation systems, based on what is often limited subsurface information, are also presented.

Library of Congress Cataloging-in-Publication Data

Hyman, Marve, 1930-
   Groundwater and soil remediation : process design and cost estimating of proven technologies / Marve Hyman and R. Ryan Dupont.
     p. cm.
   Includes bibliographical references and index.
   ISBN 0-7844-0427-5
   1. Groundwater—Purification. 2. Soil remediation. I. Dupont, R.Ryan. II Title.

TD426 .H95 2001
628.1'68--dc21   2001037323

Any statements expressed in these materials are those of the individual authors and do not necessarily represent the views of ASCE, which takes no responsibility for any statement made herein. No reference made in this publication to any specific method, product, process, or service constitutes or implies an endorsement, recommendation, or warranty thereof by ASCE. The materials are for general information only and do not represent a standard of ASCE, nor are they intended as a reference in purchase specifications, contracts, regulations, statutes, or any other legal document. ASCE makes no representation or warranty of any kind, whether express or implied, concerning the accuracy, completeness, suitability, or utility of any information, apparatus, product, or process discussed in this publication, and assumes no liability therefore. This information should not be used without first securing competent advice with respect to its suitability for any general or specific application. Anyone utilizing this information assumes all liability arising from such use, including but not limited to infringement of any patent or patents.

ASCE and American Society of Civil Engineers—Registered in U.S. Patent and Trademark Office.

Photocopies: Authorization to photocopy material for internal or personal use under circumstances not falling within the fair use provisions of the Copyright Act is granted by ASCE to libraries and other users registered with the Copyright Clearance Center (CCC) Transactional Reporting Service, provided that the base fee of $8.00 per article plus $.50 per page is paid directly to CCC, 222 Rosewood Drive, Danvers, MA 01923. The identification for ASCE Books is 0-7844-0427-5/01/ $8.00 + $.50 per page. Requests for special permission or bulk copying should be addressed to Permissions & Copyright Dept., ASCE.

Copyright © 2001 by the American Society of Civil Engineers. All Rights Reserved.
Library of Congress Catalog Card No: 2001037323. ISBN 0-7844-0427-5.
Manufactured in the United States of America.

Preface

Since the Love Canal incident, Americans have become increasingly concerned with the importance of cleaning up contaminated groundwater and soils. *Groundwater and Soil Remediation: Process Design and Cost Estimating of Proven Technologies* discussed technologies available for cleanup, provides a systematic approach to integrating process technologies into a design scheme, and describes cost estimating methods for remediation. This book fills a pressing need for technical descriptions of existing cleanup methods and how to apply them.

Unique features of this text include
- Application of chemical engineering techniques to civil engineering and hydrogeological subsurface cleanup problems
- Development of process design
- Use of available software for design and cost estimating
- Application of cost estimating methods for both treatment and associated work such as remedial excavation.

The approach use here is to identify and quantify process parameters with an eye toward making the most accurate cost estimates possible with limited subsurface information.

Engineers and scientists responsible for environmental restoration will profit from using this material. Emphasis is on proven techniques and how to select feasible remediation solutions. The steps for system design, control and process monitoring are described in detail, along with techniques for estimating costs for both capital investments and annual expenses.

Dedication

To the many students and colleagues that we have taught and learned so much from over the years.

LIST OF FIGURES

1-1	Criteria for detailed analysis of alternatives	30
2-1	Mass flow rate tabulation	43
2-2	P&ID piping symbols	48
2-3	An example of the use of pressure control	50
2-4	An example of the use of continuously modulating level control	51
2-5	An example of the use of on/off level control	53
2-6	An example of the use of temperature control (Used with permission, CSM Environmental Systems, Inc., Mountainside, New Jersey)	54
2-7	An example of the use of a flow control valve	55
2-8	A flow control loop	56
2-9	An example of a logic diagram	58
3-1	The solubility of metal hydroxides (A) and metal sulfides (B)	74
3-2	Pretreatment of hydrocarbon-contaminated groundwater to remove calcium and ferrous ions	77
3-3	Conventional hexavalent chromium reduction/precipitation scheme	78
3-4	An ultrafiltration system	80
3-5	Electrodialysis system	80
3-6	An RO and anion-exchange system process flow diagram	86
3-7	Examples of typical column adsorption breakthrough curves. A, Sharp breakthrough curve. B, Gradual breakthrough curve	89
3-8	Ion exchange system P&ID	92
3-9	Breakthrough curve extrapolation	98
4-1	A process flow diagram for an aqueous phase carbon adsorption system	121
4-2	P&ID for an aqueous phase carbon adsorption system	124
4-3	Experimental determination of saturated zones and MTZs	130
5-1	A flow scheme for an air stripper with polishing carbon	139
5-2	Plot of air stripper computations	148
5-3	Cross-section of an air sparging/vapor extraction system (US EPA, 1992e)	153
5-4	Example of a negative-pressure air stripping system	155
5-5	Preliminary P&ID for an air stripper	174
7-1	Biological treatment processes that can be used for contaminated soil, air, and groundwater remediation (adapted from US EPA, 1989)	204
7-2	A typical process configuration for a suspended growth bioreactor	218
7-3	Typical configuration of a fixed-film biological reactor	223
7-4	Schematic of a typical Raymond process using an infiltration gallery and recovery wells for the treatment of contaminated groundwater (US EPA, 1995a)	226
7-5	Components of the natural attenuation assessment approach	234
7-6	An example of Thiessen polygon network construction	235
7-7	Configuration of soil cores and associated geometry used for average bore hole contaminant concentrations and total mass calculations	251
7-8	Example groundwater monitoring network applied at a natural attenuation site for both compliance and natural attenuation process monitoring	260
7-9	Schematic of a typical land treatment system with optional liner and leachate collection system, indicating the typical sampling type and locations	264

7-10 Schematic of a typical biomound soil treatment system using air extraction or injection to aerate the piled soil, drip irrigation for soil water management, and an impervious cover to minimize volatile contaminant release 272
7-11 Oxygen utilization over time in soil pile 5 following shutdown of the air injection system at the PCP/TPH-contaminated soil biomound field site 279
7-12 Typical schematic of a slurry reactor for contaminated soil remediation 281
7-13 Bioventing system design and performance evaluation protocol 289
7-14 Typical soil respiration gas data collected during a field in situ respiration test 292
7-15 Typical first-order soil respiration gas data collected during a field in situ respiration test. The first-order respiration rate is less than 0.015/hr. The curved lines are 95% confidence intervals for the slope of the relationship 294
7-16 Typical regression results for linear regression analysis of field respiration data for bioventing systems 294
7-17 Zero-order linear regression results for hypothetical in situ respiration rate data ... 297
7-17 Residuals plot for zero-order linear regression results for hypothetical in situ respiration rate data 297
7-19 First-order linear regression results for hypothetical in situ respiration rate data 298
7-20 Residuals plot for first-order linear regression results for hypothetical in situ respiration rate data 298
7-21 Respiration rate/contaminant concentration relationships generated in laboratory-scale microcosm studies conducted with JP-4 contaminated soil from Hill AFB, Utah (from Ravipaty, 1996) 303
7-22 Theoretical air, water, and product flow occurring from the operation of a bioslurping system. Modified from the US Air Force (1995) 305
7-23 Configuration and product flow occurring in a bioslurping well operated in a conventional skimming mode and in a bioslurping (vacuum enhanced) mode. From the US Air Force (1995) 307
7-24 Configuration and product/air flow occurring in a bioslurping well operated in a drawdown simulation mode. From the US Air Force (1995) 308
7-25 Schematic of typical multilevel vapor probe installation in a single bore hole. From the US Air Force (1995) 309
8-1 Typical extraction well schematic for SVE (US EPA, 1991d) 335
8-2 Flow schematic for liquid-ring compressor application (Courtesy of the Nash Engineering Company) 339
8-3 Air flowrate versus soil air permeability and applied vacuum (US EPA, 1991d, Appendix E) 350
9-1 Schematic of a rotary kiln system 374
9-2 Circulating fluidized bed combustor 376
9-3 Relationship among plume opacity, stack width, and plume mass concentration .. 406
9-4 Scrubber blower horsepower 408
9-5 Contaminant concentration versus a function of desorption temperature 414
9-6 Thermal desorption costs 419
10-1 Aqueous soil washing process (From the US EPA 1990b) 437
10-2 Example of a soil washing scheme 442
10-3 Mass balance for screens 446
10-4 Mass balance for a spiral classifier and a flat-deck screen 446
10-5 Mass balance for attrition mills and hydrocyclones 447

10-6 Mass balance for flotation cells and belt filter..447
10-7 Mass balance for hydrocyclone and thickener filter system.....................................448
12-1 Unit costs of fixed-tubesheets heat exchangers (From *Chemical Engineering*,
July 27, 1981)..481

LIST OF TABLES

1-1	General response actions and associated technology types and process options	5
1-2	Treatment technologies screening matrix	6
1-3	Remediation technologies: Application and cost guide for contaminant removal	11
1-4	Remediation technologies: Application and cost guide for contaminant destruction	16
1-5	Groundwater treatment options for on-site remediation (Source: Long, G.M. Clean up hydrocarbon contamination effectively. *Chemical Engineering Progress*, May 1993 pp. 58-67). (Reproduced with permission of the American Institute of Chemical Engineers. Copyright © 1993 AIChE. All rights reserved)	22
1-6	Unsaturated soil treatment options for on-site remediation (Source: long, G.M. Clean up hydrocarbon contamination effectively. *Chemical Engineerin Progress*, May 1993, pp. 58-67). (Reproduced with permission of the American Institute of Chemical Engineers. Copyright © 1993 AIChE. All rights reserved)	22
1-7	Remediation technologies matrix (Source: *Chemical Engineering*. May 1997 pp. 104-112)	23
1-8	Groundwater treatment (Source: Marve Hyman, M, Bagaasen, L. Select a site cleanup technology. *Chemical Engineering Progress*, August 1997, pp. 22-43). Reproduced with permission of the American Institute of Chemical Engineers. Copyright© 1997 AIChE. All rights reserved)	27
1-9	Soil treatment (Source: Marve Hyman, M, Bagaasen, L. Select a site cleanup technology. *Chemical Engineering Progress*, August 1997, pp. 22-43). (Reproduced with permission of the American Institute of Chemical Engineers. Copyright© 1997 AIChE. All rights reserved)	28
2-1	Example contingency plans from the Weldon Springs project	41
3-1	Capital costs for RO treatment of 1,260 gal/min of extracted groundwater. (The California Department of Health Services, 1988)	101
3-2	Operation and maintenance costs for RO treatment of 1,260 gal/min of extracted groundwater (The California Department of Health Services, 1988)	102
3-3	Cost comparison of ion exchange (IX) and chemical precipitation for chromate removal from contaminated groundwater	103
3-4	Unit cost comparison of ion exchange (IX) and chemical precipitation for chromate removal from contaminated groundwater	104
4-1	Activated carbon x/m values	117
4-2	Influent volumes for activated carbon example	117
4-3	Influent volumes with prestripping	119
4-4	Dual-bed carbon canisters	134
5-1	Constant for correlating H with temperature	143
5-2	Parameter values for correlating H with temperature. (From *Journal AWWA*, Vol. 72, No. 12 (December 1980), by permission. Copyright © 2000, American Water Works Association)	143
5-3	Conditions for biofilters	160
5-4	Total stripping costs ($/1,000 gal) for 99% VOC removal	180
5-5	Vapor treatment costs	181

5-6	Volatility rankings	182
5-7	Blower ratings for air sparging	184
6-1	Relative oxidative power of commonly used oxidants	190
6-2	Absolute oxidation potentials of commonly used oxidants	190
6-3	Key UV oxidation system vendor experience	196
6-4	The effectiveness of UV/hydrogen peroxide, ozone oxidation technologies in the treatment of contaminants	197
6-5	Costs of wet air oxidation systems	199
6-6	Costs of ozone generators	201
7-1	Biodegradable chemical classes of concern that have been shown to be susceptible to biodegradation. (Numbers indicate sources of biodegradation information)	206
7-2	Test for determining the feasibility of in situ bioremediation (Reprinted from The Hazardous Waste Consultant. *Evaluating the Feasibility of In Situ Bioremediation*. Jan/Feb 1992; pp. 1.16-1.20. With permission from Elsevier Science)	214
7-3	Total mass and CoM calculations for benzene from sample monitoring network for which Thiessen areas have been determined	238
7-4	Possible changes in contaminant mass and mass center coordinates for a contaminant plume and the corresponding interpretation of these changes relative to plume mobility and persistence	240
7-5	Summary results for contaminant concentrations using Equation 7-43 with input data as presented in the example problem	248
7-6	Summary of estimated total residual contaminant mass based on residual product volume estimates and dissolved plume mass measured at a field site	253
7-7	Potential aromatic and aliphatic hydrocarbon assimilative capacity relationships for TEAs of importance at contaminated sites	257
7-8	General physical and operational characteristics of typical biomound systems	275
7-9	General design and application considerations appropriate for conventional versus bioventing SVE systems	287
7-10	Potential oxygen transfer rates in various soils	291
7-12	Example treatability study and full-scale bioventing system respiration rates reported from various sources	296
7-12	Costs for in situ and ex situ groundwater biotreatment (DuTeaux, 1996)	316
7-13	Estimated capital costs of an RBC system (California Department of Health Services, 1988)	317
7-14	Estimated annual operation and maintenance costs of an RBC system (California Department of Health Services, 1988)	317
7-15	Estimated capital costs for PACT treatment of 5 L/sec (75 gal/min) of extracted groundwater (California Department of Health Services, 1988)	318
7-16	Estimated annual operation and maintenance costs for PACT treatment of 5 L/sec (75 gal/min) of extracted groundwater (California Department of Health Services, 1988)	318
7-17	Cost of ex situ soil treatment based on case histories (DuTeaux, 1996)	320
7-18	Biological treatment system variables that must be controlled to ensure uninhibited biodegradation rates	328

7-19	Applicability of in situ versus ex situ technologies for contaminated site remediation	329
8-1	Applicability of soil vapor treatment methods (From the US EPA, 1989, "Soil Vapor Extraction, VOC Control Technology Assessment," EPA/450/4-89/017)	342
8-2	Multiplying factor applied to air rate for extraction wells (US EPA, 1991d, Appendix E)	351
8-3	Example of air flow rates (US EPA, 1991d, Appendix E)	351
8-4	Typical unit costs of SVE components (US EPA, 1991a)	357
8-5	Typical unit costs for SVE vapor treatment (US EPA, 1991a)	359
9-1	Advantage/disadvantage analysis of each incinerator type	368
9-2	Effectiveness of incinerators for various contaminants	370
9-3	LELs (Reprinted with permission from NFPA 325. "Fire Hazard Properties of Flammable Liquids, Gases, and Volatile Solids," National Fire Protection Association, Quincy, Massachusetts, 1994. This reprinted material is not the complete and official position of the NFPA on the referenced subject, which is represented only by the standard in its entirety)	373
9-4	Advantages and disadvantages of collection devices	383
9-5	Efficiency of abatement devices. (From Niessen, 1978, courtesy of Marcel Dekker, Inc.)	386
9-6	Example specific heats of exhaust gases (From Reynolds, J.P., Dupont, R.R., Theodore, L. 1991, "Hazardous Waste Incineration Calculations — Problems and Software," John Wiley & Sons, Inc. Reprinted by permission)	396
9-7	Example of correlating plume opacity to grain loading	405
9-8	Thermal desorption treatment costs	418
9-9	Thermal desorption costs from case histories	419
10-1	Contaminant removal efficiencies with soil washing	427
10-2	Contaminant removal efficiencies and residual concentrations	428
10-3	Surfactant characteristics	431
10-4	Physical and chemical characteristics that affect soil washing	433
10-5	Soil-contaminant technology matrix	434
10-6	Summary of process parameters and reagents for various contaminant groups	436
10-7	Particle separation techniques (From Eagle, M.C., et al. "Soil Washing for Volume Reduction of Radioactively Contaminated Soils," *Remediation*, Summer 1993. Copyright 1993 John Wiley & Sons, Inc. Reprinted by permission of John Wiley & Sons, Inc.)	438
10-8	Dewatering techniques (From Eagle, M.C., et al. "Soil Washing for Volume Reduction of Radioactively Contaminated Soils," *Remediation*, Summer 1993. Copyright 1993 John Wiley & Sons, Inc. Reprinted by permission of John Wiley & Sons, Inc.)	438
10-9	Generic soil washing major equipment list	443
10-10	Generic soil washing material balance	449
10-11	Laboratory tests for soil washing	454
11-1	Effect of waste components on stabilization/solidification	468
11-2	Costs of fixation with cement, using mobile equipment (From *Chemical Engineering*, February 1990)	470
11-3	Estimated capital costs for stabilization/solidification	472

11-3	Estimated stabilization/solidification operation and maintenance costs (for 5 yr)	473
12-1	Example of a cost estimate using Hand factors	482
12-2	Example of a detailed factored cost estimate	483
12-3	Example of typical closure and post closure costs (From Cressman, K.R. "Cost Estimating of the Closure/Post-Closure Phase," *Remediation*, Summer 1991. Copyright John Wiley & Sons, Inc., reprinted by permission of John Wiley & Sons, Inc.)	495
12-4	Selected present worth factors	496
12-5	Example of comparing present values for three alternatives	497
12-6	Example of a present value summary	498

TABLE OF CONTENTS

Chapter 1   **The Basis for Remediation Process Design and Cost Estimating............1**
       1.1   The Importance of Cost Analysis..1
       1.2   Natural Attenuation..3
       1.3   Selecting Among Competing Remediation Methods 4
              1.3.1   Listing the process options 4
              1.3.2   Comparing process options 10
              1.3.3   Defining and evaluating alternative treatment trains 28
       1.4   The Approach to Process Design and Cost Estimating 32

Chapter 2   **Process and Conceptual Design of Remediation Systems 35**
       2.1   Basic Principles...36
       2.2   Feasibility Studies and Work Plans 37
              2.2.1   Feasibility Study Alternatives 37
              2.2.2   Work Plans, Corrective Action Plans, Remedial Action
                       Plans 38
              2.2.3   Informal Studies, CERCLA Studies And Records Of
                       Decision, RCRA Studies 38
              2.2.4   The Observational Approach 39
       2.3   Treatability Studies 40
       2.4   Process Flow Diagram...42
              2.4.1   Main Parameters and Mass Balance 42
              2.4.2   Energy Balance 43
              2.4.3   Sizing and Rating of Equipment 44
       2.5   Site Plan and Preliminary Plot Plan 44
       2.6   P&ID and Sequence of Operations..46
              2.6.1   P&ID Development..46
              2.6.2   Pressure Instrumentation 50
              2.6.3   Liquid Level Instrumentation...51
              2.6.4   Temperature Instrumentation 53
              2.6.5   Flow Instrumentation..55
              2.6.6   Analysis Instrumentation..57
              2.6.7   Sequence of Operations Development.......................................57
       2.7   Logic Diagrams...57
       2.8   Computerized Controls and Process Monitoring 59
              2.8.1   Computer Functions 60
                       2.8.1.1   Adapting Computers 60
                       2.8.1.2   Data Manipulation...60
                       2.8.1.3   Adapting Computers to On/Off Control Function.60
                       2.8.1.4   Graphic Displays 61
                       2.8.1.5   Automatic Restart, Supervisory Control, and
                                   Combination Systems......................................61

		2.8.2	Remote Monitoring..62
			2.8.2.1 Computer-Based Systems 62
			2.8.2.2 Phone Dialers..62
			2.8.2.3 Computer/Fax 63
	2.9	Design Basis, Tradeoff Analysis, and Preliminary Specifications.......63	
		2.9.1	Preliminary Specifications..65
		2.9.2	Conceptual Design Report 66

Chapter 3   Metals Removal from Groundwater 69

    3.1   Basic Principles...70
           3.1.1   Chemical Precipitation Basics..70
           3.1.2   Membrane Separation Basics for Dissolved Ions 71
           3.1.3   Ion Exchange Basics...71
           3.1.4   Adsorption Basics...72
           3.1.5   Evaporation Basics 73
    3.2   Chemical Precipitation..73
           3.2.1   Alkaline Precipitation 73
           3.2.2   Sulfide Precipitation 75
           3.2.3   Precipitation with Iron 75
           3.2.4   Precipitation Examples 75
                    3.2.4.1   Iron, Manganese and Calcium Removal 75
                    3.2.4.2   Chromium 76
                    3.2.4.3   Mercury 78
                    3.2.4.4   Arsenic 79
           3.2.5   Alternatives to Conventional Clarification 79
    3.3   Membrane Separation for Dissolved Ions 79
    3.4   Ion Exchange..81
           3.4.1   Ion Exchange for Nitrates And Chromate 81
           3.4.2   Ion Exchange for Radionuclides 82
    3.5   Adsorption..82
    3.6   Forced Evaporation..83
    3.7   Main System Design Parameters 84
           3.7.1   Sizing and Rating of Major Equipment 84
           3.7.2   Conceptual and Process Design 91
           3.7.3   Controls 93
           3.7.4   Utilities Requirements 94
    3.8   Treatability Studies for Metal Removal...94
           3.8.1   Treatability Studies for Precipitation and Prediction of
                      Treated Effluent Concentrations 94
           3.8.2   Treatability Studies for Reverse Osmosis..................................96
           3.8.3   Treatability Studies for Ion Exchange and Adsorbent
                      Systems..97
           3.8.4   Treatability Studies for Evaporation 99
    3.9   Cost Estimating for Metals Removal..99
    3.10 Summary of Important Points for Metals Removal.....................106

Chapter 4	Groundwater Remediation Using Carbon Adsorption 109	
	4.1	Basic Principles of Carbon Adsorption 109
	4.2	Adsorption Isotherms 111
	4.3	Methods of Determining Adsorptive Capacity 113
	4.4	Breakthrough Curves..113
	4.5	Sizing of Carbon Beds and Duration of Bed Life 114
	4.6	Arrangements and Performance of Organic Adsorption Systems 118
		4.6.1 Prestripping 118
		4.6.2 Prefiltering and Preventing Overpressure.............................118
		4.6.3 Improving Performance with Three-Stage Adsorption 119
		4.6.4 Presoaking and Backwashing..120
		4.6.5 Lower Explosive Limit (LEL) monitoring for Breakthrough...120
	4.7	Main System Design Parameters 122
		4.7.1 Concept and Process Design 125
		4.7.2 Sizing and Rating of Major Equipment 127
		4.7.3 Controls 127
		4.7.4 Utilities Requirements 127
	4.8	Aqueous Phase Adsorption Treatability Studies................................128
	4.9	Cost Estimating 131
	4.10	Summary of Important Points for Carbon Adsorption 134
Chapter 5	Stripping of Groundwater..137	
	5.1	Basic Principles of Stripping 137
		5.1.1 Use of Polishing Carbon..138
		5.1.2 The Design Problem 140
	5.2	Packed Strippers..144
		5.2.1 Packing Depth and A/W (or G/L) Ratio................................144
		5.2.2 Packed Strippers – Pressure Drop and Cross-Sectional Area 148
		5.2.3 Packed Strippers – Computer Applications..........................149
	5.3	Alternatives to Packed Towers...150
		5.3.1 Tray Designs 150
		5.3.2 Aeration Chambers 151
		5.3.3 Cooling Towers Used as Air Strippers 151
		5.3.4 In Situ Air Stripping (In-Well Stripping and Air Sparging) 152
	5.4	Blower Arrangements and Mist Separation 154
	5.5	Turndown and Liquid Distribution 156
	5.6	Recycled Strippers 157
	5.7	Heated Strippers..158
	5.8	Emission Abatement...160
		5.8.1 Carbon Adsorption..161
		5.8.2 Regenerating Vapor-Phase Activated Carbon......................163
		5.8.3 Direct Thermal Oxidizers 164
		5.8.4 Catalytic Oxidizers 165
		5.8.5 Auxiliary Fuel Consumption and Heat Exchange................166

		5.9	Main System Design Parameters 169
			5.9.1 Concept and Process Design 169
			5.9.2 Sizing and Rating of Major Equipment....................................175
			5.9.3 Controls..177
			5.9.4 Utilities Requirements...178
		5.10	Treatability Studies for Groundwater Stripping.......................................179
		5.11	Cost Estimating for Groundwater Stripping...179
			5.11.1 Equipment Costs...179
			5.11.2 Operating Costs and Total Costs..180
			5.11.3 Emission Abatement Costs 180
			5.11.4 Software for Stripping Process Design and Cost Estimating...182
		5.12	Summary of Important Points for Groundwater Stripping................184

Chapter 6	Aqueous Chemical Oxidation 189
	6.1 Basic Principles..189
	6.1.1 Ranking of Oxidants and UV Oxidation Power Consumption 189
	6.1.2 Ultraviolet Light...189
	6.1.3 Emerging Technology Using Electrochemical Oxidation....190
	6.2 Wet Air and Supercritical Water Oxidation 191
	6.3 Fenton's Reagent 193
	6.4 UV Light with Oxidants 193
	6.5 Main System Design Parameters 195
	6.6 Treatability Studies for Aqueous Oxidation..198
	6.7 Costs for Aqueous Oxidation 199
	6.8 Summary of Important Points for Aqueous Chemical Oxidation 201

Chapter 7	Bioremediation Systems 203
	7.1 Basic Principles..203
	7.1.1 Microbial Metabolism 205
	7.1.2 System Environmental Requirements 207
	7.1.2.1 Microbial Populations 207
	7.1.2.2 Oxygen..207
	7.1.2.3 Soil Water 209
	7.1.2.4 pH 209
	7.1.2.5 Nutrients..209
	7.1.2.6 Temperature 211
	7.1.2.7 Toxicants in Waste 212
	7.1.3 In Situ Versus Ex Situ Treatment 212
	7.1.4 Bioaugmentation Versus Bioacclimation 215
	7.2 Aqueous Phase Treatment 217
	7.2.1 Ex Situ Treatment 217
	7.2.1.1 Suspended Growth Systems.......................................217
	7.2.1.1.1 Reactor configurations.......................219
	7.2.1.1.2 Design parameters..............................221
	7.2.1.2 Fixed-Film Systems..222

| | | 7.2.1.2.1 | Reactor configurations..........................223 |
| | | 7.2.1.2.2 | Denitrification reactors ....224 |

    7.2.2 In Situ Treatment 226
        7.2.2.1 Raymond Process 226
            7.2.2.1.1 Use of alternatives to air to supply electron acceptors..................................228
            7.2.2.1.2 Use of alternative injection systems...230
            7.2.2.1.3 Process limitations 231
            7.2.2.1.4 In situ biodenitrification.....................232
        7.2.2.2 Natural Attenuation 232
            7.2.2.2.1 Principles of operation 232
            7.2.2.2.2 Process design principles.....................233
            7.2.2.2.3 Determination of steady-state plume conditions 233
            7.2.2.2.4 Estimation of contaminant degradation rates 239
            7.2.2.2.5 Estimation of source mass and source lifetime.......................................249
            7.2.2.2.6 Predicting the long-term behavior of the dissolved plume 255
            7.2.2.2.7 Decision making regarding natural attenuation 256
            7.2.2.2.8 Long-term monitoring 259

7.3 Solid Phase Biological Treatment...261
    7.3.1 Ex Situ Treatment 261
        7.3.1.1 Land Treatment Systems......................................261
            7.3.1.1.1 Application-limited contaminant loading 262
            7.3.1.1.2 Rate-limiting contaminant loading......263
            7.3.1.1.3 Capacity-limiting contaminant loading 263
            7.3.1.1.4 Process configuration...........................264
            7.3.1.1.5 Process design 265
            7.3.1.1.6 Process operations.................................267
            7.3.1.1.7 Process design calculations 267
        7.3.1.2 Soil Piles 270
            7.3.1.2.1 Process configuration...........................271
            7.3.1.2.2 Process design 272
            7.3.1.2.3 Process operation 273
            7.3.1.2.4 Composting...274
            7.3.1.2.5 Process design case study 276
        7.3.1.3 Slurry Reactors 280
            7.3.1.3.1 Process design 282
            7.3.1.3.2 Process operation 283
    7.3.2 In Situ Treatment 285
        7.3.2.1 Bioventing..285
            7.3.2.1.1 Process design 287

			7.3.2.1.2	Process operation 301

- 7.3.2.2 Bioslurping.. 304
  - 7.3.2.2.1 Process design 304
  - 7.3.2.2.2 Process operation 308
- 7.4 Treatability Studies for Bioremediation Systems............................. 310
  - 7.4.1 Treatability Studies Applicable to Aqueous Phase Treatment 312
  - 7.4.2 Treatability Studies Applicable to Solid Phase Systems 312
    - 7.4.2.1 Land Treatment Systems.................................... 312
    - 7.4.2.2 Soil Piles 313
    - 7.4.2.3 Slurry-Phase Reactors 314
    - 7.4.2.4 Bioventing.. 314
    - 7.4.2.5 Bioslurping... 314
- 7.5 Cost-Estimating for Bioremediation Systems 315
  - 7.5.1 Costs for Aqueous Phase Treatment..................................... 315
    - 7.5.1.1 Costs from Case Histories................................ 315
    - 7.5.1.2 Computer programs for Process Design and Cost Estimating 316
  - 7.5.2 Costs for Solid Phase Treatment 320
    - 7.5.2.1 Land Treatment Systems.................................... 321
    - 7.5.2.2 Soil Piles 322
    - 7.5.2.3 Slurry-Phase Reactors 322
    - 7.5.2.4 Bioventing.. 324
    - 7.5.2.5 Bioslurping... 327
- 7.6 Summary of Important Points for Bioremediation............................. 327

Chapter 8   Soil Venting... 333
- 8.1 Basic Principles of Soil Venting 334
- 8.2 Inducing Vacuum.. 337
  - 8.2.1 Vacuum Blowers 337
  - 8.2.2 Internal Combustion Engines (ICEs)..................................... 340
  - 8.2.3 Passive Soil Venting 341
- 8.3 Vapor Treatment and Discharge 341
  - 8.3.1 Adsorption.. 343
  - 8.3.2 Oxidizers 344
- 8.4 Main System Design Parameters 346
  - 8.4.1 Pneumatic testing 346
  - 8.4.2 Radius of Influence of Extraction Wells and Soil Air Permeability... 347
  - 8.4.3 Volumetric Air Flow and Contaminant Mass Removal Rate... 352
  - 8.4.4 Ventilation wells 354
- 8.5 Treatability Studies for Soil Venting 354
- 8.6 Cost Estimating for Soil Venting 355
  - 8.6.1 Utilities Costs 361
  - 8.6.2 Carbon Adsorption Costs 361
  - 8.6.3 Software for Soil Venting Process Design and Cost Estimating.. 362

		8.7	Summary of Important Points for Soil Venting 363
Chapter 9		**Thermal Treatment for Soils and Sludges** **367**
	9.1	Basic Principles 367
		9.1.1	Incineration Basics 367
		9.1.2	Low-Temperature Thermal Desorption Basics 370
		9.1.3	Heat Recovery 373
	9.2	Incinerators 373
		9.2.1	Rotary kilns 373
			9.2.1.1 Features 374
			9.2.1.2 Oxygen Lancing and Oxygen Enrichment ... 375
		9.2.2	Fluidized CBCs 375
		9.2.3	Infrared Furnace Systems 378
	9.3	Thermal Desorbers 379
	9.4	Handling of Feed and of Treated Soils 380
	9.5	Air Pollution Control 380
		9.5.1	Use of Afterburners (Thermal Oxidizers) 380
		9.5.2	Recovery of Organic Fluids from Indirect-Fired Desorbers 381
		9.5.3	Abatement of Particulate Emissions and Acid Gases 382
		9.5.4	Emissions of NOx 388
		9.5.5	CO Emissions 389
	9.6	Main System Design Parameters for Thermal Treatment 390
		9.6.1	Characterization of the "Waste" for Thermal Treatment 391
		9.6.2	Vapor Pressure Considerations for Thermal Desorbers 392
		9.6.3	Examples of Design Calculations 395
		9.6.4	Contaminant Destruction Efficiency and Emission Limitations 401
		9.6.5	Limitations on Particulate Emissions and Plume Opacity Correlations 402
			9.6.5.1 Plume Opacity 403
			9.6.5.2 Emissions Source Testing 406
		9.6.6	Baghouse Design Parameters 406
		9.6.7	Wet Scrubber Power Requirements 407
		9.6.8	Design of Vertical Packed Acid Gas Scrubbers 409
		9.6.9	Venturi Scrubber Design Parameters 411
	9.7	Treatability Studies and Trial Burns 412
		9.7.1	Testing Thermal Desorption from Soils 413
		9.7.2	Trial Burns 414
	9.8	Cost Estimating for Thermal Soil Treatment 415
		9.8.1	Incineration Costs 415
			9.8.1.1 Incineration Costs per Ton 415
			9.8.1.2 Computerized Cost Estimating for Incineration ... 416
		9.8.2	Desorption Costs 417
			9.8.2.1 Desorption Costs per Ton or per Cubic Yard 417
			9.8.2.2 Computerized Equipment Sizing and Cost Estimating for Desorption 420

		9.8.3	Total Project Costs for Ex Situ Soil Remediation.................421
	9.9		Summary of Important Points for Thermal Desorption....................422

Chapter 10 Soil Washing...427
    10.1 Basic Principles of Soil Washing..428
    10.2 In Situ Soil Flushing 430
    10.3 Soil Washing and Solvent Extraction 432
           10.3.1 Aqueous Soil Washing for Particle Size Separation............435
           10.3.2 Solvent Extraction for Removing Organic Contaminants....437
    10.4 Main System Design Parameters for Soil Washing.........................440
           10.4.1 Conceptual Designs..440
           10.4.2 Mass Balances...441
           10.4.3 Treatment of Wash Water 448
    10.5 Treatability Studies for Soil Washing...453
    10.6 Cost Estimating for Soil Washing...456
    10.7 Summary of Important Points for Soil Washing..............................458

Chapter 11 Stabilization and Solidification 461
    11.1 Basic Principles for Stabilization and Solidification.........................461
    11.2 In Situ Applications and Area Mixing..462
    11.3 Microencapsulation..463
           11.3.1 Cement/Pozzolanic (Silicaceous) Solidifiers........................463
           11.3.2 Thermoplastic Agents...465
    11.4 Silicate Sorbents...466
    11.5 Main System Design Parameters 466
    11.6 Treatability Studies for Stabilization and Solidification 467
    11.7 Cost Estimating for Stabilization and Solidification........................470
    11.8 Summary of Important Points for Stabilization and Solidification...475

Chapter 12 Cost Estimating and Life Cycle Analysis..477
    12.1 Basic Principles..477
    12.2 Investment Costs..478
           12.2.1 Preliminary Estimates for Investment Cost...........................478
                  12.2.1.1 Ratioing Costs with Exponents.............................478
                  12.2.1.2 Factoring Costs of Principal Equipment 480
           12.2.2 Definitive Estimating of Investment Cost 483
                  12.2.2.1 Associated Costs..484
                  12.2.2.2 Engineering Design Costs....................................484
    12.3 Estimating Annual Expenses...485
           12.3.1 Utilities Consumption...485
           12.3.2 Operating Labor and Overhead..488
           12.3.3 Maintenance Expense 489
           12.3.4 Chemicals, Adsorbents, and Supplies....................................489
           12.3.5 Property Taxes and Insurance 490
           12.3.6 Monitoring and Reporting..490
           12.3.7 Other Direct Costs 491
    12.4 Computer Applications to Cost Estimating......................................492

  12.5 Life Cycle Analysis 493
     12.5.1 Investment, Expense, Closure and Post-Closure Costs 494
     12.5.2 Present value Factors 495
  12.6 Summary of Important Points for Cost Estimating.............................497
  Appendix 12-A Investment Costs and Yearly Expense Example 500

References 505

Index 525

Chapter 1

The Basis for Remediation Process Design and Cost Estimating

An aspect of remediation technology that is exciting to environmental and process engineers is the proliferation of new technologies that have been developed over the past decade. The site remediation manager now can choose from a wide variety of options, including both proven and emerging technologies.

Although this book emphasizes proven treatment processes, emerging technology is described in some chapters by way of example. The book does not deal with site assessment and characterization (except as applicable to natural attenuation as a remediation technique), nor with the determination of cleanup goals. The basis for design described here assumes prior establishment of the nature and extent of contamination and the cleanup target concentration for each contaminant of concern. In addition, the depth of contamination and subsurface soil stratigraphy must be known for the design of soil treatment systems, whereas the flow rate of groundwater that can be extracted from wells must be known for groundwater pump-and-treat technologies.

1.1 The Importance Of Cost Analyses

Accompanying the exciting emergence of new technologies is the development of software that directly aids in treatment process design and cost estimating. Most notable are COMPOSER GOLD, marketed by Building Systems Design (Atlanta, Georgia) and RACER/ENVEST^{TM}, marketed by Talisman Partners Ltd. (Englewood, Colorado). Cost will always be a major factor in selecting among applicable treatment processes. It is essential to the costing process that the size or rating of the main equipment be determined. The cost-estimating models in COMPOSER GOLD and RACER/ENVEST have default values for design parameters that provide essential data when the size or rating is unknown. The result is a preliminary process design that furthers the cleanup project, aids in feasibility analyses, and leads to a cost estimate from which to compare process alternatives.

Other information on various ENVEST models is provided in the cost estimating sections of Chapters 3 through 11. The costs for each process are based on data published since 1989. Such data are currently valid for the estimates needed in feasibility studies. The

Plant Cost Index published monthly in *Chemical Engineering* (McGraw-Hill, New York) indicates that costs have not increased by more than 10% since 1989.

Cost analyses in some form are needed very early when selecting among feasible remediation alternatives. The US Environmental Protection Agency (EPA) (1988) indicates that initial cost comparisons can be based on relative costs. Relative costs of groundwater treatment are usually expressed in terms of cost per volume treated (e.g., $/1,000 L), whereas costs for soil treatment are expressed in terms of cost per ton or per unit volume treated. In this manner, technologies that can be implemented and that are effective may be screened out early in a feasibility study, based on relative treatment costs.

A complete cost estimate is needed for each alternative selected from the initial screening phase as part of a formal feasibility study. This complete cost estimate can be accomplished either with traditional estimating methods or with commercially available remediation cost-estimating computer programs.

Traditional estimating methods are detailed in Chapter 12 and include the following (in decreasing order of accuracy): (a) definitive estimates, (b) factored estimates, and (c) ratios that use historical capacity or throughput data. Definitive estimates are generally applicable only after a considerable amount of design has been accomplished. Use of factored estimates or ratios provides a short cut to arriving at capital investment requirements.

Factored estimates apply a factor to the cost of principal equipment. The factor accounts for installation of equipment and for minor equipment such as common instruments, piping, supports, etc. Major equipment includes towers, vessels, heat exchangers, pumps, etc. As described by Lang (1948), the total major equipment cost is multiplied by a factor ranging from 3.1 to 4.74. The value of the factor depends on whether solids handling or fluids handling is involved.

A more accurate factoring method that applies an individual factor to each type of major equipment is described by Hand (1958). A list of Hand factors ranging from 2 to 4 is given in Chapter 12.

The use of cost ratios applies when historical data are available for the cost and capacity of an old treatment process or equipment that is similar to the new treatment or equipment being considered. Equation 1-1 can be applied:

$$Cost_n/Cost_o = (Index_n/Index_o)(Size_n/Size_o)^{exp} \qquad (1\text{-}1)$$

In this equation n indicates new and o indicates old, Index refers to a cost index that takes inflation into account, and exp is an exponent. As detailed in Chapter 12, the exponent for an entire treatment plant is in the range of 0.6 to 0.8, and 12 exponents ranging from

0.5 to 0.8 are listed for major types of equipment. A more detailed list is given in Table 25 through 49 in Perry and Green (1984).

1.2 Natural Attenuation

If passive remediation is selected for cleanup of a site, detailed consideration must be given to quantifying the nature, extent, and magnitude of the source of soil and groundwater contamination and to describing contaminant assimilation in these impacted environments by natural processes that take place without human intervention. One well developed site management approach that is described in detail in Chapter 7 is termed *natural attenuation*. Natural attenuation, as used in this text, focuses on the assessment of groundwater impacts. The process of site assessment and data reduction and interpretation is described, emphasizing the quantification of the capacity of a given aquifer system to assimilate groundwater contaminants through physical, chemical and/or biological means. Although numerous protocols describe approaches for collecting and analyzing data to verify that natural attenuation processes are taking place at a given site (Wiedemeier, et al. 1996; Wilson et al., 1994), the connection of these data with decisions regarding source removal activities, or with estimates of source lifetime, has generally not been presented in the literature.

Implementation of natural attenuation concepts from data collection through source removal and source lifetime considerations has been carried out for the US EPA and the US Air Force (Dupont et al., 1996, 1998) and forms the basis for the natural attenuation assessment process described in Chapter 7. This process involves: 1) determination of whether steady-state contaminant plume conditions exist at the site; 2) estimation of contaminant degradation rates; 3) estimation of the source mass term; 4) estimation of the source lifetime; 5) prediction of long-term plume behavior with and without source removal; 6) decision making regarding the use of natural attenuation and the impact and desirability of source removal at a given site; and 7) development of a long-term monitoring strategy if natural attenuation is selected for plume management.

Once steady-state plume conditions have been identified (i.e., when the rate of contaminant release from the source area is equivalent to the rate of contaminant assimilation by biotic and abiotic processes taking place within an aquifer), issues remain regarding the length of time required to provide complete assimilation of contaminant mass within the source area. Quantification of this source lifetime is critical because it defines the ultimate duration of monitoring required to ensure the protection of public health and the environment and to demonstrate the success of a natural attenuation remedy at a given site. This source lifetime directly impacts costs, and drives the decision regarding the need for source control and remediation.

If it is found that natural attenuation processes have indeed provided plume containment and if the mass of source material is small and is weathered rapidly under site conditions, it may not be necessary to provide additional source treatment to produce a remediation time that is acceptable to the regulatory community and/or a remediation cost that is acceptable to the site owner. On the other hand, if because of a large mass of contaminant in the source area, the dissolved plume will exist at the site for many decades, a decision may be made by the owner, or may be required by the regulatory agency, to invest in active source treatment to reduce the required monitoring period and long-term liability that would result if natural attenuation is used as a sole remedy.

The natural attenuation approach described in Chapter 7 can be used to make management decisions regarding the appropriateness of source treatment and the effect of such actions on the projected lifetime of contamination at a site. If source treatment is found to be required for the cost-effective management of the site over its entire life cycle, then various soil and groundwater remediation techniques described throughout this text should be considered for implementation. The source remediation technology that is optimal for a given site must be selected based on site-specific soil and waste constraints.

1.3 Selecting Among Competing Remediation Methods

1.3.1 Listing the Process Options

Selection of the remediation method to be used in the event that human intervention is to be implemented starts with the consideration of applicable process options. An organized method of listing potentially applicable process options is presented by the US EPA (1988) in the form of General Response Actions. Various processes are grouped as "technologies." For example, stripping and carbon adsorption are processes grouped within the technology "physical treatment" for groundwater. Process options such as chemical precipitation and ultraviolet (UV) oxidation are grouped within "chemical treatment." In developing the General Response Actions for a contaminated site, the best approach is to separate ex situ from in situ treatment methods, and groundwater from soil treatment methods. Some technologies apply to both media (e.g., bioremediation can be used for both groundwater and soil). Table 1-1 gives an example of General Response Actions for an actual site with soil contaminated with organics and metals. Information on applicable process options that can be used for General Response Actions is given by the US EPA (1994) in Table 1-2 and the US Army Environmental Center (1997).

More detailed information on process options is in the computer program ReOpt (short for remediation options) (Battelle Memorial Institute, 1995). ReOpt presents information on more than 100 processes, gives physical parameter values for more than 400 contaminants, and presents flow schematics. ReOpt is available from EnviroWin Software (Chicago, Illinois) or from Battelle Press (Columbus, Ohio). For information on process options that apply to cleanup of hydrocarbon contamination, a good source is the American Petroleum Institute (1990).

Table 1-1 General response actions and associated technology types and process options.

GENERAL RESPONSE ACTION	REMEDIAL TECHNOLOGY TYPES	PROCESS OPTIONS	SITE PROBLEM PRIMARILY ADDRESSED
No Action	None	None	
Institutional Actions	Access Restrictions	Fencing	Reduces human exposure
		Deed Restrictions	
	Monitoring	Groundwater Monitoring	Indicates if groundwater has been impacted
Containment	Capping	Synthetic Membrane	Reduces exposure through volatilization and reduces potential for leaching
		Clay	
		Asphalt/Concrete	
		Multimedia Cap	
	Horizontal barriers	Grout Injection	Reduces potential for leaching to groundwater
	Vertical Barriers	Slurry Wall	Reduces leaching and lateral migration potential
		Sheet Piling	
		Grout Curtains	
In Situ Soil Treatment	Physical Processes	Solidification	Reduces leaching potential
		Soil Venting	Removes contaminants
		Steam Stripping	
	Chemical Processes	Soil Flushing	Removes and/or destroys contaminants
	Biological Processes	Biological Degradation	Destroys organic contaminants
	Thermal Processes	Vitrification	Destroys, removes or fixes contaminants
		RF Heating	Removes contaminants
Soil Removal, Treatment, Disposal	Excavation	Backhoe	Removes contaminants
		Bucket Auger	
	Physical Treatment	Soil Washing	Removes contaminants
		Aeration/Venting	Removes volatile contaminants
	Chemical Treatment	Stabilization	Reduces mobility or toxicity of contaminants
	Biological Treatment	Aerobic Bioreactor	Destroys organic contaminants
		Anaerobic Bioreactor	
		Land Farming	
	Thermal Processes	Low Temp. Stripping	Removes volatile and semi-volatile contaminants
		Incineration	Removes and/or destroys contaminants
	Disposal	Landfill	Minimizes or eliminates exposure to sensitive populations
		Backfill	
		Asphalt Incorporation	

Table 1-2 Treatment technologies screening matrix.

NOTE: Specific site and contaminant characteristics may limit the applicability and effectiveness of any of the technologies and treatments listed below. This matrix is optimistic in nature and should always be used in conjunction with the referenced test sections, which contain additional information that can be useful in identifying potentially applicable technologies.

Technology	Development Status	Availability	Residuals Produced	Treatment Train (excludes off-gas treatment)	VOCs	SVOCs	Fuels	Inorganic	Explosives	System Reliability & Maintainability	Cleanup Time	Overall Cost	O&M or Capital Intensive
SOIL SEDIMENT AND SLUDGE													
3.1 In Situ Biological Treatment													
4.1 Biodegradation	Full	■	None	No	■	■	■	△	■	△	△	●	O&M
4.2 Bioventing	Full	■	None	No	■	■	■	△	—	■	●	■	Neither
4.3 White Rot Fungus	Pilot	△	None	No	△	△	△	△	■	△	△	●	O&M
3.2 In Situ Physical/Chemical Treatment													
4.4 Pneumatic Fracturing (enhancement)	Pilot	△	None	Yes	●	●	●	●	●	■	NA	■	Neither
4.5 Soil Flushing	Pilot	■	Liquid	No	●	●	●	■	△	●	△	—	O&M
4.6 Soil Vapor Extraction (in situ)	Full	■	Liquid	No	■	●	■	△	△	■	●	■	O&M
4.7 Solidification/Stabilization	Full	■	Solid	No	△	●	△	■	△	■	■	■	CAP
3.3 In Situ Thermal Treatment													
4.8 Thermally Enhanced SVE	Full	●	Liquid	No	●	●	●	△	△	●	■	●	Both
4.9 Vitrification	Pilot	△	Liquid	No	●	●	●	■	△	△	■	△	Both
3.4 Ex Situ Biological Treatment (assuming excavation)													
4.10 Composting	Full	■	None	No	■	●	■	△	■	■	●	■	Neither
4.11 Controlled Solid Phase Bio. Treatment	Full	■	None	No	■	●	■	△	■	■	●	■	Neither
4.12 Landfarming	Full	■	None	No	■	●	■	△	●	■	△	■	Neither
4.13 Slurry Phase Bio. Treatment	Full	●	None	No	■	●	●	●	■	●	●	■	Both
3.5 Ex Situ Physical/Chemical Treatment (assuming excavation)													
4.14 Chemical Reduction/Oxidation	Full	■	Solid	Yes	●	●	●	■	△	■	■	●	Neither

Table 1-2 Treatment technologies screening matrix (continued).

	Development Status	Availability	Residuals Produced	Treatment Train (excludes off-gas treatment)	VOCs	SVOCs	Fuels	Inorganic	Explosives	System Reliability & Maintainability	Cleanup Time	Overall Cost	O&M or Capital Intensive
NOTE: Specific site and contaminant characteristics may limit the applicability and effectiveness of any of the technologies and treatments listed below. This matrix is optimistic in nature and should always be used in conjunction with the referenced test sections, which contain additional information that can be useful in identifying potentially applicable technologies.					\multicolumn{5}{c}{Contaminants Treated}				
SOIL, SEDIMENT AND SLUDGE (continued)													
4.15 Dehalogenation (BCD)	Full	—	Vapor	No	●	■	△	△	△	—	—	—	—
4.16 Dehalogenation (Glycolate)	Full	●	Liquid	No	●	■	△	△	■	△	△	△	Both
4.17 Soil Washing	Full	●	Solid, Liquid	Yes	●	■	●	△	■	●	■	●	Both
4.18 Soil Vapor Extraction (ex situ)	Full	■	Liquid	No	●	●	●	△	△	■	●	■	Neither
4.19 Solidification/Stabilization	Full	—	Solid	No	△	△	△	■	△	■	■	■	CAP
4.20 Solvent Extraction (chemical extraction)	Full	●	Liquid	Yes	●	■	●	△	■	●	△	△	Both
3.6 Ex Situ Thermal Treatment (assuming excavation)													
4.21 High Temperature Thermal Desorption	Full	■	Liquid	Yes	●	●	●	△	△	●	■	●	Both
4.22 Hot Gas Decontamination	Pilot	●	None	No	△	△	△	△	■	●	■	■	Both
4.23 Incineration	Full	■	Liquid, Solid	No	●	●	■	△	■	●	■	△	Both
4.24 Low Temperature Thermal Desorption	Full	■	Liquid	Yes	■	●	■	△	■	●	■	■	Both
4.25 Open Burn/Open Detonation	Full	■	Solid	No	△	△	△	△	■	—	■	■	Both
4.26 Pyrolysis	Full	△	Liquid, Solid	No	●	■	●	△	△	●	■	△	Both
4.27 Vitrification	Full	●	Liquid	No	●	●	●	■	■	●	●	△	Both
3.7 Other Treatment													
4.28 Excavation, Retrieval, and Off-Site Disposal	NA	■	NA	No	●	●	●	●	●	■	■	△	Neither
4.29 Natural Attenuation	NA	■	None	No	■	■	■	△	△	■	△	■	Neither

Table 1-2 Treatment technologies screening matrix (continued).

NOTE: Specific site and contaminant characteristics may limit the applicability and effectiveness of any of the technologies and treatments listed below. This matrix is optimistic in nature and should always be used in conjunction with the referenced text sections, which contain additional information that can be useful in identifying potentially applicable technologies.

	Development Status	Availability	Residuals Produced	Treatment Train (excludes off-gas treatment)	Contaminants Treated					System Reliability & Maintainability	Cleanup Time	Overall Cost	O&M or Capital Intensive
					VOCs	SVOCs	Fuels	Inorganic	Explosives				
GROUNDWATER, SURFACE WATER, AND LEACHATE													
3.8 In Situ Biological Treatment													
4.30 Co-metabolic Treatment	Pilot	△	None	No	■	■	●	△	●	△	●	●	O&M
4.31 Nitrate Enhancement	Pilot	△	None	No	■	■	■	△	●	●	●	●	Neither
4.32 Oxygen Enhancement with Air Sparging	Full	■	None	No	■	■	■	△	●	●	●	■	Neither
4.33 Oxygen Enhancement with H2O2	Full	■	None	No	■	■	■	△	●	△	●	●	O&M
3.9 In Situ Physical/Chemical Treatment													
4.34 Air Sparging	Full	■	Vapor	Yes	■	△	■	△	△	■	■	■	Neither
4.35 Directional Wells (enhancement)	Full	△	NA	Yes	●	●	●	●	●	●	●	—	Neither
4.36 Dual Phase Extraction	Full	■	Liquid, Vapor	Yes	■	△	●	△	△	●	●	●	O&M
4.37 Free Product Recovery	Full	■	Liquid	No	△	■	■	△	△	■	■	■	Neither
4.38 Hot Water or Steam Flushing/Stripping	Pilot	●	Liquid, Vapor	Yes	●	●	■	■	△	△	■	●	CAP
4.39 Hydrofracturing (enhancement)	Pilot	—	None	Yes	●	●	●	●	●	■	■	●	Neither
4.40 Passive Treatment Walls	Pilot	△	Solid	No	■	■	●	■	■	—	△	—	CAP
4.41 Slurry Walls (containment only)	Full	■	NA	NA	■	■	●	—	●	■	■	●	CAP
4.42 Vacuum Vapor Extraction	Pilot	△	Liquid, Vapor	No	■	●	●	△	△	■	●	●	CAP
3.10 Ex Situ Biological Treatment (assuming pumping)													
4.43 Bioreactors	Full	■	Solid	No	■	■	■	△	●	●	NA	■	CAP

Table 1-2 Treatment technologies screening matrix (continued).

NOTE: Specific site and contaminant characteristics may limit the applicability and effectiveness of any of the technologies and treatments listed below. This matrix is optimistic in nature and should always be used in conjunction with the referenced test sections, which contain additional information that can be useful in identifying potentially applicable technologies.

	Development Status	Availability	Residuals Produced	Treatment Train (excludes off-gas treatment)	VOCs	SVOCs	Fuels	Inorganic	Explosives	System Reliability & Maintainability	Cleanup Time	Overall Cost	O&M or Capital Intensive
GROUNDWATER, SURFACE WATER, AND LEACHATE (continued)													
3.11 Ex Situ Physical/Chemical Treatment (assuming pumping)													
4.44 Air Stripping	Full	■	Liquid, Vapor	No	■	●	●	△	△	■	NA	■	O&M
4.45 Filtration	Full	■	Solid	Yes	△	△	△	■	●	■	■	■	Neither
4.46 Ion Exchange	Full	■	Solid	Yes	△	△	△	■	△	■	●	■	Neither
4.47 Liquid Phase Carbon Adsorption	Full	■	Solid	No	■	■	■	●	■	■	NA	△	O&M
4.48 Precipitation	Full	■	Solid	Yes	△	△	△	■	—	■	NA	●	Neither
4.49 UV Oxidation	Full	■	None	No	■	■	■	△	■	△	NA	●	Both
3.12 Other Treatment													
4.50 Natural Attenuation	NA	■		No	■	■	■	△	△	■	△	●	Neither
3.13 AIR EMISSIONS/OFF GAS TREATMENT													
4.51 Biofiltration	Full	●	None	NA	■	●	■	△	●	△	NA	●	Neither
4.52 High Energy Corona	Pilot	△	None	NA	■	■	■	●	△	△	NA	●	I
4.53 Membrane Separation	Pilot	△	None	NA	■	■	●	△	●	△	NA	●	I
4.54 Oxidation	Full	■	None	NA	■	■	■	△	■	■	NA	■	Neither
4.55 Vapor Phase Carbon Adsorption	Full	■	None	NA	■	■	■	●	■	■	NA	■	Neither

Rating Codes
■ Better
● Average
△ Worse

I Inadequate Information
NA Not Applicable

Source: US EPA (1994)

An Internet source for information on process options can be found at http://www.frtr.gov/matrix2/top_page.html. This source also provides a matrix that is helpful in screening process options. Other Internet sources are listed in Katz (1997). John Wiley and Sons (New York) provides to subscribers a library of remediation processes data in RIMS2000 (Remediation Information Management System), which can be accessed at www.enviroglobe.com. The Global Environment and Technology Foundation (Annandale, Virginia) maintains a database with profiles of environmental technologies, including remediation processes, called TechKnow^{TM}, at the GNET^{TM} web site (www.gnet.org).

1.3.2 Comparing Process Options

For Comprehensive Environmental Response, Compensation and Liability Act (CERCLA) or Superfund sites, implementability screening is accompanied by screening based on effectiveness and cost (USEPA, 1988). When there are many options to evaluate, relative costs (unit costs) may be used for screening instead of total estimated costs to implement each remediation option.

Resources that aid directly in the comparison of process options include the US Army Environmental Center (1997), the US EPA (1994), Long (1993), Roote et al. (1997), and Hyman and Bagaasen (1997).

The US Army Environmental Center (1997) describes 43 treatment processes and gives the applicability, limitations, data needs, and unit costs for each option. Performance data are given for some processes. A matrix is presented by the US EPA (1994) that gives the following information for proven processes: conditions that favor the use of each process, conditions that are unfavorable, unit costs, and cost drivers. The matrix (Tables 1-3 and 1-4) covers both groundwater and soil treatment. The MITRE Corporation prepared the matrix (August 1994) for the Air Force Center for Environmental Excellence (Brooks Air Force Base, Texas). Sources used for this guide were remedial equipment vendors; remediation service companies; the Remediation Technology Design, Performance, and Cost Study (July 1992) by the MITRE Corporation; and the Means Site Work and Landscape Cost Data (1993) by the R.S. Means Company, Inc. Abbreviations in the matrix include DNAPL (dense non-aqueous phase liquid); DRE (destruction and removal efficiency); gpm (gallons per minute); HDPE (high-density polyethylene); O&M (operations and maintenance); PCBs (polychlorinated biphenyls); POL (petroleum, oils, and lubricants); ppm (parts per million, which is equivalent to mg/L or mg/kg); and ppm$_V$ (parts per million on a volume basis, which is generally applied to gas mixtures).

Table 1-3 Remediation technologies: Application and cost guide for contaminant removal.

Media	Remediation Technology	Conditions Favorable for Use
Groundwater or Surface Water	Liquid phase carbon adsorption	Contamination < 10 ppm
		Presence of semi-volatile halogenated and non-halogenated contaminants
		Flow rate < 10 gpm if contamination > 10 ppm
	Air stripping	Volatile organic contaminants > 10 ppm
	Free product removal by pumping	Measured thickness of organic layer > 6 in.
		Water table depth <50 ft below ground surface
	Phase separation (oil-water)	Contamination > 2,000 ppm
		Flow rate > 100 gpm
	Air sparging	Volatile contaminants present
Soil	Soil vapor extraction	Volatile contaminant concentrations > 1,000 ppmv in soil gas
		Presence of low permeability surface cap
		Presence of contamination > 30 ft below
	Excavation	Ex-situ treatment planned, such as thermal, soil washing or biological treatment
		Off-site treatment available
		Contamination < 20 ft below ground
	Soil washing	Thermal treatment prohibited
		Soil cannot be disposed of off-site
Gas	SVE exhaust Condensation	Gas flow rate < 200 scfm
		High contaminant concentrations
		Collection efficiencies > 80-90 % are not required
	Air stripper exhaust Vapor phase carbon adsorption	Application on trial test SVE units
		Short term (< 1 month) emission control required
		Contaminant concentrations < 100 ppmv
• Each medium affected by a completed risk exposure pathway should be remediated • The majority of contaminant mass is likely to be located in soil • Groundwater remediation should be coordinated with source remediation in the unsaturated zone	• These technologies may be considered to be proven technologies; innovative technologies are not included • Technology selection can be guided by performance of nearby remediation projects at similar sites. IRPIMS may be used to locate sites having similar media, stratigraphy and contamination of concern	• These generalizations are based on projects demonstrating acceptable performance and cost effectiveness; they are not presented as rigid guidelines because each project needs to be evaluated individually.

Source: US EPA (1994)

Table 1-3 Remediation technologies: Application and cost guide for contaminant removal (continued).

Media	Remediation Technology	Conditions Unfavorable for Use
Groundwater or Surface Water	Liquid phase carbon adsorption	Suspended solids > 50 ppm
		Oil, grease content > 10 ppm
		High volatile organic content
		Presence of humic and fulvic acids
	Air stripping	Presence of non-volatile organics
		Iron content > 10 ppm
		Iron content > 10 ppm
		Hardness > 800 ppm as $CaCO_3$
	Free product removal by pumping	Viscous free product that is difficult to pump
		Thin free product layers
		Water table depth > 100 ft below ground surface
	Phase separation (oil-water)	Presence of emulsions
	Air sparging	Low permeability aquifer
		Presence of free product > 6 in. thick
Soil	Soil vapor extraction	Water table < 10 ft below ground surface
		Clay content > 20 %
	Excavation	Presence of structures and utilities
		Very volatile or toxic contaminants
		Noise sensitive environments
	Soil washing	Presence of > 30 % silt and clay
		Presence of a sensitive aquifer that may be affected by residual washing chemicals
Gas	SVE exhaust Condensation	Gas flow rate > 200 scfm
		Dense or viscous condensate
	Air stripper exhaust	Application on air strippers
		Flow rates > 200 acfm
	Vapor phase carbon adsorption	Applications to water-saturated gas streams
• Each medium affected by a completed risk exposure pathway should be remediated • The majority of contaminant mass is likely to be located in soil • Groundwater remediation should be coordinated with source remediation in the unsaturated zone	• These technologies may be considered to be proven technologies; innovative technologies are not included • Technology selection can be guided by performance of nearby remediation projects at similar sites. IRPIMS may be used to locate sites having similar media, stratigraphy and contamination of concern	• These generalizations are based on projects demonstrating acceptable performance and cost effectiveness; they are not presented as rigid guidelines because each project needs to be evaluated individually

Source: US EPA (1994)

Table 1-3 Remediation technologies: Application and cost guide for contaminant removal (continued).

Media	Remediation Technology	Unit Cost Range
Groundwater or Surface Water	Liquid phase carbon adsorption	Capital cost: 10-30 gpm: $200 per gpm 30-500 gpm: $130 per gpm Operating cost: $20-50 per pound contaminant removed $20-$50 per pound of contaminant removed
	Air stripping	Capital cost: $250-$400 per gpm throughput up to 100 gpm Operating cost: $20-50 per pound contaminant removed
	Free product removal by pumping	$3,000-$5,000 for a single well $1,500 per well for additional wells in multi-well systems
	Phase separation (oil-water)	$10-$20 per gpm capacity of separator
	Air sparging	Capital cost: $15 per foot for injection wells $5,000-$25,000 for air injection pump
Soil	Soil vapor extraction	$15-$25 per scfm capacity for extraction skid with no emission controls (See contaminant destruction by thermal treatment for emission control costs)
		$40-$75 per foot for extraction wells
	Excavation	$2-$5 per cubic yard for excavating and loading
		$1-$3 per cubic yard for backfilling and compacting
		Treatment costs additional
	Soil washing	$100-$500 per ton of soil treated
Gas	SVE exhaust Condensation	$15,000-$20,000 for a 200 scfm unit
	Air stripper exhaust Vapor phase carbon adsorption	Capital cost: < $1,000 for units 200 scfm or less $3-$4 per scfm capacity for larger units
		Operating cost: $40-$$100 per pound of contaminant removed
• Each medium affected by a completed risk exposure pathway should be remediated • The majority of contaminant mass is likely to be located in soil • Groundwater remediation should be coordinated with source remediation in the unsaturated zone	• These technologies may be considered to be proven technologies; innovative technologies are not included • Technology selection can be guided by performance of nearby remediation projects at similar sites. IRPIMS may be used to locate sites having similar media, stratigraphy and contamination of concern	• These costs are typical of successful projects conducted by the Air Force and private industry • These costs may be used for budget planning and as a rough check of contractor proposals • These costs are typical of those charged by companies specialized in each technology • A useful measure of merit for evaluating costs is to compute the project cost per pound of contaminant • Cost effective projects typically run < $200/lb • Projects where the costs are orders of magnitude higher should be scrutinized

Source: US EPA (1994)

Table 1-3 Remediation technologies: Application and cost guide for contaminant removal (continued).

Media	Remediation Technology	Major Cost Drivers
Groundwater or Surface Water	Liquid phase carbon adsorption	Carbon regeneration
		Residuals disposal
	Air Stripping	Instrumentation for automated operation
		Power consumption
		Air reheat
		Power consumption
		Offgas treatment (options listed below)
	Free product removal by pumping	Product treatment or disposal (excluding recovery credits)
	Phase separation (oil-water)	Equipment
	Air sparging	Trial test
		Implementation
Soil	Soil vapor extraction	Equipment
		Process monitoring
		Trial test if no nearby SVE applications
	Excavation	Field implementation
		Treatment or disposal of contaminated material
	Soil washing	Number of extraction stages required
		Waste stream management or decontamination
Gas	SVE exhaust Condensation	Equipment Compressor power
	Air stripper exhaust Vapor phase carbon adsorption	Equipment Carbon replacement
• Each medium affected by a completed risk exposure pathway should be remediated • The majority of contaminant mass is likely to be located in soil • Groundwater remediation should be coordinated with source remediation in the unsaturated zone	• These technologies may be considered to be proven technologies; innovative technologies are not included • Technology selection can be guided by performance of nearby remediation projects at similar sites. IRPIMS may be used to locate sites having similar media, stratigraphy and contamination of concern	• Reviews of project cost estimates can focus on these areas to expedite the reviews • Regulatory requirements for monitoring and preparing project documentation can be major cost drivers for any project, and they are not addressed in this guide because of the wide range of variability in regulatory requirements among state and local agencies

Source: US EPA (1994)

Table 1-3 Remediation technologies: Application and cost guide for contaminant removal (continued).

Media	Remediation Technology	Additional Comments
Groundwater or Surface Water	Liquid phase carbon adsorption	Best suited for low volume, low concentration applications such as effluent polishing
		Removal efficiencies of 100% can be obtained
		On-site regeneration usually not coat effective
	Air stripping	Tray strippers have less visual impact than packed towers and tray strippers may be less prone to fouling
		Units designed for removal efficiencies $\approx$ 99%
	Free product removal by pumping	Should be initiated immediately upon discovery of free product layer
		Single phase pumping less costly than two phase pumping which requires water treatment
	Phase separation (oil-water)	Effluent concentration seldom < 10 ppm
	Air sparging	Small scale (one or two well) pilot test recommended
		Sparging may spread contamination to clean areas, such as basements or utility lines
		May be used with SVE
Soil	Soil vapor extraction	Emission control equipment probably necessary; contaminant destruction by thermal treatment the preferred alternative
		Operation is generally not cost effective at removal rates <10 lb/day
		Air flow promotes biodegradation
		Can be used with air sparging
	Excavation	
	Soil washing	Due to the complexity of this technology, a compelling reason for use should exist
		Treatment of numerous waste streams required
Gas	SVE exhaust Condensation	Both contaminants and water will condense, water will require treatment prior to discharge
		Recovered product may be partially oxidized, unfit for reuse, and may plug the condenser
	Air stripper exhaust Vapor phase carbon adsorption	On-site carbon reactivation is generally not cost effective; vendors provide carbon replacement service
		Removal efficiencies of 100% can be obtained but saturated gases impede performance
• Each medium affected by a completed risk exposure pathway should be remediated • The majority of contaminant mass is likely to be located in soil • Groundwater remediation should be coordinated with source remediation in the unsaturated zone	• These technologies may be considered to be proven technologies; innovative technologies are not included • Technology selection can be guided by performance of nearby remediation projects at similar sites. IRPIMS may be used to locate sites having similar media, stratigraphy and contamination of concern.	• These comments and performance estimates are provided as helpful hints end observations based on successful projects

Source: US EPA (1994)

Table 1-4 Remediation technologies: Application and cost guide for contaminant destruction.

Media	Remediation Technology	Conditions Favorable for Use
Groundwater	Intrinsic remediation or natural attenuation	Contaminant mass < 2,000 lb
		No receptors at risk
	Biotreatment: In situ	Presence of water soluble organic contaminants
	Ex situ	For in situ treatment, aquifer must have permeability > 10^{-2} ft/d
		Contaminant mass ranging from 1,000 to 8,000 lb
Soil	Biotreatment: In situ Ex situ (composting) Bioventing	Moist, permeable soil, neutral to basic pH, Temperature > 40°F
	Thermal treatment: Low temperature	High contaminant concentrations and presence of free product
	High temperature	Water content < 20 %
		Contaminant mass > 2,000 lb for on-site treatment
		Rapid remediation required
Gas (SVE exhaust, air stripping exhaust, air sparging emissions)	Thermal treatment: Catalytic	Emission control stipulated by regulatory agencies
	Flame Reactive bed	Contaminant concentrations > 1,000 ppmv favors use of flame units
		Concentrations from 100 to 5,000 ppmv can be treated in catalytic oxidizers
• Each medium affected by a completed risk exposure pathway should be remediated • The majority of contaminant mass is likely to be located in soil • Groundwater remediation should be coordinated with source remediation in the unsaturated zone	• These technologies may be considered to be proven technologies; innovative technologies are not included • Technology selection can be guided by performance of nearby remediation projects at similar sites. IRPIMS may be used to locate sites having similar media, stratigraphy and contamination of concern	• These generalizations are based on projects demonstrating acceptable performance and cost effectiveness; they are not presented as rigid guidelines because each project needs to be evaluated individually

Table 1-4 Remediation technologies: Application and cost guide for contaminant destruction (continued).

Media	Remediation Technology	Conditions Unfavorable for Use
Groundwater	Intrinsic remediation or natural attenuation	Presence of halogenated organics or heavy metals
		Presence of free product
	Biotreatment: In situ Ex situ	Presence of halogenated organics Presence of free product Presence of inorganic contaminants
Soil	Biotreatment: In situ Ex situ (composting)	Presence of free product Presence of halogenated organics or inorganics
	Bioventing	Saturated soil or water content > 50%
		Rapid remediation required
	Thermal treatment: Low temperature High temperature	High clay content
Gas (SVE exhaust, air stripping exhaust, air sparging emissions)	Thermal treatment: Catalytic Flame Reactive bed	High particulate or water droplet loading requires filtering or separation
• Each medium affected by a completed risk exposure pathway should be remediated • The majority of contaminant mass is likely to be located in soil • Groundwater remediation should be coordinated with source remediation in the unsaturated zone	• These technologies may be considered to be proven technologies; innovative technologies are not included • Technology selection can be guided by performance of nearby remediation projects at similar sites; IRPIMS may be used to locate sites having similar media, stratigraphy and contamination of concern	• These generalizations are based on projects demonstrating acceptable performance and cost effectiveness; they are not presented as rigid guidelines because each project needs to be evaluated individually

Table 1-4 Remediation technologies: Application and cost guide for contaminant destruction (continued).

Media	Remediation Technology	Unit Cost Range
Groundwater	Intrinsic remediation or natural attenuation	No capital or O&M costs other than long term monitoring
	Biotreatment: In situ Ex situ	$40-$170/yd^3 for ex situ $13-$50/yd^3 for in situ
Soil	Biotreatment: In situ Ex situ (composting) Bioventing	$15-$50 per cubic yard
	Thermal treatment: Low temperature High temperature	$50-$150 per ton for POL only (low temperature treatment)
		$300-$600 per ton for halogenated organics (high temperature treatment)
		$70-$1,500 per ton if PCBs present
		$6,000 per ton if process-related dioxins are present in soil
Gas (SVE exhaust, air stripping exhaust, air sparging emissions)	Thermal treatment: Catalytic Flame Reactive bed	Capital cost: $65-$100 per scfm throughput for thermal equipment cost
		$60-$90 per scfm throughput for acid gas emission control equipment
		Operating cost:
		≈ $50 per scfm throughput annual O&M cost for thermal unit
		$250-$400 per scfm throughput annual O&M cost for thermal unit with scrubber
• Each medium affected by a completed risk exposure pathway should be remediated • The majority of contaminant mass is likely to be located in soil • Groundwater remediation should be coordinated with source remediation in the unsaturated zone	• These technologies may be considered to be proven technologies; innovative technologies are not included • Technology selection can be guided by performance of nearby remediation projects at similar sites. IRPIMS may be used to locate sites having similar media, stratigraphy and contamination of concern.	• These costs are typical of successful projects conducted by the Air Force and private industry • These costs may be used for budget planning and as a rough check of contractor proposals • These costs are typical of those charged by companies specialized in each technology • A useful measure of merit for evaluating costs is to compute the project cost per pound of • Cost effective projects typically run < $200/lb • Projects where the costs are orders of magnitude higher should be scrutinized

Table 1-4 Remediation technologies: Application and cost guide for contaminant destruction (continued).

Media	Remediation Technology	Major Cost Drivers
Groundwater	Intrinsic remediation or natural attenuation	Monitoring

	Biotreatment:
	  In situ
	  Ex situ
		Trial test
		Monitoring

Soil	Biotreatment:
	  In situ
	  Ex situ (composting)
	  Bioventing
		Trial test
		Field implementation

	Thermal treatment:
	  Low temperature
	  High temperature
		Contaminant type determining whether high or low temperature treatment is required
		On-site or off-site location of treatment unit
		Need for air emission controls

Gas (SVE exhaust, air stripping exhaust, air sparging emissions)	Thermal treatment:
	  Catalytic
	  Flame
	  Reactive bed
		Gas flow rate
		Presence of halogens requiring acid gas cleaning

• Each medium affected by a completed risk exposure pathway should be remediated • The majority of contaminant mass is likely to be located in soil • Groundwater remediation should be coordinated with source remediation in the unsaturated zone	• These technologies may be considered to be proven technologies; innovative technologies are not included • Technology selection can be guided by performance of nearby remediation projects at similar sites. IRPIMS may be used to locate sites having similar media, stratigraphy and contamination of concern	• Reviews of project cost estimates can focus on these areas to expedite the reviews • Regulatory requirements for monitoring and preparing project documentation can be major cost drivers for any project, and they are not addressed in this guide because of the wide range of variability in regulatory requirements among state and local agencies

Table 1-4 Remediation technologies: Application and cost guide for contaminant destruction (continued).

Media	Remediation Technology	Additional Comments
Groundwater	Intrinsic remediation or natural attenuation	Halogenated organics may degrade slowly or not at all without amendments
	Biotreatment: 　In situ 　Ex situ	Trial test is recommended to estimate performance
Soil	Biotreatment: 　In situ	Trial test is recommended, especially if microorganisms are added to soil
	Ex situ (composting)	Nutrient requirements need to be determined
	Bioventing	Performance depends on soil pore structure
		Low ppm levels may not be obtained
	Thermal treatment: 　Low temperature 　High temperature	Off-site treatment at high range of costs, on site at low range of costs
		Soils with water content > 25% require drying
		High temperature treatment units achieve DREs >99.99% and often require acid gas scrubbing and pollution control systems
		Low temperature unit performance is usually > 95% DRE
Gas (SVE exhaust, air stripping exhaust, air sparging emissions)	Thermal treatment: 　Catalytic 　Flame 　Reactive bed	Performance >95% DRE usually attained
		Base metal catalysis may be more cost effective than precious metal catalysis if halogens are present
		Units available that convert from flame to catalytic operation as concentration decreases
		Influent concentration generally kept <25% of lower explosive limit by dilution
• Each medium affected by a completed risk exposure pathway should be remediated • The majority of contaminant mass is likely to be located in soil • Groundwater remediation should be coordinated with source remediation in the unsaturated zone	• These technologies may be considered to be proven technologies; innovative technologies are not included • Technology selection can be guided by performance of nearby remediation projects at similar sites. IRPIMS may be used to locate sites having similar media, stratigraphy and contamination of concern	• These comments and performance estimates are provided as helpful hints end observations based on successful projects

The matrix identifies the following as "information necessary to use this guide":

- **An estimate of contaminant mass.** This estimate is not the same as the volume of contaminated media. Instead it is the weight of spilled jet fuel, landfilled solvent or whatever material contaminates a site.
- **A conceptual site model or site map showing the source and the approximate extent of contamination.** The contaminant concentration at a source is several orders of magnitude higher than detection limits and regulatory

action levels. If no source can be identified, the need for remediation should be reevaluated. Features within approximately 500 yd of a site should be known.
- **Site characteristics such as depth to groundwater, the measured thickness of any free product layer, groundwater flow direction, and subsurface geology.**
- **A list of completed exposure pathways identified through a risk assessment.**
- **The range of contaminant concentrations in environmental media.**

Because remediation of a site may include a combination of process options, the matrix includes a remediation strategy which should be developed prior to technology.

A remediation strategy may include any combination of these options. No containment measure should be considered permanent, and removal and destruction are often used together.

The matrix includes this disclaimer: "Innovative technologies are not represented in this guide and may be an acceptable or preferred alternative to technologies listed herein. Indications for use of these selected technologies and their costs are generalizations only. Site-specific data and regulatory requirements should be evaluated fully to determine the appropriate remedial technology and associated costs."

Long (1993) gives tables (Tables 1-5 and 1-6) that are useful for comparing groundwater and soil treatment processes for cleanup of petroleum contamination. In these tables the initial contaminant levels are in the left column. In Table 1-5 the top row indicates treated effluent contaminant levels, expressed as various regulatory discharge limits, whereas in Table 1-6 the top row indicates three ranges of water permeability.

Roote et al. (1997) present a matrix (Table 1-7) that is useful for comparing groundwater treatment options. Hyman and Bagaasen (1997) present tables that compare process options for both groundwater and soil remediation based on effectiveness, cost and speed of cleanup. Included are the following comparison tables:

Table 1-8: Groundwater treatment
- Ex situ removal of metals
- Ex situ treatment for organic contaminants
- In situ treatment for organic contaminants

Table 1-9: Soil treatment
- Ex situ treatment for organic contaminants
- Ex situ treatment for metal contaminants
- In situ treatment for organic contaminants
- In situ treatment for metal contaminants

Table 1-5 Groundwater treatment options for on-site remediation (Source: Long, G.M. Clean up hydrocarbon contamination effectively. *Chemical Engineering Progress*, May 1993 pp. 58-67). (Reproduced with permission of the American Institute of Chemical Engineers. Copyright © 1993 AIChE. All rights reserved.).

Inlet Concentration	Benzene-Toluene-Xylene Discharge Limits			
	Low	Medium	High	Very High
Water Flow Rate: 3-30 gal/min	(<1 µg/L)	(1-5 µg/L)	(5-50 µg/L)	(>50 µg/L)
Low (<500 µg/L)	CBL	AS/CBV, CBL, UVO, B, SZB	AS/CBV, CBL UVO, SZB	AS/CBV, CBL
Medium (500-5,000 µg/L)	CBL, AS/CBL/CBV	AS/CBV, UVO, B, SZB	AS/CBV, CBL, UVO, SZB	AS/CBV, ISO
High ($>5,000$ µg/L)	AS/CBL/CBV, AS/CBL/CC	AS/CBV, AS/CC, B, SZB	AS/CBV, AS/CC, UVO, SZB	AS/CBV, AS/CC, ISO
Water Flow Rate: >30 gal/min				
Low (<500 µg/L)	CBL	AS/CBV, CBL, UVO, B, SZB	UVO, B, SZB	AS/CBV, B
Medium (500-5,000 µg/L)	AS/CBL/CBV	AS/CBV, UVO, B, SZB	AS/CBV, B, SZB, UVO	AS/CBV, ISO, B
High ($>5,000$ µg/L)	AS/CBL/CC	AS/CC, UVO, B, SZB	AS/CC, UVO, B, SZB	AS/CC, ISO, B

Legend:
AS = Air Stripping   CC = Catalytic Combustion   B = Bioreactor
ISO = In situ Oxidation   CBV = Carbon Bed, Vapor   CBL = Carbon Bed, Liquid
UVO = Ultraviolet Oxidation   SZB = Saturated Zone Bioremediation
/ Denotes combination of treatment technologies, primary followed by secondary treatment.

Table 1-6 Unsaturated soil treatment options for on-site remediation (Source: long, G.M. Clean up hydrocarbon contamination effectively. *Chemical Engineering Progress*, May 1993, pp. 58-67). (Reproduced with permission of the American Institute of Chemical Engineers. Copyright © 1993 AIChE. All rights reserved.).

Contaminant Level	Saturated Hydraulic Conductivity (cm/s)		
	Low	Medium	High
Fuel: Gasoline, Naphtha	($<10^{-5}$)	(10^{-5}-10^{-2})	($>10^{-2}$)
Low (100-500 mg/kg)	SVE, LT, TD	UZB, SVE, LT	UZB, SVE, LT
Medium (500-5,000 mg/kg)	SVE, LT, TD	UZB, SVE, LT	UZB, SVE, LT
High ($>5,000$ mg/kg)	SVE, LT, TD	UZB, SVE, LT	SVE
Fuel: Diesel, No. 2 Fuel Oil, Kerosene, JP-1, Mineral Spirits			
Low (100-500 mg/kg)	SVE, LT	UZB, SVE, LT	UZB, LT, TD
Medium (500-5,000 mg/kg)	SVE, LT, TD	UZB, SVE, LT	UZB, LT, TD
High ($>5,000$ mg/kg)	SVE, LT, TD	UZB, LT, SVE, TD	LT, TD
Fuel: No. 4 or No. 6 Fuel Oil, Crude Oil, Tar, Residue Oils			
Low (100-500 mg/kg)	LT, I	LT, I	UZB, TD, I
Medium (500-5,000 mg/kg)	SVE, LT, TD, I, B	SVE, LT, TD, 1, B	SVE, TD, 1, B
High ($>5,000$ mg/kg)	SVE, LT, TD, I, B	SVE, LT, TD, 1, B	SVE, LT, 1, B

Legend:
B = Bioreactor   UZB = Unsaturated Zone Bioremediation   I = Incineration
TD = Thermal Desorption   SVE = Soil Vapor Extraction   LT = Land Treatment

Table 1-7 Remediation technologies matrix (Source: *Chemical Engineering*, May 1997, pp. 104-112).

REMEDIATION TECHNOLOGIES MATRIX	Contaminants & pollutants treated	Overall cost	Minimum contaminant concentration achievable	Maximum contaminant concentration applicable	Long-term effectiveness & permanence	Reduction of toxicity, mobility & volume	Time to complete cleanup	Health & safety risk	Commercial availability
GROUNDWATER									
Ex Situ Biodegradation									
• Activated sludge	HSO, NHVO, NHSO, P	1	3	1	4	3	4	1	3
Aqueous treatment system	HSO, NHVO, NHSO, P	0	3	2	4	3	4	1	1
Ex Situ Physical, Chemical Treatment									
• Air stripping	HVO, HSO, NHVO, NHSO	4	3	1	4	3	1	1	4
• Carbon adsorption (liquid phase)	HVO, HSO, NHVO, NHSO, P	2	3	1	4	3	4	1	4
UV oxidation	HVO, NHVO	0	2	4	4	2	4	1	4
In Situ Biodegradation									
Oxygen enhancement with H_2O_2	HVO, NHVO	4	3	2	4	3	3	3	1
Methanotrophic biodegradation	HVO, NHVO	1	3	1	4	3	1	3	1
Nitrate enhancement	HVO, NHVO	4	3	2	4	3	3	3	1
Oxygen enhancement with air sparging	HVO, NHVO	1	3	3	4	3	1	3	1

4 - Outstanding
3 - Good
2 - Average
1 - Marginal
0 - Unsatisfactory
I - Inadequate Info.
• = Conventional technologies and processes

HVO — Halogenated volatile organics
HSO — Halogenated semivolatile organics
NHVO — Nonhalogenated volatile organics
NHSO — Nonhalogenated semivolatile organics

P — Pesticides
IN — Inorganics
C — Corrosives

Table 1-7 Remediation technologies matrix (Source: *Chemical Engineering*, May 1997, pp. 104-112) (continued).

REMEDIATION TECHNOLOGIES MATRIX	Contaminants and pollutants treated	Overall cost	Minimum contaminant concentration achievable	Maximum contaminant concentration applicable	Long-term effectiveness and permanence	Reduction of toxicity, mobility and volume	Time to complete cleanup	Health and safety risk	Commercial availability
				GROUNDWATER					
				In Situ Physical, Chemical Treatment					
• Slurry walls	HVO, HSO, NHVO, NHSO, C	NA	NA	3	1	4	NA	3	4
Passive treatment walls	HVO, HSO	1	3	3	1	3	NA	3	1
Water or steam flushing or stripping	HVO, HSO, NHVO, NHSO	1	NA	1	NA	NA	NA	1	1
Hydrofracturing	HVO, NHVO	1	NA	4	NA	NA	NA	1	1
Pneumatic fracturing	HVO, HSO, NHVO, NHSO	1	NA	4	NA	NA	NA	1	2
Dynamic underground stripping	HVO, NHVO	1	NA	4	NA	NA	NA	1	1
				AIR EMISSIONS, OFFGAS					
• Carbon adsorption (vapor phase)	HVO, HSO, NHVO, NHSO, P	1	3	2	4	3	1	1	3
• Catalytic oxidation	HVO	1	3	2	4	3	1	1	2
Biofiltration	HVO, NHVO	1	3	2	4	3	1	1	2

4 - Outstanding
3 - Good
2 - Average
1 - Marginal
0 - Unsatisfactory
I - Inadequate Info.
• = Conventional technologies and processes

HVO   Halogenated volatile organics
HSO   Halogenated semivolatile organics
NHVO Nonhalogenated volatile organics
NHSO Nonhalogenated semivolatile organics
P   Pesticides
IN   Inorganics
C   Corrosives

Table 1-7 Remediation technologies matrix (Source: *Chemical Engineering*, May 1997, pp. 104-112) (continued).

REMEDIATION TECHNOLOGIES MATRIX	Contaminants and pollutants treated	Awareness of design & construction engineers	System reliability and maintainability	Regulatory and permitting barriers	Institutional acceptability	Labor intensity	Skill and training required to operate	Automation potential
GROUNDWATER								
Ex Situ Biodegradation								
• Activated sludge	HSO, NHVO, NHSO, P	1	1	1	2	1	1	1
Aqueous treatment system	HSO, NHVO, NHSO, P	2	3	1	2	4	0	1
Ex Situ Physical, Chemical Treatment								
• Air stripping	HVO, HSO, NHVO, NHSO	4	2	2	4	1	1	1
• Carbon adsorption (liquid phase)	HVO, HSO, NHVO, NHSO, P	4	2	4	4	1	1	0
UV oxidation	HVO, NHVO	4	2	3	1	4	1	0
Ex Situ Biodegradation								
Oxygen enhancement with H_2O_2	HVO, NHVO	1	1	1	1	1	1	1
Methanotrophic biodegradation	HVO, NHVO	1	1	1	1	1	1	1
Nitrate enhancement	HVO, NHVO	1	1	1	1	1	1	1
Oxygen enhancement with air sparging	HVO, NHVO	2	1	1	1	1	1	1

4 - Outstanding
3 - Good
2 - Average
• = Conventional technologies and processes
1 - Marginal
0 - Unsatisfactory
I - Inadequate Info.

HVO   Halogenated volatile organics
HSO   Halogenated semivolatile organics
NHVO   Nonhalogenated volatile organics
NHSO   Nonhalogenated semivolatile organics

P   Pesticides
IN   Inorganics
C   Corrosives

Table 1-7 Remediation technologies matrix (Source: *Chemical Engineering*, May 1997, pp. 104-112) (continued).

REMEDIATION TECHNOLOGIES MATRIX	Contaminants and pollutants treated	Awareness of design & construction engineers	System reliability and maintainability	Regulatory and permitting barriers	Institutional acceptability	Labor intensity	Skill and training required to operate	Automation potential
GROUNDWATER								
In Situ Physical, Chemical Treatment								
• Slurry walls	HVO, HSO, NHVO, NHSO, C	4	1	1	1	1	1	1
Passive treatment walls	HVO, HSO	2	1	1	1	1	1	1
Water or steam flushing or stripping	HVO, HSO, NHVO, NHSO	2	1	1	1	1	1	0
Hydrofracturing	HVO, NHVO	1	1	1	1	1	1	1
Pneumatic fracturing	HVO, HSO, NHVO, NHSO	2	1	1	1	1	1	1
Dynamic underground stripping	HVO, NHVO	1	1	1	1	1	1	1
AIR EMISSIONS, OFFGAS								
• Carbon adsorption (vapor phase)	HVO, HSO, NHVO, NHSO, P	4	3	3	4	1	1	2
• Catalytic oxidation	HVO	3	2	3	2	1	1	2
Biofiltration	HVO, NHVO	2	3	2	2	1	1	4

4 - Outstanding
3 - Good
2 - Average
1 - Marginal
0 - Unsatisfactory
I - Inadequate Info.
• = Conventional technologies and processes

HVO — Halogenated volatile organics
HSO — Halogenated semivolatile organics
NHVO — Nonhalogenated volatile organics
NHSO — Nonhalogenated semivolatile organics
P — Pesticides
IN — Inorganics
C — Corrosives

26

Table 1-8 Groundwater treatment (Source: Marve Hyman, M, Bagaasen, L. select a site cleanup technology. *Chemical Engineering Progress*, August 1997, pp. 22-43). (Reproduced with permission of the American Institute of Chemical Engineers. Copyright© 1997 AIChE. All rights reserved.).

	Effectiveness	Cost	Required Duration of Treatment
Ex situ removal of metals from groundwater			
Chemical Precipitation	Good	Low	
Reverse Osmosis	Moderate	Medium	Dependent upon contaminant extraction rate from contaminated aquifer
Ion Exchange	Most Effective	High	
Forced Evaporation	Very, Very Good	High	
Ex situ treatment of organic contaminants in groundwater			
Carbon Adsorption	Very Good	High	
Stripping - Air	Good for VOC	Medium	Dependent upon contaminant extraction rate from contaminated aquifer
Stripping - Steam	Very Good for VOC and SVOC	Very High	
Aqueous Oxidation	Medium	Medium	
Bioreactors	Medium	Medium	Years (Hours for Hydraulic Contact Time)
Bioponds	Low	Low	Years (Days for Hydraulic Contact Time)
In situ treatment of organic contaminants in groundwater			
Stripping - Air Sparging	Very Good for VOC	Medium	Months
Bioremediation	Good	Low	Months

Notes for Table 1-8 are as follows:

- The cleanup time for ex situ removal of metals from groundwater depends on how quickly contaminant concentrations in the aquifer can be reduced to required levels, which is a function of water extraction rate, aquifer hydraulics, and contaminant mobility.
- Carbon adsorption is very good for nonpolar organics, except for small molecules (e.g., vinyl chloride is not adsorbed).
- The cleanup time for ex situ removal of organics from groundwater depends on the extraction rate from wells and on groundwater hydraulics. It is usually on the order of many years. Contact time in adsorbers, strippers, and UV oxidizers is minutes.
- For biological processes, larger molecules biodegrade more slowly than smaller molecules among similar hydrocarbons, and many chlorinated organics do not biodegrade sufficiently.

Notes for Table 1-9 are as follows:

- Washing with water plus surfactants is noted as "limited use" because it is effective for coarse-grained soil fractions only and sometimes its main purpose is to reduce the volume needing further treatment.
- In situ bioremediation and flushing are generally limited to homogeneous sand formations.
- For biological processes, larger molecules biodegrade more slowly than smaller molecules among similar hydrocarbons, and many chlorinated organics do not biodegrade sufficiently.

Table 1-9 Soil treatment (Source: Marve Hyman, M, Bagaasen, L. select a site cleanup technology. *Chemical Engineering Progress*, August 1997, pp. 22-43). (Reproduced with permission of the American Institute of Chemical Engineers. Copyright© 1997 AIChE. All rights reserved.).

	Effectiveness	Cost	Required Duration of Treatment
Ex situ treatment of organic contaminants in soil			
Venting at Ambient Temperature	Good for Volatiles	Low	Several Weeks
Thermal Treatment - Incineration	Most Effective	Highest	Immediate
Thermal Treatment - Desorption	Very Good	Medium	Immediate
Bioremediation	Good	Low	Months
Washing - Water + Surfactant	Limited Use	Medium	Immediate
Washing - Solvent	Good	High	Immediate
Fixation - Cement or Pozzalans	Limited Use for Some Organics	Low	Days (for Curing)
Fixation - Thermoplastics	Very Good	Medium	Immediate
Ex situ treatment for metal contaminants in soil			
Washing - Water or Acid	Limited Use	Low	Immediate
Fixation - Cement or Pozzalans	Very Good Immobilization	Low	Immediate
Vitrification	Best Immobilization	High	Immediate
In situ treatment of organic contaminants in soil			
Venting	For Volatiles	Low	Several Months
Heating or Thermal Desorption	For Volatiles and Semivolatiles	High	Fast
Bioremediation - Infiltrate with Nutrients and H_2O_2	Generally limited to homogeneous, coarse grained formations	Medium	Dependent Upon Contaminant
Bioremediation - Bioventing		Low	Degradation Rate
Flushing		Medium	Medium-Fast
Electrokinetics	Low	Medium	A Few Months
Ex situ treatment for metal contaminants in soil			
Flushing	Limited Use	Low	Fast
Fixation - Shallow, with Backhoe	Low	Low	Fast
Fixation - with Auger	Very Good	High	Medium-Fast
Electrokinetics	Good	Medium	A Few Months

Source: Hyman and Bagassen (1997)

Flushing is generally limited to homogeneous sand formations where contaminants will not be washed down into an aquifer.

1.3.3 Defining and Evaluating Alternative Treatment Trains

Hyman and Bagaasen (1997) also describe a four-step approach to selecting among alternatives that are each comprised of multiple treatment processes. The four steps are used in the computer program RAAS (Remedial Action Assessment System) (Battelle Memorial Institute, 1996), and are as follows:

Step 1: List potentially applicable process options and perform initial screening.
Step 2: Formulate alternative treatment trains that are technically feasible and place the process options in each train in a logical order.
Step 3: Estimate the quantitative effectiveness of each alternative.
Step 4: Choose from the technically feasible alternatives.

Certain criteria set forth by the US EPA (1988) for evaluating and comparing alternatives are used for Step 4. These criteria are used in feasibility studies for CERCLA or Superfund sites, and are useful for any remediation site where multiple choices apply. The criteria are given in Figure 1-1.

The first two — overall protection and meeting Applicable or Relevant and Appropriate Requirements (ARARs) — are primary objectives. Any alternative being considered in a feasibility study that does not meet these two criteria can be immediately screened out of further consideration. Among the ARARs are chemical-specific requirements that can be used to quantitatively evaluate the effectiveness of an alternative. Other quantitative criteria are reduction of toxicity, mobility, or volume, and cost.

As explained in Hyman and Bagaasen (1997), using such evaluation criteria is practiced in the third step of the four-step approach for selecting among alternatives.

Most remediation treatment includes a combination of process steps. If more than three treatment methods can be implemented and are effective, the number of possible alternatives is large. Each process option alone or in combination with one or more of the other options may be defined as an alternative to evaluate for selection. For example, groundwater with dissolved hydrocarbon contaminants may be treated ex situ with:

- Bioremediation using activated sludge
- Bioremediation using a fixed-film bioreactor
- Air stripping
- Steam stripping
- Carbon adsorption
- Aqueous oxidation with UV light and peroxide
- Aqueous oxidation with ozone
- Aqueous oxidation with peroxide and ozone

If just three of these process options appear implementable and effective, the processes (labeled 1, 2, and 3) can be ordered in various sequences, and the possible alternative treatment trains are as follows:

1. 1 alone
2. 2 alone
3. 3 alone
4. 1 followed by 2
5. 1 followed by 3
6. 2 followed
7. 2 followed by 3
8. 3 followed by 1
9. 3 followed by 2
10. 1, 2 and 3

Overall Protection of Human Health and the Environment	Compliance with ARARs
• How Alternative Provides Human Health and Environmental Protection	• Compliance with Chemical-Specific ARARs • Compliance with Action-Specific ARARs • Compliance with Location-Specific ARARs • Compliance with Other Criteria, Advisories, and Guidelines

Long-Term Effectiveness and Permanence	Reduction of Toxicity, Mobility, and Volume through Treatment	Short-Term Effectiveness	Implementability	Cost
• Magnitude of Residual Risk • Adequacy and Reliability of Controls	• Treatment Process Used and Materials Treated • Amount of Hazardous Materials Destroyed or Treated • Degree of Expected Reductions in Toxicity, Mobility, and Volume • Degree to Which Treatment is Irreversible • Type and Quantity of Residuals Remaining After Treatment	• Protection of Community During Remedial Actions • Protection of Workers During Remedial Actions • Environmental Impacts • Time Until Remedial Action Objectives are Achieved	• Ability to Construct and Operate the Technology • Reliability of the Technology • Ease of Undertaking Additional Remedial Actions, if Necessary • Ability to Monitor Effectiveness of Remedy • Ability to Obtain Approvals from Other Agencies • Coordination with Other Agencies • Availability of Offsite Treatment, Storage, and Disposal Capacity • Availability of Necessary Equipment and Specialists • Availability of Prospective Technologies	• Capital Costs • Operating and Maintenance Costs • Present Worth Costs

State Acceptance[1]	Community Acceptance[1]

Figure 1-1 Criteria for detailed analysis of alternatives.

[1] These criteria are assessed following comment on the RI/FS report and the proposed plan.

11. 1, 3 and 2
12. 2, 1 and 3
13. 2, 3 and 1
14. 3, 1 and 2
15. 3, 2 and 1

With four process options used to form even more alternatives, a computerized system designed to aid in evaluation and selection becomes attractive. The logic employed in the RAAS computer program (Battelle Memorial Institute, 1996) is useful for defining alternative treatment trains manually or with the computer program. RAAS aids in developing cleanup goals, uses ReOpt (Battelle Memorial Institute, 1995) to develop the list of process options in Step 1, and estimates the time it would take for a treatment process to achieve a cleanup goal. Some screening of processes is carried out in Step 1.

Step 1 (listing applicable process options) selections are based on whether the concern is with groundwater or soil, and whether ex situ or in situ treatment (or both) can be applied. Step 1 also depends on contaminant type, as follows:

Inorganics

- Soluble (e.g., chromate ion) or relatively insoluble (e.g., chromic ion)
- Partitioning properties (whether the contaminant adsorbs strongly on fine soil particle surfaces)

Organics

- Volatile, semivolatile or nonvolatile
- Partitioning properties
- Biodegradability

Molecular Size

- Whether membrane separation may be effective
- Whether biodegradability is low

Step 1 also depends on potential exposures that could impact human health. For example, a process that transfers volatile organics to an air stream could affect neighboring downwind property.

Step 2 (formulating technically feasible alternative treatment trains) screening first involves specific site conditions that may preclude the use of certain process options if they cannot be implemented or are ineffective. Examples include the depth of contamination, the uses of downgradient water wells, the presence of fine-grained soils, and contaminant concentrations and mobility. Between competing effective technologies, those with higher relative costs can be excluded.

Step 2 includes ordering the remaining applicable processes within alternative treatment trains. For example, aqueous phase carbon adsorption is frequently used in conjunction with air stripping for removing volatile organic compounds (VOCs) from groundwater. If VOC concentrations are high, the stripper should be ahead of the carbon in a treatment train. Strippers have unlimited capacity for removal of organic mass. However, the limited capacity of each batch of aqueous phase carbon favors its use for polishing treatment.
Other examples of the ordering logic used in Step 2 are as follows: In situ processes precede or preclude extraction or excavation. Destruction processes (e.g., biodegradation, incineration) precede fixation. Excavation precedes capping.

Step 3 (estimating effectiveness) is done with the aid of mathematical models that simulate the processes. Included are the computation of secondary stream flows (e.g., stripping air, soil-wash water, vapor extraction air, spent carbon) and their concentrations, contaminant concentration or volume reductions, and immobilization.

Step 4 (choosing from among the alternatives) evaluations can be based on some of the criteria shown in Figure 1-1, especially:

- Compliance with ARARs
- Long-term effectiveness and permanence
- Reduction of toxicity, mobility or volume
- Short-term effectiveness
- Cost

Also, Step 4 includes another look at the ability to implement a given remediation option. The secondary stream flows and concentrations must meet the substantive requirements of state control agencies, such as air and water pollution control authorities.

1.4 The Approach to Process Design And Cost Estimating

It is the goal of this book to provide insights into the design and cost of a variety of technologies commonly used for site remediation that are effective under a variety of groundwater and soil cleanup conditions. Chapters 3 through 11 provide background understanding of processes and their practical implementation to solve site remediation problems. For each of the main treatment processes commonly deployed, methods and examples are then given for evaluating process design parameters and for estimating costs. Finally, Chapter 12 gives chemical engineering approaches to both preliminary and definitive cost estimating for process treatment schemes as well as approaches that have been developed specifically for remediation problems.

Because individual sites usually have unique design requirements and because treatment costs vary over wide ranges, it is essential that process design for each situation be customized and extensively detailed. This approach provides the best pathway to definitive cost estimating and to final mechanical, electrical and structural designs and construction. Chapter 2 emphasizes a systematic method for process design that includes these steps:

- Developing a conceptual design.
- Producing a process flow diagram that shows each major treatment step and all emissions and effluents.
- Adding to the process flow diagram a mass balance for streams entering and exiting each treatment step and for each contaminant of concern, plus all water and air flows.
- Adding to the process flow diagram the temperature and pressure for streams entering and exiting each treatment step, and the size, rating or capacity of each major equipment item.
- Developing a Piping and Instrumentation Diagram (P&ID) that shows control loops and all pipes, valves, sensors, controllers, alarms, interlocks and automatic shut-downs.

Chapter 2

Process and Conceptual Design of Remediation Systems

The process and conceptual design for a treatment system includes the following items and documents:

- Process flow diagram
- Site plan
- Preliminary plot plan
- P&ID
- Sequence of operation
- Preliminary specifications
- Design basis
- Tradeoff analyses
- One-line electric schematic
- Controls logic diagram
- Ladder logic diagram

This chapter covers feasibility studies, work plans, and treatability studies that precede the production of these items, their development, and how they are related. The keys to finally implementing construction of a project are the development of the detailed mechanical design and the detailed instrumentation/electrical design that follow process design.

The steps needed to complete these documents and detailed designs are no different than the steps taken for the design of a chemical processing plant. What distinguishes remediation treatment systems from many chemical processing plants is the need for flexibility in configuring the process and laying out the equipment. No matter how many samples and treatability tests are completed, a complete characterization of contaminated groundwater or soil is not practical. Because of this, a treatment system must be adaptable to groundwater or soil characteristics that may not have been identified during investigations carried out before design. Heterogeneity in the subsurface environment and changes that occur during treatment also cause a contaminated site to change as treatment progresses.

The amount of water that can be extracted from wells cannot be predicted accurately. Even if pump tests are conducted for wells, there may be future plugging or changes in the elevation of the water table. The content of groundwater chemical constituents may also change unpredictably. Hydrogeological modeling of the mass transfer between soil and water can furnish some predictive capability but is not sufficient and cannot be totally relied on when designing pump-and-treat systems. For soil that is to be treated ex situ, the extent of contamination is never completely known until the soil is excavated.

For these reasons, pilot tests are often extensive, and final remediation is frequently preceded by interim remediation. These steps sometimes further the design of groundwater extraction systems, as well as remedial treatment systems.

2.1   Basic Principles

The design of soil and groundwater remediation systems often requires flexibility in their capability to be expanded and to be operated at highly variable flow rates. Two examples of flexibility needs are as follows:

1. **The capability to expand**–Space, access, and connections should be provided in the design to allow for additional equipment or future larger component replacements.
2. **The capability to turn down**–Proper treatment must be achieved even when throughput is well below design rates. Some equipment does not operate effectively at low flow: Centrifugal pumps may overheat, carbon beds and packed strippers may suffer from channeling, or incinerators may clog with slag buildup. The design may have to be refined so that the system will function effectively during times of low throughput. Alternatively, surge capacity may be included that will permit semi-batch operations, so that the process runs near the design flow rate for intermittent periods of time.

Another challenge is the frequent need for practical, economic designs for treatment units with low flow rates or with short life cycles. This is difficult because the engineering, permit requirements, instrumentation, procurement, and monitoring require about the same effort for these plants as they do for larger, more permanent plants.

Even though the remediation industry is young, much of this challenge has been met by the development of semi-standard designs and of vendor packages. And unlike many chemical or refining plants, which may suffer from processing interruptions, most remediation units can be stopped any time and restarted when convenient. With this philosophy in mind, unattended operation with fail-safe automated shutdowns and remote or occasional monitoring can readily be built into the design.

Because of the hazardous nature of some contaminants that remediation units are designed to control, fail-safe designs often need certain instrumentation redundancy and secondary containment. Proposed vendor applications should be examined in detail to ensure that provisions for redundancy are possible for pressure, temperature, and level alarms and shutdowns. Custom designs should include such provisions as discussed later in this chapter in the section on "P&ID and the Sequence of Operations."

The mechanical design (specifications and detailed dimensions of the layout of equipment and connecting piping with field instruments) can be started with the site plan, the preliminary plot plan, P&ID, plus civil/structural analysis. Development of the mechanical design and the plot plan proceed iteratively as specific information is received from vendors on main components and pressure drops are calculated.

The P&ID and the electrical/logic and control diagrams, along with electrical loads derived from process flow diagram information or vendors' data, are used to complete the detailed instrumentation/electrical design.

As the process and conceptual designs proceed, the preliminary specifications can be developed and later refined to become detailed specifications. Two main categories of detailed specifications are needed:

- Equipment specifications
- Installation specifications

The equipment specifications define requirements for fabrication and performance of the main components of the remediation system. The installation specifications usually cover requirements for other items such as foundations, piping, electrical components, civil engineering items, instruments, painting, and insulation. The mechanical design, instrumentation/electrical design, civil design, and finished specifications are needed for the installation to proceed.

The conceptual design is the basis for communicating the objectives and approach used in the project to meet the remediation needs of the parties responsible for the site, as well as the permitting authorities or control agencies. It is also used for scheduling the project and for determining preliminary estimates of the capital investment costs and annual expenses of the system. The process flow diagram, site plan, and preliminary plot plan, along with process calculations, are the essential items making up the conceptual design. A preliminary P&ID with a sequence of operation or design basis helps make a more complete conceptual design. Tradeoff analyses help refine the design basis and can materially affect the process flow diagram. The conceptual design, or at least a preliminary process flow diagram, is usually first developed during a feasibility study and work plan, which precede the process design.

2.2 Feasibility Studies and Work Plans

2.2.1 Feasibility Study Alternatives

A feasibility study considers the likely remediation methods that have the potential for reaching cleanup goals. Often, for each method, more than one process applies, and the processes are linked in a treatment train. For example, groundwater may go through an oil/water separator, metals removal, filtration, air stripping, another filtration step, and carbon polishing. Soil might go through a thermal desorber for organics removal and next be subjected to fixation of metals with Portland cement.

Usually more than one method or alternative may be effective. Instead of stripping, a UV/peroxide unit or a biotreatment system, incineration instead of low-temperature thermal desorption, or in situ vitrification instead of ex situ thermal treatment plus fixation with cement might be considered applicable at a given site. (See Section 1.3 regarding various guidance documents that aid in choosing which methods to consider.)

2.2.2 Work Plans, Corrective Action Plans, Remedial Action Plans

A feasibility study identifies options, screens out the least attractive ones, and analyzes in detail the remaining options. When what is considered the best option is chosen, a work plan (or remedial action plan or corrective action plan) is developed. This plan details some basic criteria for design, details how the process will work and where equipment can be physically located, identifies the monitoring requirements, and schedules the implementation of the final design, construction, operation, and maintenance of this chosen alternative.

The work plan (or its equivalent) details for the owner or responsible party the physical progress that will be made and when and how progress and effectiveness will be monitored. It also advises control agencies of these same items and is a vehicle with which the owner can make a commitment to implement the selected remediation strategy. The agencies can accept and enforce that commitment.

2.2.3 Informal Studies, CERCLA Studies and Records of Decision, Resource Conservation and Recovery Act (RCRA) Studies

Most feasibility studies are informal and cover what is reasonable from an objective perspective. Usually a limited number of options are screened, and a budget is estimated for each of the most effective options. An informal study concludes with a recommended alternative and a schedule for implementation.

For CERCLA (or Superfund) sites, a formal feasibility study is completed in accordance with "Guidance for Conducting Remedial Investigations and Feasibility Studies Under CERCLA, EPA/540/G-89/004" (October 1988). Numerous steps, screening criteria, and a comparative analysis of alternatives are required, as detailed in Chapter 1. No recommendation is made. The EPA considers the results of the study and, in conjunction with the lead state control agency, develops a formal proposed plan. This plan describes the alternatives considered and the proposed remediation scheme, with specific requirements for cleanup goals, and sets forth a schedule for implementation. After subjecting the proposed plan to a public hearing, a Record of Decision is issued and serves as a work plan for completion of a given project.

For RCRA facilities, a formal Corrective Measures Study is required, with numerous steps and with some of the same screening criteria as used for CERCLA feasibility studies. The EPA guidance for Corrective Measures Studies is given in the "RCRA Corrective Action Plan (May 1994), available from National Technical Information Service (NTIS) (Order No. PB-94963657).

An important feature of a CERCLA feasibility study or a RCRA Corrective Measures Study is the requirement that cost estimates for the alternatives finally evaluated be accurate within +50% and -30%. The more complete the conceptual design for each alternative, the better is the chance of achieving this cost-estimating accuracy. Certainly a process flow diagram with mass and energy balances and a preliminary plot plan are needed for each alternative, and a preliminary site plan applicable to all alternatives is needed so that this level of cost-estimating accuracy can be achieved.

2.2.4 The Observational Approach

It is often a lengthy process to adequately characterize the nature and extent of contamination below the ground surface, and to conduct necessary treatability studies. Because of this, a work plan or equivalent document is often issued, and implementation of a remedial design forges ahead without complete characterization or without running all of the desired treatability assessments. If the remediation scheme is successful, cleanup is accomplished sooner, and the costs of some sampling and analyses or the costs of some treatability tests are saved.

With an observational approach, implementation is begun with a greater degree of uncertainty. The uncertainty is recognized and is managed by setting up contingency plans for handling potentially unfavorable results.

An example of how a formal observational approach scheme works is presented in the Department of Energy's "Principles of Environmental Restoration" program (DOE/EH/(CERCLA)-002, February 1997), formerly the "Streamlined Approach for Environmental Restoration" (SAFER, DOE/EH-94007658). Detailed information on the program is available from Department of Energy's Office of Program Initiatives (EM-47) (call 301-903-7791). This SAFER scheme has two main elements:

1. Data quality objectives (DQOs), which define what the data collection effort is to achieve and the type and quality of data needed for problem resolution
2. An observational approach, which is the framework for managing uncertainty and for planning decision-making.

Some of the objectives of SAFER, besides explicitly recognizing and managing uncertainty, include linking decision-making needs directly to data collection; learning and applying new data as remediation proceeds; and ensuring participation and consensus from stakeholders. The key to the whole process is converging on an early remedy.

The SAFER framework has three main components:

1. Planning
2. Assessment and selection
3. Implementation

The planning steps culminate in defining what site conditions may arise that change the view of the problem (i.e., the deviations from probable conditions expected) and in developing initial decision rules. Preliminary contingency plans are made for those deviations considered reasonable. The types of decision rules include sampling strategy, cleanup goals, remediation technology selection, monitoring of cleanup attainment, and detection of deviations. The last two decision rules, monitoring and detection of deviations, are used during the implementation steps.

During the assessment and selection steps, remedial alternatives are evaluated, decision rules are refined, additional data are collected (if needed), and the remedy and contingency alternatives are selected.

The implementation steps start with designing the remedy, the contingency plans, and a monitoring plan. An example of the development of contingency plans is given in Table 2-1 from the Weldon Springs project, the site of an old quarry used for the disposal of radioactive waste and pesticides and where explosives had been used for quarry operations.

After a remediation system is operational, potential deviations from the probable conditions are monitored. If there are no deviations, remediation is completed when the decision rule for remediation is satisfied (i.e., when the cleanup goal has been met). If a deviation is detected, the contingency plan is implemented, conditions are monitored, and decision rules are refined if needed.

2.3 Treatability Studies

Treatability studies consist of either bench-scale or pilot tests, designed to evaluate the operating performance of a given remediation system using actual samples of contaminated media from the site, or on a limited scale at the site under actual field conditions. Groundwater treatability studies provide the opportunity to test both the wells and the treatment process. Soil venting pilot tests help determine the radius of influence of soil vapor extraction wells and provide data on the initial vapor concentration that can be extracted and on the system requirements for adequate treatment of this contaminated vapor stream. For testing an ex situ process on shallow soils, excavating a trench for treatability samples provides the opportunity to better define the extent of near-surface contamination and to test the treatment process.

Other goals include determining the following:

- Optimum values of treatment parameters
- General operating conditions
- Characteristics and quantities of products of treatment

Table 2-1 Example contingency plans from the Weldon Springs project.

Contaminants/ Media	Probable Conditions	Reasonable Deviations	Contingency Plans
VOC vapors	Less than 5 ppm$_v$	Higher levels	-Mechanical ventilation -Covers
Radon vapors	Greater than 3 picocuries/L progressively lower	Higher levels	-Water sprays -Reduced area -Surface seal
Radioactive dust	Less than respiratory action levels	Higher than action levels	-Water sprays -Surface seal -Covers
Nitroaromatic dust	Less than respiratory action levels	Higher than action levels	-Water sprays -Surface seal -Covers
Gamma radiation	Exposure less than 10 milirem/hr	Higher levels	-Limit working area -Remote excavation
Explosives	Concentrations less than 10%	-Higher levels -Hot spots	-Water sprays -Oil/water spray
Groundwater	-Gradient reversal -Little chance for migration	-Preferential pathways -Rapid transport	-Interceptor wells -Cut-off wall -Shut-down of municipal well(s)

- Determining treatment effectiveness is usually a prime goal of treatability studies.
- Chemical usage
- Utilities consumption
- pH requirements
- Emissions

Evaluating each of these items can be essential in developing an optimal process design.

Bench-scale tests are often conducted using laboratory glassware. Both batch equilibrium and dynamic flow-through tests are done. Water flow-through rates are usually of the order of milliliters per minute. Bench-scale air flow-through rates are usually of the order of 1 L/min.

Pilot-scale tests are conducted for both batch processes and flow-through processes. Flow-through rates are usually on the order of liters per minute. Scaling up the design to full-scale flow rates is usually a straightforward step from pilot data. However, scaling up from bench-scale data may be risky. It frequently is effective to complete one or more bench-scale tests first, then use bench-scale data to develop pilot test conditions. Examples of remediation treatability studies are given in Chapters 3 through 11. Treatability studies are not used for all technologies or with all waste types. In cases in which sufficient experience exists, treatability studies are not needed.

2.4 Process Flow Diagram

2.4.1 Main Parameters and Mass Balance

A process flow diagram is a schematic representation initially showing:

- Temperatures of flowing streams
- Pressures of flowing streams
- Total flow rates (volumetric and mass flow)
- Compositions of flowing streams (individual concentrations and mass flow rates)

Some form of such a flow schematic is needed for conceptual design. As detailed mechanical design work proceeds, additional information can be shown on the diagram. With mass flow rates for each component included, the document becomes useful for specifying equipment, applying for permits, and for cost estimating.

Flow conditions and composition information for each flow stream can be shown in a box next to each major line or in a table on the diagram. An example of part of a flow sheet and a typical table is shown in Figure 2-1. The mass flow rates (lb/day in this example) for each line are determined by multiplying each concentration by the volumetric flow rate and converting it to the desired units. The example for constituent "ABC" is as follows, with the 22,000 µg/L concentration expressed as 22 mg/L, as shown in Equation 2-1.

$$(22 \text{ mg/L}) (50 \text{ gal/min}) (3.78 \text{ L/gal})/[(453,700 \text{ mg/lb}) (1440 \text{ min/day})]$$
$$= 13.2 \text{ lb/day} \qquad (2\text{-}1)$$

When this example diagram is completed, the steam line would be given a stream number, and the steam flow conditions would be added to the table. The tower overhead line would be added with its flow conditions. The mass balance would require that the amount of ABC in the overhead equal the difference between the amount in Stream 11 and Stream 12 (i.e., 13.2 lb/day - 0.27 lb/day). If the air stream, not included in the table as a constituent of concern, was accounted for, the total lb/hr for the tower overhead must equal the total net lb/hr of constituents stripped out of the groundwater plus the air flow rate. For example, if the air rate is 10,000 lb/day, a mass balance (on a dry basis) shows that the total overhead rate must be 10,103.7 lb/day. See Equations 2-2 and 2-3.

Net constituents stripped = (Stream 11 benzene + ABC + Z) - (Stream 12 benzene + ABC + Z)
= (32.4 + 13.2 + 58.7) lb/day - (0.0027 + 0.27 + 0.33) lb/day = 103.7 lb/day   (2-2)

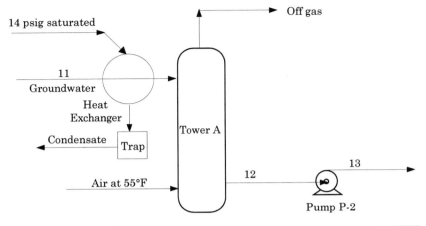

LINE NO.		11	12		13		
SERVICE		FEED	P-2 SUCTION		EFFLUENT		
LB/HR		25,000	24,896		24,896		
GAL/MIN @ 60°F		50	50		50		
TEMPERATURE		55°F	135°F		135°F		
PRESSURE		30 psig	0.2 psig		21 psig		
CONSTITUENTS		µg/L	lb/day	µg/L	lb/day	µg/L	lb/day
	Benzene	54,000	32.4	4	0.0027	4	0.0027
	ABC	22,000	13.2	400	0.27	400	0.27
	Z	98,000	58.7	490	0.33	490	0.33

Figure 2-1 Mass flow rate tabulation.

Overhead flow = Constituents stripped + Air = 103.7 lb/day + 10,000 lb/day
= 10,103.7 lb/day   (2-3)

Software for process simulation is available. Such software aids in developing flow diagrams and completes mass and energy balances for a variety of unit processes. An example is Enviropro Designer™ from Intelligen, Inc. This program is useful for wastewater treatment and air pollution control system design.

2.4.2 Energy Balance

The sum of the mass rates feeding into a treatment device or auxiliary equipment must equal the sum of the mass rates exiting the system. When there are temperature changes or changes in state, the law of conservation of energy applies, and energy input must equal the energy content of exiting streams. The flow sheet can also show temperatures and/or enthalpies derived from energy balances. An example of a partial energy balance is as follows:

Stream 11 is warmed to 135° F with 14-psig saturated steam in auxiliary equipment — a heat exchanger. Find the duty (in British Thermal Units, Btu/hr) of the heat exchanger and the amount of steam flow. The enthalpies of water and steam from a steam table (with a zero-base enthalpy at 32° F) are as follows:

	135° F Water	55° F Water	Saturated steam	At 14-psig condensate
Btu/lb	102.9	23.1	1163.2	216.1

The heat exchanger duty based on Stream 11 is shown in Equation 2-4.

25,000 lb/hr (102.9 Btu/lb - 23.1 Btu/lb) = 1,995,000 Btu/hr   (2-4)

The amount of steam required, assuming that the exchanger is well insulated, is shown in Equation 2-5.

(1,995,000 Btu/hr)/(1163.2 Btu/lb - 216.1 Btu/lb) = 2106 lb/hr   (2-5)

2.4.3 Sizing and Rating of Equipment

The rating of equipment can be calculated for heat exchange duty or obtained from vendor information as detailed design work proceeds. For example, the horsepower rating of a pump motor or a blower motor depends on the efficiency of the pump or blower. The kilowatt rating of a UV light/peroxide reactor depends on the manufacturer's design. The diameter of a stripper and the depth of packing required can be either calculated or obtained from vendors. The square-foot transfer area of a heat exchanger can be calculated or obtained from vendors. The diameter of a sand bed filter can be calculated, but a vendor's standard size (possibly somewhat larger than the calculated size) will likely be chosen.

The process flow diagram is enhanced if the sizes and the ratings are added as such information becomes known. This diagram, along with the P&ID and plot plan, can fairly completely define how the system will operate and what its main physical dimensions will be.

2.5   The Site Plan and Preliminary Plot Plan

The site plan is drawn to scale and usually shows the property boundaries, adjacent and nearby roads, and the following items:

- A. **The outline of the plot where treatment equipment will be** – If the plot is to be fenced, this is designated.
- B. **Buildings** – Major buildings existing on the site, plus minor existing buildings near or within the plot, should be shown. New buildings associated with the treatment project should be shown.

C. **Road or driveway access to the plot** – Any gates or security check points should be shown. In making this layout, thought should be given to access and turnaround areas for both emergency vehicles and supply trucks.
D. **Elevations of the plot grade and of distinctive areas near the plot and on the access road, or contours showing elevations** – Surface drainage features should be indicated.
E. **Any existing berms, dikes, levees, canopied areas, drainage piping, sewers, culverts, etc., near or on the plot, near wells, or near proposed new piping or conduit runs**
F. **Existing or proposed wells or trenches associated with the project** – Wells should be clearly designated whether they are for extraction or injection. A typical wellhead detail and dimensions of any boxes or vaults for well heads and valves should accompany the plan.
G. **Piping and conduit runs connecting well heads** – The piping and conduit should be designated as above ground or buried. If buried, the depth below grade should be indicated; if this depth varies, a profile or sectional views should accompany the plan. Existing or interfering nearby structures and utilities (whether above ground or buried) should be highlighted, and required clearances or notes should be added, resolving potential interferences.
H. **Locations of utility interconnects and utility runs associated with the project** – Utility runs may include natural gas, fuel oil, electric power, process sewers, sanitary sewers, storm drains, fire water, potable water, process or cooling water, compressed air, steam, and heat transfer oil. Any piped nitrogen or oxygen should be shown. Fuel tanks, such as for propane and fuel oil, should be indicated and designated as above ground or buried. Secondary containment requirements for fuel oil tanks should be drawn or described with notes on the plan. If water lines can potentially freeze, notes explaining freeze protection should be added.
I. **Storage tanks for process fluids or water, with secondary containment requirements** – If nitrogen or oxygen is to be stored on the site, the locations of such storage facilities should be designated.

The preliminary plot plan focuses on and enlarges Item A (treatment equipment location) in this list. It shows to scale the plan view of where major equipment will be placed. The dimensions may not be known, so in some cases it may be necessary to block out an area of reasonable size and note what components will be within the blocked area. If all or part of the plot is to be fenced, bermed, or diked, these features should be noted. The locations where main process pipes and utilities leave or enter the plot should be designated. The locations for doors on buildings and any storage sheds should also be shown.

As the plot is laid out, thought must be given to operator access, forklift or crane access, and emergency vehicles access.

2.6 P&ID and the Sequence of Operations

A P&ID schematically shows every pipe, valve, and instrument in the system, as well as the manner in which control loops are linked from sensors/transmitters to controllers to control valves. Also, all alarms, shutdowns, interlocks, and nominal pipe diameters and valve diameters are shown. Each pipe wall thickness is shown, either as an actual thickness or as a pipe schedule. Pipe insulation thicknesses are shown. For simple treatment systems (e.g., pumping from a well through two carbon vessels), the P&ID and process flow diagram can be combined into one control flow diagram. Such a diagram might show the few valves involved, a high-pressure switch, pressure gauges, and the piping/valve diameters, in addition to the flow rate and composition, all on one sheet.

Given a proper P&ID, preliminary plot plan, and drawings from the major equipment manufacturers, an experienced designer or draftsperson can produce the finished, dimensioned, detailed drawings suitable for construction. (The main remaining information needed for construction includes specifications, drawings showing buildings and civil work, and electrical diagrams and drawings.)

2.6.1 P&ID Development

Normal progression in P&ID development is as follows:

A. **Main lines connecting the equipment are drawn**, showing the block valves, check valves, and sample valves. Semi-standard design features are included, as desired at this time. Examples include a pressure gauge, check valve, and block valve on the discharge of each centrifugal pump; manifolding to allow the switching of lead and follow vessels in adsorbent systems; and block and bypass valves for equipment that can be momentarily bypassed.

B. **Pressure controls are added**, including any relief valves and high- or low-pressure alarms and switches. Two types of modulating pressure controls are frequently used:

   1. Pressure-regulating valves, in which either the downstream or upstream pressurized fluid is used directly to position the valve
   2. Pressure control loops, in which a pressure sensor/transmitter signals to a controller, and the controller in turn sends a signal out that corrects the position of a continuously modulating control valve

   On/off pressure control is rarely used in treatment applications, except for the air receiver vessel on an air compressor unit. The control system switches the compressor off when rising pressure reaches a preset level and switches the compressor on again when falling pressure goes below a lower preset level.
   Pressure gauges or pressure recorders are added to the system. Or a controller can be specified that indicates or records pressure. Instruments that indicate or transmit differential pressures are added.

C. **Level controls are added**, including any overflow piping and high- and low-level alarms and switches. In conjunction with pumps, a modulating control loop with a sensor/transmitter, controller, and control valve is used, or on/off switches are used. Level indicators or level recorders are added to the system, or controllers are specified accordingly.
D. **Temperature controls are added**, including emergency cooling systems and high- and low-temperature alarms and switches. Almost all treatment plant temperature control systems are modulating control loops, each with a sensor that directly signals a controller (such as a thermocouple) or a sensor plus a transmitter, with a controller and a control valve. On/off thermostats are rarely used in treatment applications, except for building heating/ventilation systems. Thermometers or temperature recorders are added to the system. Or a controller can be specified that indicates or records temperature.
E. **Flow controls are added**, including high- and low-flow alarms and switches. Two types of modulating flow control are used:

   1. A flow control valve, which automatically adjusts internally to maintain a preset adjustable rate.
   2. Flow control loops, in which a flow sensor/transmitter signals to a controller, and the controller in turn sends a signal out that corrects the position of a control valve.

F. **Automatic analyzer indicators are added**, such as pH meters and gas concentration indicators, sometimes with high and low switches. In some instances, analyzer controllers are used, either to modulate a control valve or to switch pumps off and on based on the system reaching specific set points.
G. **The diagram is reviewed**, and any remaining valves are added. Every valve is then studied individually, and a decision is made as to what type of valve should be indicated for each one. The most common valves used in treatment plants are ball valves, globe valves, gate valves, and butterfly valves. A different valve symbol is used on the drawing for each type of valve used.
H. **Each automatic control valve is studied individually**, and a decision is made as to what failure position applies (closed, open, or remain in the last position). The valve action subsequent to signal failure should be indicated for each control valve.
I. **The pipes and valves are sized preliminarily**, and the sizes are indicated on the diagram. (The sizes are checked and adjusted as needed in a later issue of the diagram, after the mechanical layout drawings are done and piping pressure drops have been calculated.) Copps (1995) gives equations for determining optimum piping diameters based on life cycle analyses. Adams (1997) suggests Equations 2-6 and 2-7 for determining a liquid line nominal diameter, d (in inches), at flow rate Q (in gpm).

For pipes 2-in. and smaller: $Q = d/2(d+2)^3$ (2-6)
For pipes greater than 2-in.: $Q = 1.2(d+2)^3$ (2-7)

Alternatively, the preliminary sizing can be done based on the following rules:

1. Pressurized liquid lines having a nominal diameter of 1- to 6-in. are sized so that the velocity is approximately 6 ft/sec. Eight-in. lines are sized at 8 ft/sec, and 10-in. lines are sized at 10 ft/sec. (These velocities result in a reasonable economic balance between increased investment costs with larger diameters and reduced pumping horsepower costs as a result of lower friction head losses. Using larger diameters at lower velocities might result in a net present worth savings if the plant life cycle is long. However, these suggested velocities are needed if settling in the horizontal piping runs is to be avoided and suspended solids are present in sizes ranging up to 200 μm.
2. Centrifugal pump suction lines are sized at a velocity that is 2 ft/sec less than each velocity given above for pressurized lines.
3. Air ducts and stacks are sized at a velocity of 50 ft/sec.

Valves are assumed to be the same size as the connecting pipe unless indicated otherwise on the diagram for special instances. Special instances include the following: (1) Connecting piping and a valve are changed only at a pump, blower, or vessel to match the manufacturer's fitting size. (2) A valve is made smaller because it is in throttling service or it is a control valve designed to reduce pressure or modulate flow. (3) A valve is used to bypass an equipment component on occasion. If a valve is a different size than the connecting pipe, its size is noted next to the valve. The pipe sizes are usually shown next to a piping run or included in an oval attached to the piping run. The oval may contain this type of information: 4"-116-40-CS, in which 4-in. is the nominal pipe diameter; 116 is the line number, which can be arbitrarily assigned to each run if desired; 40 is the schedule that designates a standard pipe wall thickness; and the letters designate the material as given in a legend. In this example, CS is carbon steel. Figure 2-2 shows an example of a pipe run with varying valve sizes and with insulation.

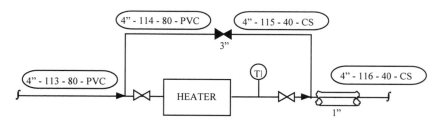

Figure 2-2 P&ID piping symbols.

J. **For each insulated run of pipe, an insulation symbol is added** with the insulation nominal thickness noted nearby. If heat tracing is involved, an appropriate symbol is included.
K. **Each instrument is given two to four letters that symbolize the type of instrument it is**, contained in the upper half of a circle (also known as the bubble). Usually, the lower half of each circle is assigned an arbitrary

48

instrument number at some stage in P&ID development. The best way to assign instrument symbols is to use, wherever possible and where not ambiguous, the symbols listed by the ISA (the Instrument Society of America, the former name of the International Society for Measurement and Control). A legend should be given either on the P&ID or accompanying it.

The main pattern used for instrument symbols and some examples are illustrated as follows:

1. The first letter indicates function, such as:
   P = Pressure
   L = Level, usually liquid level
   T = Temperature
   F = Flow
   A = Analyze
2. The second letter indicates type, such as:
   I = Indicator
   R = Recorder
   C = Controller
   A = Alarm
   S = Switch
   T = Transmitter
   E = Sensor (element)
3. A third letter is not always used; it indicates whether the function value is high or low or whether an indicator or recorder is also a controller or transmitter. Examples include:
   PAH = High-pressure alarm
   FSL = Low flow switch
   LSL = Low level switch
   TSH = High temperature switch
   FIC = Flow indicator/controller
   TRC = Temperature recorder-controller
   TIT = Temperature indicator-transmitter
4. A fourth letter is rarely used; sometimes it designates an extremely high or low condition, such as in these examples:
   LSLL = Low low level switch, used in conjunction with an LSL, at a somewhat lower level setting
   TSHH = High high temperature switch, used in conjunction with a TSH, at a higher temperature setting

Other logical four-letter combinations may be used, as long as their meaning is clear in the legend. For example, dPIT or PDIT could mean that the instrument indicates and transmits differential pressure.

L. **Signals from transmitters and from controllers are shown on the P&ID** as a connecting dashed line for electric or electronic signals and a double-hatched solid line for pneumatic signals. If an electronic signal is transduced to a proportional pneumatic signal, the signal lines are joined by a circle with the symbol I/P or I-P.

M. **Instruments that will not be mounted in the field**, such as on a control panel in an on-site control room or a remote office, **are shown** with a horizontal line across the diameter of the instrument schedule. Sometimes a double line is used to indicate mounting on a small field control panel hanging on or adjacent to operating equipment. Or the legend may assign another use for the double line and indicate LB or LP for a local board or local panel.

2.6.2 Pressure Instrumentation

Suppose the desire is to control the pressure from a multistage centrifugal high-pressure pump in a system feeding groundwater to a wet air oxidization reactor. Figure 2-3 shows a downstream pressure control loop that continuously modulates the pressure. The desired pressure is fed in by manually positioning the set point adjustment at pressure recorder-controller PRC-17. This controller receives the pressure signal from pressure indicator-transmitter PIT-17 and sends out a signal with a strength proportional to the deviation from a set point.

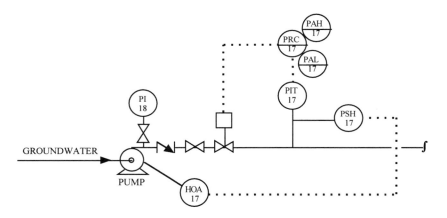

Figure 2-3 An example of the use of pressure control.

Suppose the desired pressure is 270 psig and the reactor is rated for 300 psig. The high-pressure alarm PAH-17 in the figure could be set at 280 psig. The high-pressure switch PSH-17 could be set at 290 psig. If the pressure ever reaches this amount, PSH-17 will open an electric circuit connected to the pump motor controls. In this example, when PSH-17 trips, the hand-off-automatic switch HOA-17 will cause the pump motor to shut off. HOA-17 is a three-position switch: if the switch is set in the hand position, the pump motor is on; in the off position, the motor is off; and in the auto position, the motor shuts off in the event of an automatic signal.

Note that PAH-17 and PAL-17 are triggered by switches that are part of PRC-17. PSH-17 could also be triggered by another switch as a part of PRC-17. Such an arrangement would cost less than using the independent PSH shown. However, PSH-17 should be independent of the controller, so that it will work in the event that the controller or the transmitter feeding the controller fails.

2.6.3 Liquid Level Instrumentation

An example of a continuously modulating level control loop is shown in Figure 2-4. The bottom of a stripper is a sump providing suitable suction head for pump P-2. The liquid level is sensed and transmitted by LT-16. (A summary of liquid level measuring/transmitting devices is given by Parker [1995].) The desired level is fed in by manually positioning the set point adjustment at level indicator-controller LIC-16. Level transmitter LT-16 signals LIC-16, which in turn sends out a signal with a strength proportional to the deviation from the set point.

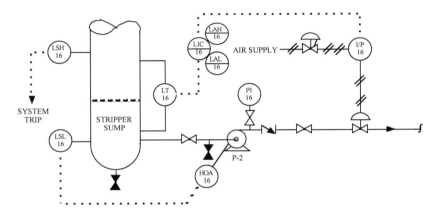

Figure 2-4 An example of the use of continuously modulating level control.

LIC-16 is depicted as an electronic controller; dashed lines show signals in and out. Conventional electronic controllers work with a signal strength in the range of 4-20 milliamps. A 12-milliamp signal (i.e., in the middle of the range) from LT-16 might indicate that the level in the stripper sump is momentarily at the desired set point.

The control valve is depicted with a pneumatically energized diaphragm operator, with the incoming pneumatic signal fed from transducer I/P-16. This transducer is needed because the controller sends out an electronic signal. Conventional pneumatic control valves and instruments work with a signal strength in the range of 3-15 psig. A 12-milliamp signal from the electronic controller in this example is transduced by I/P-16 to a signal of 9 psig (in the middle of the range).

Instrument air from a compressor-filter-dryer system (not shown in the figure) is fed to the transducer at a pressure somewhat greater than 15 psig, perhaps 20 psig. A downstream pressure-regulating valve is show in the figure, which can be manually adjusted to supply air at 20 psig.

The high- and low-level alarms, LAH-16 and LAL-16, are triggered by switches that are part of LIC-16. High- and low-level switches LSH-16 and LSL-16 are independent of LIC-16 and LT-16. Thus, a failure at LIC-16 or LT-16 will not affect the proper functioning of the level switches.

In the event of an extremely high level, LSH-16 causes an electric circuit to open, resulting in a complete shut-down of the stripper system or, at the least, a shutdown of any pumps feeding the stripper. In the event of an extremely low level, LSL-16 would signal hand-off-automatic switch HOA-16 to shut down the motor driving centrifugal pump P-2. This is important because a loss of suction head can damage a working centrifugal pump.

It would be good design practice to have two parallel signals from LSL-16. One signal would shut down pump P-2 as described. The parallel signal would cause a complete shutdown of the entire stripping system.

Not shown in the figure are flashing lights and/or audible alarms that would notify the operator that a shutdown switch has tripped. Such an alarm is a shutdown notification, not a warning. It could be shown as an alarm bubble connected to the shutdown switch bubble. However, it simplifies the P&ID to leave out such notification alarm bubbles and to merely state in a general note that alarms will notify if any shutdown switches are tripped.

An example of an on/off liquid level control system is shown in Figure 2-5. With this type of control, the high- and low-level switches (the LSH and LSL) are used for normal conditions, not to shut down the stripping system. Instead of holding the liquid level at one location (as with modulating control), the level is purposely allowed to fluctuate between two preset locations.

When the liquid level rises to the location where LSH-14 is tripped, pump P-11 is shut down. Pump P-11 stays shut down while P-1 continues to function, drawing down the liquid level in the tank until LSL-14 is tripped, at which point P-11 automatically starts.

If this on/off pumping scheme with P-11 fails to function properly and the level rises extremely high, LSHH-14 shuts down P-11. With this possibility in mind, it would be prudent to design the system so that manual reset is required before LSL-14 is allowed to function normally. This procedure could require the operator to attempt to correct the fault that caused the high high level to occur.

If a failure occurs and the level drops extremely low, LSLL-14 shuts down pump P-1 immediately, because P-1 must have appropriate suction head. LSLL-14 is used to shut down the entire system (requiring a manual reset to restart the system) until the fault is investigated.

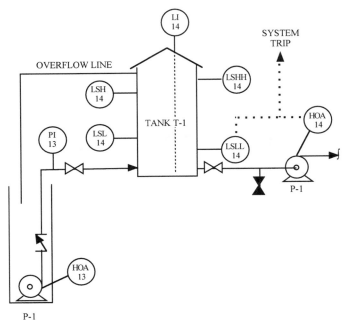

Figure 2-5 An example of the use of on/off level control.

2.6.4 Temperature Instrumentation

Figure 2-6 shows a catalytic oxidizer in a soil venting system. For simplification, not all the valves, burner controls, nor the flame arrestor is shown. In this example, the oxidizer manufacturer has recommended that at least 650° F be maintained ahead of the catalyst block. A natural gas-fired burner is controlled with temperature recorder-controller TRC-1 accordingly. This controller receives a signal proportional to the temperature from temperature transmitter TT-1A, which in turn receives a temperature signal from temperature sensor TE-1A. This sensor is located in thermowell TW-1A. TRC-1 sends a signal to control valve TCV-1 in the gas burner train, with a strength proportional to the deviation from the set point. In this example, the set point is 665° F in order to be certain that the actual temperature does not dip below 650° F, as recommended by the manufacturer.

Note that TRC-1 is a two-pen recorder. One pen records the temperature ahead of the catalyst, as described above, and another pen (TR-1) shows the final discharge temperature, after the catalyst. The manufacturer's design in this example limits the discharge temperature to 1,200° F. This maximum discharge temperature poses a problem. The flammable vapor content of the soil gas fluctuates. At a high vapor concentration (which occurs each time a new well is brought into service), the temperature rise through the catalyst can be excessive. The burner control valve

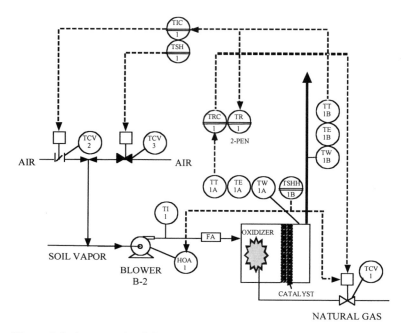

Figure 2-6 An example of the use of temperature control (Used with permission, CSM Environmental Systems, Inc., Mountainside, New Jersey).

TCV-1 may automatically be at its minimum opening, and the discharge temperature keeps rising. Either of these two alternative automatic control systems for admitting dilution air can be used for safe operation:

1. Continuously or semi-continuously analyze the flammable vapor content, and then apply an analyzer-controller system to control the positioning of the air dilution control valve.
2. Use another temperature controller to monitor and control the discharge temperature by continuously controlling the position of the dilution air inlet valve.

The figure shows the second choice using TIC-1. If the discharge temperature starts to rise above 1,100° F, dilution air control valve TCV-2 starts opening. TIC-1 modulates the opening of TCV-2 to hold the discharge temperature at or below 1,100° F.

If TCV-2 opens all the way and the temperature rises to 1,125° F, TSH-1 is triggered and causes TCV-3 to open all the way. TCV-3 is another air dilution valve, but it is on/off (fully open or closed) rather than modulating. If the discharge temperature falls to a preset value, TCV-2 quickly closes all the way. CSM Environmental (New York) uses this type of control in some of their catalytic oxidizer systems.

If TCV-1 and TCV-2 are open all the way (or there is a malfunction) and the temperature continues rising above 1,125° F, TSHH-1 shuts down the entire system at 1,150° F. In this event, the natural gas control valve closes all the way, and blower B-2 is shut down.

2.6.5 Flow Instrumentation

Figure 2-7 shows flow control valve FCV-15. Over a range of pressures, this type of valve holds the flow rate constant. It can be adjusted to a desired flow rate. No external control loop is involved. The flow rate can be monitored with a separate flow indicator or flow recording system. The flow monitoring system shown in the figure includes an indicating flow totalizer. A totalizer can, for example, show the total gallons processed since the date the instrument was manually reset to zero.

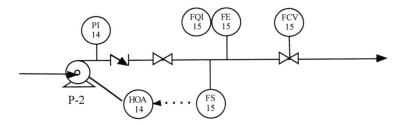

Figure 2-7 An example of the use of a flow control valve.

As with all centrifugal pumps, P-2 should operate above the minimum flow rate prescribed by the pump manufacturer. Flow switch FS-15 is set to shut down P-2 at that flow rate. The electric circuitry must be set up such that the shutdown signal from FS-15 is momentarily bypassed or is time-delayed when starting up.

Another method of flow control is shown in Figure 2-8. This is an example of a conventional flow control loop. Flow sensor FE-18 is a device within the pipe or attached to the pipe that emits signals proportional to either the flow rate or to the square of the flow rate. (Information on selecting flow sensors and meters is given by Dolenc, 1996.) Flow transmitter FT-18 sends such a signal to FRC-18. If needed, FRC-18 includes a square-root extraction module that computes a signal proportional in strength to the flow rate. FRC-18 sends a signal to the control valve that is proportional in strength to the deviation from the set point.

The figure shows some important features in control valve system designs that are often used for all control valve applications. The control valve in this example is marked F.O (fails open). In the event of a signal failure or power failure, the valve

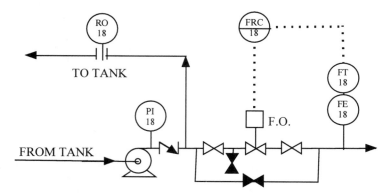

Figure 2-8 A flow control loop.

operator, energized by a backup battery, drives the valve to the full open position. In other examples, it may be desirable to use fail-close control valves, marked F.C. Or a pneumatic control valve may be marked A.C. (air closes), which means the valve spring drives the valve to the full open position in the event of a signal failure or loss of instrument air. A.O. (air opens) means the valve spring drives the valve toward being closed. The P&ID legend should show what designations are used. Either no designation or a specially defined designation could be used to indicate that control valves stay in the last position should there be a failure of signal or of power.

The control valve manifold shows blocks, a bypass, and a bleed. In the event that the control valve needs inspection or repair while the process unit continues to operate, the bypass globe valve can be adjusted manually to stay near the desired flow rate while the block valves on either side of the control valve are shut. Before disconnecting the control valve, fluid is drained from the 0.75-in. bleed valve into a container to minimize potential spills when the control valve is temporarily removed.

Note that the pump discharge manifold in this example includes a restrictive orifice, RO-18. This is an alternative method of ensuring that the pump is operating above the manufacturer's recommended minimum flow rate. If the main pump discharge flow is blocked, such as when the control valve closes all the way, the RO ensures that minimum flow is maintained, with the flow path being from the pump discharge back into the tank.

This configuration wastes some pumping energy but is easier to maintain and to start up than one with a flow switch. Waisvisz (1987) gives a method of sizing the orifice hole diameter. Or pump vendors will provide correctly sized orifices with their pumps.

Another method of ensuring minimum flow is to have the control valve designed to not close all the way.

2.6.6 Analysis Instrumentation

The signal from an automatic analyzer can be transmitted to a controller that sends out a signal proportional to the deviation from the set point. Alternatively, an on/off function can be performed based on preset values for the analyzer. For example, a pH analyzer-controller system can be used to adjust the stroke of a reciprocating metering pump for injection of acid or base. Or an oxygen detector in an incinerator flue gas stream can cause a shut-off, preventing organics-contaminated soil from being fed to the incinerator in the event of low oxygen concentration.

2.6.7 Sequence of Operations Development

The Sequence of Operations documentation is a terse set of notes on the P&ID or accompanying it, which serves to accomplish the following:

- Explain process steps
- Define control interlocks
- Give reset/start up conditions

The following is a typical Sequence of Operations description that might be developed for a groundwater extraction/steam stripping system:

1. The stripper is directly heated to 175° F with low-pressure steam injection.
2. The overhead is condensed with water-cooled heat exchanger E-2.
3. Two liquid phases are recovered in accumulator D-1. The floating phase is decanted each week to a drum that is picked up by a recycling firm. The water phase is treated with activated carbon and returned to the steam boiler.
4. If the low-temperature switch TSL-3 is triggered (at 145° F), the following equipment/streams shut down automatically:
    - Groundwater recovery pumps P-1A and P-1B
    - The transfer pump from the oil/water separator
    - Steam flow
    - Stripper bottoms pump P-2
    - Condensate recovery pump P-4
5. System restart requires manual starting of all pumps after liquid levels are established and checked by the operator. Steam flow will restart automatically if the boiler is functioning.

2.7 Logic Diagrams

The main diagram that logically follows the development of P&IDs is a logic diagram, from which a ladder diagram can be prepared, showing how the logic is applied in electric circuits.

Logic diagrams show how signals from sensors, transmitters, or controllers are processed and what the results will be electronically from an automatic decision. An example is given in Figure 2-9 for the pumps shown in Figure 2-1.

In the example, when the three-position manual switch is in the auto position, the motor for steam stripper feed pump P-1 shuts down if there is either low steam pressure (PSL-1) or low groundwater flow (FS-1).

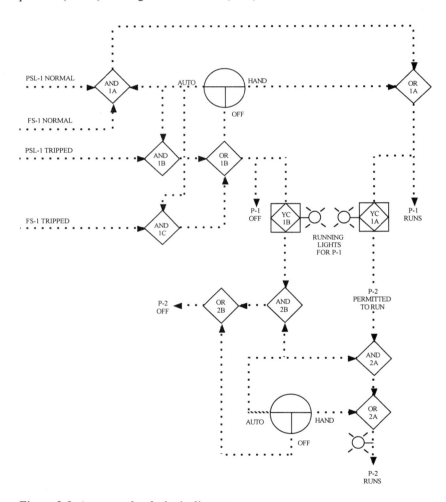

Figure 2-9 An example of a logic diagram.

The decision is made at "AND GATES" (AND-1B and AND-1C), which represent relays in series that open the electric power circuit connected to the motor with either low steam pressure or low flow when the three-position switch is in the auto position. An "OR GATE" (OR-1B) also controls the shutdown of the pump motor if the three-position switch is set to the off position. OR-1A lets the motor for pump P-1 run if the three-position switch is set in the hand position or auto position.

This last condition lets pump P-1 run only if AND-1A has closed contacts for PSL-1 (steam pressure is adequate) and for FS-1 (flow is adequate). Note that to start pump P-1 when the three-position switch is on auto, the electric circuit must be arranged such that the contacts for FS-1 stay closed for a period of time until the groundwater flow is established by pump P-1.

The diagram also shows that pump P-2 is interlocked with pump P-1, i.e., it cannot run when its three-position switch (not shown) is on auto unless P-1 is running and a signal is coming from YC-1A. YC-1A also signals a running light to go on when P-1 is running

Again referring to the example, acid injection pump P-2 will run if P-1 is on and the three-position switch for P-3 is in the auto position. This condition is controlled by an AND gate (AND-2A). P-2 is off when P-1 is off (these two pumps are interlocked) or when the three-position switch is in the off position as controlled by an OR gate (OR-2B). Also, OR-2A permits P-2 to run with the three-position switch in either the auto or hand position.

As mentioned, a ladder diagram showing how the logic can be applied in electric circuits can be prepared in conjunction with the logic diagram.

2.8 Computerized Controls and Process Monitoring

Conventional modulating controllers receive a signal from a sensor or transmitter and automatically derive the deviation from the set point. They then transmit a signal to a control valve. The signal to the control valve is proportional in strength to the amount of deviation from the set point. Many controllers can be tuned into three different modes to improve control precision:

- Proportional control or gain
- Integral control, bias control, or reset action
- Derivative or rate action

Three-mode tuning, with proportional, integral, and derivative features, is called PID control. By adjusting the gain, the strength of the signal from the controller (and the corresponding amount of control valve movement) can be multiplied. By using the integral control (adjusting the amount of bias), a constant average deviation (a constant offset) from the set point can be eliminated. By adjusting the amount of derivative action, the amount of control valve movement can be made proportional to the rate at which the deviation from the set point is changing.

Conventional modulating controllers usually also indicate, or record with a pen on a chart, the value of the controlled parameter, such as the amount of pressure, fluid level, temperature, flow rate, or analyzed strength.

A conventional modulating controller can also be used to open or close electric contacts at preset discreet points. Typically, this capability is used as an alarm or shut-down function in the event of extreme pressure, fluid level, temperature, flow rate, or analyzed strength.

2.8.1 Computer Functions

2.8.1.1 Adapting Computers

A computer or a programmable logic controller (PLC) can be designed to perform all of the functions of a number of conventional modulating controllers---sending out signals to position control valves, indicating or recording values, and causing the opening or closing of electric contacts. The modulating control functions can be tuned by manually setting the amounts of gain, bias, and/or derivative action. A computer can also be programmed to automatically adjust these three modes of control to minimize deviations from the set point for each control loop.

2.8.1.2 Data Manipulation

Computers and, to some extent, PLCs can be programmed to perform calculations, to perform data reduction and reporting in the form of tables and graphs, and to log data. They can be programmed to report the average value of a parameter such as temperature or flow rate over the past 24 hours or the past 1 hour, etc., or the cumulative flow over a variety of time periods. They can be programmed to sense and record which extreme deviation first caused multiple alarms or a shutdown to have been triggered. This capability helps equipment operators determine the basic cause of process conditions deviating too far from set points.

2.8.1.3 Adapting Computers to On/Off Control Functions

Conventional on/off controls usually employ a sensor/switch device, which triggers an electrical relay, solenoid valve, or motor starter action. Some typical examples are the opening or closing of a line feeding a chemical additive and the start up or shutdown of a pump or compressor. By using an adjustable time-delay relay in the electric circuit, manual or automatic intervention can be used to perform corrective action or to dampen the frequency of on/off actions.

A computer or PLC can perform on/off control and time-delay functions, as well as carry out alarm, shutdown, and modulating control functions. Modulating control can be accomplished with either analog or digital signals. On/off functions, alarms, and shutdowns are accomplished with digital signals. The control system is purchased with a prescribed capability to handle limited numbers of analog input signals, analog output signals, digital input signals, and digital output signals. Sometimes the system is designed such that it can readily be expanded to handle additional analog or digital signals.

2.8.1.4 Graphic Displays

Conventional controllers and status lights (e.g., colored lights indicating that a pump is running or shutdown) can be placed on a graphic control panel. Symbols and connecting lines are drawn on the graphic panel so that the display represents a simplified process flow diagram. A computer can be programmed to produce a similar display. However, unlike conventional graphic panels, a computer can be used to focus on a particular section of the plant or on a particular control loop or set of loops. That section can be expanded visually to display more details on a screen or to display a portion of the P&ID.

2.8.1.5 Automatic Restart, Supervisory Control, and Combination Systems

Conventional control systems can be designed so that the plant will automatically start up when an extreme deviation that caused an automatic shutdown has been corrected. Interlocks can help ensure that multiple parameters affected by the shutdown are restored to suitable conditions for successful plant operation. However, with a computerized control system, it is easier to set up the interlocks and to adjust the parameter values corresponding to suitable operating conditions.

Conventional controllers are sometimes integrated with supervisory controllers. The output signal from a supervisory controller is used to automatically adjust the set point of the conventional controller. For example, when a preset temperature is reached, a signal from a supervisory temperature controller is cascaded to a flow controller. The set point of the flow controller is automatically adjusted depending on the supervisory signal strength. Such supervisory or cascade control setups can be readily accomplished with a single computer instead of a number of individual supervisory and conventional controllers.

Necessary computer programming or customized PLC programming can become quite complex when modulating control, data reduction, or sophisticated graphic displays are included with the control functions. With additional complexity, troubleshooting and revising the control system become more complex. Often the number of people capable of operating the system or accessing the data is reduced when complexity increases. A somewhat simple system to set up is the use of a PLC for most of the digital signals and the use of conventional modulating controllers for control valve loops and a limited number of alarm or shutdown functions. With this approach, the PLC can be readily programmed based on electrical ladder diagrams. Making revisions to the control system and troubleshooting are relatively easy when using this approach. Either the PLC or the controllers can be used to monitor or react to extreme deviations, and redundant sensor/switches can be used for shutdowns in the event of a PLC or controller malfunction.

Using such a combination of a PLC and individual controllers will save significant programming and troubleshooting costs, and it may not cost much more for hardware than use of a PLC for all of the main control functions. The impact on hardware costs is minor if the PLC is augmented with external devices that display data and permit the plant operators to dial in set point adjustments.

A common combination is the use of a PLC for control, alarm, and shutdown functions, plus a desktop computer for data manipulation, logging, and displays. Again, the PLC can be readily programmed based on electrical ladder diagrams. A slightly more complex combination would include use of the desktop computer to perform start up or supervisory control functions.

2.8.2 Remote Monitoring

2.8.2.1 Computer-Based Systems

A desktop computer can be connected remotely to a PLC or a computer in the field via a modem and telephone lines or a radio transmission system. The desktop computer can be used for manipulating and logging data, observing displays of data, monitoring alarms and shutdowns, and performing start up or supervisory control functions. One desktop computer can serve any or all of these functions for a number of remote remediation sites.

2.8.2.2 Phone Dialers

Using an automatic phone dialer is the most inexpensive method of signaling a change in status (e.g., alarm or shutdown) to a remote location of a processing unit. It is a one-way, non-computerized monitoring system for status switches only, not for control and not for monitoring of trends in parameter values. A typical phone dialer can transmit four prerecorded messages from an audio cassette player. Examples of messages are as follows:

- "This is a report from the remediation unit at location No. 4 in Centerville. The level in the auxiliary fuel tank storing liquid propane fuel has fallen below 10 %."
- "Treatment Unit No. 2 has undergone an automatic shut-down."
- "The high-pressure alarm at the remediation system carbon canisters has tripped on."

Such a system can also be used to automatically deliver a telephone message that a security fence has been breached (i.e., a burglar alarm has been tripped). Setting up the system is simple wherever telephone lines are available. The messages can readily be set up and later changed as desired. Models are available with battery backup that will work in the event of a power failure.

An automatic phone dialer and answering machine can be purchased for approximately $200. One answering machine can serve multiple remote remediation sites.

2.8.2.3 Computer/Fax

A blending of computer application and simplified remote monitoring is accomplished with an on-site computer and modem connected to a remote facsimile (fax) receiver. A disk-oriented computer can be programmed to perform relay-control functions (such as switching equipment off and on, triggering alarms, and storing data), transmit data, and signal a change in status.

The Para-Fax system, originally marketed by Paragon Environmental Systems (Escondido, California), can be used to transmit a routine daily report summarizing values of parameters such as flows, temperatures, and pressures. The system will also transmit notification of an alarm or shutdown at any time. The user can input the treatment unit, reporting frequency, the phone/fax number to be automatically dialed for normal reporting, the phone/fax number to be automatically dialed for alarm notification, parameter identification, scaling of units, alarm limits, and the frequency at which parameter values will be scanned. The unit can be powered by solar batteries or by 120-volt power. Connection can be accomplished via conventional phone lines, or, at an extra price, a cellular phone.

2.9 Design Basis, Tradeoff Analyses, and Preliminary Specifications

A record should be kept of the basis for design decisions and equipment selection. For example, on/off level control may have been selected over modulating level control for certain reasons, such as cost or reliability; or a low-profile air stripper at high cost may have been selected over a conventional stripper because of a height limitation where a remediation system is being installed.

Regulations and permits can affect the design of a treatment system. The regulatory analysis done during the feasibility study focuses mainly on remedial action objectives, and it may not reveal detailed regulations regarding control of air emissions, disposal of effluents, and safety requirements. Most feasibility studies do not take into account municipal regulations enforced by community planning departments and building inspectors. Although CERCLA sites are exempt from local permitting requirements, substantive state permitting conditions must be met. Detailed regulatory analysis should be done as early as possible during the process design of the selected remediation alternative, so unanticipated regulatory and permitting requirements can be identified at that time.

Regulatory analysis and definitions of permitting conditions should start as soon as the process design diagrams and a preliminary plot plan are prepared. The water control authorities should be contacted to determine discharge limitations and sampling requirements during start up and long-term operations. The air and water pollution permitting requirements, if applicable, should also be determined. The local fire department or hazardous materials control authority may have numerous requirements that will affect design, including requirements for secondary containment, canopies, and explosion-proof electrical components. The community planning department and/or building inspection department may impose height limitations, dictate finished appearance, restrict access, impose noise limits, and require permits for construction.

The utilities available may affect equipment specifications. For example, 460-volt, three-phase power may not be available for high-horsepower motors anticipated during design. Natural gas pressure in the local mains may not be high enough for the oxidizer that has been selected, and it may be necessary to add a gas booster compressor not anticipated in the original design. Utility suppliers should be contacted early during process design. Sometimes the utility supplier needs to evaluate requested service characteristics before determining if they can meet system requirements or if they need to modify their system. Such a study can delay the remediation implementation schedule if not planned for in advance.

Some of the decisions made regarding permits and utilities supply can form part of the design basis.

When alternatives are equally acceptable except for operating and maintenance costs, a tradeoff analysis can be applied to support the decision. Tradeoffs apply when at least one alternative has a higher investment cost and a lower yearly operating and maintenance cost than another alternative. If the yearly savings eventually make up for an increased investment (within the project lifetime), the higher investment may be attractive. A decision on whether to choose that alternative is usually based on one of the following three criteria:

- **Return on investment.** This can be expressed as the yearly savings divided by the increase in investment. If expressed as a percentage, this value is multiplied by 100. The yearly savings may be estimated as the gross savings minus the cost of amortizing the increase in investment over the life of the project. For example, suppose the following conditions hold in considering the purchase of an improved UV light treatment component that will reduce power consumption: The increase in investment is $20,000, and the project life is 5 years. The annual savings in electric power is $7,000; other utilities, labor, maintenance, chemicals, property taxes, insurance, etc., do not change significantly. Salvage value at the end of the project is negligible for either alternative.

    The simple return on investment is ($7,000/$20,000) X 100%, or 35%. If corrected for amortization of the increase in investment at $4,000/yr, the net savings is ($7,000 - $4,000)/yr. The return on investment is ($3,000/$20,000) x 100%, or 15%. Management guidance is needed to judge whether a 15% return is attractive enough to warrant the increase in investment. If the projected return is borderline, the decision may be based on whether there are adequate capital funds budgeted to cover the extra investment.

- **Payback.** Payback is the number of years it will take for the accumulated savings to equal the extra investment. Numerically, it is the inverse of simple return on investment. For the example above, with a 5-year operating life, the simple payback before taxes is $20,000/($7,000/yr), or 2.8 yr. A non-tax paying entity would probably decide in favor of this alternative – after 2.8 years the entire $7,000/yr would have a positive effect until the end of the fifth year. If the savings affect the income taxes paid by the owner of the facility and the tax bracket is 38%, the net savings are $7,000/yr x (1 - 0.38), or $4,340/yr. The simple payback after taxes is $20,000/($4,340/yr), or 4.6 yr. For a tax-paying owner, the payback is borderline. A discounted payback calculation would take into account the value of money (which may, for example, be what the owner experiences in interest rates). The discounted after-tax payback in this example would probably be more than 5 years. The decision would favor the lower-cost investment.

- **Life-cycle analysis.** The present value of the savings is calculated and accumulated over the 5-yr project life, taking into account the time value of money (the discount rate). The present value of the first year's $7,000 savings is $7,000. The present value of the second year's savings is somewhat less because the owner cannot invest those savings the first year and collect interest or some other cash return until the second year's savings are actually realized. The present value of postponed dollar amounts can be calculated as described at the end of Chapter 12. The present values of the third, fourth, and fifth years' savings are progressively less than the second year's. If the total present value of the 5 years' worth of savings in this example is more than $20,000, it pays to use the proposed alternative with the higher investment cost.

If a life-cycle analysis indicates that accumulated savings at their present value are greatly in excess of the increase in investment, it would be unfortunate if the capital budget did not allow the higher investment. Either the budget should have flexibility, or it should have a built-in allowance for potential favorable tradeoffs.

2.9.1 Preliminary Specifications

If the conceptual design is at the stage at which the remedial alternative has been chosen, it is a good idea to draft preliminary equipment and installation (erection) specifications at this point. Developing specifications that define critical system characteristics results in a more precise conceptual design and subsequently more accurate cost estimates than if these critical items are not addressed early in the design phase of the remediation project. Such critical system variables may include required residence or holdup times, and temperature and pressure limits. Other relevant system requirements include: utilities available or needed, materials of construction, delivery times, construction constraints, soil strength, site grading and drainage requirements, winterization requirements, fireproofing and fire protection elements, gas monitoring requirements, pipeline and conduit routings, wind and seismic resistance, access, secondary containment, level of safety (i.e., the type of personal protective equipment to be worn by workers), and construction time limits.

Allied with the development of preliminary equipment specifications is the grouping of certain components into vendor packages. Examples include the following groupings:

- Soil venting water knockout vessel, filter, air dilution system, blower, flame arrestor, and oxidizer — all skid-mounted, pre-piped, and with controls
- Stripper tower with packing or trays, mist separator, sump, blower, and pump
- Vapor phase carbon adsorption/regeneration system with multiple parallel vessels, regeneration condenser and accumulator — all skid-mounted, pre-piped, and with automatic valves, timers, and controls

Using a vendor package has several advantages over making individual component purchases. The vendor may already have packaged similar systems, so the selection of many components and engineering has been done except for relatively minor customizing efforts. The vendor's experience with prior installations may also lead to a system that has a better chance of performing effectively than a completely custom configuration. The vendor's performance guarantee can cover the complete packaged system, so that in the event of a component failure the vendor cannot blame another supplier.

2.9.2 Conceptual Design Report

Along with the consideration of vendor packages and tradeoffs, a conceptual design report can be developed. A complete conceptual design report for a remediation system would include the following items:

- Background information on the site and the contamination problem
- Regulatory aspects and a summary of prior documentation related to requirements for remediation, emphasizing any feasibility study or treatability study reports and a work plan or EPA record of decision
- A summary of how the proposed remediation system will work in order to meet specific cleanup target levels, how it will be monitored, and where it will be located, accompanied by a process flow diagram and a site plan
- A summary of what procurement packages are envisioned, accompanied by a schedule for detailed design, procurement, construction, start up, operation, and closure
- A summary of cost estimates for design, equipment, installation, operation, maintenance, closure and monitoring
- If the remediation planning is based on an observational approach, certain remediation steps will be taken to start the cleanup and to gather data that may be used for designing potential further steps. And contingency plans should be described. The contingency plans should state what actions will be taken to respond to identified deviations from site conditions and cleanup conditions that had been anticipated using limited site characterization data
- Appendices that include cost-estimating details and backup information; calculations and analyses of chemicals consumption, utilities consumption, operator requirements, and monitoring/reporting requirements through the project life cycle; tradeoff analyses; preliminary P&ID with a sequence of operations; and preliminary specifications

Development of such a report can comprise 30% of a formal design effort. This conceptual design report directly leads to detailed design, which includes drawings showing dimensions, and detailed specifications.

Chapter 3

Metals Removal from Groundwater

This chapter discusses the removal of inorganic compounds — mainly metals — from groundwater. Inorganics of concern for groundwater remediation may be divided into these groups: heavy metals, other metals and inorganic complexes, and radionuclides. The inorganics of primary interest are those that are regulated, for example under Title 40 of the Code of Federal Regulations. However, the non-regulated inorganics can have a major impact on the unit operations for the treatment of groundwater. Metals such as iron, manganese, calcium, and magnesium are commonly removed from groundwater before air stripping to prevent scaling of equipment and pipes. Ion exchange resins may require more frequent regeneration when calcium and magnesium are present because most ion exchange resins are not selective and will remove both the regulated and non-regulated ions.

Inorganic contamination of groundwater is mainly the result of industrial, commercial, and government defense operations. Historically, soils were used as a medium for disposal of liquid and solid wastes, and many facilities constructed for disposal of liquid and solid wastes leaked or leached contaminants to the environment. Over time, water infiltration has resulted in the migration of contamination through the soils and into groundwater systems. The enactment of a series of laws in the 1970s (especially the Resources Conservation and Recovery Act, RCRA, of 1976 and the Clean Water Act) required industries and government facilities to install more waste treatment and management systems to control heavy metal discharges into the environment.

Examples of processes that contributed substantially to soil and groundwater contamination with inorganics include electroplating, metals finishing, minerals extraction and refining, nuclear weapons production, chemical processing, and electronics manufacturing.

Current practice for removal of inorganics from groundwater involves the pumping of the groundwater to the surface for treatment. This is known as "pump-and-treat." Once the groundwater is at the surface, inorganics are removed by any of the processes discussed in this chapter:

- Chemical precipitation
- Membrane separation
- Ion exchange
- Adsorption

Forced evaporation is also used in limited applications. Emerging processes, such as plant uptake (Kim, 1996) and in situ permeable barriers (Smyth, Cherry, and Jowett, 1995; US EPA, 1995b) are not within the scope here.

3.1 Basic Principles

3.1.1 Chemical Precipitation Basics

Chemical precipitation is commonly used in wastewater treatment to remove hardness and heavy metals. In general, the process involves the addition of a chemical agent to an aqueous waste stream in a stirred reaction vessel, either batch-wise or with steady flow. Most metals can be converted to insoluble compounds by chemical reactions between the agent and dissolved metal ions. Chemical agents provide precipitating ions that yield highly insoluble species of compounds of concern. In some cases, alteration of the pH is essential to the process.

The insoluble compounds (precipitates) are removed by settling and/or filtering. A goal of most precipitation processes is the creation of solid particles that can be efficiently removed by sedimentation. Processing includes coagulation and flocculation to agglomerate fine particles, increasing their size, which permits settling. It is important that the sludge formed from the settled solid precipitate be readily removable and reducible in volume with a sludge dewatering system.

Major precipitation processes used for groundwater remediation include: (1) alkaline precipitation and (2) sulfide precipitation. Iron coprecipitation and electrochemical coprecipitation are also practiced. Metals that can be treated by chemical precipitation include: arsenic, barium, cadmium, chromium, lead, mercury, selenium, silver, nickel, zinc, and strontium. Precipitation is also used to remove non-contaminant metals, such as calcium and iron, which can cause scaling in other remediation equipment.

Coagulation is used to neutralize electrostatic charges on colloidal and microparticles so that they can collide and agglomerate. This enhances the removal of the solids. Inorganic coagulants are salts, usually of iron or aluminum, which under alkaline conditions form hydrated colloidal flocs of their hydroxides. Organic polymers (polyelectrolytes) are the coagulant of choice for wastewater treatment (Shelley, 1997). The polymers are used in smaller dosages than are salts and produce a smaller volume of drier sludge, which is important if the sludge is a hazardous waste. Polymers are added in small concentrations to enhance particle growth.

Flocculation of coagulated particles or precipitates uses slow mixing to promote growth of particles into a floc. The larger particles have a higher settling velocity, making them easily and economically separable from solution by gravity or by flotation.

The flocculated particles are removed from the water by solid-liquid separation equipment. If the concentration of suspended solids is less than 50 ppm, filtration using granular media (e.g., a sand bed) may be sufficient for separation. At higher concentrations, a gravity settler or a dissolved air flotation system is usually used.

Hydroxide precipitation and sulfide precipitation are common industrial wastewater treating processes. These processes consist of the following main steps:

1. Addition of an alkaline agent or sulfide, plus addition of coagulants and flash mixing
2. Addition of flocculants and slow mixing to promote particle growth
3. Settling of the particles

The clarified water overflows as effluent. Further treatment is required if organic contaminants are present. If neutral or near-neutral pH is required for the effluent, acid may need to be added. The sludge (precipitate) is removed from the bottom of the clarifier and is dewatered (e.g., by further settling in a thickener or by centrifugation). The solids may undergo filtration or drying before disposal in a Subtitle D landfill or at a permitted Subtitle C Treatment, Storage and Disposal Facility if the dewatered solids are classified as a hazardous waste.

3.1.2 Membrane Separation Basics for Dissolved Ions

Reverse osmosis (RO) and electrodialysis reversal (EDR) are technologies that use porous membranes that can function in several ways: volume reduction, purification of liquid phase, or concentration and recovery of contaminants. Unlike filtration systems that separate out suspended particulates, RO and EDR are processes that apply to dissolved ions. For RO and EDR, pretreatment is required to filter out suspended solids, and sometimes pH is adjusted with the addition of relatively small amounts of acid to prevent scaling on the membranes. Pretreatment methods to minimize scaling and membrane fouling are given by Kucera (1997). If rapid scaling can be avoided, the membranes can last for years.

Membrane configuration details and the advantages of various designs are described by Koros (1995).

Compared to chemical precipitation, membrane separation systems are simple to operate. However, Nyer (1992) states that RO is better suited to removing anions (e.g., nitrates, sulfates) and other naturally occurring inorganics, rather than particular metals.

3.1.3 Ion Exchange Basics

When solutions are passed through cation exchange media, cations in the media (usually sodium or hydrogen ions) are displaced by certain cations from the solutions. With anion exchange media, anions in the media (usually chloride or hydroxide ions) are displaced by certain anions from the solution. Ion exchange for groundwater remediation is virtually always carried out by passing the water downward under pressure through a fixed bed of a granular medium or spherical beads. Ion exchange media most often used for remediation are zeolites (both natural and synthetic) and synthetic resins. The synthetic resins can be categorized as follows:

- Strong-acid cation exchangers
- Weak-acid cation exchangers
- Strong-base anion exchangers
- Weak-base anion exchangers

Details of the resins, including their industrial uses and manufacturers, are given by McNulty (1997).

Ion exchange media can be used and then regenerated, either in place or at an off-site regeneration facility, or the media may be used only once and disposed of. Cation resins are regenerated with sodium chloride solution (brine) or with acid, whereas anion resins are regenerated with caustic soda, ammonia, or sodium chloride solutions. Regeneration is not always practiced. Depending on the economics of each situation (including waste disposal costs), sometimes it is cost-effective to discard spent ion exchange media without any regeneration. Ion exchange and adsorption systems that are not regenerated are the simplest schemes for metals removal, requiring minimal operator attention. However, these systems produce much more waste resin (or other media) than do regenerated systems. Waste media must be handled and disposed of properly.

When regeneration is practiced, spent regenerant and rinse water removed from the regenerated resin form a waste concentrate. The concentrate may require neutralization. If direct discharge of neutralized waste regenerant is not permitted, it can be evaporated.

Ion exchange media that selectively remove toxic metals or radionuclides are preferred for groundwater treatment over media that remove a broad spectrum of cations or anions. Selective ion exchange is discussed in Section 3.4.

3.1.4 Adsorption Basics

Adsorbents that do not exchange ions can be applied for removal of metals from groundwater. Metal ions adsorb on the pore surfaces of certain media without displacing other ions. Alumina can remove anionic forms of arsenic. Charcoal or activated carbon removes dissolved mercury cations. Bone char removes dissolved plutonium in a process that includes both physical adsorption and chemical reaction. This is similar to chemisorption. Some of the plutonium reacts chemically with calcium phosphate in bones.

In some instances, the adsorbent is not regenerated. Adsorbents such as charcoal, bone char, and some natural products may be used for removing extremely small amounts of selected contaminants, and it not always cost-effective to invest in and operate a regeneration system. If the adsorbent's life expectancy is sufficient, it may be more cost-effective to dispose of it as a solid waste without regeneration, and generation of a liquid waste — concentrated regenerate — is avoided.

3.1.5 Evaporation Basics

Forced evaporation is a process in which energy, usually from steam heating or from heat of compression, is applied to convert water from the liquid phase to the vapor phase. The water vapor that is formed may be discharged directly to the atmosphere or recovered in an almost pure form by condensing it.

Most evaporation schemes conserve energy by heat exchanging hot water vapor or hot effluent with influent. The objective of evaporation is to drive off the water and concentrate the nonvolatile inorganic contaminants as a brine. The technology is expensive in terms of operating and capital costs.

Evaporation is rarely used directly for remediation of groundwater. Solar evaporation from outdoor basins and forced evaporation are more often used to concentrate secondary wastes from groundwater treatment processes, such as ion exchange with regeneration and RO.

3.2 Chemical Precipitation

3.2.1 Alkaline Precipitation

The alkaline agents used most frequently are hydrated lime (calcium hydroxide), magnesium hydroxide, and caustic soda (sodium hydroxide). (Alkaline phosphate precipitation of divalent metals is possible, but it is usually not used for groundwater treatment because of calcium interference.)

Many nonmonovalent metal cations form hydroxides that are relatively insoluble. For each cationic metal, there is an optimum pH, usually between 9 and 11, at which the hydroxide solubility is the lowest. The upper part of Figure 3-1 (adapted from Battelle, 1995) shows the solubility of a number of cationic metals as a function of pH. The optimum pH (the lowest point on the curve for a given metal) varies from metal to metal and their oxidation state (valence). If the optimum pH value is exceeded, metals resolubilize. For example, the optimum pH for zinc hydroxide precipitation is approximately 9.2; for nickel hydroxide, the optimum is 10.2. (Note that the solubility is higher if a metal is chelated with an organic ligand, and the curves cannot be used in that event).

If magnesium is used, a buffered pH of 9 is attained. Overdosing with magnesium hydroxide will not cause the pH to be higher than 9, which is often the limit allowed for effluent discharge. Precipitated sludge dewaters faster and is denser than when lime or caustic soda is used. Compared with the use of caustic soda or lime to form hydroxides, the use of magnesium hydroxide is safer because it is less corrosive and is a weaker base. If a pH higher than 9 is required, a two-step process using magnesium hydroxide followed by a stronger base should be considered. Such a two-step process with magnesium hydroxide and caustic soda is less costly for chemicals than the use of caustic soda alone.

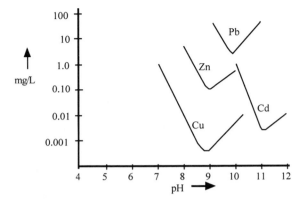

Source: Battelle (1995) and Wikoff and Prescott (1988).   Figure 3-1A

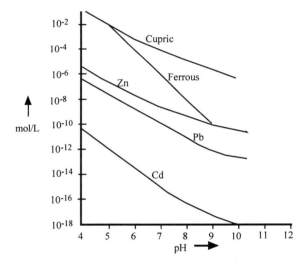

Source: Stumm and Morgan (1996), and Wikoff and Prescott (1998).   Figure 3-1B

Figure 3-1 The solubility of metal hydroxides (A) and metal sulfides (B).

Metal precipitation using a carbonate, such as soda ash, is effective for certain metals (i.e., cadmium, lead, and strontium). Precipitation with polythiocarbonate is effective for many heavy metals. Carbonate precipitation operates at lower pH levels than hydroxide precipitation, and it produces denser, more efficiently filtered solids. One strontium removal process uses lime, soda ash, and coagulants consisting of water-soluble metal complexes. Sonification is used to promote settling. After removal of the precipitate, the high-pH treated water is neutralized with gaseous carbon dioxide.

Comparisons of properties and prices of alkaline agents are given by Hairston (1996). In the design of reaction tanks, the typical residence time is 5 min, except longer times are required for chromic ion and arsenic precipitation (Rast, 1998).

3.2.2 Sulfide Precipitation

If sulfide is added instead of an hydroxide or in conjunction with an alkaline agent, the metal sulfides that form are generally even more insoluble than hydroxides (*see* Fig. 3-1). Also, unlike with hydroxides, if the pH is raised beyond the 9 to 11 range, the solubility of sulfides generally continues to decrease. Sulfide precipitation is usually accomplished using sodium sulfide (Na_2S), sodium bisulfide (NaHS), or ferrous sulfide (FeS), or by injecting hydrogen sulfide (H_2S), a very toxic gas. The primary disadvantage of sulfide precipitation is the potential release of toxic hydrogen sulfide gas at lower pH levels and the potential release of excess sulfide in the effluent. Sulfide precipitates are difficult to dewater unless sludge is recycled through a contact clarifier or (as described in Section 3.2.4) with cross-flow ultrafiltration.

3.2.3 Precipitation with Iron

Iron coprecipitation is a combination of precipitation and adsorption processes. The precipitated floc that is formed provides surfaces for adsorption. One coprecipitation process involves air oxidation of ferrous ion. In other examples, Nyer (1992) describes the use of iron additions for the precipitation of arsenic (using lime) and mercury (using sulfide).

Electrochemical coprecipitation, as described by Nyer (1992), operates at neutral pH. A relatively small amount of acid is used intermittently to clean sacrificial electrodes that produce ferrous ion in concentrations in excess of solubility. The ferrous hydroxide precipitate adsorbs heavy metal contaminants from the groundwater.

3.2.4 Precipitation Examples

3.2.4.1 Iron, Manganese, and Calcium Removal

Iron and manganese are naturally present in groundwater in a reduced state (low cationic valence). Although they are not toxic, as are many other heavy metals, removal is desirable because they form scale when exposed to air during remediation processes such as air stripping, which is used for removal of VOCs. Iron or manganese removal as a pretreatment step for an air stripping system should be considered if their concentration is more than 2 or 3 mg/L. It may be necessary to add an oxidizing agent (e.g., chlorine, peroxide, ozone, or permanganate ion) to improve the precipitation process, especially if iron or manganese is complexed by organics in the groundwater.

For iron removal, oxidation of the ferrous to ferric ion is practiced. Ferric hydroxide is more insoluble than ferrous hydroxide. Nyer (1992) suggests using 12% to 15% hypochlorite or aerating for 30 min at a pH of 7 to 7.5 to accomplish this step. (If aeration is applied with VOCs present, the off-gas must be controlled.)

Calcium removal should be considered to avoid calcium carbonate scaling in an air stripper if the Langelier saturation index (LSI) is positive. The LSI is defined in Equation 3-1.

$$LSI = pH - pH_{sat} \quad (3-1)$$

in which pH_{sat} is the pH of the water when saturated with calcium carbonate. If iron and calcium are not first removed, Dilzell (1996) suggests using a calcium carbonate antiscalent if the LSI is positive and using an iron dispersant if the iron concentration is 5 mg/L or more.

Figure 3-2 is a schematic showing a conceptual design for removal of calcium and ferrous cations as a pretreatment for feed to an air stripper.

Pretreatment is accomplished by the addition of caustic soda as an alkaline agent and potassium permanganate as an oxidizing agent. These chemicals are mixed with the groundwater in a small compartment; a polymer is injected, further mixing is used; and the flocs are separated by gravity in a very large clarifier compartment. Sludge is removed from the clarifier compartment intermittently, pumped into a thickener, and then dewatered with a filter press.

3.2.4.2 Chromium

Chromate (hexavalent chromium ion) is a highly toxic form of chromium. It is not directly removed by precipitation. It can be removed with membrane separation techniques, by ion exchange, or reduction to chromic ion followed by precipitation.

The precipitation of chromic hydroxide requires that the chromium be present in the trivalent form. A conventional process for reducing hexavalent chromium (chromate ion) to chromic ion and precipitation of chromic hydroxide is shown in Figure 3-3.

The pH is lowered to between 2 and 3 by adding a reducing agent, such as sulfur dioxide, ferrous sulfate, or sodium bisulfite plus acid. After carrying out the reduction step at a low pH, caustic (or lime or soda ash) is added in a neutralization tank. The pH is raised to greater than 8 to precipitate chromic and other metal hydroxides (Nyer, 1992). Rast (1998) states that a typical residence time is 20 min.

Freeman (1989) suggests that phosphate precipitation may be an effective method for removing trivalent chromium from a mixed solution. Freeman describes a process using ferrous sulfide for reduction at a lowered pH and then adding hydroxide to raise the pH to between 8 and 11. The optimum pH is 8.5 if chromium is the only metal of concern. Freeman recommends maintaining a pH greater than 8 to prevent evolution of hydrogen sulfide.

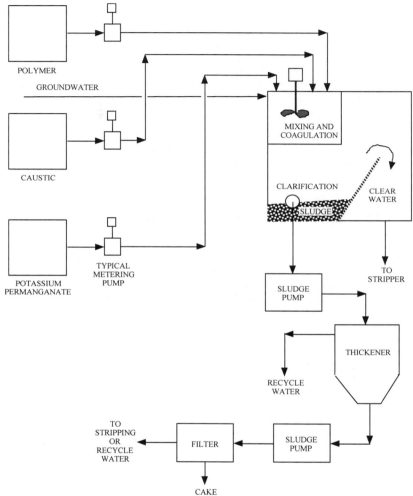

Figure 3-2 Pretreatment of hydrocarbon-contaminated groundwater to remove calcium and ferrous ions.

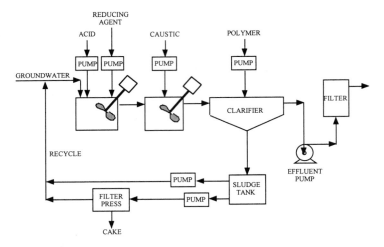

Figure 3-3 Conventional hexavalent chromium reduction/precipitation scheme.

A process developed by Environmental Research and Development, Inc. (Idaho Falls, Idaho) uses ferrous additions with sulfide to reduce chromate and precipitate chromic hydroxide all in one step at a neutral pH (Wikoff and Prescott, 1998). The precipitate must be recirculated such that crystal growth takes place, enabling separation of the solids.

Koolik (1992) describes a scheme using ferrous sulfate to reduce chromate and coprecipitate chromic hydroxide with amorphous iron oxyhydroxide that forms when ferrous sulfate is added to water. Koolik notes that chromic hydroxide is least soluble at a pH near 8.5.

The conversion of chromate ion to chromic ion with the reduction step is an important advantage of chromate reduction/precipitation processes over other chromium-removal processes, such as membrane separation or ion exchange. Chromic ion is relatively nontoxic, as well as being much less soluble than chromate ion.

3.2.4.3 Mercury

Mercury cations cannot be removed adequately with hydroxide precipitation. Sulfide precipitation and sodium borohydride precipitation are effective, however (Freeman, 1989). Sodium borohydride is a reducing agent for precipitating metals from solution as the insoluble elemental metals at a pH of 8 to 11. Sodium borohydride is used after first dissolving it in caustic soda solution. Alternatives to mercury precipitation include activated carbon adsorption and ion exchange. With activated carbon, the adsorbent cannot be regenerated. The spent carbon adsorbent may be difficult to dispose of as a RCRA waste.

3.2.4.4 Arsenic

Arsenic in groundwater is usually present as arsenate, which cannot be adequately precipitated with caustic. Nyer (1992) recommends adding eight parts of iron to each part of arsenic at a pH between 5 and 6 and precipitating with lime at a pH between 8 and 9. Freeman (1989) reports that ferric chloride and hydrated lime have been added to cyanic leachate from mine sites to precipitate arsenic as a calcium ferrocyanate complex. Rast (1998) states that a typical residence time is 30 min for arsenic precipitation. (An alternative to arsenic precipitation is adsorption on alumina.)

3.2.5 Alternatives to Conventional Clarification

A key unit operation for chemical precipitation is the solid-liquid separation step, which is commonly done using clarifiers, with plate-and-frame filter presses for dewatering the precipitated sludge. Conventional clarifying requires hours of settling time, which requires the volume of clarifiers to be very large. Plate-and-frame filter presses require operator attendance. A variety of schemes have been developed to reduce settling time and clarifier volume and to simplify filtration. A prime example is the use of slanted plates or tubes as impingement surfaces on which small particles of precipitate agglomerate.

Ultrafiltration, a membrane separation process effective for undissolved substances, can be used in place of clarification. Figure 3-4 shows a semibatch ultrafiltration schematic.

The membrane is configured as a cylinder, with flow of groundwater, treatment agents, and precipitated sludge in the central passageway. The membrane pores are sized such that at moderate pressures (usually less than 50 psig), water molecules and dissolved ions pass through into the annular space around the membrane. The clear water discharged from the annular space can be further treated or can be disposed of as effluent. The precipitated sludge is recycled through the sludge drum. A batch or semibatch operation can continue this recycling until a relatively high solids content is built up in the sludge. Then flow through the drum is stopped temporarily for sludge removal. Continuous systems require a larger membrane area and/or multiple stages. The high-velocity flow through the inner bore of the membrane helps prevent clogging of the pores. Ultrafiltration is less sensitive to fluctuations than conventional clarifying/settling systems.

3.3 Membrane Separation for Dissolved Ions

Membrane separation is accomplished by applying pressure to groundwater on one side of a semipermeable membrane. For RO, the pressure must be in excess of the osmotic pressure; it generally is in the range of 400 to 1000 psig. For electrodialysis, pressures are lower than 100 psig; the groundwater is pumped through a membrane stack with alternating cation-permeable membranes and anion-permeable membranes. When direct current is applied to the stack, cations migrate toward the cathode, and anions toward the anode, and they are separated as shown in Figure 3-5.

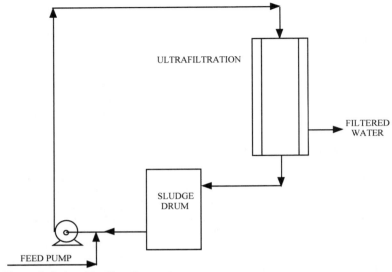

Figure 3-4 An ultrafiltration system.

With electrodialysis polarity reversal (EDR) the polarity of the electrodes is reversed approximately four times per hour. Simultaneously, the flow pathways are interchanged for the purified water versus the retained concentrating stream. With EDR, electrodialysis membranes do not fail rapidly from scaling, and the addition of acid, which is normally needed to prevent scaling, can be eliminated.

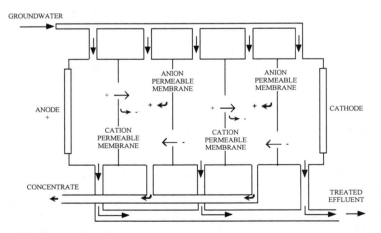

Figure 3-5 Electrodialysis system.

With these membrane separation technologies, usually up to 20% of the groundwater is retained from flowing through the membrane and is a concentrate containing 95% to 99% of the dissolved ions originally in the raw groundwater. The concentrate (retentate) can be evaporated to form a semisolid cake, which must be disposed of as a solid waste. An important goal is to lower the percentage volume of retentate as much as possible, with corresponding increased concentration in the retentate. The limiting factor is scaling of the membranes if retentate concentrations become excessive.

RO is frequently used either as a stand-alone technology or as a pretreatment step for ion exchange. Both RO retentate and ion exchange spent regenerants have to be disposed of as concentrated wastes. The permeate from an RO unit can be passed through ion exchange beds for final purification. By removing approximately 95% of the dissolved ions with RO pretreatment, the frequency required for regenerating ion exchange beds is greatly reduced. Consequently, the volume of regenerate waste concentrate and the consumption of regeneration chemicals are reduced. Overall operating costs can be less when RO is used in conjunction with ion exchange in instances in which RO alone cannot produce effluent that meets the required discharge standards.

3.4 Ion Exchange

Ion exchange, otherwise known as deionization or demineralization, is practiced in a number of industrial applications. Principal uses for the technology include demineralization of feedwater for high-pressure steam boilers and purification of tap water for laboratory use and electronics manufacturing. The technology is rarely used for groundwater and wastewater cleanup, except for recovery of valuable metals from dilute process effluents. It does have a place in groundwater remediation for removal of nitrates and for removal of metals in cases in which toxic metals and radionuclides are involved. For these instances, ion exchange media that are selective are preferred.

3.4.1 Ion Exchange for Nitrates and Chromate

In some applications, certain ion exchange resins and certain adsorbents can be applied for removing specific ions from groundwater. The advantage with such systems is that resin disposal and replacement costs or regeneration chemicals and waste disposal costs can be minimized if the natural ions in the groundwater that are not considered to be contaminants do not use up exchange sites in the resin. For example, if nitrate ions are the only contaminant of concern, an anion resin can be chosen that will remove nitrate ions and can be regenerated with sodium chloride.

There are a number of synthetic resins that have been developed that selectively remove certain metal ions, such as chromate. Some are regenerable, and there are a limited number of regeneration facilities in the United States that will regenerate chromate-laden resins.

3.4.2 Ion Exchange for Radionuclides

Certain synthetic anion resins will remove uranium and technetium complex anions while not removing many nonmetal anions. Synthetic cation resins, synthetic zeolites, and natural zeolites are available that selectively remove cesium and strontium cations.

Ion exchange media and adsorbents cannot distinguish between radioactive isotopes and their nonradioactive counterparts. This feature can be used to great advantage when dealing with something as toxic as strontium-90. The groundwater cleanup goal may be so stringent that even conventional ion exchange cannot produce acceptable effluent. In this event, isotopic dilution can be applied. For the strontium-90 example, nonradioactive strontium can be injected into the influent. Some ion exchange resins have the property of reducing the concentration of a particular ion down to a certain level, even though the influent concentration varies over a wide range. If the influent total strontium concentration is increased so that the strontium-90 portion is one-third of its original portion of total strontium, the effluent may contain approximately one-third as much strontium-90 concentration as it would without the nonradioactive dilution. The disadvantage of resorting to isotopic dilution is that the adsorbent becomes spent much sooner than it would without the dilution.

3.5 Adsorption

Adsorption processes can be categorized by the adsorption medium used, as follows:

A. Activated carbon (charcoal)
  - Granular
  - Powdered
B. Inorganic noncarbon media
  - Iron hydroxide and oxide
  - Activated alumina
  - Activated silica
C. Reactive media
  - Bone char (contains calcium phosphate, which is reactive, and activated carbon)
  - Iron filings
D. Proprietary sorbents
  - Alcoa SORBPLUS Adsorbent^{TM} (Aluminum Company of America, Alcoa Center, Pennsylvania)
  - Algasorb (Resource Management & Recovery, Las Cruces, New Mexico)

The main parameters for design, the approach to treatability studies, and cost-estimating techniques for adsorption are the same as for ion exchange. For sorbents to be practical for remediation, selectivity for toxic metals is important, and rejection of common ions, such as calcium, magnesium, and sodium, is a major consideration for choosing a sorbent. For proprietary sorbents, basic information must be obtained from the manufacturer.

Alcoa SORBPLUS Adsorbent™ is a magnesium-aluminum oxide developed for removing low-concentration toxic metals from metal-plating wastewater. The metals must be chelated or must be anionic complexes. This adsorbent is selective for chelated nickel, copper, or cobalt; for hexavalent chromium (chromate); and for metal-complexed cyanides.

Algasorb is a unique adsorbent in that it is derived mainly from a natural organic substance, harvested algae cells. The cells are encapsulated in polymeric silica gel, forming a hard, granular product. Algasorb will selectively remove heavy metals, including copper, cadmium, chromium, gold, lead, mercury, nickel, zinc, manganese, and iron. It can be regenerated with dilute mineral acid.

3.6 Forced Evaporation

Forced evaporation is rarely used for direct remediation of groundwater. More often, it plays a role in volume reduction of concentrated streams, such as from RO retentate or ion exchange regeneration waste. Evaporators produce a very highly concentrated waste brine that is sometimes supersaturated with some solid crystals; these evaporators are not designed to produce dry crystals or solid powder. The waste brine can be treated in a crystallizer that is designed to remove all free water.

The scheme for forced evaporation with the lowest investment cost uses a submerged fuel gas burner. The burner tip is under the surface of the water, and virtually all of the heat energy is converted to warming the water and vaporizing it.

Vapor recompression is a highly energy-efficient forced evaporation scheme. The vaporization occurs at temperatures slightly greater than 100° C (212° F) in a falling-film evaporator. The water vapor is compressed and returned to one side of the metal heat-exchange surface in the falling-film section of the unit, where it condenses and falls to the outlet. The heat of compression and the heat of condensation are energy sources that raise the temperature of circulating brine sufficiently for rapid evaporation to take place without inputting steam or using a burner.

Multiple-effect evaporation usually uses steam for heating in a process carried out in stages, with increasing vacuum and correspondingly decreasing temperature in each stage. The energy input is minimized by transferring heat from stage to stage. Optimum vacuum levels and the number of stages can be estimated from Durand (1996). The properties of the groundwater are critical in final selection of an evaporation system. If foaming, scaling, or corrosion are concerns, they can typically be handled with appropriate pretreatment and/or the use of additives. The critical design parameters involved in determining heat transfer coefficients include heat capacity, heat of vaporization, density, and thermal conductivity, for which values are known for water. The characterization data needed for specific groundwater treatment schemes include hardness, dissolved solids content, and related properties that affect pretreatment and brine concentrate handling.

3.7 Main System Design Parameters

This section summarizes important points that have a bearing on system design, with emphasis on the evaluation of the main parameters. It is assumed that treatability testing has been accomplished, either at the laboratory bench or with pilot-scale equipment. The following items must be known first: groundwater flow rate, influent hardness and contaminant concentrations, and desired effluent concentrations. If organic contaminants, as well as metals, are to be removed, it is assumed that free-phase (undissolved) organics have been removed in an oil/water separator and that dissolved organics will be treated after metals removal.

3.7.1 Sizing and Rating of Major Equipment

This section describes two examples, each with 60 gpm (8 ft^3/min) of influent groundwater.

Example 1. A chemical precipitation system, including a flash mixing compartment and a multimixer flocculation compartment, followed by a clarifier

Given: The flow rate is 8 ft^3/min. Laboratory bench-scale experiments indicate that a flash mixing time of 1 min and a flocculation time of 20 min are adequate.

Assumptions: Clarifiers are typically designed for an overflow rate near 600 gal/day/ft^2 of surface area, a maximum weir loading rate (overflow rate) in the range of 10,000 to 20,000 gal/day/ft, and a residence time of at least 4 hr.

Find: Volume of flash mixing compartment, flocculation compartment, and clarifier; surface area and depth of clarifier.

Step 1: Multiply the volumetric flow rate times the residence time for the first two compartments, as shown in Equations 3-2 and 3-3.

Volume of the flash mixing compartment = 8 ft^3/min (1 min) = 8 ft^3 (3-2)
Volume of the flocculation compartments = 8 ft^3/min (20 min) = 160 ft^3 (3-3)

Step 2: Use the allowable overflow rate for the clarifier area, as shown in Equation 3-4.

The surface area at the top of the clarifier
= [60 gal/min (1440 min/day)]/(600 gal/day/ft^2) = 144 ft^2 (3-4)

Step 3: Check weir loading, as shown in Equations 3-5 through 3-8.

The diameter of the clarifier, d, is calculated assuming a circular configuration.

Area = 3.14 d^2/4   (3-5)
d = [4(Area)/3.14]$^{0.5}$ = [4(144 ft^2)/3.14]$^{0.5}$ = 13.54 ft   (3-6)
Circumference = 3.14 d = (3.14) (13.54 ft) = 42.54 ft   (3-7)

Weir loading = (60 gal/min)(1440 min/day)/(42.54 ft) = 2031 gal/day/ft   (3-8)

This loading is well below the maximum of 10,000 to 20,000 gal/day/ft and is acceptable.

Step 4: Multiply the volumetric flow rate times the residence time for the clarifier volume required; then divide the volume by the area to obtain depth. See Equations 3-9 and 3-10.

Volume of clarifier = 8 ft^3/min (4 hr) (60 min/hr) = 1,920 ft^3 (3-9)

Approximate depth of upper section of clarifier = volume/plan area
= (1,920 ft^3)/(144 ft^2) = 13.3 ft   (3-10)

This example shows how large a conventional clarifier can be for a relatively small flow rate, making the investigation of alternatives worthwhile.
The design of an integral mixing-settling system can be aided with the use of AEA LEXSET software available from AEA Technology Engineering Software (Bethel Park, Pennsylvania).

Section 3.9 (Cost Estimating for Metals Removal) gives the basis for sizing a sand bed filter system that might be used for filtering water, as well as the basis for sizing a rotary vacuum drum filter that might be used for dewatering sludge from the clarifier.

Example 2. The ion exchange system is shown in Figure 3-6, including two resin beds in series (with RO modules for pretreating and an evaporator for handling waste concentrate from the RO modules). It should be noted that although the mechanism for removal of metal ions is different between ion exchange and adsorption processes, the process design steps are similar.

Given: The flow rate is 8 ft^3/min. Vendor data and experience are assumed to have indicated that RO will remove almost 96% of contaminant C and more than 97% of the other dissolved ions, producing 83% permeate (50 gal/min) and 17% concentrate (10 gal/min). Contaminant C is at 0.45 mg/L, which is equivalent to 147,000 mg/day at 60 gal/min in the groundwater. RO is expected to produce concentrate (that will be subject to evaporation) with almost 2.6 mg/L (2,588 µg/L) of contaminant C (140,876 mg/day). The remaining 6,124 mg/day of contaminant C is in the permeate fed to the first ion exchange bed. The permeate is fed to the ion exchange system, and the concentrate is fed to the evaporator. Pilot experiments have shown that contaminant C will break through a bed of the selected ion exchange resin before contaminant D breaks through. The other main dissolved constituents, A and B, are not contaminants but are naturally occurring minerals in the groundwater that do not load the resin. The total mass flow rate and concentrations of A, B, C, and D, all highly soluble constituents, affect the operation of the evaporator and determine the solids content of the waste-concentrated brine that will be finally evaporated.

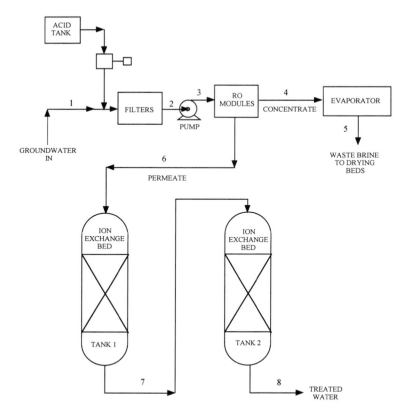

Figure 3-6 An RO and anion-exchange system process flow diagram.

Pilot data indicate that k_p, the partitioning coefficient (equilibrium concentration on the resin divided by concentration in the water), is 1.1×10^6 mg/kg per mg/L for contaminant C, and higher for D. The resin has a bulk density of 25 lb (11.3 kg)/ft^3, and the vendor recommends an operational flow rate of approximately 3 gal/min/ft^3 of resin (equivalent to 24 bed volumes/hr) and a bed depth of at least 4 ft. Dynamic pilot column tests indicate that at least a 2-min residence time is needed to ensure that the concentration of contaminant C in the effluent will not exceed the desired amount of 2.0 µg/L (0.002 mg/L). The ultimate loading on the resin in the first bed in a series of multiple beds is desired to be at least 500 g (500,000 mg) of contaminant C.

Find: Size of resin beds and resin tanks

Step 1: Determine the time period for the lead resin bed to reach a loading of 500,000 mg of contaminant C.

With the 50 gal/min of RO permeate containing 6,124 mg/day of contaminant C, the permeate concentration going into the first ion exchange resin bed is 0.0225 mg/L. The loading on the resin per day and the number of operating days before the first bed must be replaced are calculated, as given in Equations 3-11 and 3-12, based on the allowable effluent concentration of 0.002 mg/L and the desired ultimate loading.

$$(0.0225 - 0.002) \text{ mg/L } (50 \text{ gal/min}) (3.78 \text{ L/gal}) (1,440 \text{ min/day})$$
$$= 5,579 \text{ mg/day} \tag{3-11}$$

$$500,000 \text{ mg}/(5,579 \text{ mg/day}) = 90 \text{ days} \tag{3-12}$$

If multiple full-sized beds are used in series, and the second one is made the lead bed when the first bed approaches saturation, the actual bed life might be a large fraction of the calculated 90 days. This is a reasonable time period for a system without in-place regeneration of resin. Without RO pretreatment removing 96% of contaminant C, the time period would be much shorter, and installing a regeneration system would be considered. The actual time period might be less than what has been calculated here, which was derived from the k_p value. Such a calculated time period applies at equilibrium conditions, not at dynamic flow-through conditions. Batch equilibrium tests for determining k_p values of candidate ion exchange media are good for initial screening. They should be followed by dynamic flow-through tests, either pilot scale or bench scale, to obtain breakthrough and saturation data for design. Such dynamic tests are discussed at the end of this subsection.

Step 2: Determine the minimum resin volume per tank and the corresponding tank dimensions.

The minimum amount of resin per tank (bed volume) and the configuration for each adsorber are calculated, as indicated in Equations 3-13 and 3-14, applying the bulk density and the vendor's recommended operational flow rate and depth.

Minimum bed volume = 50 gal/min/(3 gal/min/ft^3) = 17 ft^3 (3-13)
Cross-sectional area = volume/depth = 17 ft^3/4 ft = 4.3 ft^2 (3-14)

The tank should be approximately 28-in. in diameter to provide approximately 4.3 ft^2 of cross-sectional area. A readily available standard diameter of 30-in. would be chosen, with a vertical straight side of at least 6 ft. This length allows 4 ft for the desired bed depth and 50% extra length for expansion of the bed during backwashing. This also eliminates a custom tank, which would cost much more than a standard-sized tank. Standard-sized tanks should be selected when available to eliminate the expense of customized manufacturing. The resin vendor may also recommend a certain minimum ratio of bed depth to diameter, which could result in the selection of an even longer vessel. The resin vendor should also provide data on the backwash flow rate required for removal of resin fines and substances in the resin that impart color.

Step 3: Determine the loading per kilogram of resin.

Each kilogram of resin can hold a loading of contaminant C determined from k_p, as indicated in Equation 3-15.

$$1.1 \times 10^6 \text{ (mg/kg)/(mg/L)} (0.002 \text{ mg/L}) = 2{,}200 \text{ mg/kg} \qquad (3\text{-}15)$$

Step 4: Check the calculated minimum bed volume of 17 ft^3 for the desired 500,000 mg of capacity for contaminant C, as indicated in Equation 3-16, using the bulk density of 11.3 kg/ ft^3 and the loading calculated in Equation 3-15; and check for adequate residence time, as indicated in Equation 3-18.

$$500{,}000 \text{ mg}/(2{,}200 \text{ mg/kg}) = 227 \text{ kg} \qquad (3\text{-}16)$$
$$227 \text{ kg}/(11.3 \text{ kg/ ft}^3) = 20 \text{ ft}^3 \qquad (3\text{-}17)$$

A bed volume of 20 ft^3 is tentatively chosen. The residence time should be checked to be certain that it safely exceeds the required 2 min for the flow rate of 50 gal/min (6.7 ft^3/min). The residence time is calculated as if the bed space were empty, as indicated in Equation 3-18, so in this example the resin pore volume does not have to be considered.

$$\text{Empty-bed contact time} = 20 \text{ ft}^3/(6.7 \text{ ft}^3/\text{min}) = 3 \text{ min} \qquad (3\text{-}18)$$

To obtain at least 20 ft^3 of bed volume, a final tank diameter and length can now be chosen, based on tank diameters available from the manufacturers and the desired ratio of bed depth to diameter. If the resin vendor does not recommend a high ratio for the bed depth to diameter, it is imperative with shallow beds that the water outlet distributor be designed to produce even flow distribution across the entire cross-sectional area of the bed. Otherwise, groundwater will channel through only portions of the bed.

If a regeneration system were to be included in the design, the regenerant flow rate would be some fraction of the groundwater flow rate of 50 gal/min. The resin vendor may suggest a regenerant flow rate of 20% of the groundwater flow rate. Rinse water should be applied after each regeneration at the same flow rate as the regenerant flow rate. The total volume of rinse water required is typically less than 100 gal/ft^3 of resin. For example, a rinse water volume of 50 gal/ft^3 applied at 20% of a groundwater flow rate of 50 gal/min in a 20 ft^3 bed would result in the parameter values derived in Equations 3-19, 3-20, and 3-21.

$$\text{Rinse water volume} = (50 \text{ gal/ ft}^3)(20 \text{ ft}^3) = 1{,}000 \text{ gal} \qquad (3\text{-}19)$$
$$\text{Rinse water flow} = 0.2 (50 \text{ gal/min}) = 10 \text{ gal/min} \qquad (3\text{-}20)$$
$$\text{Time required to rinse} = 1{,}000 \text{ gal}/10 \text{ gal/min} = 100 \text{ min} \qquad (3\text{-}21)$$

The best design data for ion exchange systems and for adsorption systems are obtained from pilot-scale or bench-scale dynamic flow-through tests that are run beyond breakthrough, preferably until saturation. Samples of groundwater are pumped through columns of ion exchange media or adsorbent media. Breakthrough is when the contaminants of concern are first detected in the effluent or when their concentration in the effluent is a small fraction of the influent concentration.

Saturation is when the effluent concentration C_{eff} equals the influent concentration C_{in} or the ratio C_{eff}/C_{in} equals 1. This ratio is plotted from test results against the number of pore volumes or bed volumes. Each test run is conducted at a chosen residence time (contact time) for the fluid flowing through the column.

The data plotted from dynamic flow-through column tests may produce a sharp breakthrough curve, as shown in Figure 3-7A, or a gradual curve, as shown in Figure 3-7B. As indicated by the curve in the second figure, the ion exchange medium never reached saturation ($C_{eff}/C_{in} = 1$), even after many volumes were passed through. The effluent concentration does not have a desirably low concentration of contaminant for most of the run. However, the medium apparently has a high loading capacity for this contaminant. The amount of that capacity is not readily apparent.

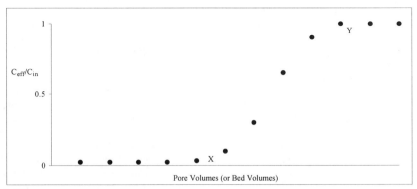

3-7A

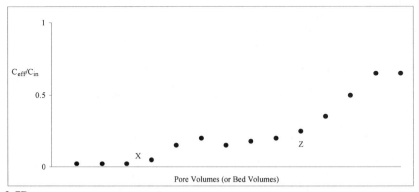

3-7B

Figure 3-7 Examples of typical column adsorption breakthrough curves. A, Sharp breakthrough curve. B, Gradual breakthrough curve.

This is in contrast to what can be quickly discerned for the medium, as indicated by the curve in the Figure 3-7A. Effective use of this medium would end with the number of volumes corresponding to point X (breakthrough), or with multiple columns in series corresponding to point Y (saturation) for the lead column.

Referring to the first figure as representing medium A for the residence times tested, medium A has less ultimate loading capacity than medium B (represented in the second figure), because A becomes saturated sooner than B. However, breakthrough is apparent (point X) for medium A at more volumes than for medium B. High-quality effluent is indicated in the left portion of either curve. This means that medium A will last longer than medium B for producing high-quality effluent for the given flow rate and the chosen residence times.

The operational capacity of full-scale ion exchange systems and of adsorption systems is derived from bench-scale testing and is taken as the capacity that corresponds to an arbitrarily chosen breakthrough percentage. The percentage depends on the desired effluent quality and may be as low as the corresponding detection limit allows. With multiple ion exchange columns in series, 50% breakthrough or 80% breakthrough, etc., may be chosen for the lead column. Close monitoring of effluent concentrations is required for continuing to operate a bed beyond the initial detection of breakthrough. Or close monitoring can be reduced if influent concentrations and flow remain steady and historical monitoring data are sufficient for predicting breakthrough accurately, based on cumulative flow volume.

If the number of bed volumes at point X for medium A is too low for practical application, tests with longer residence times can be run. The test results may show that an impractically large volume of adsorbent would be needed to attain a desirable bed life. In order to obtain both high-quality effluent and an extended bed life, a series combination could be used: a column containing medium B followed by one or more columns containing medium A. The first column will last a long time—for example, until point Z (0.2 breakthrough). The next column (with medium A) has only 20% of the groundwater contaminant of concern, so medium A will have a much longer bed life than that indicated from tests done with raw groundwater.

If only one column containing medium A is placed after one with B, the final effluent must be monitored closely, and the A column must be taken out of service when breakthrough is reached. If three columns with A are placed in series, the first A column can be left in service until it is saturated; this scenario is the most economic if the medium is discarded instead of being regenerated. The same would be true if two columns with A are placed in series, but intensive monitoring would be required.

For remediation of groundwater with toxic contaminants, a three-column series is recommended over the less costly two-column alternative. Some might argue that after attaining experience monitoring a two-column series, one could predict when breakthrough or saturation is being approached merely by tracking the volume of throughput. The monitoring frequency would be intensified only near the end of predicted bed life. However, that is a risky scenario in which toxic chemicals are involved, because sometimes breakthrough occurs early. This is because channeling of water through the bed develops, a bad batch of media is inadvertently installed, the pH undergoes an excursion, or the influent concentration at the adsorbent system has increased.

Again looking at the curve representing medium B in Figure 3-7B, the breakthrough curve is too flat to readily extrapolate it to saturation. So the bed life for medium B cannot be easily predicted. There may not be enough time or enough of a budget to test medium B until saturation ($C_{eff}/C_{in} = 1$) is attained.

At steady-state, an adsorption bed or an ion exchange bed is fully loaded (saturated) at the front of the bed, in the equilibrium zone. This zone is followed by a mass transfer zone (MTZ), in which adsorption takes place. Until initial breakthrough in a properly designed unit, there is a length of unused bed beyond the MTZ.

The length of the MTZ can be found from a total breakthrough test ($C_{eff}/C_{in} = 1$ is achieved and bed is saturated). The time at which initial breakthrough occurs is noted as t_b. The time at which $C_{eff}/C_{in} = 1$ is achieved (or estimated from extrapolating to $C_{eff}/C_{in} = 1$) is noted as t_e. Thus, the time lapse for the MTZ to exit from the bed is ($t_e - t_b$) hours. The length of the MTZ is $U(t_e - t_b)$, in which U is the velocity at which the MTZ advances (in ft/hr).

Lukchis (1973) shows that U is a constant, given by Equation 3-22.

$$U = G(Y_{in} - Y_{eff})/[\text{bed density } (x/m - x_o)] \qquad (3\text{-}22)$$

in which G is the superficial mass velocity, calculated as $(lb/hr)/ft^2$; bed density is the bulk density, lb/ft^3; x/m is the adsorbent capacity from a batch equilibrium test, weight fraction; x_o is the initial loading of contaminant on the adsorbent, weight fraction; Y_{in} is the concentration of contaminant at the bed inlet, lb/lb of feed; and Y_{eff} is the concentration of contaminant in the bed effluent until initial breakthrough occurs, lb/lb of feed.

It is important for adsorbers and ion exchange units that the bed length exceed the length of the MTZ, $U(t_e - t_b)$. Lukchis (1973) states that the preferred bed length is 1.5 to three times the length of the MTZ.

3.7.2 Conceptual and Process Design

An example of initial process design is given for an ion exchange system to remove small concentrations of anionic metals without on-site regeneration of resin. The process flow diagram is shown in the lower half of Figure 3-6 for a groundwater flow rate of 60 gal/min into the RO system that precedes the ion exchange system. Ten gal/min is rejected as RO concentrate (retentate). An example of a P&ID is given in Figure 3-8 for the ion exchange portion (at 50 gal/min of groundwater permeate). Tank 1 and tank 2 from Figure 3-6 contain ion exchange beds, and Figure 3-8 shows the initial lead and lag ion exchange columns, respectively.

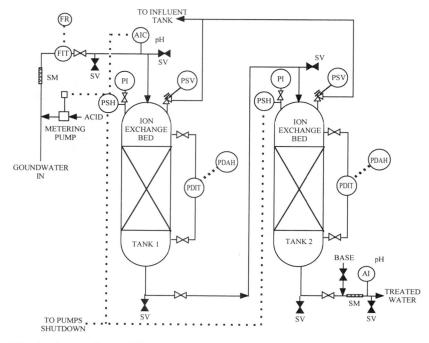

AI = Analyzer indicator (pH meter)
AIC = Analyzer indicator controller (pH controller)
BW = Backwash connection
FIT = Flow indicator transmitter
FR = Flow recorder
PDAH = High-differential pressure alarm
PDIT = Differential pressure indicator transmitter
PI = Pressure gauge
PSH = High-pressure switch
PSV = Pressure safety valve (relief valve)
SM = Static mixer
SV = Sample valve

Figure 3-8 Ion exchange system P&ID.

For the RO and ion exchange systems in Figure 3-6, the Sequence of Operation might include the following information:

- The influent bag filter at the RO system is to be changed out when the pressure drop reaches 20 psi.
- The pH of the influent is adjusted to be in the range of 6.5 to 7.0 with injection of sulfuric acid.

- Tank 1 is sized to remove contaminants C and D to less than 2 µg/L, each with 2 min of empty-bed contact time (defined in the next section) at 50 gal/min using resin X with 4 to 30-mesh resin beads.
- Tank 2 is the same as tank 1 and is a guard chamber that becomes the lead column when contaminant C in the water exiting from tank 1 is detected at concentrations exceeding 90% of the influent concentration.
- Samples of water exiting tank 1 and tank 2 should be taken for analysis each week until the calculated loading of contaminant C on the resin in tank 1 reaches 500 g. Then samples should be taken every day until breakthrough conditions for contaminant C are reached, i.e., when concentrations exceed 90% of the influent concentration.
- The high-pressure switch (PSH) trips off the influent pump if the pressure entering tank 1 exceeds 80 psig.
- Manual restart of the influent pump is required.
- High-differential pressure alarms (PDAHs) are set at 15 psi. If the influent filter alarm is triggered, change the bag within 3 days. If a resin bed alarm is triggered, check for a torn influent filter bag and consider backwashing the resin.

The process flow diagram, P&ID, plot plan, and site plan form a process design from which detailed piping, electrical, and structural designs can be developed at later phases of design.

3.7.3 Controls

Figure 3-8 is the beginning of a P&ID for the dual-bed ion exchange system. For this design example, no flow controller or pressure controller is included. The system flows at whatever rate the groundwater well pumps and RO pump can develop, or the globe valve ahead of the first ion exchange tank is throttled manually to control the flow. The high-pressure switch (PSH) shuts off the feed pumps automatically if there is a blockage.

Pressure drop should build up very gradually in the ion exchange tanks, so no automatic shutdown is needed in the event of high differential pressure. Each tank is monitored with a differential pressure indicator-transmitter (PDIT), and alarmed with a high-differential pressure alarm (PDAH). In the event of a surge in differential pressure, the pressure would rise and a relief valve would open, and/or the pressure switch would activate.

A pH control system is included in this example. Acid injection into the influent is automatic, with the pH indicator controller (AIC) setting the automatic stroke adjustment on the acid-metering pump. Injection of base into the effluent is set manually and would be done in the event of excessive acid injection. Ion exchange and adsorption systems operating without regeneration and pH control need very little instrumentation for unattended operation. Level controls for water storage tanks, pressure-monitoring instruments, and alarm/shutdown devices are sufficient.

3.7.4 Utilities Requirements

The main utility needed for chemical precipitation, membrane separation, and adsorption systems is power for pumping, instruments/alarms/interlocks, lighting, and heat tracing if freeze protection is required. For RO systems, the pumping horsepower is most significant. For the example in Figure 3-6, with 60 gal/min and a rise in pressure of 790 psi for the pump, the horsepower is given for a pump with 80% efficiency, as shown in Equation 3-23.

Horsepower = (flow rate) (pressure rise)/[1,713 (pump efficiency)]
= 60 gal/min (790 psi)/[1,713 (0.80)] = 35 hp (3-23)

The pump motor selected for this service would probably be 50 hp, a standard size motor that would allow for flow rate excursions greater than 60 gal/min.

For most electric-driven process pumps, each horsepower is equivalent to one 1 kW. This relationship accounts for the pump motor inefficiencies and power factor. The electric power load for this example is approximately 35 kW.

For evaporation, the main utility requirement for a steam evaporator would be steam, which, in turn, involves fuel consumption in a boiler. The amount of heat needed would depend on how much heat exchange is used and whether vacuum evaporation is used. If there is no heat recovery, each pound of water requires approximately 1 lb of condensing steam to be evaporated, or 500 lb/hr of steam (or approximately 500,000 BTU/hr) for each gal/min of water. If the steam is generated by a boiler at 75% efficiency, the boiler fuel consumption is 160 therm/day for each gal/min evaporated if there is no heat recovery. For example, evaporation of 60 gal/min of water with 80% heat recovery by heat exchange would consume steam and use fuel with a 75% efficient boiler, as given in Equations 3-24 and 3-25.

Steam use = (60 gal/min) (1 - 0.80) (500 lb/hr of steam)/gal/min = 6,000 lb/hr (3-24)
Boiler fuel = (60 gal/min) (1 - 0.80) (160 therm/day)/gal/min
= 1,920 therm/day at 75% boiler efficiency (3-25)

The main utility needed for a vapor-recompression evaporation system with an electric motor-driven compressor is power for the compressor motor. Again, the amount of energy consumed depends on how much heat is recovered by heat exchange.

3.8 Treatability Studies for Metals Removal

3.8.1 Treatability Studies for Precipitation and Prediction of Treated Effluent Concentrations

Treatability studies are necessary for evaluating a proposed chemical precipitation process and for determining the optimum values for design parameters. Often, the amount of chemicals needed and the pH requirements can be predicted from theory. The amount of metal ions staying in solution can be predicted from solubility constants. However, unless a complex mathematical/chemical model is used, it is worthwhile to run laboratory batch tests (jar tests) to determine the actual required chemical dosages. This is because groundwater often contains natural buffering agents or chelating agents that may significantly affect predicted metal solubility values. Also, laboratory tests evaluating settleability can help in the selection of coagulant aids and in the design of clarifiers.

Simple jar tests often help determine the optimum pH for precipitating mixtures of metals. Also, titration curves are derived from laboratory additions of an alkaline agent and then acid, showing pH versus quantity of the chemical solution added.

The solubility constant (available from a number of general chemistry texts and handbooks) is used to determine the equilibrium concentration of dissolved ions. Note that the use of solubility constants in dealing with precipitation of metal cations does not always apply when the added precipitating anion forms an anionic complex with the metal. For example, when adding hydroxide to groundwater to precipitate copper, nickel, ferrous, cadmium, zinc, or lead hydroxide beyond the optimum precipitation pH, as shown in the upper part of Figure 3-1, the solubility increases. The metals start forming complex oxygenated anions. An example of using a solubility constant is given below.

Given: Cuprous ion must be removed to the extent that the final concentration does not exceed 0.005 mg/L. The precipitation agent is sulfide, and the solubility constant for cuprous sulfide (Cu_2S) is 2×10^{-47} at 16° to 18° C (*Handbook of Chemistry and Physics*, 30th Ed., Cleveland, Ohio: Chemical Rubber Publishing Co., 1947).

Assumptions: The precipitation reaction is carried out with good contact of reagents and adequate time to approach equilibrium. No complexes are formed, and copper is precipitated as Cu_2S.

Find: Equilibrium concentration of sulfide at the target concentration for cuprous ion.

Step 1: Write the solubility equation for the precipitated compound, cuprous sulfide, as shown in Equation 3-26.

$$[Cu^+]^2 [S^=] = 2 \times 10^{-47} \tag{3-26}$$

in which the numbers in brackets are molal concentrations, usually expressed in gram-equivalent units.

Step 2: Convert the target effluent concentration for cuprous ion to molal concentration, as shown in Equations 3-27 and 3-28.

$$(0.005 \text{ mg/L})(g/1000 \text{ mg}) = 5 \times 10^{-6} \text{ g/L} \tag{3-27}$$

Molal concentration of a metal cation is g/L divided by atomic weight.

$[Cu^+] = (5 \times 10^{-6} \text{ g/L})/(63.57) = 7.86 \times 10^{-8}$ \hfill (3-28)

Step 3: Solve Equation 3-26 for the sulfide concentration in gram-equivalent units and convert to mass/unit volume units, as shown in Equations 3-29 and 3-30.

$[S^=] = (2 \times 10^{-47})/(7.86 \times 10^{-8})^2 = 0.324 \times 10^{-32}$ equivalents/L \hfill (3-29)

$(0.324 \times 10^{-32}$ equivalents/L$)$ (32 g sulfide/sulfide equivalent) (1,000 mg/g)
$= 1 \times 10^{-28}$ mg/L \hfill (3-30)

Using similar reasoning, the solubility constant for cuprous sulfide can be used to find the equilibrium Cu^+ concentration at any given sulfide concentration.

Note that concentrations derived from equilibrium constants and solubility curves are the minimum concentrations that can be attained at equilibrium, (i.e. with ideal conditions). Under actual conditions of precipitation, the concentration of a contaminant metal will usually be higher than the predicted value, because of conditional solubility. Also, any precipitate that is not removed from the treated water that remains suspended may be considered as adding to the final total concentration. This is especially true if there is any chance of an adverse subsequent pH change, because then the precipitate might dissolve. The main advantages of using equilibrium constants and solubility curves are:

- The minimum contaminant concentration that is achievable can be predicted. If this concentration does not meet treated effluent quality requirements, then another metals-removal scheme, such as ion exchange or evaporation, should be considered.
- The reagent quantities that are required can be estimated.
- When multiple heavy metals are present, determination of which metal precipitate controls the chemical dosage and pH is needed. In some instances, stepwise pH adjustments and removal of precipitates may be necessary.

In the absence of bench-scale tests, the quantity of chemicals needed for pH adjustments and for precipitation of metals can also be determined from mathematical/chemical modeling of the system. Also, if the treated groundwater is to be injected into an aquifer, it will contain some of the treatment chemicals. Modeling of the aquifer that accounts for the addition of these chemicals is needed to ensure acceptability. OLI (Morris Plains, New Jersey) has developed models for these situations.

3.8.2 Treatability Studies for RO

Membrane type, dissolved solids concentrations, pH, feed temperature, and applied pressure are the key parameters affecting performance of RO systems. Because of the variations in each waste stream, pilot-scale tests are required. The performance parameters that must be measured include pressure, flux (flow of permeate through each square meter of the membrane), ratio of permeate flow to feed flow (or percent permeate), and rejection (percent of solute that does not pass through the membrane).

For membrane separation processes such as RO, vendors have pilot units that help predict the flux, the percent of permeate, the contaminant removal efficiency or rejection, and the acid pretreatment that may be needed to prevent scaling of membranes for a particular groundwater application. Field pilot units ranging in capacity up to 30 gal/min are available. The pilot tests can be preceded by laboratory bench-scale tests with a drum sample of groundwater being passed through a flat-sheet membrane. This testing can be used to correlate flux versus pressure. More sample volume is needed to run bench-scale tests at various pressures with the membrane in a cylindrical configuration.

3.8.3 Treatability Studies for Ion Exchange and Adsorbent Systems

In evaluating the application of ion exchange or adsorption, important design parameters that must be considered include the concentrations of contaminants and non-contaminants, flow rates, media capacity, regenerant volume, and effluent concentrations. Mass balances and process parameters, such as optimum pH, can be established with treatability studies. To show the feasibility for a particular application, it is necessary to determine the contaminant capture isotherm from tests with groundwater samples.

Bench-scale tests are best performed in two steps: batch equilibrium tests are conducted in beakers with various media to find which have the best partitioning coefficients and capacity to achieve high loadings of contaminants; then flow-through dynamic tests are conducted with glass columns to find the residence time required with the most promising media. From batch equilibrium data, isotherms can be plotted that relate contaminant loading per unit mass of the media versus the final contaminant concentration in the groundwater. These batch tests can also be conducted at various pH levels to find the optimum pH. The dynamic tests with miniature glass columns are often done with at least two columns in series. Thus, removal efficiency is determined for at least two residence times, by sampling effluent after the first and second beds. By continuing the dynamic testing beyond breakthrough from the first bed, the media life can be predicted.

Bench-scale tests also show the selectivity of various media for removal of target contaminant species versus removal (or non-removal) of naturally occurring non-contaminant dissolved constituents.

A convenient way to correlate dynamic test data for a given bed residence time is to plot the first bed effluent contaminant concentration against the number of bed volumes that have passed through either a miniature glass column or a pilot column. For example, if a glass column holds 750 mL of resin and the effluent concentration reaches the cleanup target upon having 900,000 mL of groundwater flow through, the number of bed volumes is as shown in Equation 3-31.

$$900,000/750 = 1,200 \text{ bed volumes} \tag{3-31}$$

Now, calculating the bed life for a full-scale unit is straightforward. For example, if a 50-gal/min (189-L/min) system is to be installed with 20 ft^3 (150 gal, or 565 L) of resin in a bed, the bed life is calculated, as shown in Equations 3-32 and 3-33, and 3-34.

1,200 bed volumes (150 gal/bed volume) = 180,000 gal (680,000 L) (3-32)
180,000 gal/(50 gal/min) (day/1440 min) = 2.5 days (3-33)
680,000 L/(189 L/min) (day/1440 min) = 2.5 days (3-34)

When breakthrough has been achieved, regeneration testing can be conducted if regeneration of used media is being contemplated. If three beds in series are planned or a non-regenerative system is planned, and if enough groundwater sample volume is available, testing well beyond the initial breakthrough should be done.

If the slope of the breakthrough curve (effluent concentration versus bed volumes) can be established after passing through a large number of bed volumes, the curve can be extrapolated. Figure 3-9 illustrates this extrapolation.

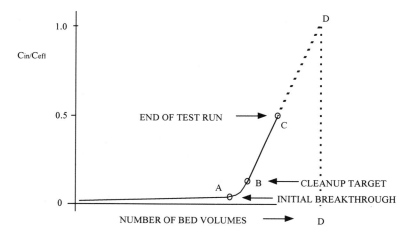

Figure 3-9 Breakthrough curve extrapolation.

Point D on the figure is the number of bed volumes at which the medium becomes saturated — the ratio of the inlet concentration to effluent concentration is 1. Thus, the number of bed volumes corresponding to saturated adsorbent can be estimated. The bed life is proportional to bed volumes, and at saturation it is usually much greater than the bed life for initial breakthrough. For example, assume there is 50 gal/min throughput with a bed volume of 150 gal; if the initial breakthrough is at 1,100 bed volumes (point A in the figure) and saturation (point D) is at 1,700 bed volumes, the bed life at any intermediate point, such as point C at 50% breakthrough, can be determined as follows:

- **At point A, initial breakthrough:** 1,100 bed volumes
- **At point D, 100% breakthrough (saturation):** 1,700 bed volumes
- **At point C, 50% breakthrough:** 50% of the way between A and D, 1,400 bed volumes. The volume of groundwater passed through the bed is (1,400 bed volumes) x (150 gal/bed volume), or 210,000 gal. The corresponding time period at 50 gal/min is 4,200 min (2.9 days).

It is important when examining test data involving bed volumes to determine whether the reported data are based on the net volume of adsorbent without counting pore space or on the gross space occupied by the adsorbent (the "empty-bed" volume). Either method of defining bed volumes is valid. The bed volume used in Equation 3-32 for the full-scale column must be figured on the same basis as that used for deriving the number of bed volumes from testing.

3.8.4 Treatability Studies for Evaporation

For the design of forced evaporation systems, treatability testing is especially important for two items: the volume reduction (with corresponding solids concentration) that can be achieved, and the amount of scaling of heat transfer surfaces that can be expected. In general, it is desired to maximize volume reduction, but scaling intensifies when corresponding solids concentrations increase. Treatability studies can include pH adjustments or the addition of sequestering agents to minimize scaling. Scaling often increases with temperature, and evaporation systems that operate under vacuum conditions may have fewer tendencies toward scaling than those operating under ambient pressure. Testing can be undertaken with a variety of evaporation systems, each of which operates at a different amount of vacuum.

A key component in a vapor recompression system used for evaporation is the falling-film heat exchanger that is often used. The system vendor may not have a complete pilot unit, including a compressor. However, for pilot testing to determine scaling tendencies, steam, instead of compression, can be used as a source of heat.

3.9 Cost Estimating For Metals Removal

Studies by Bechtel (1994) indicate the typical costs for budget planning as follows: Investment costs for adsorption systems and for ion exchange systems are approximately $700 to $1,000/(gal/min) for basic treatment equipment (including erection) if relatively low-cost adsorbents are used, such as alumina, activated carbon, bone char, and natural zeolites. Initial costs would be higher if regeneration equipment is included, or with synthetic ion exchange resin, with the costs depending on which media are selected. Investment costs for chemical precipitation and membrane separation systems are approximately $2,000/(gal/min) for capacities near 100 gal/min and range upward for smaller capacities and downward to $1,600/(gal/min) for capacities of several hundred gal/min. A case study for one such large treatment system resulted in a cost estimate as follows:

The system, including installation costs
- Basic equipment $1,400/(gal/min)
- Chemicals storage and metering pumps +10%
- Sand filter/backwash system +20%
- Instrumentation and electrical controls +30%
- Groundwater feed tank, power supply, foundations +10%

Operating and maintenance costs (per year)
- Labor at 8 hr/day (365 days/yr), replacement membranes for an RO unit, antiscalant chemicals, electricity, maintenance, and monitoring - $1,000/(gal/min)

A cost estimate of capital investment costs for a 1,260-gal/min RO unit for removing toxic metals from groundwater was published by the California Department of Health Services (1988) and is reproduced in Table 3-1. The estimate of operating and maintenance costs for running the unit at 1,230 gal/min is reproduced in Table 3-2.

Durand (1996) gives the installed cost of multiple-effect vacuum evaporators, as given in Equation 3-35.

$$\$ = 11.24A + 7{,}314 \qquad (3\text{-}35)$$

in which A is the heat transfer surface area, ft^2.

A cost comparison completed for a chromate-removal plant constructed in eastern Washington in 1996 is an interesting example of ion exchange and precipitation applications. This comparison is for treating an average of 200 gal/min of groundwater containing up to 200 µg/L of chromate so that the effluent contains less than 20 µg/L chromate after treatment. Table 3-3 shows the costs that are different between using ion exchange (without on-site regeneration) and chemical precipitation. (Note that the discounted values and total present worth are explained in Chapter 12).

These comparison costs exclude contingency and items that are approximately the same for both cases, such as site preparation, fencing, engineering, permits, wells, storage tanks, electric power installation and consumption, buildings, sampling/analytical costs, project and construction management, and general and administrative costs. For ion exchange (Case A), resin consumption is based on pilot plant data. Unit costs are shown in Table 3-4. (Note that tabulated labor rates include overhead.)

Software is available that aids both in the sizing/rating of treatment equipment and estimating costs. COMPOSER GOLD computer models, marketed by Building Systems Design (Atlanta, Georgia) includes coagulation/flocculation, sludge dewatering, neutralization, and sand bed filtration.

The RACER/ENVEST™ computer models marketed by Talisman Partners Ltd. (Englewood, Colorado) include estimating investment costs and chemicals consumption for filters, coagulation/flocculation/clarifying systems, and

Table 3-1 Capital costs for RO treatment of 1,260 gal/min of extracted groundwater. (The California Department of Health Services, 1988).

	Description	Quantity	Unit	Unit Price ($)	Total Cost (Rounded $)	Assumptions	Reference
1	Reverse osmosis unit including vessels and membranes, interconnecting piping, high pressure pumps, cleaning system, installation and start-up	1	lump sum	1,700,000	1,700,000	Four treatment trains, 60 percent product recovery, TFC magnum membrane @ 15 gpm/ft^2 flux rate. Includes 50% standby capacity.	Design and cost by CassCorp (vendor)
2	Chemical storage and feed unit for Reverse Osmosis, installed	1	lump sum	170,000	170,000	Antiscalant and pH control. Estimated @ 10 percent of RO unit cost.	Allowance based on Professional Judgement
3	Reverse osmosis feed tank installed with appurtenances and foundation	1	ea.	50,000	50,000	60,000 gal coated steel tank, 1-hr detention	Cost from Appendix A.17
4	Mixed-media filtration system for Reverse Osmosis including backwash system, chemical feed, installed	1	lump sum	345,000	345,000	400 ft^2, four cell filter @ 3 gpm/ft^2 including chemical flocculant feed. Provides 50% standby at 6 gpm/ft^2	Extrapolated from CH2M-Hill based on Professional Judgement
5	Piping and appurtenances	1	lump sum	510,000	510,000	Estimated @ 30 % of Reverse Osmosis and filtration system cost (less 25% for installation)	Allowance based on Professional Judgement
6	Instrumentation and electrical controls	1	lump sum	510,000	510,000	Estimated @ 30 % of Reverse Osmosis and filtration system cost (less 25% for installation)	Allowance based on Professional Judgement
7	Concrete foundation, not including site preparation	5,000	S.F.	4	20,000	80 ft by 60 ft, 6-in thick. Means costs of $1.92/ft^2 doubled for Stringfellow	Means Site Work Cost Data 1987, p. 96
8	Electrical supply	1	lump sum	80,000	80,000	Equipment, wiring, and labor to run electrical off of existing power line	Allowance based on Professional Judgement
	Total				$3,385,000		

Table 3-2 Operation and maintenance costs for RO treatment of 1,260 gal/min of extracted groundwater (The California Department of Health Services, 1988).

Description	Quantity	Unit	Unit Price ($)	Total Cost (Rounded $)	Assumptions	Reference
1 Membranes	200	ea.	1,550	310,000	TFC magnum membrane, 1-yr membrane life	CassCorp (vendor) Treatability Study (Appendix G)
2 Chemicals sulfuric acid (H_2SO_4)	73,000	lb	0.04	3,000	200 lb/d of 100% H_2SO_4 @ $80/ton for 93% H_2SO_4	CassCorp (vendor) Treatability Study (Appendix G)
antiscalant (FLOCON)	55,000	lb	1.32	73,000	150 lb/d (10 ppm) @ $700/55 gal	CassCorp (vendor) Treatability Study (Appendix G)
3 Electricity	3,360,000	kW-hr	0.10	336,000	Consumption based on 5.2 kW-hr/1000 gal @ 60% product recovery. Unit price based on current rate.	kW-hr/1000 gal from CassCorp
4 Labor	3,000	hr	50	150,000	8hr/d, 365 d/yr, @ $50/hr	Labor hours based on CassCorp and Professional Judgement; labor unit price from CDM, 1987
5 Maintenance materials	1	lump sum	245,000	245,000	Estimated @ 10% of system equipment capital cost (less 25% for installation)	Allowance based on Professional Judgment
6 Process lab monitoring	100	ea.	100	10,000	2 analyses/wk, 52 wk/yr for TDS and inorganics (scaling)	Analysis requirements and cost based on Professional Judgment
Total				$1,127,000		

Table 3-3 Cost comparison of ion exchange (IX) and chemical precipitation for chromate removal from contaminated groundwater.

Cost Component	Case A - IX	Case B - Precip.
Capital Investment	204	245
Operations, 5 years		
Operator labor		1,869
Resin changeout	186	
Routine	94	
Chemicals	0	200
Resin @ $150/ft^3	1,437	0
Operations Subtotal	1,717	2,069
Maintenance, 5 years		
Labor	109	120
Parts/Supplies	51	61
Maintenance Subtotal	160	181
Waste Transport, 5 yr	0.33	0.16
Waste Disposal, 5 yr	7	negligible
Total 5-Year Expense		
Undiscounted	1,884	2,250
Discounted at 5%/yr	1,631	1,948
Total Costs,		
Undiscounted	2,088	2,495
Total Present Worth	1,835	2,193

neutralization equipment. The two types of filters included in the computer models are: (a) granular media, such as sand beds to remove suspended solids from water pumped through filter vessels; and (b) rotary drum vacuum units that are used to dewater sludges.

Computerized remediation models such as COMPOSER GOLD and ENVEST are primarily used for cost estimating but aid in process design. ENVEST examples illustrate this feature.

Granular media filters use downflow and they sometimes have multiple layers, each with a different size range for the granules. For example, coarse anthracite coal may overlie a layer of sand. The user of the ENVEST program must specify the rate in gal/min and can choose the hydraulic loading and number of vessels. If the hydraulic loading is not specified, 5 gal/min/ft^2 is used. If the number of vessels is not specified, two are used, each large enough to handle all of the flow. Thus, one vessel can periodically be taken out of filtering service for backwashing to remove collected solids while the other vessel is in filtering service. The program selects standard vessel diameters of 3, 4, 5, 6, 7, or 8 ft, which correspond to the cross-sectional area that most closely meets the hydraulic loading for flow rates up to 250 gal/min.

Rotary drum vacuum filters include a drum on a horizontal axis with a wire mesh or supported fabric surface, partially immersed in a trough. The sludge from a chemical precipitation clarifier can be fed into the trough. Vacuum applied in the drum draws water through the filtering surface of the drum. As the drum slowly rotates, the cake

Table 3-4 Unit cost comparison of ion exchange (IX) and chemical precipitation for chromate removal from contaminated groundwater.

Parameter		Units	IX	Precipitation
Processing Rate		gpm	200	200
Operating Period		yr	5	5
Total Quantity Processed		gal	1,051,200,000	1,051,200,000
Unique Capital Costs		$	464,000	396,000
Labor Rates:	Design and Engineering	$/hr	77.53	77.53
	Environmental Compliance	$/hr	69.55	69.55
	Field Support	$/hr	55.60	55.60
	Project Management	$/hr	87.98	87.98
	Project Controls	$/hr	61.68	61.68
	Hydrogeologists	$/hr	62.76	62.76
	Quality Assurance	$/hr	71.53	71.53
	Procurement	$/hr	52.34	52.34
	Health & Safety	$/hr	64.26	64.26
	Drivers	$/hr	36.55	36.55
	Operators	$/hr	42.68	42.68
	Administrative	$/hr	35.25	35.25
	Sample Support	$/hr	62.76	62.76
	Data Management	$/hr	67.62	67.62
	Sample Management	$/hr	62.76	62.76
	Pipe Fitters	$/hr	45.54	45.54
	Millwrights	$/hr	45.54	45.54
	Electricians	$/hr	48.08	48.08
	Instrument Technicians	$/hr	49.26	49.26
Labor-Operations:	Design & Eng.-Operations	hr/yr	832	915
	Environmental Compliance	hr/yr	48	53
	Field Support-Operations	hr/yr	208	229
	Project Management	hr/yr	104	115
	Project Controls	hr/yr	96	106
	Hydrogeologists	hr/yr	96	106
	Quality Assurance	hr/yr	96	106
	Procurement	hr/yr	96	106
	Health & Safety	hr/yr	104	115
	Drivers-Operations	hr/yr	104	115
	Operators	hr/yr	2,624	8,760
	Administrative	hr/yr	312	344
	Sample Support	hr/yr	288	317
	Data Management	hr/yr	208	229
	Sample Management	hr/yr	96	106
Labor-Maintenance:	Design & Eng.-Maintenance	hr/yr	96	106
	Field Support-Maintenance	hr/yr	96	106
	Drivers-Maintenance	hr/yr	96	106
	Pipe Fitters	hr/yr	192	212
	Millwrights	hr/yr	96	106
	Electricians	hr/yr	96	106
	Instrument Technicians	hr/yr	192	212

Table 3-4 Unit cost comparison of ion exchange (IX) and chemical precipitation for chromate removal from contaminated groundwater (continued).

Parameter		Units	IX	Precipitation
Processing Rate		gpm	200	200
Operating Period		yr	5	5
Total Quantity Processed		gal	1,051,200,000	1,051,200,000
Consumable Costs:	Acid	$/gal	2.76	2.76
	Caustic	$/gal	1.72	1.72
	Fe	$/lb	0.29	0.29
	Polymers	$/lb	2.50	2.50
	Resin	$/ft^3	250	250
Consumption Rates:	Acid	gal/day	0	30
	Caustic	gal/day	0	28
	Fe	lb/day	0	130
	Polymers	lb/day	0	20
	Resin	ft^3/day	10.5	0
Maintenance Parts & Supplies		$/yr	23,200	19,800
Truck Operating Costs		$/hr	35.20	35.20
Distance From Plant to Permitted Landfill		miles	14	14
Average Truck Speed		mph	35	35
Total Time For Loading/Unloading/Decon		hrs.	1	1
Waste Shipments Schedule		per yr	4	1
Waste Generation Rates		ft^3/day or lb/day	10.5	2
Density, of Precipitation Waste (Sludge)		lb/ft^3	N/A	104
Disposal Cost		$/yd^3	20.00	20.00

that forms on the filtering surface passes out of the trough into a drying zone and then gets scraped off.

The ENVEST model report suggests that the drum be sized such that each square foot of filter surface collects 1 lb/hr of solids (dry basis) for each percent of solids in the sludge. The range allowed for suspended solids content of the sludge is 4% to 10%. (If the clarifier sludge is less than 4% suspended solids, the filter should be preceded by a cone-bottomed thickener tank, which allows settling time for the solids and decanting of some of the water). The model will size the drum for solids at 5 lb/hr/ft^2 if the user does not specify otherwise.

The model will include the addition of ferric chloride as a coagulant aid at the rate of 4% of the dry solids rate, unless the user specifies an additional percentage or specifies lime instead of ferric chloride. The program computes the cost of chemicals over the period of time specified by the user, as well as the investment cost.

The ENVEST model for coagulation/flocculation includes flash mixing of additives, coagulation, and clarifying all in one combined treatment clarifier, rather than in separate components. The program selects the diameter of the unit closest to the standard diameters available for combined clarifiers corresponding to an overflow rate in the range of 330 to 620 gal/day/ft^2. The standard diameters are 9, 12, 15, 20, 25, and 30 ft followed by 10-ft increments up to 70 ft.

The user must specify the water flow rate and may choose coagulants and coagulant aids and their dosage rates. The choices available for coagulants are alum, ferrous sulfate, or ferric chloride. The alum dosage ranges from 0 to 0.835 lb/1,000 gal, and 0.334 lb/1,000 gal is assumed if the user does not specify the dosage. The user must specify the dosage if one of the other coagulants is chosen. The choices available for coagulant aids are sodium carbonate, lime, or clays. The user must specify the dosage, or the program will assume it is zero. The program computes the cost of chemicals over the period of time specified by the user, as well as the investment cost.

The ENVEST model for neutralization computes the cost for pH control, a retention/mixing tank, and chemicals. The user must specify the water flow rate and whether it is acidic or alkaline. The user may choose caustic solution, solid beads or flakes; sodium carbonate or quicklime granules; or powdered hydrated lime for neutralizing acidic water. The model assumes that the pH is less than 2 and that caustic flakes will be used if no other chemical is chosen. If the user does not specify the addition rate, the model assumes a rate of 1.5 lb/1,000 gal for caustic and 1.0 lb/1,000 gal for sodium bicarbonate or lime.

The model uses drummed sulfuric acid for neutralizing alkaline water and assumes that the pH is greater than 12. If the user does not specify the acid addition rate, the model assumes a rate of 1.5 lb/1,000 gal.

If the user does not specify the retention time in the neutralizing tank, the model assumes 5 min.

3.10 Summary of Important Points for Metals Removal

- Metals removal is carried out for removing certain non-contaminant ions to reduce water hardness, as well as to remove toxic heavy metals.
- Metals removal is carried out for certain noncontaminant ions to reduce water hardness, as well as toxic heavy metal ions.
- Chemical precipitation processes involve a number of process steps and require careful additions of the correct amounts of additives and reagents.
- The success of precipitation processes depends on the formation of a dense, readily separable solid phase.
- Conventional separation of the solid phase requires a relatively large clarifier and a sludge dewatering system.
- Innovative filtration and settling schemes are sometimes used to avoid large clarifiers and labor-intensive sludge dewatering systems.
- Mathematical modeling can aid in evaluation and design of chemical precipitation systems.
- RO, ion exchange, and evaporation produce concentrated metal removal process waste streams.

- Scaling of RO membranes occurs if influent pH is not lowered or if the retentate concentrations are too high.
- RO requires high pressures, generally 400 to 1,000 psig.
- RO is often applied as a pretreatment step for ion exchange.
- Ion exchange media can be regenerated and used repeatedly, or they can be discarded after becoming spent.
- Certain media adsorb metal ions without displacing other ions.
- Energy-efficient evaporation schemes include vapor recompression and multiple-effect vacuum evaporation.
- Breakthrough occurs when metal concentrations in the effluent are an arbitrarily chosen fraction of the influent concentration. At saturation, the column effluent concentration equals the influent concentration.
- An ion exchange or adsorption treatment unit with only one or two columns should undergo media regeneration or replacement upon detection of breakthrough unless the effluent concentrations are monitored closely.
- With more than two ion exchange or adsorption beds in series, the lead column can be operated without regeneration or replacement until it becomes saturated.
- A breakthrough curve for ion exchange and adsorbent media is a plot of C_{eff}/C_{in} versus a function of cumulative water volume passed through a bed.
- Breakthrough curves are used to predict the operational life of ion exchange or adsorbent media.
- The shape of a breakthrough curve can be used to predict the operational life of a bed beyond initial breakthrough.
- Sometimes the operational life of a bed can be maximized by using two different media in series, especially if the medium in the lead position does not have a steep breakthrough curve.
- It is important for adsorbers and ion exchange units that the bed length exceed the length of the MTZ.
- The main utility needed for precipitation, membrane separation, ion exchange or adsorption, and vapor-recompression systems is electric power for pumping or compression, controls, lighting, and freeze protection.
- The main utility needed for steam evaporation is steam (or fuel for a boiler).
- Steam required for evaporation is approximately 500 lb/hr for each gal/min of water if there is no heat recovery by heat exchange.
- Simple laboratory jar tests can be applied for determining optimum pH and quantities of chemicals required for metal removal via precipitation.
- The cleanup level that an ideal precipitation process can theoretically attain can be derived from the solubility constant for the precipitate formed.
- RO treatability studies establish the optimum pH and RO pressure and provide a measure of flux, conversion, and rejection rates.
- Treatability studies establish feasibility, provide ion exchange and adsorption isotherms for estimating the capacity of a medium for retaining metal ions, and allow the determination of requirements for media regeneration.
- Flow-through tests for ion exchange and adsorption beds determine the capacity of resins and adsorbents at breakthrough and at saturation.
- Software is available that aids both in process design and cost estimating for coagulation, clarification, dewatering, neutralization, and filtration.

Chapter 4

Groundwater Remediation Using Carbon Adsorption

Activated carbon and certain resins will adsorb organic compounds in an aqueous phase, such as groundwater, or in a vapor phase, such as air. Stripping of groundwater that contains VOCs produces vapor-phase organic compounds. Ventilation of soils that contain VOCs produces vapor-phase organics also. The chapters on stripping and on soil venting will discuss vapor-phase carbon adsorption. This chapter will deal with the direct contact of groundwater with fixed beds of granular activated carbon (GAC) or of resins that adsorb organic compounds.

Emphasis will be on using GAC, but certain resins and other substances can compete with GAC in the process of organics adsorption, either in an aqueous phase or vapor phase. For example, in The Netherlands, groundwater is being remediated by absorption into vegetable oil contained in porous particles of polypropylene. The term "adsorbent," as used in this chapter, refers to either GAC or other sorbents used to remove organics from groundwater.

Most groundwater treatment systems using aqueous phase GAC ship used carbon off-site for reactivation, a process that will not be covered in depth here. This chapter discusses mechanisms of adsorption and adsorption capacity, configuration and performance of adsorption systems, process design, and costs.

4.1   Basic Principles of Carbon Adsorption

Activated carbon is charcoal that can be produced from wood, coconut shells, fruit pits, and other natural cellulose materials, or it can be made from lignite, peat, or bituminous coal. The most common activated carbon used for remediation is derived from bituminous coal. Lignite and coconut shells are also widely used for producing activated carbon. The material is activated by heating it in a multiple-hearth furnace in a steam atmosphere where there is insufficient oxygen (air) to burn the material. The temperatures are high enough to drive off virtually all of the organic compounds, leaving almost pure carbon. Carbon that has been used to adsorb organic compounds can be reactivated in a multiple-hearth furnace, with the same process used to prepare virgin activated carbon, or in a kiln.

Reactivated carbon does not adsorb as well as virgin activated carbon. If a very pure product water is desired, replacement carbon is often virgin activated material, and the spent carbon is cycled to another user when reactivated. If it is desired to avoid cross-contamination from reactivated carbon, arrangements should be made with the vendor to avoid carbon that has been cycled from another user. Note that there is approximately 10% to 20% attrition loss each cycle. This loss is made up with virgin carbon.

Heating coal to 800° to 1,000° C (1,472° to 1,832° F) in an oxygen-limited atmosphere in the presence of steam opens up pores in the carbon. Organic molecules in a medium passing through GAC diffuse to the carbon and then diffuse through macropores within the granules. The macropores provide a pathway to micropores. Adsorption takes place on the surface of the pores. If the micropores are envisioned as somewhat spherical in shape, the high degree of adsorptivity for carbon can be explained. Mathematically, the smaller the diameter of any sphere, the larger the ratio of surface area to volume. Activated carbon contains many microspheres per unit volume of carbon and therefore provides a large surface area for adsorption sites per unit volume. A gram of carbon can have as much as 1,000 m^2 of internal surface area (McLaughlin, 1995).

Two types of activated carbon are derived from bituminous coal. Macroporous coal carbon has a large fraction of pore volume in macropores; microporous coal carbon has more of its pore volume in pores that are smaller than 500 Angstrom units. Coconut shell carbon has even more of the very smallest pores (smaller than 20 Angstrom units in diameter), which makes it more advantageous in vapor-phase applications than in aqueous phase. Macroporous coal carbon is good for adsorbing large molecules. Microporous coal carbon is excellent for semivolatile organic compounds and is good for non-oxygenated VOCs. Coconut shell carbon is good for semivolatile organic compounds and excellent for non-oxygenated VOCs.

If in place regeneration of spent carbon with steam is attempted instead of removal to a furnace or kiln, the surface area will be not be maintained. The regenerated carbon will not remove trace contaminants as well as would carbon that has been reactivated at very high temperatures. High molecular weight organics will not desorb and will build up on the carbon.

Adsorption of organics on carbon in the absence of bacteria or oxygen is mainly physical, and the organic molecules are unchanged. Sawyer, McCarty, and Parkin (1994) describe three general types of adsorption: physical, chemical, and exchange. These processes are applicable to both activated carbon and polymeric sorbents. Physical adsorption is a non-specific reaction resulting from condensation on the adsorbent surface and weak attraction forces, including van der Waals' forces. Physical adsorption is reversible, resulting in desorption of the adsorbate when its concentration drops below equilibrium levels. Chemical adsorption involves interactions between the adsorbate and specific surface substances on the adsorbent. This process is seldom reversible and requires a large energy input to displace adsorbed materials during adsorbent regeneration. Exchange adsorption involves electrical attraction and is particularly relevant when the adsorbate can be ionized under ambient groundwater conditions. Here, the adsorbate concentrates at the adsorbent surface because of electrostatic attraction, with species of greater ionic charge and a smaller hydrated radius being more strongly attracted than species of small ionic charge and a large hydrated radius.

Depending on the nature of the adsorbate and the surfaces of the carbon being used for its removal, one or all of these adsorption reactions can take place. A combination of these processes leads to hysteresis in the desorption (regeneration) process. If attrition replacements are small, this can lead to the eventual requirement for total carbon replacement because of irreversible chemisorption occurring throughout the expended carbon.

Equilibrium between the molecules in the fluid passing through the GAC and the adsorbed molecules is achieved rapidly. The adsorption process releases heat. Adsorption can be reversed, and with a heated gas or steam environment the molecules readily desorb. The amount of organics that a unit mass of activated carbon can adsorb depends on the source from which the carbon is derived, the carbon pore size, the organic compounds being treated, the concentration of these organics, and the reactor temperature. Coconut shell carbon costs more than bituminous coal carbon, but it has a higher adsorptivity and can produce a cleaner treated product if VOCs are the contaminants of concern. Correspondingly, coconut shell carbon requires more energy input per unit mass to reactivate than does coal carbon.

For a given form of carbon at a given temperature, each organic compound has an adsorptivity that correlates to that compound's concentration. Almost all groundwater adsorption systems operate at or near the natural temperature of groundwater, which is usually 13° C (55° F) or slightly above. For groundwater remediation by adsorption in the aqueous phase, variations in temperature are normally not significant. Data on adsorptivity obtained at a temperature in the range of 10° to 21° C (50° to 70° F) is almost universally applicable. (Conversely, for vapor-phase adsorption, the adsorptivity is very sensitive to temperature variations.) A graph of absorbed amount of organic per unit mass of carbon versus concentration, for a given temperature, is expressed with adsorption isotherms.

Some compounds are so poorly adsorbed — e.g., vinyl chloride and methanol — that carbon is not the adsorbent of choice. In addition, methylene chloride, ethanol, and some other low-molecular weight oxygenates are somewhat poorly adsorbed on carbon in the aqueous phase. As a general rule, organic compounds that are highly soluble in water (e.g., ketones, alcohols) have the lowest amount of the adsorbed compound per unit mass of carbon values.

The most common interfering compounds are natural organic matter (e.g., humic and fulvic substances) and oil and grease.

4.2   Adsorption Isotherms

The Freundlich equation expresses adsorption isotherms, as given in Equation 4-1.

$x/m = kC^{1/n}$ (4-1)

in which x/m is the amount of the adsorbed compound per unit mass of carbon; k is a proportionality constant; and C is the equilibrium concentration of the compound involved. If this equation is plotted on log-log paper, a straight line is obtained, with its slope equal to 1/n.

The x/m values can be expressed in a variety of units, such as mg/g, g/g, lb/lb, lb/100 lb, weight percent of carbon that is adsorbed compound, etc. C can be expressed in any convenient concentration units, such as mg/L.

Aqueous-phase x/m values for organics are determined experimentally in laboratories by dispersing varying amounts of pulverized carbon in water solutions containing a given dissolved organic compound or compounds. The carbon is stirred in the water solution for many hours so that equilibrium conditions are attained. At some level of carbon addition, the carbon becomes saturated with the adsorbed matter, and the x/m value is determined for the equilibrium concentration in the water. The equilibrium concentration is the concentration of unabsorbed organics left in solution.

Several important factors about this bench-scale isotherm test are important to understand in relation to designing a steady-flow groundwater remediation system. With the usual flow-through systems (groundwater pumped through a fixed bed of GAC), the water/carbon contact time is typically only 10 to 20 min. Equilibrium conditions may be approached within the carbon bed, but not where the water exits the bed, if adsorption is taking place. Multiple organic compounds, some of them natural and relatively harmless and some of them contaminants, are usually present in groundwater. If the adsorption isotherm has been determined for a single compound, the x/m values cannot be applied to the situation with multiple organic compounds, and x/m values are not always directly additive.

What can be readily measured in a groundwater remediation system are the influent concentrations. With proper design, the effluent concentrations are frequently below or near the analysis detection limit, and equilibrium concentrations are not known. To ensure that contamination is not significant in the effluent with a single carbon bed system, the carbon bed is most often taken out of adsorption service for reactivation upon detection of initial breakthrough, well before it becomes totally saturated.

Because of these conditions, the amount of carbon proposed to be used during design for a given set of contaminant concentrations is based on experience with similar contaminant conditions, on pilot test data, or on a fraction of published x/m isotherm values. With a single carbon bed system operated until initial breakthrough, the fraction of x/m that can be achieved has been determined from Calgon Carbon Corp. (Pittsburgh, Pennsylvania) experiments. With flow-through conditions at typical residence times, the fraction is 0.45 to 0.55, as reported by Stenzel and Merz (1989). This provides a practical approach for predicting carbon bed life based on influent contaminant concentrations and x/m values from isotherms determined at equilibrium concentrations.

4.3 Methods of Determining Adsorptive Capacity

The x/m values can be found in published data or determined experimentally. If only one organic contaminant is present in the groundwater at a significant concentration, and if concentrations of natural organic compounds are not significant (measured analytically as total organic carbon, or TOC, concentration minus the contaminant carbon concentration), published x/m values or published graphical isotherms can be used. By plotting the x/m values on log-log paper, interpolation can be readily used to find the applicable x/m value.

The x/m values used in the design must be for the type of carbon to be used. Published x/m values should usually be multiplied by 0.45 to 0.55 for application to design of a flow-through system with an empty-bed contact time of 12 to 15 min (Stenzel and Merz, 1989).

For some compounds, x/m is affected by the pH of the groundwater. Organic acids adsorb better at low pH, and amino compounds, at high pH (Stover and Thomas, 1992).

With multiple organic contaminants and no experience in adsorption applications with the contaminant concentrations involved, published x/m values help determine whether the individual compounds can be readily adsorbed. Better x/m values can be determined in a laboratory using a sample of the actual groundwater. Data that are better than standard isotherm test data can be attained by running the groundwater sample through a laboratory column packed with GAC. Another experimental approach that is sometimes preferred is to run field pilot tests with drum-size quantities of carbon. Laboratory column tests and field pilot tests are described in Section 4.8 (Aqueous Phase Adsorption Treatability Studies).

4.4 Breakthrough Curves

Within an adsorbent bed with the usual downflow arrangement, there are three zones that exist before breakthrough:

- A saturated zone extending part-way down from the top of the bed
- A mass transfer zone (MTZ) in which adsorption is taking place, immediately below the saturated zone
- A clean adsorbent zone at the bottom, where essentially no contaminant has yet come out of the MTZ

It is important to understand the distinction between the contaminant loading on the adsorbent under initial breakthrough conditions versus the loading at saturation, x/m. After initial breakthrough of contaminant in the effluent, additional contaminated groundwater can still be passed through the adsorbent bed. Under breakthrough conditions, the effluent concentration is an increasing fraction of influent concentration as more volume of groundwater continues to flow. Adsorption can continue until saturation is achieved. However, once breakthrough starts, the effluent contaminant concentration will increase as more groundwater volume is treated, up to the point of saturation. At saturation, the effluent concentration equals the influent concentration. This situation can be described graphically by a breakthrough curve for each residence time, which usually is similar to the curve in Figure 3-7A. At longer residence times, the initial flat portion of the curve is longer.

The residence time used in this book is defined as the empty-bed contact time. It is calculated as the gross volume of the bed, including solid adsorbent and void space between the granules, divided by the volumetric flow rate. By basing the residence time on the gross volume of the bed, it is independent of the void space fraction and of the size of the granules.

As more volume of groundwater flows through the bed, the saturated zone grows in depth, the MTZ moves downward, and the clean adsorbent zone shrinks in depth. When the MTZ approaches the outlet, breakthrough starts. However, adsorption still continues to take place, until the saturated zone reaches the outlet. At this condition, the effluent concentration equals the influent concentration.

It must be emphasized that initial breakthrough occurs earlier than bed saturation. In practice, multi-columns in series are used. The first bed is allowed to continue in operation to some extent beyond initial breakthrough. Allowing a first bed to continue to operate until saturation would reduce needed adsorbent reactivation frequency, but it poses the risk of excess contaminant appearing in the final effluent. This is discussed in detail in Section 4.6.3 (Improving Performance with Three-Stage Adsorption).

The design total bed depth must be significantly larger than the length of the MTZ. That zone length depends on influent concentration, contaminant molecule size and polarity, and the hydraulic loading.

4.5 Sizing of Carbon Beds and Duration of Bed Life

The cross-sectional area for a carbon bed with groundwater flowing downward should be such that the hydraulic loading rate is 1 to 5 gal/min/ft^2 for most applications. For example, if 12 gal/min is flowing through a vertical cylindrical carbon canister that is 2 ft in diameter (3.14 ft^2 cross-sectional area), the hydraulic loading rate is 12/3.14 gal/min/ft^2, or 3.8 gal/min/ft^2. Design hydraulic loading rates exceeding 5 gal/min/ft^2 can be tolerated if the pumping system and canister allowable working pressures are designed for the higher fluid pressure drop that will be developed.

The chosen hydraulic loading rate will establish the cross-sectional area for any given groundwater flow rate for the initial conceptual design. The required bed depth is calculated as the gross bed volume (determined from the desired residence time) divided by the cross-sectional area. For example, if the 12-gal/min example above uses a redesigned canister such that the hydraulic loading rate is increased, a preliminary calculation of bed depth is as follows:

Assumptions: Hydraulic loading = 5 gal/min/ft^2; desired empty-bed residence time = 20 min

Given: Flow rate = 12 gal/min

Find: Cross-sectional area and bed depth

Step 1: Determine cross-sectional area as shown in Equation 4-2.

Area = 12 gal/min/(5 gal/min/ft^2) = 2.4 ft^2 (diameter = 21 in.)   (4-2)

Step 2: Calculate bed depth from Equations 4-3, 4-4, and 4-5 by dividing bed volume by the cross-sectional area.

Q = volumetric flow rate 12 gal/min (1 ft^3/7.48 gal) = 1.6 ft^3/min   (4-3)
Bed volume = 1.6 ft^3/min (20 min) = 32 ft^3 (4-4)
Bed depth = 32 ft^3/2.4 ft^2 = 13 ft, 4 in.   (4-5)

This may not be the final design. Canisters may not be readily available with a diameter of 21 in. The pressure drop may be too high with a depth of 13 ft, or there may not be enough space for a canister that tall. (Canisters arranged in series can be used to achieve a desired bed depth.) The same bed-sizing steps might be repeated, based on readily available standard canister sizes and on information from the equipment vendor on pressure drop. The desired residence time is sometimes achieved with beds in series. Adsorbent vendors can advise on the best mesh size for the granules and on the corresponding pressure drop per unit depth at any given hydraulic loading.

At higher volumetric flow rates, it is sometimes desirable to use readily available drums or canisters. To achieve the residence time with a standard container height and to keep operating pressures low, parallel containers are used.

Stover and Thomas (1992) recommend that upflow columns be considered if very high hydraulic loadings are used, with consequent high pressure drops. Gravity downflow is feasible at hydraulic loading rates less than 4 gal/min/ft^2. However, groundwater adsorbers should have prefilters, so pumped systems are normally used to overcome filter pressure drop, piping friction losses, and bed pressure drop.

At hydraulic loading rates less than 1 gal/min/ft^2, the GAC usually used will be too coarse for proper operation with a downflow design. The groundwater may channel down through the bed in a narrow pathway and not spread over the entire cross-section. In such an event, the calculated residence time will not be achieved, and breakthrough will occur earlier than expected. Carbon consumption will be higher than it need be, because not all of the carbon will not be saturated. If finer mesh carbon granules are not used, and if the container selection is not changed to something narrower, then upflow columns should be used. By introducing the groundwater to the bottom of the bed and flowing upwards, the water tends to spread over the entire cross-section more uniformly.

The duration between adsorbent change-outs depends on the fraction F of x/m achieved, the adsorbent mass m, the contaminant concentration C, and the volumetric flow rate Q as shown in Equations 4-6 and 4-7.

Mass of contaminant adsorbed until breakthrough = (F) (x/m) (m)   (4-6)
Duration = mass of contaminant adsorbed/(C)(Q)   (4-7)

As an example, consider this hypothetical situation:

Assumptions: F is the maximum in the 0.45 to 0.55 range given in Stenzel and Merz (1989) for a system in which the carbon is changed out upon breakthrough (see Section 4.6.3 regarding operating carbon beds until saturated); downtime for maintenance is approximately 1 mo. (31 days) each year.

Given: Groundwater flow rate Q is 50 gal/min; contaminant concentration C is 0.350 mg/L; effluent contaminant concentration is must not exceed 0.001 mg/L; x/m is 12.7 lb of contaminant per 100 lb of carbon, from published isotherm data

Find: Mass m of adsorbent for a duration of 1 yr (334 days net operating time, allowing for downtime)

Step 1: Rearrange Equation 4-7 to determine the mass of contaminant adsorbed in 1 yr, as shown in Equation 4-8.

Mass of contaminant adsorbed = (duration) (ΔC) (Q)
= (334 day) (0.350 mg/L) (272,160 L/day) (1 lb/453,700 mg) = 70 lb   (4-8)

Step 2: Rearrange Equation 4-6 and solve for m, as shown in Equation 4-9.

m = Mass of contaminant adsorbed/[F(x/m)]
= 70 lb/[(0.55) (12.7 lb/100 lb adsorbent)] = 1,000 lb   (4-9)

The main factors that affect the capacity of a given mass of carbon to adsorb organics and bed life are:

- The original material from which the activated carbon is derived
- Properties of contaminant
- Concentrations of contaminants
- Presence of interfering compounds
- Residence time

Each organic compound has an x/m value for a given carbon source that depends on concentration, according to the isotherm. The higher the concentration, the higher the x/m; the carbon mass can adsorb more, but the bed life at a given groundwater flow rate decreases as influent concentrations become higher. For example, suppose x/m values for contaminant Z are as given in column C in Table 4-1 for a particular activated carbon at 15° C.

Table 4-1 Activated carbon x/m values.

A Influent Z Concentration	B Concentration of Z at Equilibrium	C x/m mg Z/g carbon	D x, for 1000 g carbon
1 mg/L	0.02 mg/L	7.5 mg/L	7,500 mg Z
2 mg/L	0.05 mg/L	11 mg/L	11,000 mg Z

Note that column B is for concentrations after hours of mixing solutions of Z with powdered activated carbon. This activated carbon will theoretically become saturated with Z at cumulative influent volumes treated, as shown in column E of Table 4-2.

Table 4-2 Influent volumes for activated carbon example.

A Influent Z Concentration	D x, for 1000 g carbon	E Cumulative Influent Volume at Saturation
1 mg/L	7,500 mg Z	7,500 mg/(1 mg/L) = 7,500 L
2 mg/L	11,000 mg Z	11,000 mg/(2 mg/L) = 5,500 L

Analyses of the groundwater should include TOC, which includes the contaminants and natural organic matter. It is the TOC that often determines what the bed life will be. Because of this, dynamic laboratory column tests or field pilot tests are the best methods of obtaining data for predicting bed life.

If suspended oil and grease or other high-molecular-weight organic compounds are in the groundwater, carbon adsorption sites will become quickly saturated. Oil and grease should be filtered out before the water enters the carbon beds. The use of organophilic clays as filtration media should be considered. One commercially available clay filter is made of amine-treated bentonite clay mixed with anthracite (Alther, 1997). The anthracite helps maintain the clay's permeability.

Plugging and pore blockage can be caused by calcium scalants and by biomass growth. Groundwater with a positive Langelier saturation index (LSI), calculated from calcium hardness and total dissolved solids concentrations, pH, and alkalinity as given in Section 3.2.4 in Chapter 3, can potentially cause calcium carbonate scaling problems. A calcified carbon bed may form that may not be fit for reactivation. Bioactivity in the carbon beds may also have to be controlled with continuous UV light pretreatment or by periodically treating the beds by injecting acid and sodium hypochlorite. However, the free chlorine residual should not exceed a few parts per million to avoid degrading carbon activity (Graham, 1992).

4.6 Arrangements and Performance of Organic Adsorption Systems

Most remediation systems use downflow of groundwater through fixed beds of GAC. However, Stover and Thomas (1992) note that upflow beds can be designed with removal and replacement of carbon while the unit is in operation; the efficiency of carbon use in this configuration is higher than with downflow units because countercurrent contact operation is more closely approached.

Carbon canisters chosen should be made of corrosion-resistant materials. If carbon steel is used, it must be lined, because water in the presence of carbon develops corrosion cells (Stenzel, 1993). The lining should be at least 35 mils (0.9 mm) thick, because carbon is abrasive.

4.6.1 Prestripping

If the groundwater contains VOCs, air stripping the groundwater first can be considered. Air stripping systems generally cost less to operate than adsorption systems. Air costs nothing, and the energy involved with stripping — for pumping the groundwater to the top of a stripping tower and for operating an air blower — is much less costly than the energy required for reactivating used adsorbents. However, if stripper offgas must be treated to comply with air pollution control regulations, an economic analysis is needed to determine whether prestripping is cost-effective.

In Chapter 5 (Stripping of Groundwater), an example is given to explain the use of carbon as a polishing step for removing the residual levels of organics remaining after a stripping step. The stripper in that example removes approximately 97% of the volatile organics, reducing the influent concentration from 13 mg/L to 0.35 mg/L before carbon treatment. If there are no nonvolatile organic compounds that have not been accounted for, prestripping increases the adsorbent life between reactivations by a factor of almost 37. The factor will be somewhat less than 37 because x/m values are less at reduced concentrations.

For example, suppose that the x/m is 110 mg/g of carbon corresponding to a concentration of 13 mg/L at the inlet to the first carbon bed and 90 mg/g of carbon corresponding to a concentration of 0.35 mg/L at the inlet. (The x/m values derived from equilibrium isotherms are approximated as x/m values based on inlet concentrations. This approximation is a good one at the beginning of the MTZ.) Up to 1,000 g of activated carbon can become saturated with quantities as shown in Column D in Table 4-3 for influent to the first carbon bed.

The absorbent life in this example is increased with prestripping by a factor of 30 (257,000 L/8,460 L).

4.6.2 Prefiltering and Preventing Overpressure

With downflow systems, adsorbents can also accomplish filtration of suspended solids from groundwater, but it is not economical to use adsorbents for filtering. The filtering should be accomplished ahead of the adsorbent containers in equipment

Table 4-3 Influent volumes with prestripping.

A Concentration at Bed Inlet	B x/m	C x, for 1000 g carbon	D Influent Volume at Saturation
13 mg/L	110 mg/g carbon	110,000 mg	110,000 mg/(13 mg/L) = 8,460 L
0.35 mg/L	90 mg/g carbon	90,000 mg	90,000 mg/(0.35 mg/L) = 257,000 L

designed such that the filtering media are readily replaceable or cleanable. Most filter systems applied to remediation are either back-flushable or have vessels that are easy to open and to securely close, with cloth socks or paper cartridges that can be changed quickly. Prefiltration should reduce turbidity to less than 2.5 Jackson Turbidity Units (Stover and Thomas, 1992).

Without prefiltering, suspended solids will clog adsorbent beds and cause high pressure drops. Even with prefiltering, some suspended solids and/or biological growth often clogs adsorbent beds, causing high pressure drops. One or more of the following alternatives should be considered for the system design to prevent overpressure:

- Install pressure gauges ahead of the filter and ahead of each adsorbent bed used in series.
- Install a pressure-relief valve that discharges to an influent tank or extraction well.
- Interlock the feed pump with a high-pressure switch, or install a high-pressure switch that trips an alarm.
- Use canisters with a pressure rating that is higher than the pumping system's pressure capability.

4.6.3 Improving Performance with Three-Stage Adsorption

Many adsorbent systems include two stages of adsorption. If it is critical that the final effluent always be below certain contaminant concentration limits, the first-stage adsorbent is removed from service when breakthrough is noted. The first-stage adsorbent is not allowed to continue in service much after initial breakthrough. This is because by the time that saturation is reached in the first stage, there might be a second-stage breakthrough, causing a violation of effluent concentration limits.

However, some adsorbent capacity is wasted with such two-stage operations. If the first stage could be allowed to reach saturation, bed life would be extended. Because of this, there is a trend toward installing three-stage systems for unattended installations that are not closely monitored for final-stage breakthrough. The first stage is allowed to reach saturation, and if there were consequent breakthrough from the second stage, the contaminants would be adsorbed in the third stage. In practice, last-stage columns are not allowed to reach initial breakthrough but can operate with adsorption taking place in a MTZ that has not migrated to the bed outlet.

With either a two-stage or three-stage system, each time the first-stage adsorbent is removed from service the second-stage adsorbent is put into first-stage service. Fresh adsorbent is then used in the last stage. Interchanging stages of service is accomplished by physically moving containers and reconnecting lines or by repositioning valves in a valve manifold.

Using three-stage adsorption helps in minimizing the frequency of changing adsorptive media while assuring quality effluent. Another method of assuring quality effluent is to continuously monitor contaminant concentrations in the effluent. An alternative is to set up water sampling ports at various bed depths. Analyses of the water samples are used to define the depth and rate of movement of the MTZ within a bed. Thus, with periodic sampling, the throughput volume of water that corresponds to breakthrough can be predicted. It is not necessary to resort to continuous or very frequent analyzing of the effluent.

4.6.4 Presoaking and Backwashing

Before going into service, fresh adsorbent should be presoaked overnight and backwashed (apply upward fluid flow with systems designed for normal downflow) with clean water. These two steps will help remove trapped air and help prevent a rise in pH that occurs with fresh carbon that has not been acid washed by the supplier. Applying upward fluid flow also helps stratify the bed and rid the bed of fines. This operation is usually accomplished in the columns used for adsorption, although it can also be accomplished in separate vessels. The backwash water containing the fines is usually routed through a filter. Backwashing can be applied to partially spent adsorbent as well as to fresh adsorbent in order to reduce pressure drop.

In general, the upward fluid velocity should be enough to expand the adsorbent bed volume 30% to 50%. Therefore, the columns should be designed such that there is free space extending above the normal bed equal in height to 50% of the bed depth. The equipment vendor should recommend the corresponding hydraulic loading rate for backwash conditions and the minimum quantity of backwash water. The hydraulic loading rate for GAC backwashing may be as high as 12 gal/min/ft^2. The quantity of backwash water for carbon may be at least 0.75 gal/lb of carbon.

4.6.5 Lower Explosive Limit (LEL) Monitoring for Breakthrough

Sometimes groundwater is contaminated with relatively high concentrations of light hydrocarbons, such as gasoline, and it is essential to ensure that the effluent does not cause explosive conditions in a sewer system. Such conditions could form if there is inadvertent breakthrough from the carbon beds.

An LEL monitoring or shutdown system can be added to help prevent explosive conditions. An LEL detection tank or drum contains no adsorbent and has a probe in the air space above the effluent liquid level. Most LEL devices can detect the volumetric hydrocarbon vapor concentration in air and read out the corresponding percent of LEL (percent of the threshold of explosivity). Many LEL detectors are set to signal an alarm or shutdown a process at 25% to 40% of the LEL.

For remediation of hydrocarbon-contaminated groundwater, the LEL detector should be set to act at the lowest possible LEL percentage setting. This setting may be as low as 2% of the LEL, because normally the concentration of hydrocarbons in the effluent will be near zero, and the corresponding vapor concentration will be nondetectable.

The use of an LEL detector is a low-cost method of monitoring for a major breakthrough. High-cost, more expensive vapor phase detectors could be applied for better sensitivity. Or an elaborate liquid phase automatic detection system could be designed.

Figure 4-1 shows a typical aqueous-phase adsorption system with an LEL monitoring drum.

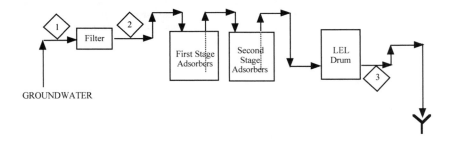

	Stream 1	Stream 2	Stream 3
Water, gpm	50	50	50
Total Petroleum Hydrocarbons, TPH			
mg/L	0.350	0.350	ND
lb/day	0.21	0.21	nil
Pressure, psig	10 to 18	10 maximum	1
Temperature, °F	60	58 to 62	54 to 66

Figure 4-1 A process flow diagram for an aqueous phase carbon adsorption system.

4.7 Main System Design Parameters

This section focuses on the evaluation of important features of system design. Groundwater flow rate, influent and desired effluent concentrations, and any treatability testing results should be known. A numerical example is given in this section for activated carbon adsorption system design.

4.7.1 Concept and Process Design

If the groundwater contaminants include VOCs, the first decision that must be made is whether prestripping is desirable, as well as whether a stripper installation would have to include offgas treatment. If no emissions controls are needed, prestripping with air will be economic except at low influent concentrations. If capital investment or equipment height must be minimized, simple aeration chambers (in which air is bubbled through shallow tanks with groundwater flowing through the tanks in series) should be considered. Another method of prestripping and meeting height limitations is to use a low-profile tray stripper.

If stripper emissions controls are required, an economic analysis based on carbon bed life should be completed. An example of the basis for such an analysis is given in Chapter 5 (Stripping of Groundwater). The analysis should be refined with an attempt to find the optimum stripper air/water ratio based on:

- Costs for controlling stripper emissions
- Costs for replacing carbon

The next decision is whether to use more than two stages of adsorption. Two stages are sufficient if the influent concentration is low or if prestripping is applied. In a different scenario with high contaminant concentrations entering the first adsorption stage and stringent contaminant concentration limits, it is too risky to allow saturation in the first stage. Addition of a third stage will usually be cost-effective.

The example considered here will be the two-stage system shown in Figure 4-1. This aqueous phase carbon adsorption system follows an air stripper that has reduced the hydrocarbon concentration to 0.35 mg/L.

From pilot testing data, the x/m was determined to be 0.10 to 0.12, equivalent to 10% to 12% based on carbon bed mass, and it is anticipated that first-stage carbon will be removed each time breakthrough is discovered. The system in this example is generally unattended except for routine visits by a technician once every 2 weeks, and a response to an alarm is possible within 6 hr. Alarms and shutdowns signals are relayed to an attended station by an automatic phone dialer. From experience with similar systems, breakthrough is expected when the mass of adsorbed organics is approximately 7% of the carbon mass. With attendance only every 2 weeks, a breakthrough may not be discovered until 14 days after it happens. This is tolerable, because it is anticipated in this example that the loading on the first-stage carbon will not exceed approximately 8%, and the second stage carbon will not yet be near breakthrough. Virtually all hydrocarbon contamination will be removed, and non-detect analytical results are anticipated in the effluent from the second stage.

The process flow diagram shows volumetric and mass flow rates and concentrations for the constituents of concern, giving preliminary estimated values for pressures and temperatures. The diagram shown in Figure 4-1 shows these parameters as exemplified for a 50-gal/min system.

The process flow diagram for the prestripper has not been included. The estimated pressures should be revised later after the piping is designed and after the canisters are finally selected and pressure-drop data are obtained from the carbon vendor. After equipment is sized, equipment ratings and capacities are added to the flow diagram. For example, the pounds of adsorbent per canister and the volume of the LEL drum should be noted.

A preliminary P&ID can be developed from Figure 4-1 by adding all instruments, interlocks, alarms, and valves, including sample valves (Figure 4-2). A complex valving manifold could be added at the adsorbers that will allow either one to be the lead absorber. As the design progresses, the piping sizes and materials should be added to the P&ID.

Along with the P&ID, a Sequence of Operations document should be developed that succinctly explains the general basis for design, how the interlocks are connected, and how responses to alarms will be handled. For this example, the sequence might include the following information:

- The carbon adsorption system takes bottoms product from the prestripper at flow rates up to 50 gal/min. At the anticipated concentrations, the life of each carbon bed will be 334 days. Either bed can serve as the lead bed, and carbon should be changed within 2 weeks of operation on detection of breakthrough.
- The filter is rated at 85 psig and should be changed if the pressure drop reaches 10 psi.
- Carbon canisters are rated at 11 psig and are fitted with a high-pressure alarm set at 10 psig. If breakthrough has not occurred, the carbon bed should be backwashed in the event of high pressure.
- The LEL detector should be set to trip at its lowest setting. It will close the effluent valve and shut off influent pumping.
- The unit has a rain canopy and secondary containment. The liquid level detector will shut off influent pumping at 1 in. of liquid level being detected within the containment.
- The autodialer has four messages that it can transmit to headquarters: (1) automatic shut off of pumping; (2) prestripper shutdown; (3) high pressure; and (4) breach of the security system at the perimeter fence.

Using the example calculations given with Equations 4-8 and 4-9, 1,000-lb carbon canisters will be installed. A carbon vendor can furnish dimensions of the standard size canister that will hold close to this amount of carbon. A preliminary plot plan drawn to scale is needed, showing likely locations of major equipment and enclosures

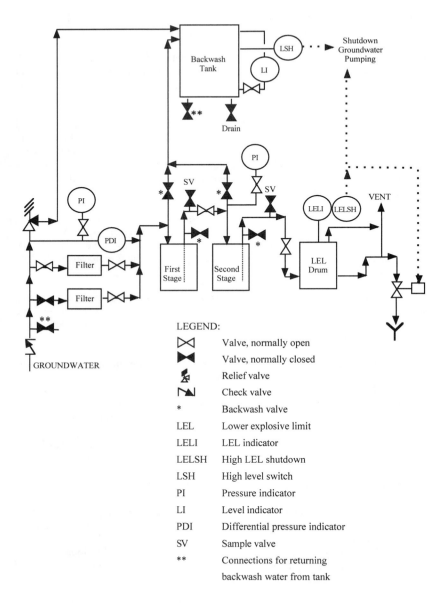

Figure 4-2 P&ID for an aqueous phase carbon adsorption system.

and allowing space for valves and access. A site plan should be developed showing where the plot fits relative to existing equipment, buildings, and nearby roads and where utilities are available. The site plan should show any needed new access driveways, fencing, underground drainage lines, groundwater extraction wells, injection wells, piping connected to wells, and water storage tanks.

The process flow diagram, P&ID, plot plan, and site plan form a process design from which detailed piping, electrical, and structural designs can be developed in later phases of design.

4.7.2 Sizing and Rating of Major Equipment

This section uses as an example the sizing of a media filter system, two downstream adsorbers in series, and an LEL drum.

For this example, it has been decided that the prestripping/carbon adsorption system will include a pair of bag filters upstream of the stripper and a pair of dual-media (sand plus anthracite) filters between the stripper and the first adsorber. For the concentration of suspended solids in the groundwater fed to the stripper, a simple bag filter may be deemed adequate for lasting more than 2 weeks each between bag changeouts. However, relatively large amounts of calcium carbonate scale can form when air strips carbon dioxide from the groundwater, depending on the groundwater chemistry. A large-capacity filter, such as a media filter, may be required to prevent the need for frequent operator attendance.

For this example, it has also been decided to avoid the maintenance of automatic valves that would be needed if the media filters were to be switched automatically between filtering service and backwashing. The plant operator will backwash either filter manually during periodic visits to the site.

Because of the need for periodic backwashing, a storage tank must be added to the flow scheme to receive the backwash water. The particulate matter in the backwash water is allowed to settle in the tank. Before starting a backwash operation, the operator decants the previous backwash water out of the tank and transfers it to the stripper sump. After some months of operation, sludge that has built up in the bottom of the tank must be pumped out and removed from the site. The Sequence of Operation should include a description of the manual operations required for backwashing.

Vendors of sand bed filters would be contacted for sizing of the filter vessels. Sizing can be done using a hydraulic loading rate of 2 to 6 gal/min/ft^2. The 50-gal/min example given above is continued here for a hydraulic loading of 3 gal/min/ft^2 for the sand bed filter, using Equation 4-10.

Area = 50 gal/min/(3 gal/min/ft^2) = 16.7 ft^2 (4-10)

This cross-sectional vessel area corresponds to a diameter of 4.6 ft; vessels with a diameter of 5 ft would be selected.

The pressure rating for the filter vessels depends on the pump selection made for the stripper sump. This pump must develop enough pressure to overcome pressure drops in a loaded filter vessel and two adsorbers in series, plus friction losses in the connecting piping. If the shutoff head of the selected pump exceeds the pressure rating of the filter vessels, protection against vessel overpressure must be included in the design.

The volume of the backwash storage tank depends on the amount of backwash volume recommended by the filter vendor. The tank should hold at least twice the recommended volume plus the volume of sludge anticipated to be collected between sludge pumpout events.

An example of sizing carbon adsorbers will now be developed using 50 gal/min (6.7 ft^3/min) with 5 min of residence time in each of two adsorbers in series (10 min of total residence time). The hydraulic loading rate usually should not exceed 5 gal/min/ft^2 if a low pressure drop is desired, but loadings at 10 gal/min/ft^2 and higher have been used in vessels rated for high pressure. Stover and Thomas (1992) recommend hydraulic loading rates in the range of 2 to 10 gal/min/ft^2. A velocity greater than 10 gal/min/ft^2 might be reached, in which the carbon could not adsorb the contaminants fast enough to prevent early breakthrough; whereas at very low velocities, Stover and Thomas indicate that diffusion of contaminant molecules through the stagnant film surrounding the granules may limit adsorber efficiency. The sizing of adsorbers is done as shown in Equations 4-11 and 4-12.

Area = 50 gal/min/(5 gal/min/ft^2) = 10 ft^2 (4-11)

A 4-ft diameter canister will be assumed until a carbon vendor is selected and a standard canister size is chosen.

Bed volume = 6.7 ft^3/min (5 min) = 33.5 ft^3 (4-12)

Considering the bulk density of aqueous-phase GAC, this volume of carbon would weigh approximately 1,000 lb. Standard canisters are available for this quantity of carbon. The bed depth would equal the bed volume divided by the actual square feet of cross-sectional area in the canisters procured. The canisters should have a vertical straight-side length that is 50% larger than the bed depth. This extra length allows for expansion of the bed during backwashing.

The pressure drop through a GAC bed is proportional to $\mu QL/d_p^2 D$, where μ is water viscosity, Q is the volumetric flow rate, L is bed depth, d_p is the geometric mean particle diameter, and D is the canister inside diameter.

The volume that should be selected for the LEL drum depends on how the liquid level will be controlled in the drum. If the final effluent were being pumped from the drum at a fully attended unit with continuously modulated level control, the drum volume might be chosen to provide several minutes of holdup time after a high-level or low-level alarm. This time would give the operator a chance to correct the level control action without shutting down the system. With the example in Figure 4-2, the drum effluent piping includes an upward loop that holds the liquid level even with the elevation of the horizontal piping in the loop. An arbitrarily small amount of holdup time is needed, because LEL detectors react very quickly in the vapor phase above the liquid level. For a 1-min holding time at 50 gal/min, the drum volume would be 50 gal. A standard 55-gal drum would be selected.

Provisions should be made for periodic removal of spent adsorbent. If the installation is not designed for ready replacement of entire canisters, then either vacuuming out spent adsorbent should be anticipated or a slurry removal system should be installed. Slurry removal requires that the canisters have a bottom nozzle and a line to a receiving tank set up such that the adsorbent can be dewatered by gravity drainage.

4.7.3 Controls

Figure 4-2 is the beginning of a P&ID. It includes the dual-media filters and backwash storage tank not included in the original concept, and it shows all of the valves, automatic shutdown switches, and indicators for proper operation and control of the treatment system.

For this design example, it has been assumed that any pressure buildup in the adsorbers would be gradual. If any pressure surges were to be anticipated, a high-pressure shutdown switch would be included to shut off pumping. Also, some designs might include pressure relief valves for the adsorbers, discharging to the prestripper sump, to an influent tank, or, as a last resort, to an extraction well.

If pumping from the LEL drum were desired, liquid level controls and high- and low-level shutdown switches would be included.

4.7.4 Utilities Requirements

The main utility needed for aqueous-phase adsorption is power for pumping, instruments/alarms/interlocks, lighting, and (if freeze protection is included) heat tracing. If pneumatic controls or a continuous backwash sand filter were used, a compressed air system would also be needed.

Pumping from groundwater extraction wells and the prestripper sump will account for almost all of the electric power requirements. For centrifugal pumps most commonly used, each horsepower is equivalent to approximately 1kW. This relationship accounts for the pump motor inefficiencies and power factor. The motor horsepower consumed is calculated as shown in Equation 4-13.

hp = gal/min (pressure rise)/[1,713 (pump efficiency)] (4-13)

For an example of 50 gal/min, if a centrifugal pump with 0.65 efficiency is chosen for pumping from the stripper sump, with a 25-psi pressure rise, the horsepower is calculated as shown in Equation 4-14.

50 (25)/[1,713 (0.65)] = 1.1 hp (4-14)

Allowing approximately 500 W (0.5 kW) for lighting and controls, the total load for the adsorption system is 1.5 kW if no air compressor is used. Additional power is needed for well pumps. (See Section 5.9.2 on sizing and rating of groundwater strippers for the prestripper air blower power requirements.)

4.8 Aqueous-Phase Adsorption Treatability Studies

Adsorption isotherms can be derived from laboratory jar tests conducted under equilibrium conditions using samples of the actual groundwater to be remediated. This is often worthwhile for groundwater cleanup because groundwater may contain multiple contaminants and natural organic matter that affect the adsorptivity of activated carbon.

Either bench-scale or pilot flow-through column tests will produce much more data needed for design than equilibrium isotherm tests because of nonequilibrium conditions that exist in adsorption columns under dynamic flow-through conditions. For example, a typical groundwater remediation system operating with 15 min of residence time might actually have half the bed life that would be predicted from equilibrium isotherm data. Flow-through testing results can be used to relate bed life to influent concentrations rather than equilibrium concentrations associated with isotherms. This is because equilibrium conditions are approached with the influent near the beginning of the saturated zone.

Stover and Thomas (1992) list these parameters that can be evaluated with pilot tests:

- Residence time
- Bed depth
- Pretreatment requirements (These could include filtration, or a consideration of whether prestripping should be considered.)
- Breakthrough characteristics
- Pressure drop
- Adsorptivity (x/m isotherm values)
- Carbon usage (the mass of carbon that becomes spent divided by the volume of water)

Pilot tests can also include an evaluation of whether natural organic matter (approximated as the TOC minus total organic contaminant concentrations in the influent and effluent) is being adsorbed on the carbon. Sometimes natural organic matter removal is essentially zero, but frequently it controls carbon life.

A column that is 5 cm (2 in.) in diameter is convenient to use for laboratory flow-through pilot tests and will not impair the experiment with excessive wall effects. The depth of packing should be selected to match the residence time expected for the full-scale design. Such a column test is dynamic, with a realistic residence time, rather than static, with a very long residence time. This test provides data that are directly applicable to full-scale design and expected performance.

If the groundwater sample has a low organic concentration, column testing takes a long time before breakthrough occurs. Crittendon et al. (1991) describe how to accelerate column tests. Small columns (e.g., 1-cm inside diameter) with fine, crushed carbon are used, and results are scaled up using mathematical modeling for application with larger-diameter particles. Usually, such small-scale column tests can be accomplished within 5 days, using less than 750 L (200 gal) of groundwater sample.

Unless rapid small-scale column tests are run in a laboratory, it is better to run either bench-scale or pilot column tests in the field, pumping from a groundwater well, because a very large volume of water is usually needed to achieve breakthrough. An example will demonstrate the advantage of a field test with pumping from a well into a bench-scale unit. If a 5.08-cm diameter column is packed 76.2 cm deep with GAC, the bed volume is 18,484 mL. The corresponding flow rate for this example at 15 min residence time is as shown in Equation 4-15.

$$18{,}484 \text{ mL}/15 \text{ min} = 1{,}232 \text{ mL/min } (0.326 \text{ gal/min}) \tag{4-15}$$

At this rate, a drum sample of groundwater would be depleted in less than 3 hr — not nearly enough time for achieving breakthrough with most groundwater contamination situations. And, as with ion exchange resins used for adsorbing metals, the bed life (or number of bed volumes) corresponding to a saturated carbon bed is even longer than for breakthrough.

The time period required to reach breakthrough in a pilot experiment may be too long to be practical. It might be desirable to sample the bed after a period of operation to determine the depth of the saturated zone and of the MTZ. From such measurements, the time to reach breakthrough and the time to reach saturation could be calculated, and the experiment could be ended without waiting for breakthrough. However, it is usually not practical to remove samples of GAC with adsorbed VOCs and measure the concentrations on the carbon. A better approach is to withdraw water samples for analysis while probing the bed at various depths. The depth and the end of the MTZ can be determined with this method, and the time or groundwater volume corresponding to breakthrough can be predicted.

An actual pilot system is described here to help clarify application of these principles. A 30-in. deep pilot carbon bed was tested at a 15 min residence time with groundwater containing carbon tetrachloride and chloroform. The water flowing through the carbon bed was sampled with a movable vertical probe at various depths. Figure 4-3 shows that for carbon tetrachloride, the thickness of the saturated zone was 4 in. (point A in the figure).

On the day of the sampling, 2,058 bed volumes had passed through the carbon column. The number of bed volumes to saturate 30 in. of carbon depth instead of 4 in. is estimated, as shown in Equation 4-16.

$$2{,}058 \text{ bed volumes } (30/4) = 15{,}400 \text{ bed volumes} \tag{4-16}$$

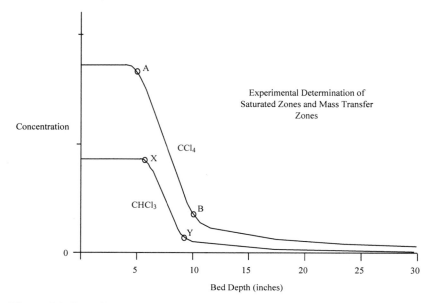

Figure 4-3 Experimental determination of saturated zones and MTZs.

The corresponding carbon usage rate, at a wet bulk density of 28 lb of GAC/ft^3, is 0.24 lb of carbon/1,000 gal of water. Equation 4-17 shows how usage is calculated.

$$(28 \text{ lb/ft}^3) (V \text{ ft}^3)/[(15,400 \text{ V ft}^3) (7.48 \text{ gal/ft}^3)] = 0.00024 \text{ lb/gal} \quad (4\text{-}17)$$

in which V is a bed volume, ft^3.

The cleanup target concentration, at point B in the figure, will be reached when a little more than 2,058 bed volumes have passed through it.

Examination of the curve for chloroform shows that the thickness of the saturated zone was 5 in. (point X in the figure). The number of bed volumes to saturate 30 in. of carbon depth with chloroform is estimated, as shown in Equation 4-18.

$$2,058 \text{ bed volumes } (30/5) = 12,350 \text{ bed volumes} \quad (4\text{-}18)$$

The corresponding carbon usage rate, at a wet bulk density of 28 lb of GAC/ft^3, is 0.30 lb of carbon/1,000 gal of water.

The figure also shows that the MTZ for chloroform (between points X and Y in the figure) is 4 in. thick. The bottom of the saturated zone can be expected to move toward the outlet of the bed as the saturated zone gets thicker. For either contaminant, the thickness of the MTZ can be expected to remain at approximately the same over time, but the MTZ will advance toward the outlet of the bed.

It is of value to compare the adsorptive capacity of the carbon derived from data obtained in this actual dynamic pilot test with published equilibrium isotherm x/m values. Although isotherm data correspond to contaminant concentrations in treated water, the inlet concentration is used for predicting flow-through carbon bed performance. Inlet concentration and flow rate are known measures of how much contaminant will potentially be adsorbed. The saturated zone was estimated (from analyses of the water and the volume of water treated) to have adsorbed 0.047 units mass of chloroform per 100 units mass of carbon in that zone. The carbon isotherm showed that at the actual measured inlet chloroform concentration of 0.024 mg/L, the carbon could adsorb 0.1 units mass of chloroform per 100 units mass of carbon. In practice, with flow-through dynamic conditions and 15 min residence time, the portion of carbon in the saturated zone in this example adsorbed 47% as much as the equilibrium isotherm predicted.

That the carbon under dynamic flow-through conditions adsorbed approximately half of what could be predicted from the isotherm data may be expected. The isotherm data relate adsorptivity to equilibrium concentration — what is left in the water after hours of contact time. With downflow conditions, the top of the saturated zone approaches equilibrium with the inlet concentration, but there is not enough contact to achieve equilibrium throughout the carbon bed.

A very good, simple field pilot test can be conducted using a filter and two drums of carbon in series connected by hoses. Two drums are set up in series with a prefilter, pressure gauge, and hose connections. Groundwater is pumped through the system, typically for a period of weeks, until significant breakthrough of contaminants is noted in samples taken from between the drums. The second drum is a guard chamber, used to ensure that breakthrough does not appear in the final effluent. That effluent may need to meet stringent permit conditions for disposal.

The two-drum pilot experiment can be easily expanded to test a resin at the same time and check its performance against carbon. Which adsorbent has contaminant breakthrough first and the adsorptive capacity of each adsorbent can be determined. Some aqueous-phase resins can be regenerated with steam. The experiment can be expanded further to determine the x/m value after regeneration.

A flow rate of approximately 4 gal/min is suitable for 55-gal drums. If 8 to 10 gal/min can be extracted from a groundwater well, two drums in parallel can be used, followed by a final drum of activated carbon acting as a guard chamber. One of the parallel drums can be filled with a resin suitable for adsorbing organics, and the other parallel drum can be filled with activated carbon. This setup allows simultaneous testing of both carbon and a potentially competing resin.

4.9 Cost Estimating

The purchase costs for the main equipment (vessels, pumps, tanks, filter systems) for aqueous phase carbon adsorption systems can be readily obtained from vendors of filters, pumps, tanks, and carbon. There are a number of carbon vendors who supply canisters and handle shipping and reactivation of used carbon.

Stenzel (1993) indicates that the costs for purchasing two-stage skid-mounted systems (excluding installation costs) are $50,000 for canisters with 2,000 lb of carbon each, handling up to 60 gal/min, and $110,000 for canisters with 10,000 lb each, handling up to 250 gal/min.

A typical steel bag filter housing for handling 50 gal/min, non-American Society of Mechanical Engineers code rated, costs $700. A bank of four bag filter housings pre-assembled with valves costs $5,000.

A 55-gal drum fitted with piping connections on the top lid and internal effluent piping, containing 180 lb of activated carbon, typically costs more than $500. When the carbon is spent, other costs amounting to more than $500 are involved for laboratory analyses, hazardous waste manifesting, shipping, and reactivation charges in order to replace the drum. Off-site carbon reactivation costs typically include the following items:

    A. Basic reactivation charge
    B. Replacement of carbon lost by attrition – (10% to 20%)
    C. Profiling charges by an analysis laboratory
    D. Shipment charges
        • Manifesting as hazardous waste
        • Shipping hazardous waste to off-site facility
        • Return shipment of non-hazardous material

Items A and B amount to almost $1/lb in bulk quantities. If Item B is done with reactivated carbon, the cost is slightly less than with virgin carbon. Item C is often a one-time charge of several hundred dollars for the initial shipment. If the groundwater contaminant species do not change to over time, Item C does not have to be repeated. For Item D, shipments of 1,000 lb cost approximately $0.50/lb; the exact amount depends on distance and individual state taxes. Shipments of 20,000 pounds at a time could be done in bulk trailers at a much lower cost; one West coast supplier estimated in 1999 a total shipping cost of $0.30/lb. One Pennsylvania supplier estimated in 1999 a total cost for A, B, C, and D of $1.50/lb for a 1,000-lb batch and $0.60/lb for a 20,000-lb batch.

Stenzel (1993) estimates that yearly maintenance costs (excluding carbon purchases) are 5% to 10% of installed capital costs, depending on water characteristics and the frequency of carbon exchanges and of backwashes.

Stenzel and Merz (1989) indicate that installed costs in 1989 for a single-stage 20,000-lb carbon unit were $417/gal/min for a 300-gal/min system and, for a two-stage 20,000-lb carbon (each stage) system, were $1,750/gal/min for a 200-gal/min system. (Note that based on the Plant Cost Index published monthly in *Chemical Engineering*, McGraw-Hill, New York, costs did not increase more than 10% between 1989 and 1998.) This system included a spent carbon transfer tank, so carbon return can be done in the same bulk trailer used to deliver fresh carbon. They report that annual costs for two-stage systems, with recycling of carbon via thermal reactivation, ranged from $0.48 to $2.52/1,000 gal (3.785 m^3) treated. Near the higher end of this range, they indicate the following breakdown:

- 15% — Amortization of the investment
- 7% — Maintenance
- 65% — Carbon usage in an example using 200,000 lb/yr
- 11% — Carbon freight

Near the lower end of this range, Stenzel (1993) indicates that the breakdown is:

- 28% — Amortization
- 16% — Maintenance
- 4% — Power costs for pumping
- 7% — Carbon freight

The remaining 45% is for carbon usage in an example using 60,000 lb/yr for a 400-gal/min system. Capital costs for this two-stage system with 20,000-lb adsorbers are estimated to be $563/gal/min.

The RACER/ENVEST™ computer models marketed by Talisman Partners Ltd. (Englewood, Colorado) include cost estimating of liquid carbon adsorption systems. Similar information is in Rast (1997), when accompanied with the database "ECHOS" cost books. The primary choices the program user has are as follows:

- Dual-bed unit, with two adsorbers in series or in parallel
- One disposable adsorber
- Permanent canisters

The model computes investment costs, including the adsorbers, pumps, influent and effluent piping, and an 8-in. thick concrete slab (12-in. if the total weight of adsorbers is 25 tons or more).

Secondary choices include the following:

- Carbon can be virgin coal-derived, virgin coconut shell-derived, or reactivated.
- Dual-bed canisters each can be as shown in Table 4-4.
- Disposable canisters can be:
    - 5 gal/min, 85 lb
    - 15 gal/min, 165 lb
    - 20 gal/min, 250 lb, up to 200 gal/min, 6,000 lb
- Permanent canisters can be polyethylene-lined steel and can be:
    - 25 gal/min, 330 lb
    - 35 gal/min, 660 lb
    - 50 gal/min, 1,050 lb
    - 75 gal/min, 1,650 lb
    - 100 gal/min, 3,000 lb
    - 200 gal/min, 6,000 lb

Using the program, investment costs were estimated to be $670/gal/min for a 60-gal/min system with two adsorbers in series. The program can also estimate operating and maintenance costs for an unattended system, including electric power for pumping and rework of pump motors (assumed to be once each 18 months).

Table 4-4 Dual-bed carbon canisters.

Flow Rating (gal/min)	Diameter (ft)	Carbon (lb)
50	-	1,760
65	4	2,000
75	-	3,300
175	7.5	10,000
350	10	20,000

4.10 Summary of Important Points for Carbon Adsorptions

- Aqueous phase carbon and certain resins can often remove very high proportions of organic contaminants from groundwater.
- Used carbon can be reactivated in a deprived-oxygen high-temperature furnace or kiln — usually off-site.
- Most activated carbon used for groundwater is derived from bituminous coal. Lignite- and coconut shell--derived carbon are also used. Carbon suppliers have some control over pore size distribution. A preponderance of micropores favors adsorption of VOCs. Macropores adsorb large molecules.
- The capacity of carbon for adsorption of individual organic compounds (x/m) is a function of their concentration. Carbon capacity can be obtained from isotherm plots, equations of the isotherms, or tabulated points obtained from isotherms.
- Isotherm data are taken at equilibrium conditions. The adsorptivity for carbon with dynamic flow-through remediation systems with 15 min residence time is approximately half the x/m derived from published isotherm data if influent concentration is used as the independent variable instead of effluent or equilibrium concentration.
- Published isotherm data are not as useful when multiple organic compounds, especially natural organic matter, are present. Better adsorption data useful for design are obtained using actual groundwater in laboratory column tests or field pilot tests.
- A method of accelerating laboratory column tests is useful for shortening the time required to obtain adsorption data that can be used for modeling and scaling up to design conditions.
- Within an adsorption bed, there are a saturated zone that grows in depth and a MTZ (where adsorption takes place) that moves toward the discharge.
- A method of probing pilot adsorption beds for water analyses is useful for shortening the time required to obtain adsorption data directly useful for design. Analysis of water sampled at various depths locates the MTZ, from which eventual breakthrough can be predicted.

- Adsorption continues after breakthrough until the effluent concentration equals the influent concentration.
- The volume of groundwater that can be treated and the corresponding duration of bed life increase with increased residence time (bed contact time).
- The design hydraulic loading rate is often in the range of 1 to 5 gal/min/ft^2. Systems can be designed to operate up to 10 gal/min/ft^2.
- The design bed cross-sectional area is equal to the volumetric flow rate divided by the design hydraulic loading rate. The design bed volume is equal to the empty-bed residence time multiplied by volumetric flow rate.
- The mass of contaminant that a bed can adsorb is (F) (x/m) multiplied by the mass of adsorbent. If x/m is derived from equilibrium isotherm data, then F for carbon is approximately 0.5. If x/m is derived from flow-through tests, then F is 1.
- The duration between adsorbent changeouts (bed life) is the mass adsorbed divided by (Q x ΔC), in which Q is the volumetric flow rate and ΔC is the reduction in concentration. If the mass is based on published isotherm data, interfering compounds may shorten the calculated duration.
- Oil and grease are interfering compounds that can be filtered out first.
- Carbon bed life between reactivations can be extended by prestripping VOCs.
- Groundwater should be prefiltered for removal of suspended solids before activated carbon treatment.
- Adsorbent canisters should be monitored for pressure. If the system is unattended, the canisters should be protected from overpressure with a relief valve or pressure switch, unless fed by a low-head centrifugal pump.
- Adsorbent canisters need extra length to accommodate bed expansion during backwash.
- Two adsorption beds in series are usually used, with the lead bed changed out soon after breakthrough.
- Duration of the lead bed life can be extended to saturation if three stages are used or if the second stage is closely monitored for breakthrough.
- Fresh adsorbent should be presoaked, backwashed, and started in service in the last stage.
- Repeated backwashing can be applied to reduce pressure drop buildup.
- Major breakthrough can be detected by monitoring the vapor space in a tank with no adsorbent with an effluent liquid level maintained in the tank.
- Process design usually includes a flow diagram with flow rates and concentrations, a P&ID, a plot plan showing major treatment equipment, and a site plan showing features surrounding the plot, utility interconnections, wells, and off-plot piping runs.
- Capital investment costs in 1993 for systems with two carbon beds in series were estimated to be approximately $600/gal/min for large-capacity systems (20,000-lb adsorbers rated at 400 gal/min) and higher for small-capacity systems.

Chapter 5

Stripping of Groundwater

By intimately contacting groundwater with air or steam or other gases, dissolved VOCs and sometimes semivolatile organic compounds can be transferred to the vapor phase and are thereby stripped from groundwater. Such stripping can be accomplished with distillation columns containing packing or trays, with aeration tanks, or with cooling towers. The most common device used is the packed column, with air most commonly used as the stripping gas.

This chapter deals in detail with packed air strippers and with alternative types of strippers, blower arrangements and mist elimination, turndown and liquid distribution, recycled strippers, heated strippers, abatement of overhead emissions, in situ stripping, and cost estimating of stripper systems.

5.1 Basic Principles of Stripping

With packed or tray strippers, groundwater is fed near the top of a vertical tower over the top of the packing (or onto the top tray), and air or steam is passed upward through the falling water. The tower is usually circular or rectangular in cross-section. Round towers of plastic construction are most commonly used. A variety of materials besides plastics are used with both shapes (e.g., stainless steel, lined carbon steel, concrete).

The groundwater usually requires filtration and possibly other pretreatment steps to avoid scaling and biological fouling within the tower. Dilzell (1996) describes methods of controlling scaling caused by calcium, magnesium, and iron compounds, including acidification to achieve a pH of 6.5 to 7.0, or injection of phosphonate and polymeric dispersants. Chapter 3 describes methods of removing metals that cause scaling. Bacteria in the groundwater can be inactivated with continuous ex situ application of UV light. Dilzell describes methods of controlling bacteria in wells using pH shocking with acid injection to reduce the pH to less than 3.5 for several hours or using gamma radiation from a cobalt-60 source.

Another pretreatment operation needed in cases in which undissolved liquid organic product is pumped with the raw groundwater is separation of the product. Any entrained dense non-aqueous phase liquids (DNAPLs) such as chlorinated solvents, will sink below the groundwater in a gravity separator. Other organic liquids, such as hydrocarbons, can usually be decanted from the groundwater in an oil-water separator and/or removed with air flotation. Design parameters for oil-water separators are discussed in Section 5.11 (Cost Estimating for Groundwater Stripping). The design of dissolved air flotation systems is given by Yeh (1996).

Usually design work begins after a hydrogeologic investigation of the site has been accomplished. The investigation determines the maximum flow rate at which groundwater can be extracted and fed to the strippers, the degree to which the flow rate might vary, and the concentration of each contaminant in the recovered groundwater. A regulatory analysis is usually needed to determine what the groundwater cleanup goal is — the maximum concentration desired for each contaminant in the treated groundwater. For this chapter, it is assumed that the design groundwater flow, in terms of gal/min or ft^3/min, and the stripper inlet concentration (C_{in}) for each contaminant are known.

The corresponding effluent from the stripper will have a concentration (C_{eff}) that is known or, in the event that polishing treatment will be applied, is derived from an economic or feasibility analysis of the polishing treatment system.

5.1.1 Use of Polishing Carbon

Polishing treatment might be accomplished by passing the effluent from the bottom of the stripper through an adsorbent or an alternative treatment system. For example, consider a polishing system consisting of a series of two GAC adsorbent vessels, with the GAC to be periodically removed from the first vessel for off-site reactivation. It is desired in this example that the stripper remove enough of the total petroleum hydrocarbons (TPH) so that the first adsorbent vessel does not have to be taken out of service more often than once per 120 days. For a given mass of GAC in the vessel, there will be a corresponding mass, X, of TPH that the carbon can adsorb. Then, the stripper effluent TPH concentration (mg/L) can be derived from setting X equal to the flow rate in L/d times the effluent TPH concentration times 120 days.

In practice, more than one preliminary design iteration between selecting the mass of carbon and deriving the TPH concentration may be needed. For groundwater flow rates up to 50 gal/min, a GAC mass of up to 1,000 lb is often selected. For higher flow rates, 2,000 lb or larger amounts may be appropriate.

A design example is given in Figure 5-1, showing a packed tower for removing gasoline contaminants followed by two canisters of GAC.

Each canister holds 1,000 lb of GAC, so the GAC mass for the calculations is 1,000 lb. The final canister is a guard chamber that contains carbon. The mass of carbon in the final canister is not considered in these calculations, but it may be considered in calculating residence time. From experience with adsorption of gasoline components,

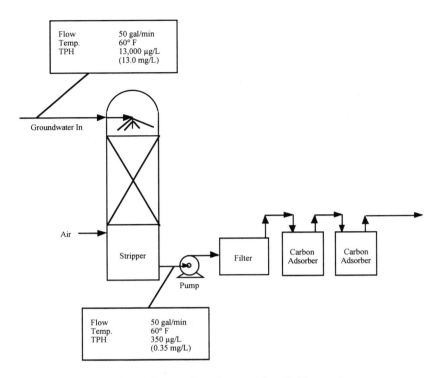

Figure 5-1 A flow scheme for an air stripper with polishing carbon.

breakthrough and removal of the first canister from service should occur when the mass of adsorbed TPH equals 7% of the mass of carbon in the bed. With proper adsorption system design, the concentration of aromatic compounds and TPH in the final effluent will be close to nondetectable. Then the mass of carbon needed per canister is 0.07 x 1,000 lb, or 31,800 g. It is desired to change carbon no more often than once per year, with 11 mo. (334 days) of net operating time each year. The corresponding stripper effluent TPH concentration for a flow rate of 50 gal/min (272,160 L/day) is derived, as given by Equation 5-1.

(272,160 L/day) (C_{eff}) (334 day) = (1,000 mg/g) (31,800 g)   (5-1)

Solving for the stripper effluent concentration yields C_{eff} is 0.35 mg/L.

Note that the stripping efficiency related to the inlet TPH concentration of 13,000 µg/L (13.0 mg/L) is derived, as given in Equation 5-2.

$(13.0 - 0.35)/13.0 = 0.973$ (5-2)

Such a stripping efficiency, 97%, is readily attained for volatile compounds (such as gasoline hydrocarbons) with air strippers, with reasonable amounts of packing depth or of trays and air/water (A/W) flow ratios.

5.1.2 The Design Problem

This analysis leads to formulation of the air stripper design problem. The known values are for these parameters:

- Groundwater design flow rate, gal/min or ft^3/min
- C_{in}/C_{eff} and the corresponding stripping efficiency, for each contaminant of most concern

What needs to be determined are:

- The A/W ratio and the corresponding depth of packing or number of trays
- Air pressure drop through the packing and tower cross-sectional area

All of these items depend on the packing or tray design chosen. Packings are available in a variety of shapes and sizes. Traditional packings include rings, saddles, and tellerettes, ranging in size for most strippers from 1 to 3 in. and made of plastics, stainless steel, or ceramics. Modern packings mostly used are special plastic shapes such as the 1.75-in. Tri-Pak, marketed by Jaeger Products (Houston, Texas), and the 2-in. Lanpak, marketed by Lantec (Agoura Hills, California). These traditional and modern packings are randomly dumped in the stripping tower on a support grid mounted internally above the air inlet.

Structured packings that are set in place in layers or blocks are also available. Structured packings have higher separation efficiencies and lower pressure drops than do random packings for a given depth of packing. They are, however, more sensitive to liquid maldistribution (Helling and DesJardin, 1994). Also, Sloley and Martin (1995) indicate that if suspended solids are present, structured packing is the worst choice versus modern random packings. It is important to note that with air strippers, solids fouling often occurs in the stripping tower even if the groundwater is prefiltered. The fouling that occurs upon air contact with the groundwater can be from these processes: microbial growth, forming bacterial sludge; calcium carbonate scaling when carbon dioxide is stripped from the water; and ferric oxide or manganic oxide precipitation.

The air pressure drop through packing depends on the packing factor. The depth of packing depends on the mass transfer coefficient (K_L) for contaminant molecules passing from the liquid phase across the liquid film on the packing surfaces to the adjacent gas film and on the packing interfacial wetted surface area (a). These last two parameters are described as the product $K_L a$. The $K_L a$ value may be available for a particular packing material from the manufacturer, as is the value for the packing factor. Or the manufacturer may know the value of the gas film mass transfer coefficient, and chemical engineering correlations (given in Section 5.2.1) can be used to relate the liquid film mass transfer coefficient to the gas film mass transfer coefficient. Another method of determining pressure drop and $K_L a$ consists of running pilot tests to measure pressure drop directly and then calculating $K_L a$ using the correlations given in Section 5.2.1.

It is important here to define some terms:

1. <u>A/W ratio and G/L ratio</u>. By expressing the air flow rate ft^3/min and the water flow rate in ft^3/min, the A/W ratio is dimensionless. A/W is numerically equal to the dimensionless volumetric G/L ratio, in which G is the gas (air) flow rate divided by the cross-sectional area of the column, and L is the liquid (water) flow rate divided by the cross-sectional area of the column. However, for certain correlations, it is better to express the G/L ratio as moles of air per unit of time and per unit of area divided by moles of water per unit of time and per unit of area. Because air and water have different molecular weights and densities, the molar G/L does not equal the volumetric G/L. In the correlations below, molar G/L should be used; the air flow rate is the water flow rate (ft^3/min) times volumetric A/W. The molar G/L equals 0.000762 times A/W (dimensionless) for strippers operated near 1 atm and 60° F.

2. <u>Henry's constant</u>. The depth of packing and A/W required to achieve a given reduction in contaminant concentration depend on how volatile the contaminant is. A measure of volatility is the Henry's constant (H). At a given temperature, the partial pressure of a component i in the vapor phase (p_i) is proportional to the concentration in the liquid phase (C_i). This is a method of expressing Henry's Law. The proportionality constant is H, as shown in Equation 5-3.

Partial pressure of component i = p_i = H (C_i) (5-3)

The units of H can be units of pressure divided by units of liquid phase concentration. If the partial pressure units are atm, and if the liquid phase concentration units are mole fraction, then the units of H are atm.

The concentration in the vapor phase, expressed as mole fraction, is equal to the partial pressure of the component divided by the total pressure. Therefore, H relates vapor phase concentration (as well as partial pressure) to liquid phase concentration.

If the vapor phase and liquid phase concentrations are expressed in mole fraction units, then H is dimensionless. The numerical value of dimensionless H is equal to the numerical value of H expressed in atm multiplied by 0.219/T, in which T is the temperature in Kelvin. Using this relationship, at 20° C (293 K), multiply dimensionless H by 1,338 to obtain H in atm. The following multipliers, adapted from Chidkopkar (1996), can be used to obtain H in atm:

- Atm-m^3 water/mol compound, multiply by 55,600
- Atm-L/mg, multiply by (55,600) x (molecular weight)

If H is unknown, it can be approximated from Wilson (1995) in units of atm-m^3/mol as (Vp)(MW)/S, in which Vp is the vapor pressure, atm; MW is the molecular weight, g/mol; and S is the solubility, mg/L.

Henry's Law is accurate when the liquid phase concentration is low, which holds true in most problems involving hydrocarbons or chlorinated solvents dissolved in groundwater. Values of H have been published at various temperatures for almost all VOCs and semivolatile organic compounds. It is important to use consistent units. The published values vary the units for H from source to source. In this chapter, the units for H are atm. If the design temperature T (Kelvin) for the stripper is different than the temperature T_{publ}, corresponding to the published value for H, then the design value for H can be calculated from the following correlation from Haarhoff and Cleasby (1990).

$$H = H_{publ}\, 10^{A(1/T_{publ} - 1/T)} \tag{5-4}$$

in which the constant A is numerically equal to the heat of solution divided by the universal gas constant R, has units of Kelvin, and has the values shown in Table 5-1.

If H is not expressed in atm, Equation 5-5 can be used.

$$H = H_{publ}(T_{publ}/T)\, 10^{A(1/T_{publ} - 1/T)} \tag{5-5}$$

Kavanaugh and Trusell (1980) relate H to temperature as a function of the heat of solution, as given in Equation 5-6, and give the correlation constants for 11 organic compounds.

$$\log H, \text{atm} = -A/T + \text{constant} \tag{5-6}$$

in which T is the temperature measured in Kelvin, and the values in Table 5-2 apply.

When A values differ from those given by Haarhoff and Cleasby (1990), the Haarhoff values are probably more accurate, having been derived from more recent data. An example of determining H at two temperatures is as given in Equations 5-7, 5-8, and 5-9 for benzene at 20°C (293 K) and at 50°C (323 K).

$$\log H = -1{,}850/293 + 8.68 = 2.366 \text{ at } 293 \text{ K} \tag{5-7}$$
$$H = \text{antilog } 2.366 = 233 \text{ atm at } 293 \text{ K} \tag{5-8}$$

Table 5-1 Constant for correlating H with temperature.

Compound	Constant
Benzene	1,850
Tribromomethane	1,910
Bromomethane	1,720
Trichloromethane	1,730
Chloromethane	1,250
1,1-Dichloroethane	1,900
1,2-Dichloroethane	1,820
Hexachloroethane	2,320
Tetrachloroethane	1,800
Toluene	1,820
1,1,1-Trichloroethane	1,640
Trichloroethylene	1,910

Table 5-2 Parameter values for correlating H with temperature. (From *Journal AWWA*, Vol. 72, No. 12 (December 1980), by permission. Copyright © 2000, American Water Works Association).

Compound	A	Constant
Benzene	1852	8.68
Trichloromethane	2013	9.10
Carbon tetrachloride	2038	10.06
Chloromethane	1248	6.93
1,2-Dichloromethane	1822	7.92
1,1,1-Trichloroethane	1992	9.39
1,1-Dichloroethane	1902	8.87
Trichloroethylene	1716	8.59
Tetrachloroethylene	2159	10.38
Difluorochloromethane	1470	8.18

At the higher temperature,

$$H = 233 \text{ atm } (10^{1850(1/293 - 1/323)}) = 899 \text{ atm at } 323 \text{ K} \tag{5-9}$$

The same value can be derived using Kavanaugh's correlation.

3. <u>Stripping factor</u>. The term R will be used to simplify the presentation of equations that relate depth of packing (Z) and A/W (G/L) ratio. The stripping factor R is dimensionless and is defined by Equation 5-10.

$$R = (H/P)(G/L) \tag{5-10}$$

in which H should be in atm when P, the total pressure, is in atm. For most air strippers, the total pressure is very close to 1 atm. Note that 1 atm is equal to 760 mm (29.92 in.) of mercury, which is equal to 10,369 mm (approximately 400 in.) of water column (w.c.). The pressure in a typical packed air stripper does not exceed 500 mm (approximately 20 in.) w.c. above 1 atm. Therefore, for most strippers, P can be taken as equal to 1 atm.

5.2 Packed Strippers

5.2.1 Packing Depth and A/W (or G/L) Ratio

The packing depth Z depends on the G/L ratio. The higher the G/L, the smaller the depth Z can be. If the air exiting the top of the stripper can be discharged directly to the atmosphere without the need for an emission abatement system, a relatively high G/L should be selected. Such an approach leads to reduced costs for packing, for the tower, and for tower supports. Air is free, and the costs of purchasing and powering a larger blower are relatively low.

Volumetric A/W ratios as high as 200 or 250 have been used. On the other hand, if emission abatement is required, it is more economical to select the lowest A/W (and corresponding G/L) that corresponds to a reasonable depth of packing Z. Minimizing the air volume that an emission abatement system must treat is important economically. Capital investment costs for an abatement system are comparable to stripper system investment costs, or higher. And the operating costs, which are mainly proportional to the air rate, are much higher than stripper system operating costs.

Looking again at the stripper example in Figure 5-1, packed strippers with gasoline hydrocarbons contamination usually have a volumetric A/W ratio in the range of 30 to 65. The lower value for A/W is a logical choice if there is a groundwater polishing treatment system. Without a polishing system, the higher value for A/W would be preferred to ensure meeting the treatment goal.

If the stripping factor R is 1.0, an infinite depth of packing would be required for complete stripping, so R must be greater than 1.0. And for most designs, an R value greater than 5 for the main contaminant of concern would imply an impractical amount of air. However, R values are sometimes in the range of 5 to 100 for some components.

An approach often used for sizing of strippers is first to select a value of G/L such that R is in the range of 1.5 to 5 for a key contaminant. For the C_{in}/C_{eff} value for that contaminant, Z is calculated from Equation 5-11, the correlation for packing height.

$$Z = HTU \, (NTU) \qquad (5\text{-}11)$$

The definition and calculation of the terms HTU (height of the transfer unit) and NTU (number of transfer units) are given with Equations 5-12 and 5-13. Then, for any value of G/L and Z, there is a value of C_{in}/C_{eff} for each of the other contaminants that are determined from the correlation for NTU.

If the calculated C_{in}/C_{eff} values for some of the key components are too low to be acceptable, the G/L (and corresponding R values) is increased, while Z is held constant. If G/L becomes too large (R greater than 5 for most volatile organics), Z is increased, and the effluent concentrations are checked again.

A section of packing of a given height will reduce the concentration of a volatile component by a factor. A section of packing below the first section will further reduce the concentration by the same factor. For example, if a component in groundwater transported downward through a section is reduced from 100 units to 10 units in that section, the factor is 10; the concentration exiting from the next section down will be one unit; and from the third section, 0.1 units, etc. HTU characterizes the transfer of dissolved contaminant to the vapor phase.

HTU is proportional to the molar liquid load per unit of cross-sectional area of the column. If the volumetric liquid load used for deriving HTU is expressed in units of ft^3 of water per second and the tower cross-sectional area is expressed in ft^2, then HTU will have units of feet. In metric units, if the water flow rate is in gram moles per second, the cross-sectional area should be expressed in m^2, and L is in mol/(sec-m^2), with units for HTU now being meters. The cross-sectional area can be selected, as discussed in Section 5.2.2. HTU is evaluated from Equation 5-12.

$$\text{HTU} = (L/55{,}555 \text{ gmol/m}^3)/K_L a \qquad (5\text{-}12)$$

K_L is the liquid mass transfer coefficient, with dimensions of m/sec; and a is the interfacial surface area of the packing, with dimensions of m^2/bulk m^3 of packing.

$K_L a$ depends on the packing selected, and values for this parameter or for the related parameter $K_G a$ may be obtainable from the packing manufacturer. K_G is the gas mass transfer coefficient, (gmol/sec)/(atm-m^2). For older, traditional packing types, $K_G a$ can be derived from values in Table 4 of Kavanaugh and Trussell (1980). For VOCs, K_L is approximately equal to $K_G H$ divided by the molal density of water, 55,555 gmol/m^3.

$K_L a$ value may also be derived from $K_G a$ values where $K_L a$ is known for a different packing by multiplying that known $K_L a$ value by the ratio of the $K_G a$ values for the two different packings.

K_L can be derived from diffusivities and surface tension data. Nyer (1992) recommends that in the absence of test data, $K_L a$ be derived from the correlation of Onda, Takeuchi, and Okumoto (1968), with application of a safety factor.

Another method of determining K_La values is to obtain pilot data using the selected packing and samples of the groundwater. The NTU is calculated from Equation 5-13 using the measured influent and effluent concentrations. The HTU is calculated by rearranging Equation 5-11 and using the actual depth Z employed in the pilot column. Using the actual pilot test liquid loading, the K_La is derived as shown in Equation 5-14 by rearranging Equation 5-12.

$\text{NTU} = (R/(R-1)) \ln[((C_{in}/C_{eff})(R-1) + 1)/R]$ (5-13)
$K_La = (L/55,555 \text{ gmol/m}^3)/\text{HTU}$ (5-14)

An example with two A/W ratios illustrates how the units (dimensions) are managed. A pilot stripping column is 0.254 m (10 in.) in diameter (0.05063 m² cross-sectional area), with 7 m of packing depth Z. An experimental run with 9 gal/min (1.2 ft³/min) of groundwater and 35 ft³/min of air reduced the influent concentration of 10,000 µg/L to 240 µg/L in the effluent. The volumetric A/W ratio was 35/1.2, or 29.17. Find the NTU, HTU, and K_La at that A/W ratio. Then, using that K_La, predict the effluent concentration if the water flow rate is raised so that the A/W ratio is 22.

Given: A/W is 29.17 for case A, with an air flow of 35 ft³/min; A/W is 22 for case B; the tower cross-sectional area is 0.05063 m²; the packing depth is 7 m; the effluent concentration for case A is 240 µg/L; H is 205 atm; the groundwater temperature is 60° F; the air temperature is 60° F; and the density of air at these conditions is 0.076 lb/ft³.

Assumptions: The total pressure is 1 atm, and K_La remains constant.

Find: NTU, HTU, and K_La for case A (with A/W at 29.17), and effluent concentration for case B (with A/W at 22)

Step 1: Determine G/L using Equations 5-15, 5-16, and 5-17. Note that G/L is 0.000762 (A/W).

G = [35 ft³/min (0.076 lb/ft³) (453.7 g/lb)/(29 g/gmol)]/[60 sec/min (0.05063 m²)]
= 13.7 gmol/sec-m² (5-15)
L = [9 gal/min (0.0037854 m³/gal (55,555 gmol/m³)]/[(60 sec/min (0.05063 m²)]
= 623 gmol/sec-m² (5-16)
G/L = 13.7/623 = 0.022 (5-17)

Step 2: Determine stripping factor R from Equation 5-18.

R = (H/P) (G/L) = (205 atm/1 atm) (0.022) = 4.1 (5-18)

Step 3: Determine NTU and HTU from rearranging Equation 5-11 as shown in Equations 5-19, 5-20, and 5-21.

$$C_{in}/C_{eff} = 10,000/240 = 41.7 \tag{5-19}$$
$$NTU = (4.1/3.1) \ln[(41.7\,(3.1) + 1)/4.1] = 4.57 \tag{5-20}$$
$$HTU = Z/NTU = 7\text{ m}/4.57 = 1.53\text{ m} \tag{5-21}$$

Step 4: Determine $K_L a$ from Equation 5-14 as shown in Equation 5-22, using the HTU from Step 3.

$$K_L a = [(623 \text{ gmol/sec-m}^2)/(55,555 \text{ gmol/m}^3)]/1.53\text{ m} = 0.00733/\text{sec} \tag{5-22}$$

Step 5: Now, lower the A/W ratio from 29.17 to 22 and determine the new C_{eff}. Determine the new values for L, HTU, and R as shown in Equations 5-23, 5-24, and 5-25.

$$L = (29.17/22)\,623 \text{ gmol/sec-m}^2 = 826 \text{ gmol/sec-m}^2 \tag{5-23}$$
$$HTU = (L/55,555 \text{ gmol/m}^3)/K_L a = (826/55,555)/(0.00733\text{ m}) = 2.03\text{ m} \tag{5-24}$$
$$R = (22/29.17)\,4.1 = 3.1 \tag{5-25}$$

Step 6: Rearrange Equation 5-11 to obtain NTU and Equation 5-13 to obtain C_{in}/C_{eff}, and derive C_{eff} from the given value of 10,000 µg/L for C_{in}, as shown in Equations 5-26, 5-27, 5-28, and 5-29.

$$NTU = Z/HTU = 7\text{ m}/2.03\text{ m} = 3.45 \tag{5-26}$$
$$\ln[((C_{in}/C_{eff})(2.1) +1)]/3.1] = NTU\,(2.1/3.1) = 2.337 \tag{5-27}$$
$$C_{in}/C_{eff} = 14.8 \tag{5-28}$$
$$C_{eff} = C_{in}/14.8 = (10,000\text{ µg/L})/14.8 = 676\text{ µg/L} \tag{5-29}$$

The more difficult it is to accomplish the transfer to vapor, the higher the NTU needed to reach a given effluent concentration. For a given L and packing, HTU is somewhat constant, and C_{in}/C_{eff} depends on R (for each component), which is determined by G/L, and on the NTU. As R (or corresponding G/L and A/W) is arbitrarily changed while checking corresponding effluent concentrations, it is convenient to plot for each contaminant of most concern the relationship between computed packing depth Z, A/W, and the concentrations or concentration reductions. Such a plot makes it easy to determine C_{in}/C_{eff} values at any given Z, as shown in Figure 5-2.

Figures 6 and 7 in Kavanaugh and Trussell (1980) can also be used for stripper design. The figures are plots for Equation 5-14, with contaminant removal efficiency substituted for C_{in}/C_{eff}, and they apply to any contaminant. However, when using such plots that relate NTU to R, remember that the A/W ratio (and hence the air rate) is different for each compound at a given R. For this reason, a good strategy is to prepare a plot of Z or NTU and concentration reduction versus volumetric A/W ratio for a given L for each contaminant.

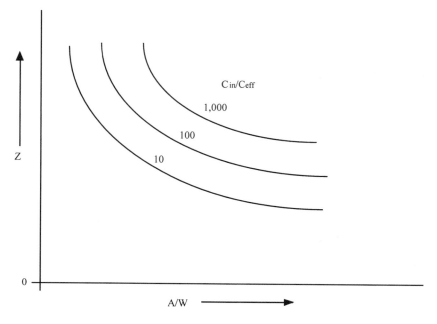

Figure 5-2 Plot of air stripper computations.

In the design of strippers, the contaminants of most concern will usually be those with the highest toxicity (most stringent cleanup level, C_{eff}), the highest influent concentration (C_{in}), and/or the least volatility (low H). For example, with a hypothetical mixture of toluene and other aromatic hydrocarbons, benzene (highest toxicity), toluene (highest concentration), and naphthalene (lowest H) may be thecontaminants of most concern. The NTU (with the corresponding Z) and A/W ratio must be high enough so that the effluent concentrations of these three contaminants are below their required cleanup levels at the chosen G/L.

After Z is determined with this procedure, it should be multiplied by a safety factor, to arrive at the depth of packing to be used for design. Kavenaugh and Trussell (1980) recommend a safety factor of 1.5. Unless experience has been gained with a similar groundwater stripper, or pilot tests are run, a factor of at least this magnitude should be used. With pilot data available, the factor can be lower, but it should be greater than 1 because of variations in influent concentrations and K_La values.

5.2.2 Packed Strippers — Pressure Drop and Cross-Sectional Area

There is a limit to the gas (or air) flow rate for any given tower cross-sectional area beyond which water will be carried by the gas over the top of the tower, a condition that chemical engineers call "flooding." The gas pressure drop, a function of gas rate G, liquid rate L, packing factor, liquid viscosity and density, and gas density, must be below the value that corresponds to flooding. The correlation between these variables is given as Figure 9 in Kavenaugh and Trussell (1980), in Eckert (1975), and in Perry's "Handbook for Chemical Engineers."

With traditional packings (e.g., rings, saddles, and tellerettes) the air pressure drop through the packing should be in the range of 0.25 to 0.50 in. w.c./ft of packing depth (200 to 400 N/m^2/meter of packing depth). This range is well below what the pressure drop would be at flooding conditions. Modern packings (e.g., Tri-Pak, Lanpak) and structured packings have lower pressure drops. A procedure for determining the cross-sectional area is to select a reasonable pressure drop, obtain G from the pressure drop correlation, and take the area as the influent water flow rate times (G/L)/G. The selection of pressure drop might be arbitrarily set at 0.25 in. w.c./ft for traditional packings or derived from a value suggested by the packing vendor or from experience. For the purpose of specifying an air blower, the total pressure drop through all of the packing depth, plus ductwork friction losses, plus the pressure drop through the mist separator and any abatement devices, and a safety factor, must be taken into account.

With modern packings, the cross-sectional area is frequently chosen to correspond to a liquid loading of approximately 1 $m^3/min/m^2$ (20 $gal/min/ft^2$ or slightly greater).

5.2.3 Packed Strippers — Computer Applications

Because a number of iterations may be required to arrive at reasonable values for packing depth and A/W ratio with acceptable effluent concentrations, applying computer programs can make the process easy to design. Air stripping computer programs can be obtained free from the US EPA and (for Lanpak packing) from Lantec (Agoura Hills, California), or at a small cost from D. Schoeler (Ames, Iowa).

Typically, computer programs are applied by selecting an arbitrary A/W ratio and checking that the computed outlet concentrations are acceptable. If these concentrations are acceptable, the corresponding computed packing depth and pressure drop are used, with safety factors, as a basis for design.

To use most air stripping programs, input data must include values for the volumetric liquid flow rate and cross-sectional area (or diameter), or a value for the liquid loading, (in gpm/ft^2 or $m^3/min-m^2$), so the tower cross-sectional area must be chosen first. A reasonable starting value would be that which corresponds to a liquid loading in the range of 20 to 25 $gal/min/ft^2$, or 1 $m^3/min-m^2$. D. Schoeler's air strip program documentation indicates that typical liquid loadings are in the range of 22 to 44 $gal/min/ft^2$.

5.3 Alternatives to Packed Towers

5.3.1 Tray Designs

Tall stripping towers are traditionally designed with packing or trays to provide good contact between water falling downward through the tower and air rising upward. Different types of trays are distinguished mainly by the method the air contacts the water. Three types of designs are:

- Trays with bubble caps
- Sieve trays
- Trays with air diffusers

Bubble caps. The trays can be designed with distillation bubble caps or valve caps that reverse the air flow under water. The water passes horizontally across each tray over a weir to a downcomer that conveys the water to the next tray below.

Sieve trays. An alternative design uses sieve trays, which are plates that are perforated with a large number of small holes. Most tray towers use sieve tray designs. Weirs and downcomers for these systems are similar to those for designs with caps. Sieve trays are much less costly than bubble trays with caps, but they do not have high turndown ratios. Turndown is the capability to operate satisfactorily below design flow rates. With sieve trays, if the air flow rate falls much below a proper design rate, water weeps downward through the perforations instead of across the trays, so the A/W contact is inadequate to reach required treatment performance. However, with proper design and sustained air flow rates, sieve trays provide an excellent means of economically contacting groundwater with air.

Air diffusers. Another design for achieving A/W contact in tray towers has submerged perforated plastic pipes for diffusing air into water that is flowing across each tray.

Tall towers with many trays have much higher air pressure drops than packed towers. The pressure drop at each tray is at least a few inches w.c. because the air has to rise through a head of water equal to the weir height plus at least the height of the crest over the weir. Additionally, there is air pressure drop in going through tray holes and, if caps are used, in reversing direction around caps.

Traditional tall tray towers typically have many trays. Depending on tray efficiency, equilibrium contact between water and air often requires several trays.

Low-profile tray strippers are usually rectangular in cross-section and use only several trays, each of which has a relatively long horizontal water flow path. This configuration can provide the equivalent of one or more equilibrium contacts on a single tray. Thus, not many trays are required to achieve a desired effluent concentration. However, the A/W ratio for a low-profile stripper is usually much higher than for a tall packed scrubber to achieve a comparable contaminant concentration reduction. If emission abatement is required and there are height limitations, costs of a low-profile unit, including abatement, should be compared against two or more packed towers with series flow, including abatement. Low-profile tray strippers foul less than do packed strippers. They are usually constructed with bolted flanges at each tray for ready maintenance access.

North East Environmental (West Lebanon, New Hampshire) offers a computer program for selecting the air rate and the number of trays for their line of low-profile strippers.

5.3.2 Aeration Chambers

Another type of low-profile stripper is an aeration tank with no trays. Air is injected near the bottom of the tank through perforated piping. More equilibrium contacts can be attained if two or more aeration tanks are arranged such that the water flows in series. A critical feature in the design of aeration tanks is the need to produce very fine air bubbles that disperse through all of the water. Small bubbles have a desirably large ratio of surface area per unit volume. The bubble surface is where mass transfer of dissolved organics into the vapor phase takes place.

5.3.3 Cooling Towers Used As Air Strippers

The most economical ex situ air stripper for high groundwater flow rates is a cooling tower. Cooling towers are used in processing plants, steam-turbine electric generating plants, and in air conditioning systems to reduce the temperature of circulating cooling water. The warm water returning from condensers or coolers is distributed across the top surface of the cooling tower and is allowed to trickle down through plastic or wood fill to a basin. Air is admitted through the sides of the tower and turns upward, exiting at the top. Usually, very large side-mounted forced-draft fans or top-mounted induced-draft fans are used. Some of the water evaporates, producing a cooling effect and concentrating any minerals dissolved in the water. To prevent excessive mineral concentration buildup and scaling, a small fraction of the circulating water is blown down (bled off) and discarded. A cooling tower used in stripping service would not have a circulation loop or high amounts of evaporation. Instead of bleeding off a small fraction as blowdown, water would flow through it once, down from the top and out from the basin.

Because cooling towers are designed with high A/W ratios and intimate contact of water with air, they do an excellent job of stripping volatile contaminants from groundwater. They are not normally used in cases in which emission abatement is required. The top surface area is entirely open to the atmosphere, and capturing the air for feeding into an abatement device often is not practical. Also, cooling tower fans are not usually designed for overcoming the pressure drop associated with abatement systems.

5.3.4 In Situ Air Stripping (In-Well Stripping and Air Sparging)

In situ stripping can be accomplished by injecting air into dual-screened wells and through sparge points. With dual-screened wells, either pressurized air systems are used in conjunction with a vacuum blower, as practiced at numerous sites in Germany, or all-vacuum systems are used, as practiced at some sites in the United States and France. With these systems, an aboveground vacuum blower induces air, with VOCs, that is released from the water in the upper part of the well to flow into an emissions abatement system.

The German UVB system is described by Schrauf, Sheehan, and Pennington (1993, 1994) as useful in coarse-grained soils. Compressed air is injected in a screened interval that traverses the water table. A convection cell is formed in situ with oxygenated water flowing out and downward from this upper screen, and then into a lower screened interval with a well pump.

An all-vacuum system is NoVOCs, (Cichon, Mantovani, and McKeon, 1996). Another process that uses air-lift pumping principles within a well with injected air is the Density Driven Convection system developed by Wasatch Environmental (Salt Lake City, Utah).

A sparging system is different than in-well stripping. Air sparging is most effective when volatile organics are mainly in the upper part of an aquifer, such as when the contaminants have infiltrated downward from soil above the water table. Sparging systems bubble air through the aquifer, rather than in wells. The air is usually injected at depths up to approximately 25 ft below the water table, with wells or sparge points. With either wells or sparge points, usually less than 1 ft of length near the bottom is screened. With this limited screening, narrow, conventional wells are drilled and utilized. As an alternative to using wells, sparge points are usually constructed of pipe that is less than 1-in. in diameter, fitted with a pointed tip, and driven into the ground rather than using drilling techniques. Sparging systems can also be constructed using horizontal or slanted borings. KVA Co. (Falmouth, Massachusetts) markets some sparge nozzles that are made of plastic pipe that is approximately 5-cm long and is drilled with very small air diffusion holes.

A shallow soil venting system using extraction trenches or vadose zone extraction wells is used, including an abatement system if emissions must be controlled (Figure 5-3). In general, air sparging is applicable to permeable aquifer soil formations that are not very deep.

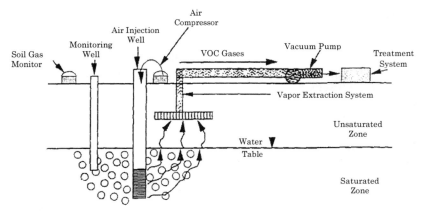

Figure 5-3 Cross-section of an air sparging/vapor extraction system (US EPA, 1992e).

The injected air strips organic compounds from the soil pores and from light non-aqueous phase liquids (LNAPLs), as well as from groundwater. The technique is also effective in stripping residual NAPLs that sometimes are trapped in soil pores in the upper part of an aquifer. The LNAPLs float near the top of an aquifer, where sparged air does the most stripping. Dense NAPLs move through the unsaturated zone, penetrate the capillary fringe, and continue to migrate downward through the saturated zone until they encounter a confining layer, where they pool and/or move along the slope of this layer. Air sparging can strip the NAPL that remains in the upper part of the saturated zone (whether it is dense or light NAPL) as the injected air moves upward through the aquifer and contacts this NAPL.

Air sparging usually can achieve the desired aquifer cleanup goal better than with pumping groundwater and treating it ex situ. And it is less likely that contamination will rebound after cleanup is attained because sparging can remove NAPLs that are often difficult to pump out. Air sparging can be used in place of or to augment a pump-and-treat system.

Sparging can be used as a plume containment strategy, as well as a direct cleanup method. A row of sparge points near the downgradient edge of a contaminant plume can prevent plume spread.

Conventional pump-and-treat systems with an ex situ stripper may cost less to install than air sparging for aquifers that are too deep for driving sparge points or where horizontal wells are not feasible. Each sparger typically has a radius of influence of only up to approximately 25 ft (7.5 m) (more in some stratified soils).

Sparging systems may have the added expense of soil venting wells or trenches and vacuum-producing equipment.

According to Marley et al. (1996), the radius of influence is typically 10 to 25 ft (3 to 7.5 m) and is less in highly permeable soils. This is expected, because the air will tend to flow primarily vertically unless a lower-permeable stratum is encountered as the air rises. In order to have adequate volumetric air flow, the soil hydraulic conductivity should be at least 0.001 cm/sec.

With either in-well stripping or sparging, biodegradation, oxidation of metals, and calcium carbonate formation may occur. Biodegradation may be beneficial, but bacterial cell mass is formed that can cause plugging. Oxidized metals, especially iron and manganese, and calcium carbonate can all cause plugging problems as well.

Implementation usually should be done in phases, using up to 20 standard ft^3/min (scfm) of air with a very few wells or points. If this investigation proves promising, a pilot system using up to 100 scfm is a logical second phase.

A typical sparge point handles as little as 5 ft^3/min, and the soil vapor extraction system handles approximately four times as much air as the total volume of sparge air. If wells are used instead of trenches for vapor extraction, these wells can be either concentric with the sparge wells or sparge points, or drilled independently.

Marley and Droste (1995) indicate that the air pressure generally must be 1 to 2 psi in excess of the hydrostatic head at the top of the injection well screen with coarse-grained soils. With fine-grained soils, injection pressures may be a factor of two or more greater than the hydrostatic head. Too high a pressure (0.8 psi/ft of normally consolidated overburden) could create fractures that may or may not be beneficial. Marley and Droste (1995) describe measurements that should be made during pilot testing. Pulsing the air with either pilot or full-scale installations reduces energy consumption and may create convective water movement.

Clarke, Wilson, and Norris (1996) describe how mathematical models aid the design and operation of air sparging systems and argue in support of using short pulses at high air flow rates. Because the air tends to flow in channels, not evenly, when sparging at a steady air flow (without pulsing), the contaminants must diffuse or disperse to an air channel for stripping to continue.

5.4 Blower Arrangements and Mist Separation

Figure 5-4 illustrates an example of a negative-pressure air stripper, with a vacuum blower inducing airflow.

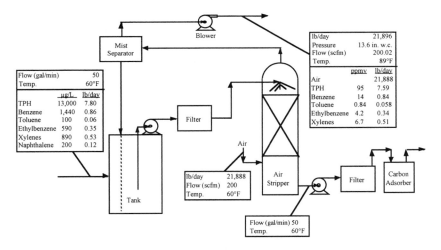

Figure 5-4 Example of a negative-pressure air stripping system.

The more common arrangement is with the blower forcing air into the tower, with the stripper at a positive pressure. Both arrangements are suitable. Advantages of the negative-pressure design illustrated in the figure include:

- There is less likelihood of contaminated air leaks to the surroundings.
- Stripping is slightly improved.. With a lower total pressure, the stripping factor R is increased. (Some strippers are designed to operate at high vacuum, in which event high R values can be attained at relatively low A/W ratios.)
- The heat of compression in the blower raises the off-gas temperature, with a consequent decrease in relative humidity. This results in improved adsorbent life if carbon adsorption is used for emission abatement, provided that the off-gas temperature is not increased by more than 20° to 30° F.
- There is no depression of water level in the bottom section of strippers that discharge effluent by gravity to a system at atmospheric pressure.

This last item is important if a high water level is needed to provide adequate head to a siphon breaker for discharge of stripper effluent to a sewer. The amount of depression is the in. w.c. discharge pressure that the blower must develop and is equal to the sum of the pressure drops through the following: packing or trays, mist separator, overhead ductwork, and abatement device (if used). This sum might amount to 20 in. w.c., and the design height of a tower with a conventional forced-air blower arrangement may accordingly have to be increased approximately 2 ft.

The air leaving the top of a stripper is saturated with water vapor, and more drops form if the overhead ductwork is exposed to cold ambient temperatures. With a negative-pressure design, these drops can erode the blower blades unless a mist separation device is located as close to the blower suction as possible. The advantage with most positive-pressure designs is that the mist separator can be located within the tower. Rather than installing a separate vessel for mist separation, as shown in the figure, the tower design height is increased slightly, and a section of baffles or mesh or packing is installed above the groundwater inlet. The mist separator coalesces fine water droplets carried by the air leaving the top of the packing or trays. The coalescing action forms water drops large enough to move back into packing or top tray by gravity.

With a negative-pressure design, a low-vacuum switch (or, equivalently, a high absolute pressure switch) should be installed at the tower overhead duct. With a positive-pressure design, a low-pressure switch should be installed at the blower discharge. With either arrangement, the switch can be automatically activated on a blower failure, and it can be interlocked so as to shut down groundwater pumps and the vapor emission abatement system.

It is vital that the blower be shut down whenever there is no water flow through the packing or across the trays for more than a few minutes. Otherwise, water will rapidly evaporate from the wetted surfaces, leaving all of the dissolved minerals as a scale.

5.5 Turndown and Liquid Distribution

Recovery of groundwater from wells is variable, so most strippers used for groundwater treatment must have good turndown capability, i.e., must be able to function at groundwater flow rates below the design flow rate. With packed strippers, the design should provide for keeping the downflowing water spread through the entire cross-section. Otherwise, at low flow rates, the water will channel through pathways that are most open, whereas upflowing air might find other pathways. A groundwater inlet distributor is needed above the packing that wets the top area as evenly as possible without groundwater impinging against the interior tower wall. Towers with a very large depth of packing need one or more distributors part way down the packing. Distributor designs are described by Bonilla (1993) and by Sloley and Martin (1995). Klemas and Bonilla (1995) define distribution quality, graphically correlate maldistribution versus quality, and present a mathematical model of maldistribution.

Two main types of distributors are available. One type is parallel troughs with notched overflow weirs. Two tiers of troughs oriented at right-angles to each other comprise a good design. The other type is simply a solid-cone water spray nozzle. Spray nozzle vendors can supply nozzles that have a preset cone angle. The tower designer can choose the distance between the nozzle and the top of the bed so that the entire cross-section is evenly wetted without impinging against the interior tower wall.

Spray nozzles cost less than trough distributors, but have 5 to 20 psi more pressure drop and have potential problems with nozzle-hole erosion and plugging. Some plugging tendencies can be avoided by prefiltering the groundwater. Sloley and Martin (1995) indicate that nozzle run lengths of 2 to 3 yr can be attained if the ratio of minimum particle size to nozzle minimum free path length is 10:1. Any prefiltering device used should be designed accordingly.

Nozzles have very poor turndown capability. The spray pattern cannot be maintained at water flow rates or pressures much below design values. If good turndown capability is necessary and a spray nozzle is chosen instead of distributor troughs, a recycled stripper, as discussed in the next section, should be considered. In any event, spray nozzles are not practical for liquid redistribution part way down the packing.

If a large packing height is used without liquid redistribution, at the least a deflector should be provided that catches water running down the inside tower wall and redirects it toward the center of the tower. A donut-shaped, flat ring sealed against the inside tower wall is normally used in cylindrical towers.

5.6 Recycled Strippers

The air from the top of a stripper (off-gas) can be treated and recycled. With this arrangement, an air emissions permit might be avoided, and monitoring of treated air quality can be reduced. The extent of treatment must be sufficient such that any residual contaminant in the air stream (at concentration y) is not concentrated enough to cause mass transfer into the stripper effluent (which should be below the target concentration x). The value of y should be less than H(x), where H is expressed in concentration units.

A more common form of recycle is the pumping of some of the stripper effluent back into the stripper feed stream. Recycling some of the treated effluent from the bottom of the stripper has several advantages over once-through systems. An example is where the groundwater flow rate varies from 20 to 50 gal/min. The feed pump and other feed equipment and the tower are designed for 50 gal/min. When groundwater is flowing at 50 gal/min into the tower, the recycle flow rate is zero. When groundwater is flowing at 20 gal/min into the tower, the recycle flow rate is 30 gal/min. At groundwater rates of 20 and 50 gal/min, the recycle rate automatically adjusts so that 50 gal/min is maintained at all times at the tower inlet.

The advantages include:

- Constant wetting of the packing or trays, thereby avoiding scaling if the blower keeps running when groundwater is not flowing
- Constant flow through the inlet spray nozzle, thereby avoiding liquid distribution turndown problems
- Constant flow through other devices, such as a groundwater heater, thereby avoiding potential scaling in the heater at low groundwater flow rates
- Constant hydraulic conditions in the stripper, so it can always function as designed
- Less chance of ice formation in lines and in the tower during cold weather when groundwater flow stops

The main disadvantage with recycling is that there is some energy expended in pumping recycled fluid back to the top of the tower. Also, without recycling, better stripping is achieved if the air rate is constant and the groundwater flow rate drops, because the A/W ratio is correspondingly higher.

5.7 Heated Strippers

Unheated air strippers are most commonly used and can do an adequate job of removing volatile compounds from water. However, higher-molecular-weight compounds and semivolatile compounds cannot be effectively removed unless the groundwater is heated to raise H. Heating can also be used for removing volatile compounds if it is desired to minimize the air rate or maximize the removal efficiency. In addition to raising H, heating may be needed during cold weather if the groundwater is stored in an uninsulated storage tank.

Either the gas or the water fed to the stripper, or both, can be heated. In general, it costs less to heat water than to heat gas. Water has a higher specific heat value, and heat transfer coefficients are generally higher for liquids than for gases.

A common design for heated systems that use air for stripping includes a steam heater for the water fed to an air stripper. However, the best designs use steam instead of air as the stripping medium. The steam heats the water (thereby increasing H), as well as functions as the stripping gas. Another advantage with steam is that it condenses as it rises through the stripper; very little stripping gas exits from the top of the tower. The tower overhead stream with such a design is mainly vapors stripped from the water, so any abatement system is greatly reduced in size and does not have to handle large volumes of air. If desired, the contaminant vapors can often be condensed without refrigeration, and a liquid product can be recovered. The liquid product is gravity separated from the condensed steam in an accumulator vessel. The water withdrawn from the accumulator can be treated with aqueous phase activated carbon for return to the steam boiler or pumped back to the top of the stripper. Steam strippers often operate under a partial vacuum, which enhances volatilization and increases stripping factor R values. Bravo (1994) suggests that steam stripping is mainly applicable to the removal of soluble, semivolatile organics.

In order to save heating costs, the stripper feed water can be heat exchanged with treated effluent leaving the bottom of the stripper tower. However, as noted in Bravo (1994), the heat exchanger will probably be subject to scaling from dissolved minerals in this configuration. If it is desired to maintain stripping while the exchanger is out of service for cleaning, the design must provide for more steam usage. Also, more steam is needed for startup than at steady state.

If no exchanger is included, more fouling occurs in the stripper. Sieve trays are more tolerant of fouling than is packing, and they are used more often in steam strippers than in air strippers. Bravo (1994) also recommends that if packing is to be used in a steam stripper to minimize pressure drop, the pressure drop should be determined using the Robbins (1991) correlation. Temperatures can be reduced, and fouling is less severe if strippers are operated at partial vacuum. Although the stripping factor R tends to be less at lower temperatures (because H is less), R is greater at reduced pressures. Bravo (1994) recommends that R be between 3 and 6 and suggests methods of estimating H at elevated temperatures based on activity coefficients and on solubility data.

Two types of steam strippers are generally available. With the first type, steam is injected directly, below the bottom of the packing or below the bottom tray. With the second type, some of the effluent from the bottom of the tower is cycled through a reboiler, which could be a fired boiler but more often is a steam-heated exchanger.
A system with direct steam injection is simpler than one with a reboiler, and it costs less to install because one less heat exchanger is involved. However, condensate cannot be recovered from direct steam injection systems, because the condensate goes out the bottom of the tower with the treated effluent. In fact, the effluent volumetric flow rate will be increased by the amount of steam condensate that forms.

With a steam-heated reboiler, there is no direct contact of the steam supplied to the reboiler with the water being stripped, and very pure condensate can be recovered. This condensate is valuable steam boiler feedwater and does not have to be treated to prevent scaling in the steam boiler. However, reboilers are seldom used for groundwater stripping because of potential scaling of the heat exchange tubing. The design of various types of reboiler systems has been reviewed by McCarthy and Smith (1995). Two types of reboilers are the kettle (horizontal tubes and with a pump-around circuit) and the thermosiphon (vertical tubes, no pump-around). The kettle-type reboiler has liquid level control, and only steam from the space above the liquid level is transported to the tower where it acts as stripping gas. The thermosiphon-type reboiler circulates a considerable amount of liquid water back into the tower, along with some steam that acts as stripping gas. Approximately 20% of the water that enters the reboiler each pass is vaporized. Thermosiphon reboilers are simpler than kettle reboilers, but they require extra space to be built into the bottom section of the stripping tower from which still water can be pumped out as treated effluent.

5.8 Emission Abatement

With steam strippers, the off-gas emissions can often be abated by condensing the vapors at ambient temperatures. With air strippers, condensers are rarely used because cryogenic temperatures would be needed. Such refrigerated condensing systems are complex, and provision must be made for controlling icing on condenser tubes. If the vapors from a stripper are not condensed and off-gas must be treated, one of two types of abatement devices is usually used: a vapor phase carbon adsorption system or a thermal oxidizer. It should be noted that the treated off-gas can be either discharged to the atmosphere or recycled as stripping gas.

Biofilters, which pass the off-gas through a moist, biologically active, porous medium, are rarely used because of their large area and the need to sustain temperatures at a somewhat constant level greater than approximately 15° C (59° F). With biofilters, organic destruction efficiencies are lower than with carbon adsorbers or oxidizers. An optimum design might include biofiltration followed by a dehumdification step and carbon adsorption. Thus, exhaustion of activated carbon would be minimized. Pennington (1996) indicates biofilters are suitable for water-soluble organics in humid off-gas with the conditions shown in Table 5-3.

Table 5-3 Conditions for biofilters.

Parameter	Range of Values
Oxygen content	Greater than 18%
VOC concentration	Less than 2,000 ppmv
Molecular weight	Low
Operation	Continuous

Williams (1996) indicates that air velocity should be 1 to 2 ft/sec, and that a typical area is 0.25 ac. Bohn (1992) compares costs per million ft^3 of air treated as follows:

    $20   Activated carbon systems with regeneration
    $130   Oxidizers
    $8   Biofilters

For controlling emissions of halogenated VOCs, Keller and Dyer (1998) describe a number of technologies besides condensation, carbon adsorption, oxidation, and biofiltration.

- **Pressure- and temperature-swing adsorption.** Synthetic polymeric adsorbents are regenerated by applying vacuum and heating, with nitrogen purging of the adsorption bed.
- **Membrane separation.** The off-gas permeates through membranes that are 10 to 100 times more permeable to organics than to air. Such systems cost more than condensation systems at concentrations greater than 10,000 ppmv.
- **Rotor concentration.** Carbon or zeolite adsorbent is regenerated with hot gas that is then treated with a relatively small adsorber or oxidizer.
- **Flameless thermal oxidation.** The off-gas is passed through a heated ceramic matrix.
- **Biotrickling filtration.** The off-gas flows co-currently with recirculated water down through a vertical packed tower. The packing has a coating of biofilm. Contaminants are biodegraded to carbon dioxide, water, and acid gases. The pH of recirculated water must be controlled because the acid gases would otherwise deactivate the bacteria in the biofilm. Most halogenated organics have slow biodegradation rates, so this technology applies to dilute air streams, as is often the case with off-gases from strippers.

For systems that use air stripping of halogenated VOCs, data from an economic evaluation by Keller and Dyer (1998) indicate that the lowest net present cost for a 10-year life at a discount rate of 12% is a catalytic oxidizer, without heat recovery, followed by wet scrubber. The oxidizer destroys the organic compounds, and, as described in Section 5.8.3, the scrubber controls acid gases that are generated by the oxidizer.

5.8.1 Carbon Adsorption

Granular activated carbon will effectively adsorb organic vapors at temperatures up to approximately 61° C (130° F). Vapor phase carbon granules are similar to those used for aqueous phase adsorption, but they are coarser and have a higher fraction of smaller micropores. Coarse granules with a high percentage of void space between the granules are preferred in order to minimize pressure drop through the carbon beds. For each organic compound, vapor phase adsorption isotherms are higher than the aqueous phase adsorption isotherms described in Chapter 4.

Most stripper blowers are designed to develop heads on the order of only inches of water column, so the carbon beds must be shallow. In order to obtain enough bed residence time for efficient adsorption of organic compounds, broad shallow beds in horizontal vessels or multiple short vertical canisters in parallel service can be used. Typical bed depths are 1 m, with face velocities in the range of 0.2 to 0.4 m/sec (40 to 80 ft/min) (Stenzel and Merz, 1989).

The adsorption capacity of vapor phase carbon is given for 283 organic compounds by Yaws, Bu, and Nijhawan (1995). Much of these data are derived from a correlation involving temperature, refractive index, and partial pressure.

The colder the temperature, the more efficient is the vapor removal by adsorption. However, at greater than approximately 40% relative humidity, the carbon has a progressively decreased capacity for adsorbing organics. (Synthetic polymeric adsorbents that compete with carbon for adsorption service are generally not affected by humidity.) Because the overhead discharge of a stripper is at 100% relative humidity, it is often economical to increase the carbon bed life by reducing the relative humidity ahead of carbon treatment. If the reduction of relative humidity is achieved by heating the overhead discharge, some potential carbon adsorption efficiency is lost because the temperature of the gas has been raised.

At greater than approximately 50% to 70% relative humidity, water molecules compete effectively with organic vapor molecules for carbon adsorption sites. If a system is installed to reduce the relative humidity of the off-gas, the optimum target is usually 45%. For most unheated stripping operations, raising the overhead air temperature by 25° F will lower the relative humidity from 100% to approximately 45%. At cooler than 90° F, activated carbon will typically remove approximately 90% of many organic compounds. The actual percentage removed will depend on the individual Freundlich isotherms, bed residence time, and bed inlet concentration.

With a heated stripper, the overhead air temperature should first be cooled, with the condensate that forms collected and recycled to the influent. Then the air stream temperature should be heated by approximately 25° F just before entering a carbon adsorber.

The best conditions for carbon adsorption include low temperature and low relative humidity. These conditions can be attained by refrigerating the air exiting from the top of a stripper, removing the liquid condensate that forms, and heating the air approximately 25° F. Such a scheme requires relatively high investment and operating costs for the refrigeration system and is not often used. It should be noted that with immiscible organics, the condensate will have both a water phase and a liquid organic phase that can be separated, or both phases that comprise the condensate can be recycled through the stripper. Note that recycling this relatively small amount of organics does not impair effective stripping, because the bulk of the volatile organics are removed by adsorption on the carbon.

It is easier to achieve higher removal percentages at higher inlet concentrations than at lower concentrations with any adsorption system. However, many air strippers have very dilute overhead concentrations because concentrations are low in the groundwater to start with. Stringent regulations may require that a certain minimum percent removal of organics from the off-gas be achieved. A series of carbon beds may be needed to meet the regulatory limits for organics in the air finally discharged to the atmosphere.

Because of the heat of adsorption, the off-gas temperature is increased by passing through activated carbon. With most air strippers, the organic concentrations are so dilute that this temperature increase is negligible. Heated strippers (or any stripper with a low G/L ratio) may produce an off-gas so rich in organic vapor concentrations that temperature control is needed after carbon adsorption. If the temperature exiting from a first-stage carbon bed is too high, a second-stage bed will not effectively remove the remaining organic vapors, and an interstage gas cooler will be needed.

If the air taken overhead from a stripper goes through vapor phase carbon and the effluent from the bottom of the stripper goes through aqueous phase carbon polishing beds, all of the organic contaminants captured end up adsorbed on carbon. It is logical to ask, "Why bother with the stripper? Why not use all aqueous phase carbon to start with?" The reason is that vapor phase carbon can adsorb more organics per pound of carbon than can aqueous phase carbon. By first stripping almost all of the contaminants in a vapor phase adsorption system, the aqueous phase polishing carbon will last longer before breakthrough occurs. Also, vapor phase carbon is easier to regenerate than is aqueous phase carbon, because aqueous phase carbon adsorbs nonvolatile organics as well as volatile organics.

5.8.2 Regenerating Vapor Phase Activated Carbon

A few states have regional furnaces for activation of virgin carbon and for reactivation. These facilities are used for reactivating both aqueous phase and vapor phase carbon. A typical reactivation unit uses a multi-hearth fired furnace. Steam is injected so that pores open up. Oxygen is limited, and the carbon does not burn. Almost all of the organic compounds, not just highly volatile ones, are vaporized and removed from the carbon at the temperatures used. The organics are combusted in an oxygen-rich afterburner.

The most common method of regenerating vapor phase carbon on-site is to steam it. Aqueous phase carbon might be regenerated with steam but not as thoroughly as with vapor phase carbon. At least two parallel carbon beds are used if continuous adsorption is desired, which is the case with most air stripping operations. While one bed is being steamed to regenerate the carbon, at least one other bed is operating in the adsorption mode.

It should be noted that up to 20% of the working capacity of the carbon can be lost because of retention of organics in the small pores (Stenzel and Merz, 1989).

The steam leaving the carbon bed contains VOCs. This mixed stream is passed through a condenser into an accumulator vessel. Water is separated from immiscible organic liquids by gravity settling in the accumulator, and the water may be returned to the stripper or to the steam boiler after passing through aqueous phase activated carbon.

Methods other than using steam for applying heat in the absence of oxygen can be applied for regenerating carbon. A carbon system with regeneration can serve as a concentrator for an oxidizer. A portion of the hot exhaust from the oxidizer is recycled through the spent carbon as the regenerating medium. The gases from the carbon regeneration process are much richer than the original air stream being treated. The oxidizer is thereby reduced in size and consumes much less fuel than an oxidizer installed without a carbon system preceding it.

Vacuum can also be applied for removing VOCs from carbon. Some of the best regeneration schemes use vacuum in conjunction with heat. A regeneration system with electric heating under vacuum conditions with nitrogen used to sweep out the organic vapors provides a more thorough job of regeneration than can typical steam heating systems.

Some adsorption/regeneration systems use one hollow wheel filled with activated carbon instead of multiple vessels with fixed beds. The wheel slowly rotates, with one section of the carbon exposed at any given time to the flowing air stream for adsorption of organic vapors in the air. At the same time, another section of the wheel is undergoing regeneration.

When halogenated VOCs are present, regeneration of carbon or of other adsorption media can generate corrosive acids. Equipment must be constructed of appropriate materials accordingly.

5.8.3 Direct Thermal Oxidizers

Oxidizers combust organic vapors with heating and ignition in the presence of excess oxygen (air), and they do not involve using carbon. Unless the air stream has a relatively high concentration of flammable organic vapors, auxiliary fuel such as natural gas or propane is used to sustain combustion. Because large air volumes are used in air strippers to remove relatively small amounts of organic compounds dissolved in water, the concentration of organics in the air stream is usually so lean that auxiliary fuel must be used. Direct thermal oxidizers are organic vapor incinerators that generally operate at temperatures greater than 1,200° F (649° C), with intimate contact of the burner flame with the air stream containing organic vapors. Some of the air stream is used as combustion air.

In addition to good mixing throughout the combustion chamber, efficient conversion of organic compounds to carbon dioxide and water vapor depends on having adequate residence time and temperature. A residence time of at least 0.3 sec at greater than

1,200° F (649° C) will usually result in more than 95% destruction efficiency. State or local air pollution control agencies often set the minimum allowable temperature and sometimes the minimum residence time that must be maintained in an oxidizer. In some jurisdictions, 1,400° F (760° C) and 0.5 sec are minimum requirements. With these conditions, hydrocarbon destruction efficiencies greater than 95% are possible. For non-halogenated VOCs, Keller and Dyer (1998) indicate that 98% destruction efficiency is attained with a 0.75-sec residence time at 1,600° F (871° C); for halogenated VOCs, 2 sec at 1,800° to 2,000° F (982° to 1,093° C) is needed for a 98% destruction efficiency.

Straitz (1995) gives the outside diameter and height of thermal oxidizers that are designed for a 0.6-sec residence time at 1,600° F. Examples are as follows: for 100 scfm of off-gas, 14 in. by 16 ft; for 200 scfm, 16 in. by 19 ft; for 400 scfm, 20 in. by 19 ft, and for 1,000 scfm, 30 in. by 16 ft.

When an oxidizer is used for emission abatement, a flame arrestor should be installed in the ducting connecting the stripper overhead with any oxidizer. Such devices prevent backward propagation of flame from the oxidizer to the stripper in the event of a blower failure.

If chlorinated solvents are being stripped, hydrogen chloride gas and chlorine will be emitted from either direct thermal or catalytic oxidizers. Because hydrogen chloride gas forms a hydrochloric acid solution with water, it may be required to water scrub the oxidizer exhaust. The scrubbing solution may have to be neutralized with caustic. Theodore (1996) gives the design basis for packed-column wet scrubbers. Spray towers, trayed columns, and Venturi scrubbers are also used for control of acid gas emissions. Frequently, air quality control agencies require that overall emissions of hydrogen halides and halogens be less than 1 lb/hr (75 g/min) or be abated by at least 99%.

5.8.4 Catalytic Oxidizers

Catalytic units use a fixed or fluidized bed of catalyst for combustion of organic vapors in the air stream from a stripper. The air stream is first heated to a temperature greater than 600° F (316° C). Because catalytic oxidizers use auxiliary fuel to heat the air stream to only approximately half the temperature that direct thermal oxidizers do, potential savings from using catalysts can be significant. This is especially true with air from an air stripper, which typically has very low organic vapor concentrations.

Most units use platinum impregnated on a fixed substrate, the same catalyst system used for most automobile exhaust converters. The preheating does not ignite the organic vapors but prepares them for reaction within the catalyst bed. The conversion of organic compounds to carbon dioxide and water vapor takes place while the air stream passes through the catalyst. Because the reactions are exothermic, the air stream rises in temperature through the catalyst bed.

The temperature rise is distinct, is somewhat proportional to the off-gas organics concentration, and can be used to monitor performance. Too high of an influent gas concentration could cause an excessive temperature rise detrimental to the catalyst, but that is unlikely with dilute concentrations usually encountered with air stripper off-gas. Catalytic oxidizers can readily achieve 95% destruction efficiency with 600° F (316° C) preheat. Either a gas burner with auxiliary fuel or electric heating can be applied for preheating. Higher destruction efficiencies, greater than 99%, are possible with higher preheat temperatures and large catalyst beds.

The catalyst must maintain adequate activity. If chlorinated solvents are being stripped, the usual platinum catalysts will become deactivated. At low chloride concentrations, this deactivation may not be significant, even after a long period of time. US-manufactured catalysts are now available that do not deactivate in the presence of even high chloride concentrations (Buck and Hauck, 1992). A Danish firm is marketing a catalyst that is effective with chlorinated and brominated hydrocarbons (Ondrey, 1995). HCl, Cl_2, HBr, and Br_2 that form can be removed with a caustic scrubber. Keller and Dyer (1998) indicate that halogen-tolerant catalysts can treat gas streams with up to 5,000 ppmv of equivalent chlorine or bromine content; temperatures must be higher or space velocity must be lower than with non-halogenated VOC destruction. Space velocity is defined as actual cubic feet per hour or actual cubic meters per hour divided by catalyst volume in cubic feet or cubic meters, and it is typically 7,500 to 15,000/hr with halogenated VOCs.

Certain heavy metals will poison catalysts. Lead, especially in the form of tetramethyl lead, which is volatile and was used in some gasoline blends, will quickly poison a platinum catalyst if not converted first to lead oxide in a preheating flame. Platinum catalysts can be deactivated by phosphorus, arsenic, cadmium, alkali metals and iron. According to Bar Ilan et al. (1994), catalyst manufacturers can restore some activity by removing these substances with wet chemical cleaning and with baking off of soot and other organics.

5.8.5 Auxiliary Fuel Consumption and Heat Exchange

A rapid method of estimating auxiliary fuel consumption with an oxidizer for an air stripping system is illustrated below. Fuel costs usually greatly outweigh other operating costs if the stripping system is automated so that operator attendance is only occasional. Oxidizers are well insulated, with relatively minor radiation losses to the atmosphere. Because of this, fuel consumption can be approximated from the amount of heat needed to raise the temperature of the stripper overhead air. The approximate amount of heat needed to be supplemented by the auxiliary fuel is given by Equation 5-30.

Air heating duty = (air flow rate) (ΔT) C_p − (fuel value of the organic vapors)   (5-30)

in which ΔT is the heat rise provided to the influent stripper off-gas; the air flow rate is in pounds per unit of time, and C_p is the average specific heat of air over the temperature range involved. Moisture content has been neglected; for heated strippers, the heating rate for water vapor should be accounted for. For the temperature range of most oxidizer systems, C_p is 0.24 Btu/lb of air/°F of temperature rise (for water vapor, C_p is approximately 0.5 Btu/lb of air/°F). As noted for Equation 5-16, the density of air is 0.076 lb/standard ft^3 for 60° F standard temperature. If the air flow rate is known in standard cubic feet, the density is multiplied by air flow rate to obtain units in pounds. Using the example given in Figure 5-4 for 200 scfm (21,888 lb/day) of air at 60° F, the air heating duty for a catalytic oxidizer with preheating to 650° F is calculated as follows:

Given: The air mass flow rate is 21,888 lb/day; the temperature rise is from 60° to 650° F; and the gasoline TPH vapor rate is 7.59 lb/day, from Figure 4-6.

Assumptions: The heating value of the gasoline hydrocarbons involved is approximately 20,000 Btu/lb; no heat recovery is used.

Find: Heat duty required from auxiliary fuel

Step 1: Determine heat duty to increase the air temperature, without considering the fuel value of the organic vapors, as given by Equation 5-31.

(21,888 lb/day) (650° F - 60° F) (0.24 Btu/lb/°F) = 3,099,300 Btu/day   (5-31)

Step 2: Determine net duty, using the assumed heating value for the organic vapors, as given by Equation 5-32.

3,099,300 Btu/day − [20,000 Btu/lb (7.59 lb/day)]
= 3,099,300 Btu/day - 151,800 Btu/day = 2,947,500 Btu/day net   (5-32)

The temperature rise in this example is 650° F - 60° F, or 590° F. For a direct thermal oxidizer operating at, for example, 1,300° F, the temperature rise would be 1240° F. The air heating rate could be calculated from Equation 5-30 and would be approximately (1,240/590) (3,099,300 Btu/day), or 6,513,800 Btu/day, minus 151,800 Btu/day fuel value of TPH vapor. The net heating duty is more than double the amount calculated in Equation 5-32 for the catalytic unit.

With either a direct thermal oxidizer or a catalytic oxidizer, a heat exchanger can be added to the system to save auxiliary fuel. The most common heat exchanger uses metal surfaces or tubes to separate air from the stripper overhead and the oxidizer exhaust. Recuperative heat recovery is accomplished by heat transfer between the two gas streams. The amount of heat saved is approximately proportional to the amount of heat exchange surface. Exchangers are rated by the percent of gross air heating saved, without taking into account the fuel value of the organic vapors. Typical standard designs provide 50%, 65%, or 70% in savings.

Keller and Dyer (1998) recommend that the exhaust gases be maintained at least 100° F above (38° C above) the acid dew point temperature to avoid exchanger corrosion.

Straitz (1995) gives the percentage of heating saved as equal to $(t_2 - t_1)/(T-t_1)$ times 100%, in which t_2 is the stripper off-gas temperature after warming in the heat exchanger, t_1 is the temperature leaving the stripper, and T is the oxidation temperature.

Another type of heat recovery is attained using regenerative oxidizers that provide more than 90% in fuel savings. These devices have at least two extra chambers surrounding a firebox that contains an auxiliary fuel burner. The extra chambers are filled with gravel or ceramic packing that provides a mass with high heat capacity and allows air flow through the mass. The hot exhaust flowing from the firebox through the solids heats the mass for a relatively short period of time. Valves are then automatically repositioned so that stripper overhead air passes through the chamber containing the heated mass before the air enters the firebox. More than 90% of the air heating duty is thereby accomplished before the air contacts the auxiliary fuel burner flame. The burner duty is minimal using these systems. Smith Environmental (Ontario, California) recommends in an interview by *Chemical Engineering* (September 1996) that regenerative oxidizers be avoided if vapor concentrations exceed 20% of the LEL. Overheating at 10% to 20% of the LEL should be avoided by either admitting dilution air or reducing the depth of the ceramic packing material. With typical air stripping, concentrations do not exceed 10% of the LEL. With high organic vapor concentrations, either no heat recovery or a recuperative heat exchanger should be used instead of a regenerative unit.

Renco (1994) compares regenerative heat recovery with conventional recuperative heat exchange, and recommends regenerative systems for flows greater than 10,000 scfm with flammable vapor concentrations below 10% of the LEL. Virtually all stripping towers operate below this air rate, but they usually have such low vapor concentrations. For large stripping systems, it would pay to compare the costs and savings of regenerative units with recuperative units. Klobucar (1995) compares the two types of heat recovery and gives a method of accounting for operating costs, taking into account fan electrical energy consumption, as well as fuel consumption. An example is shown with pressure drop through a regenerative unit at a 25-in. w.c. versus a 12-in. w.c. through a recuperative unit. Klobucar also compares such systems with a hybrid emission abatement system using vapor phase activated carbon or zeolite adsorption with hot-air absorbent regeneration. The hot air exhaust contains the organic vapors at perhaps 10 times the concentration of the main air stream and is fed to a direct thermal oxidizer with recuperative heat recovery.

If there is a chance that organic vapor concentrations can approach the LEL in a heat exchanger, controls should be added that automatically open a dilution air damper or a heat exchange bypass damper at 10% to 20% of the LEL.

5.9 Main System Design Parameters

This section summarizes, by way of an example, important parameters that impact air stripper system design and evaluation. A packed air stripper with carbon polishing of the liquid effluent is used in the example for removing gasoline hydrocarbons from groundwater. As shown in the Chapter 4 on carbon adsorption, a stripper effluent gasoline TPH concentration of 0.35 mg/L corresponds to a reasonably long carbon life. Aqueous phase carbon will reduce the effluent concentration from this level to near nondetectable levels. With this polishing action by the carbon, and considering that the air emission abatement equipment and operations will cost less if air volume is minimized, the low end of the A/W range for gasoline strippers will be chosen — namely, A/W is 30.

5.9.1 Concept and Process Design

The influent TPH is given as 13,000 µg/L, or 13.0 mg/L. The stripping efficiency for TPH would be (influent - effluent)/influent, or $(13.0 - 0.35)/13.0 = 0.973$. The next steps illustrated are as follows:

- Mass balance
- Calculation of volumetric air flow rate
- Calculation of vapor concentrations
- Initial development of P&ID, plot plan, and site plan
- Use of an air stripping computer program

Mass balance example. Figure 5-4 shows the system illustrated in Figure 5-1, with the contaminant removal efficiencies and corresponding effluent concentrations shown. In addition to the TPH concentrations, the diagram shows concentrations of contaminants of most concern to pollution control authorities, the aromatic compounds. Note that the sum of the aromatics concentrations is not equal to the TPH concentration because TPH includes nonaromatics that are not shown.

In this example for gasoline hydrocarbons, the stripping factor R could be approximated by using H for octane. Use of an air stripping computer program would give the corresponding depth of packing for the chosen A/W ratio and the effluent concentrations for the individual aromatic compounds.

The mass flow rates, g/day or lb/day, are equal to the volumetric water flow rate times the concentration. The water rate is 50 gal/min (272,160 L/day). Because 1 g equals 1,000 mg, the influent TPH mass rate at 13 mg/L is 3,580 g/day (7.80 lb/day); the effluent at 0.35 mg/L is 95.3 g/day (0.21 lb/day). The tower overhead vapors contain 7.59 lb/day of TPH, determined by subtracting the effluent mass flow rate from the influent mass flow rate. Figure 5-4 shows these values for TPH and values for the same parameters determined in identical fashion for each of the aromatic compounds of interest.

Volumetric air flow rate. Figure 5-4 also shows the air flow rate, on a dry weight basis — that is, air humidity is not taken into account, and it is assumed that the incoming air is saturated. At air temperatures up to 25° C, the moisture vapor content of air cannot exceed 2% by weight. The calculated values are still accurate enough for designing the air blower and ductwork. (As discussed earlier, for vapor phase carbon adsorption, the relative humidity of the air being treated is important for designing carbon systems.) For an influent flow rate of 50 gal/min (6.68 ft^3/min) and an A/W of 30, the air flow rate is 6.68 ft^3/min (30), or 200 ft^3/min. If the incoming air is the same temperature as the groundwater, as in this simplified example, there would be some cooling effect because of evaporation if the incoming air were not saturated.

For these calculations, it is convenient to express air volumes in standard cubic foot, and air flow rates in standard cubic foot per minute as scfm or SCFM. A standard cubic foot is a cubic foot of air measured at 1 atm (29.92 in. of mercury column, in. Hg or 760 mm Hg) and at standard temperature. (An actual cubic foot of air, acf, is the volume at its actual pressure and temperature). The chemical engineering standard temperature of 60° F, equivalent to 520° R, is used in examples throughout this book.

Figure 5-4 could be enhanced to include values of acfm as well as scfm for air flow rates. Assuming that the incoming air is at sea level and at 1 atm, there would be 29.9 in. Hg of pressure entering the tower just below the bottom of the packing. The incoming air temperature is given as 60° F, so the incoming acfm equals the scfm rate of 200 at a standard temperature of 60° F. The total pressure drop through the tower is assumed to have been computed in this example to be 5 in. w.c., equivalent to 0.38 in. Hg. The diagram shows a negative-pressure stripping system, i.e., the air blower induces atmospheric air into the tower. The gauge pressure in the tower overhead duct leading to the blower suction would be -0.38 in. Hg. The absolute pressure would be 29.9 in. Hg - 0.38 in. Hg, or 29.5 in. Hg. For the temperature and pressure ranges within which groundwater strippers operate, the ideal gas pressure-temperature-volume relationship holds quite accurately. The ideal gas law is used to relate actual and standard conditions as follows:

acfm = scfm (29.9 in. Hg)/[(absolute actual pressure) (actual temperature)]/520   (5-33)

At 29.5 in. Hg and 60° F, the 200 scfm effluent flow rate converts to actual cubic feet per minute as follows:

200 scfm (29.9/29.5) (520/520) = 203 acfm (5-34)

Figure 5-4 shows blower discharge conditions to be 89° F (549° R) and 13.6 in. w.c. (1 in. Hg) gauge pressure (30.9 in. of Hg absolute pressure), enough pressure to force the off-gas through ductwork and an abatement system in this example. The 200 scfm flow rate converts to actual cubic feet per minute in the blower effluent, as given in Equation 5-35.

200 scfm (29.9/30.9) (549/520) = 204 acfm (5-35)

Vapor concentrations. Figure 5-4 shows for the tower overhead air/vapor mixture the vapor concentrations in parts per million by volume (ppmv) on a dry basis. These concentrations are determined by calculating the scfm for each contaminant, dividing by the total scfm (air plus TPH), and multiplying by one million. Again, ideal gas relationships are assumed to hold. The mass flow rates (lb/day) are known from Figure 5-4. The volumetric flow rate is the mass rate divided by density of vapor (lb/scf) for each contaminant. The density of a vapor is proportional to its molecular weight. A convenient method of finding density is to divide the molecular weight by 379 (the volume, scf, for 1 lbmol of any gas at 60° F and 1 atm pressure). Another method is to multiply the density of air (0.076 lb/scf for 60° F standard temperature) by the vapor/air molecular weight ratio. The molecular weight of air is 28.8, so the scfm for a contaminant vapor can be determined from Equation 5-36 using the density calculated from Equation 5-37 or Equation 5-38:

scfm = (lb/day)/[(density) (1,440 min/day)] (5-36)
density = 0.076 (MW/28.8) lb/scf; or (5-37)
density = MW/379 lb/scf (5-38)

The molecular weight of gasoline, given in Johnson et al. (1990), varies from approximately 95 to 110, depending on what fraction is light hydrocarbons. As the gasoline ages ("weathers"), the lighter hydrocarbons evaporate preferentially over the higher-molecular-weight hydrocarbons, and its molecular weight increases.

Given: The TPH mass flow rate in tower off-gas, determined by subtracting the effluent mass rate from the influent mass flow rate and shown in Figure 5-4, is 7.59 lb/day. The air flow rate corresponding to an A/W of 30 is 200 scfm.

Assumptions: Groundwater is contaminated from an old leaking underground storage tank. The gasoline components that have dissolved in the groundwater have weathered to a molecular weight of 105. Note that weathered NAPL typically has a higher molecular weight than dissolved TPH, because some components such as benzene have a higher solubility and lower molecular weight than the bulk of weathered NAPL. As is typical of gasoline air strippers, almost all of the TPH volatilizes into the vapor phase, so the average off-gas molecular weight of the stripped gasoline components is almost 105.

Find: Total scfm, and ppmv of TPH in the off-gas.

Step 1: The density of gasoline vapor from Equation 5-37 for a molecular weight of almost 105 is almost 0.275 lb/scf. The air flow in scfm is calculated from Equation 5-36 as shown in Equation 5-39, with the sum shown in Equation 5-40.

(7.59 lb/day)/(1,440 min/day)/(0.275 lb/scf)
= 0.019 scfm of stripped gasoline components (TPH vapor)   (5-39)
Total scfm = 200 + 0.019 = 200.02   (5-40)

Step 2: Multiply TPH vapor fraction by 1 million to arrive at the parts per million by volume, as shown in Equation 5-41.

ppmv = (0.019/200.02) (10^6) = 95   (5-41)

For benzene, with a molecular weight of 78, the density from Equation 5-37 or 5-38 is 0.206 lb/scf. The standard cubic feet per minute and vapor concentration for benzene will be calculated next.

Given: Air flow rate is 200 scfm.

Assumption: Use the lb/day rate for tower overhead going to abatement in Figure 5-4.

Find: The benzene standard cubic feet per minute flow rate and the volumetric concentration

Step 1: Applying Equation 5-36 gives the volumetric flow rate, as shown in Equation 5-42.

(0.84 lb/day)/(1,440 min/day)/(0.206 lb/scf) = 0.00283 scfm   (5-42)

Step 2: Multiply benzene vapor fraction by 1 million, as shown in Equation 5-43.

ppmv = (0.00283/200.003) (10^6) = 14   (5-43)

The standard cubic feet per minute rates and concentrations shown in Figure 5-4 reflect the values calculated for TPH and benzene and the values calculated similarly for the other aromatics.

Initial P&ID and layout. A preliminary P&ID is developed by adding to the flow schematic all instruments, interlocks, alarms, and valves. (*See* Figure 5-5 and Section 5.9.3).

For example, each centrifugal pump discharge should include a pressure gauge, check valve, and either a block valve or throttling valve. Each filter designed for intermittent backwashing should be manifolded for filtering flow and backwash flow, and each nonbackwashable filter should be manifolded with a standby filter or with block-and-bypass valves. As the design progresses, the size of each piping run should be added to the diagram along with the pipe wall thickness (or schedule) and material. Any heated, insulated lines should be designated with the insulation thickness. Valves and instrumentation should be included for monitoring of stripper pressure drop and the pressure drop across all filters.

Along with the P&ID, a Sequence of Operations document should be developed that briefly describes the basis for stripper and polishing carbon system design, how the interlocks are connected, how responses to alarms will be handled, and what controls must be manually reset for startup after an automatic shutdown has occurred.

As soon as the stripper cross-sectional area, carbon canister diameters, and tank sizes have been selected, a preliminary plot plan is developed. This plan shows feasible locations of major and auxiliary equipment such as pumps and filters, drawn to scale, allowing space for valves and access. A site plan is developed showing where the plot fits relative to existing equipment; buildings; roads; plot fencing, gates and driveway access; where utilities will be interconnected and routed; drainage systems; groundwater extraction wells, injection wells, and connecting piping routings; and water storage tanks.

The process flow diagram, P&ID, plot plan and site plan form a process design from which detailed piping, electrical, and structural designs can be developed.

Air stripping computations. The objectives of stripping computations are to check that a selected cross-sectional area (or diameter) and A/W ratio correspond to a reasonable pressure drop and packing depth, while meeting desired effluent criteria. A computer program (see Section 5.2.3) that correlates concentration reduction or effluent concentration with packing depth at various A/W ratios (or various corresponding stripping factor R values) makes this task easier. Or Equations 5-11, 5-12, 5-13, and 5-14 can be applied. A plot of Equation 5-13 can be made, or the equivalent plots in Figures 6 and 7 in Kavenaugh and Trussell (1980) can be used to speed the calculation of NTU. The computer programs or such plots show at a glance the effect on required packing depth caused by a change in the A/W ratio.

For this example, C_{in}/C_{eff} for TPH is (13.0 mg/L)/(0.35 mg/L), or 37, which corresponds to 97.3% removal. The A/W ratio was chosen to be 30; G/L is 0.000762

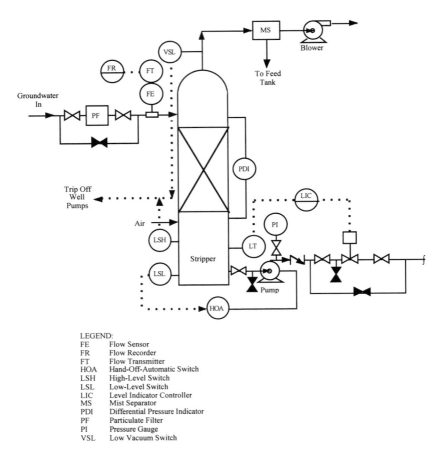

Figure 5-5 Preliminary P&ID for an air stripper.

LEGEND:
FE   Flow Sensor
FR   Flow Recorder
FT   Flow Transmitter
HOA   Hand-Off-Automatic Switch
LSH   High-Level Switch
LSL   Low-Level Switch
LIC   Level Indicator Controller
MS   Mist Separator
PDI   Differential Pressure Indicator
PF   Particulate Filter
PI   Pressure Gauge
VSL   Low Vacuum Switch

(A/W), or 0.023. A diameter of 2 ft is chosen for the tower (corresponding to 3.14 ft^2 cross-sectional area) so that a standard-diameter stripper can be purchased and so the liquid loading (50 gal/min/3.14 ft^2) is close to 20. Using an H value of 200 atm at 60° F as being representative of dissolved gasoline TPH, the stripping factor R is (200 atm/1 atm) (G/L), or 4.6. The NTU from Equation 5-13 or from Figure 6 in Kavenaugh and Trussell (1980) is 4.3. A type of packing is chosen, and from pilot data, from an air stripping computer program, or from the packing manufacturer's information, HTU is 3.5 ft. From Equation 5-11, the depth of packing is as shown in Equation 5-44.

$$Z = \text{HTU (NTU)} = 3.5 \text{ ft } (4.3) = 15 \text{ ft } (4.6 \text{ m}) \qquad (5\text{-}44)$$

5.9.2 Sizing and Rating of Major Equipment

The objective here is to determine the tower shell length, air blower design conditions, and blower horsepower.

Tower shell length. Applying a safety factor, the design packing depth is given in Equation 5-45.

$$1.5\ Z = 1.5\ (15\ \text{ft}) = 22.5\ \text{ft}\ (6.9\ \text{m}) \tag{5-45}$$

The total tower shell length must be at least this depth plus allowances for each of the following items:

Sump for stripped liquid holdup. Consider that the liquid loading is 50 gal/min/3.14 ft^2 and that there are 7.48 gal/ft^3. If the sump residence time is to be approximately 5 min, then the sump length is as given in Equation 5-46.

$$\text{Sump length} = 5\ (50/3.14)/7.48\ \text{ft} = 10.7\ \text{ft}\ (3.3\ \text{m}) \tag{5-46}$$

Space between sump and bottom of packing. This space must accommodate the air inlet and the packing support grid — a minimum of 2 ft.

Space above the packing for the water inlet liquid distributor, or spray nozzle and spray cone. Assume 2 ft for this example.

Space for water redistribution midway in packing. Assume 2 ft for this example.

Mist elimination section. If included inside the top of the tower above the water inlet, assume approximately 2.5 ft.

Depression of sump water level with forced-draft strippers that discharge to systems at atmospheric pressure. This is as discussed in Section 5.4.

In this example of a negative-pressure stripper with a mist separator located at the vacuum blower suction, the last two items do not apply. The total tower length is 22.5 ft of packing depth plus sump length plus the sump to bottom of packing length plus space for water inlet liquid distributor plus space for water redistribution midway in packing, or 39.2 ft (12.0 m).

Blower design. The objective is to define the following:

- Pressures at blower suction and discharge and corresponding pressure rise (blower head)
- Standard cubic feet per minute and actual cubic feet per minute at blower suction

- Blower brake horsepower and motor horsepower rating

For a negative-pressure stripper, as in this example with a vacuum blower in the off-gas system, the stripper air inlet is at 1 atm absolute pressure. The negative pressure at the blower suction must be numerically equivalent to the sum of the pressure drops through the packing, ducting, and mist separator. Evaluation of these three pressure drops is as follows:

For this example, assume that an air stripper computer run indicates a packing pressure drop of 1 in. w.c.

The ducting pressure drop depends on velocity. Usually the duct diameter is chosen so that the air velocity is approximately 3,000 ft/min. The pressure drop is given by the equation from Wright (1945), as shown in Equation 5-47.

$$h' = 2.74(V/1000)^{1.9}/D^{1.22} \tag{5-47}$$

in which h' is in. w.c./100 ft of duct equivalent length; V is the air velocity, ft/min; and D is the duct diameter, in. The equivalent length is the total length of straight duct runs plus the number of 90°-elbows multiplied by $2D$. For this example, at 200 ft^3/min, a velocity of 3,000 ft/min corresponds to a duct diameter between 3 in. and 4 in. The larger standard size, 4 in., is chosen with a corresponding velocity V of 2,293 ft/min and h of 2.4 in. w.c./100 ft of equivalent length. Assume that straight duct runs total 60 ft and that there are four 4-in. 90°-elbows, the equivalent length is 60 ft plus 32 ft. Allowing 1 in. w.c. for other losses, the ducting pressure drop is (2.4 in. w.c./100 ft) (60 + 32) ft + 1 in. w.c., for a total of 3.2 in. w.c.

The pressure drop through the mist separator must be obtained from the manufacturer. In the absence of this information, allow 4 in. w.c. for a preliminary design.

If there is no emission abatement device, the blower discharge is at 1 atm absolute pressure. The blower head requirement would be the sum of all pressure drops listed above. There will be an additional head requirement if a damper is included in the ducting or if the blower discharge ducting or stack is of a significant length. If there is a catalytic oxidizer or vapor phase activated carbon used for abatement, allow approximately 12 in. w.c. additional pressure drop for a preliminary design. The blower horsepower required is proportional to the air flow rate and the head h. It is also slightly affected by air moisture content (which at ambient temperatures is less than 2% and can be neglected).

For a conventional blower arrangement with a forced-draft blower feeding air into the stripper, the blower suction is at 1 atm absolute pressure. The blower discharge pressure must overcome all the pressure drops, including those listed above plus that of the abatement equipment. The pressure rise required through the blower is the same as the head h for the vacuum blower example.

The air flow in scfm entering the stripper is the same as the air flow in acfm in this example, because the air is at standard conditions of 1 atm and 60° F (520° R). Neglecting possible changes in air moisture content, the air flow rate in scfm is constant through the tower, ducting, mist separator, and blower. The actual air flow rate is an important design parameter for sizing ducting and for the blower manufacturer, and it is related to the air flow rate expressed in units of scfm as shown in Equation 5-48.

acfm = scfm (1 atm/absolute actual P) (absolute actual T)/(standard T)   (5-48)

For this example with the blower suction at -8.2 in. w.c, the absolute pressure in in., Hg is (29.92 - 8.2)/(specific gravity of Hg, 13.6), or 29.32 in. Hg. The actual volumetric air flow rate at the blower suction at 60° F is shown in Equation 5-49.

acfm = 200 (29.92 in. Hg/29.32 in. Hg) (460 + 60)/520 = 204 acfm   (5-49)

The head h is 8.2 in. w.c. plus 12 in. w.c. for emissions abatement, or 20.2 in. w.c. total. The brake horsepower, bhp, is given by Equation 5-50.

bhp = scfm (h) (0.000157)/eff   (5-50)

in which eff is the blower fractional efficiency as obtained from the blower manufacturer's data. For this example at 200 scfm and h equal to 20.2 in. w.c., assuming that a 65%-efficient centrifugal blower is chosen, the bhp is given by Equation 5-51.

200 (20.2) (0.000157/0.65) = 0.98 bhp   (5-51)

The blower motor initially selected for this service should be rated at least at 1 hp, the closest larger standard motor rating. An even larger motor should be selected, depending on the blower manufacturer's horsepower requirements at the upper end of the blower's air flow rate range. Each blower model has a range of blower-wheel diameters and rotational speeds. The air flow rate value used for determining the motor horsepower requirement should correspond to the largest wheel ever expected to be installed and the highest rotational speed anticipated.

5.9.3 Controls

Figure 5-5 is the beginning of a P&ID for the stripper portion of the example groundwater remediation system. Additional diagrams would apply to groundwater pumping, storage, pretreatment, and the polishing carbon adsorption system. The figure includes a non-backwashable filter, level control in the stripper sump with redundant level switches that activate in the event of controller failure, and a low-vacuum switch that activates in the event of vacuum blower failure. If a conventional forced-draft blower feeding air into the stripper were used instead, the blower discharge duct would be fitted with a low-pressure switch.

Not shown on the diagram, but a good practice to consider including, is an interlock that would shut down the blower if influent flow stops. This is important, because continuing air flow would dry the tower internals, leading to severe scaling and plugging from groundwater dissolved solids precipitation. This automatic blower shutdown circuit should include a time delay that would allow the blower to operate for a short time after the interlock is tripped. During that short time, residual groundwater that is migrating down the tower will be stripped, thereby avoiding contamination of the stripper sump.

5.9.4 Utilities Requirements

The main utility requirement for an unheated stripper without emission abatement is electric power for the blower and pumps. If pneumatically energized pumps or controls are used, additional electric power will be required for an air compressor. If vapor phase activated carbon is used, the saturated stripper off-gas should first be heated approximately 25° F. This heating would achieve the optimum off-gas relative humidity and would require either additional electric energy or fuel gas. If vapor phase activated carbon with on-site steam regeneration is used for emission abatement, then steam use will be an additional major utility requirement. If a gas-fired oxidizer is used, then fuel to heat the oxidizer will be an additional major utility requirement. The amount of fuel gas use depends on the degree of heat recovery used and the fuel value of the organic vapors in the off-gas.

The centrifugal blowers and pumps most commonly used in remediation applications are driven by induction motors, and each horsepower is approximately equivalent to 1 kW of motor power consumption. This relationship accounts for the pump motor inefficiencies and power factor. The total number of kilowatts needed for a blower and pumps is then the sum of the motor horsepower. The blower horsepower equation (Equation 5-46) indicates for the 200 ft^3/min example that 1 bhp applies. A groundwater feed pump feeding from an atmospheric storage tank would need enough head to lift the groundwater to near the top of the stripper and to overcome pressure drops through the filter and piping (and spray nozzle if used), approximately 45 psi pressure rise through the pump. The brake horsepower required for a 70% efficient feed pump at 50 gal/min is calculated, as shown in Equation 5-52.

Brake horsepower = gal/min (pressure rise)/[1,713 (pump efficiency)]
= 50 (45)/[1713 (0.70)] = 1.9 hp (5-52)

A similar calculation applies to pumping from the stripper sump. For example, if that were a 65% efficient pump requiring a 25 psi pressure rise, the corresponding brake horsepower would be 1.1 hp. The total horsepower requirement for the blower, feed pump, and stripper sump pump would be 1 hp + 1.9 hp + 1.1 hp = 4 hp. If the usual induction motors are used, the corresponding electric power load would be approximately 4 kW.

The off-gas heating duty if vapor phase carbon is used is calculated, as shown in Equation 5-53.

Heating duty = (air flow rate) (temperature rise) (C_P) (5-53)

in which the air flow rate is in pounds per unit of time, and C_P is the average specific heat of air over the temperature range involved. For 200 scfm of air (density of 0.076 lb/scf at 60° F standard temperature) undergoing a 25° F rise, the duty per hour is as shown in Equation 5-54.

(200 scfm) (60 min/hr) (0.076 lb/scf) (25° F) (0.24 Btu/lb-°F) = 5,470 Btu/hr (5-54)

A kilowatt is equivalent to 3,413 Btu/hr. If an electric resistance heater is used, the power consumption would be (5,470/3,413) kW, or 1.6 kW.

5.10 Treatability Studies for Groundwater Stripping

Pilot testing of groundwater air strippers is sometimes done at vendors' shops or by packing manufacturers. For packed strippers, such testing can be used to find the HTU and the K_La for the packing. The HTU is a function of the groundwater flow rate per unit cross-section of tower area. If K_La is established, the HTU can be derived theoretically for other flow rates and cross-sectional areas. The HTU is derived from pilot data as the depth of packing divided by the NTU. The equations given in Section 5.2.1 can be applied to find the relationship between NTU and H, the G/L ratio used in each pilot run, and the concentration reduction as measured in the pilot tests.

5.11 Cost Estimating for Groundwater Stripping

5.11.1 Equipment Costs

The investment costs per gal/min of capacity for stripping systems depend on the volatility of the contaminants and whether auxiliary equipment (e.g., water storage tanks, filters, emission abatement, etc.) is included. As given in Attachment 2 to US EPA (1994), the cost of basic air stripping equipment ranges from $200/gal/min for small units (10 to 30 gal/min) down to $130/gal/min for units rated at 30 to 500 gal/min. Addition of off-gas emission abatement equipment has a major impact, sometimes doubling or tripling the cost of a stripping system, depending on the A/W ratio used.

Table 8-2 in Chapter 8 gives costs of sparging and associated soil vapor extraction equipment, well costs, and some installation costs. Attachment 2 to US EPA (1994) gives sparging costs for air injection pumps of $5,000 to $25,000 and for air injection well installation at $75/ft.

5.11.2 Operating Costs and Total Costs

Attachment 2 to the US EPA (1994) states that operating costs for air stripping range from $20 to $50/lb of contaminant removed.

Total air stripping costs from Figure 2 of the US EPA (1991a) are summarized as given in Table 5-4, based on dimensionless values of H.

Table 5-4 Total stripping costs ($/1,000 gal) for 99% VOC removal.

Value of Dimensionless Henry's Constant	Operating flow Rate = 70 gpm	Operating flow Rate = 700 gpm
H = 0.02 to 0.1	$0.50	$.020
H = 0.1 to 0.2	$0.45	$0.16
H over 0.2	$0.42	$0.13

DuTeaux (1996) tabulates three case histories for groundwater stripping as follows:

- Total cost of $0.75/1,000 gal (in 1994) with a 500-gal/min stripper reducing 15 ppm TCE and 6.7 ppm PCE to less than 1 ppm
- A 250-gal/min air stripper (in 1993) with two-stage carbon polishing, thermal oxidizer, and caustic scrubber for chlorinated hydrocarbons averaging 60 ppm:
  - $1.7 million — stripper, GAC, oxidizer, scrubber, wells
  - $1.0 million — oxidizer heat exchanger, pumps, compressors, control center
  - $1.3 million — other capital investment costs
  - $1,240,000/yr — operating cost, including contractor, utilities, sampling/analysis, project management
- A 250-gal/min air stripper (in 1992) with fluidized bed catalytic oxidizer for 10,000 ppm TCE:
  - $3.19/1,000 gal for air stripping
  - $1.70/1,000 gal for catalytic oxidation of off-gas

5.11.3 Emission Abatement Costs

Table 8-3 in Chapter 8 gives the reported capital costs and alternative rental costs for carbon adsorption and catalytic oxidation systems. Table 8-3 also gives electric power requirements for operating emission abatement equipment. Attachment 2 to US EPA (1994) gives vapor treatment costs, as shown in Table 5-5.

Straitz (1995) compared the costs of various gasoline vapor emission control methods. The installed costs of a carbon adsorption unit with regeneration are 2.5 times the cost of a $100,000 oxidizer installation. Operating costs for the carbon unit in this example were $23,000/yr, of which $13,000 was recouped as the value of the recovered gasoline. Operating costs for the oxidizer would depend on concentrations of gasoline vapor, because the vapor is fuel for the burner. Air stripper off-gas is usually very low in organic vapor concentration, so auxiliary fuel costs are very high.

Table 5-5 Vapor treatment costs.

Treatment Process	Capital Investment Costs	Annual Operating Costs
Carbon Adsorption	<$1,000 for <200 scfm $3 to $4/scfm >200 scfm	*
Condensation	$15,000 to $20,000 for 200 scfm	no data
Oxidizer	$65 to $100/scfm	$50/scfm
Acid Gas Scrubber	$60 to $90/scfm	$200 to $350/scfm

*Operating costs for vapor phase carbon adsorption are $40 to $100/lb of contaminant removed.

In the example in Section 5.8.5, the direct thermal oxidizer duty is 65 therm/day, minus only 1.5 therm/day for the fuel value of the gasoline vapor. The corresponding auxiliary fuel cost would be approximately $13,800/yr at an on-stream factor (fraction of actual operating time per year) of 0.85 with fuel costing $0.70/therm. This cost could be reduced to approximately $5,000/yr with a 65% efficient heat recovery exchanger.

In Straitz (1995), the costs of thermal oxidizers with no heat recovery were given as $7,700 (plus approximately 10% for a flame arrestor) for 100 scfm of stripper off-gas; $9,400 for 200 scfm; $11,100 for 400 scfm; and $14,300 (plus approximately 20% for a flame arrestor) for 1,000 scfm. Natural gas (methane) auxiliary fuel consumption (assuming negligible fuel value in the off-gas) was given in units of standard cubic foot per hour, scfh, or equal to the off-gas scfm times 2.0 for an 1,800° F oxidizer temperature. This corresponds to a temperature rise of 1,740° F. If the temperature rise is, for example, 1,340° F, the flow rate of fuel in units of scfh would be the off-gas scfm times 2.0 times (1,340/1,740). The fuel cost per hour is the flow rate in units of scfh times the natural gas cost per scf, which is approximately equal to the flow rate in units of scfh times cost per therm divided by 95. Propane fuel consumption is given in units of scfh or equal to the off-gas times 0.82. The propane cost/hour is the flow rate in units of scfh times the propane cost per standard cubic foot.

5.11.4 Software for Stripping Process Design and Cost Estimating

The computerized estimating scheme COMPOSER GOLD marketed by Building Systems Design (Atlanta, Georgia) and the RACER/ENVEST™ computer models marketed by Talisman Partners Ltd. (Englewood, Colorado) include cost estimating for packed air strippers, air sparging systems, and vapor phase carbon adsorption. Use of ENVEST models is described as an example of how these estimating systems are used. The primary parameter values for air stripping that the ENVEST program user must furnish are:

- The number of towers and whether they are in series or parallel
- Influent flow rate in gal/min
- Whether the key contaminant volatility is low, moderate, high, or very high.

Volatilities are ranked based on values for H, as given in Table 5-6.

Table 5-6 Volatility rankings.

Relative Volatility	Corresponding H (atm-m^3)
Low	<0.0012
Moderate	0.0012 to .0144
High	0.0145 to .027
Very high	> 0.028

If the user does not set the packing height, the program computes a packing height based on these values for H. Other options allow the user to specify associated tanks, transfer pumps, piping, and chemical feed pumps for which the program will estimate costs.

The program computes a rounded value for the tower diameter using a maximum liquid loading of 20 gal/min/ft^2 if the user does not set the diameter.

Another ENVEST model estimates the cost of an oil-water separator that would be used upstream of a stripper in the event that the fluids pumped from wells contains free-phase organic products. Unless otherwise specified by the program user, the separator vessel is sized for a minimum groundwater residence time of 10 min.

The ENVEST model for in situ air sparging requires these input data:

- Soil type — coarse sand, silty sand, silt, or clay
- Aerial extent of groundwater contamination plume to be sparged
- Depth to groundwater
- Depth to the base of the contamination plume (not exceeding 65 ft more than the depth to groundwater)

Other factors affecting the estimated costs that the model takes into account are the number of weeks for startup, the number of weeks of operation, and the safety level for workers' personal protective equipment. A more accurate estimate can be produced if these secondary parameters are evaluated by the user and fed as input data, instead of the program using default values:

- **Number of sparge points.** If not included in input data, the program computes a default value by assuming well spacings.

40 ft	coarse sand
20 ft	silty sand
15 ft	silt
10 ft	clay

- **Drilling method for sparge wells.** A hollow stem auger is assumed for depths up to 150 ft, and a mud rotary for depths greater than 150 ft; air rotary is available as an option. The program defaults to 2-in. diameter sparge points, with the following borehole diameters:

8 in.	hollow stem auger
6 in.	mud rotary
6 in.	air rotary

- **Construction material for sparge wells and monitoring wells.** Default choices include schedule 40 polyvinyl chloride (PVC) for depths up to 85 ft and schedule 80 for depths greater than 85 ft. Stainless steel and other materials are user options. Connecting piping is assumed to be schedule 40 2-in. PVC aboveground; another ENVEST model for piping is available for underground installations.
- **Whether sampling while drilling is to be done.** If yes, the model assumes that hydrogeologist labor will correspond to drilling at 20 ft/hr (instead of 40 ft/hr without sampling) plus 2-hr/sparge point. Split-spoon sample collection every 5 ft is assumed. For laboratory analysis costs, the user can use the ENVEST sampling and analysis model.
- **Depth of sparge well.** The default value is contamination depth plus 4 ft, with a screen interval of 2 ft.
- **Whether drill cuttings are to be contained in drums.**
- **Air injection blower type.** An oil-free reciprocating compressor or rotary lobe blower, with four user options, as shown in Table 5-7.

Table 5-7 Blower ratings for air sparging.

Water Column, ft	Blower Rating
H < 11	98 scfm, 3.2 hp, 5 psi
11 < H < 23	170 scfm, 10.3 hp, 10 psi
23 < H < 35	163 scfm, 15 hp, 15 psi
35 < H < 65	426 scfm, 84 hp, 30 psi

The number of blowers is calculated assuming that the air flow rate per sparge point is at the high end of the 3 to 10 ft^3/min range.

Operating costs for a sparge system are based on the assumption that a crew for sampling and maintenance will visit the site once per week for the first month and once per month thereafter.

For estimating the cost of applying carbon adsorption to the stripper overhead air or to vapor extraction air, another ENVEST program is used. Values for the following items must be given:

- Air flow rate in units of ft^3/min
- Whether dual carbon beds are used, and whether the carbon canister is permanent
- The program user can exercise options involving the following:
- Use of packaged dual bed systems with 2,000-lb canisters and a 10-hp blower, or a variety of canisters ranging from 110 to 6,300 lb of carbon
- Number of canisters
- Whether carbon is virgin, reactivated, coal derived, or coconut shell derived and replaced with new disposable adsorbers
- Carbon replacement schedule (if this is not specified, the program assumes every 3 mo.)
- Use of a blower ranging from 50 to 8,000 ft^3/min, 5 to 12 in. w.c. pressure, 1/3 to 25 hp
- Use of an electric air heater rated for hazardous service for reducing relative humidity, ranging from 7.5 to 20 kW

ENVEST models are available for estimating other related costs, including sampling and analysis; blower, heater, and motor maintenance; electric power usage; clearing and grubbing; concrete slab; fencing and signage; and electric power service. The ENVEST model for soil vapor extraction can be used in conjunction with the air sparging model if vapor extraction must be used with a given air sparging application.

5.12 Summary of Important Points for Groundwater Stripping

- Target effluent concentrations from strippers that are followed by aqueous phase carbon polishing can be determined from an economic analysis or can be based on selection of a reasonable carbon life.

- The main design problem for packed towers is to determine the volumetric A/W ratio and corresponding depth of packing, and the tower cross-sectional area and pressure drop.
- A key parameter for each volatile contaminant of concern is H, a measure of volatility that increases rapidly with temperature.
- Another key parameter is G/L, the ratio of the molar gas loading per unit cross-sectional area to the molar liquid loading per unit cross-sectional area.
- A simplifying computational parameter is the stripping factor R, for each contaminant of concern. R is dimensionless and is H, in units of atm/total pressure, multiplied by G/L.
- R must be greater than 1 for each contaminant being stripped and is usually in the range of 1.5 to 5 for the key contaminant.
- The depth of packing required is equal to the height of the transfer unit, HTU, multiplied by the number of transfer units, NTU.
- A section of packing will reduce the concentration of a volatile component by a certain factor. Each successive section will further reduce the concentration by the same factor.
- HTU is proportional to the liquid loading per unit of cross-sectional area and inversely proportional to $K_L a$, the liquid film mass transfer coefficient times the specific area of the chosen packing.
- K_L is approximately equal to K_G multiplied by H and divided by the molal density of water; K_G can be obtained from packing manufacturers or from published tables. Or K_L can be determined from diffusivities and surface tension data, such as with the Onda correlations, or from pilot data.
- If pilot data are not available, a significant safety factor should be applied to the calculated HTU (or to the corresponding calculated packing depth).
- Enlarging the tower cross-sectional area reduces the HTU, the overall depth of packing, and the pressure drop.
- NTU is related to stripping factor R and the reduction in concentration for each volatile component.
- A convenient calculational procedure for packed air stripper design is to select the A/W ratio or stripping factor R, and then determine NTU; packing depth is then equal to HTU times NTU.
- The calculations can be made easier by plotting packing depth or NTU against A/W for various concentration reductions, by using published plots for NTU versus R or by using an air stripping computer program.
- Published correlations are available that relate the gas loading per unit cross-sectional area and pressure drop for traditional packings, from which the cross-sectional area can be selected for a given G/L.
- For modern packings, the cross-sectional area is often chosen so that the liquid loading per unit area is less than 1 m^3/min for each m^2 of area (or approximately 20 gal/min for each ft^2).

- Computerized correlations allow rapid checking of the effect on required packing depth and pressure drop when adjusting A/W and tower cross-sectional area for multiple components.
- Stripping towers may use trays instead of packing. Sieve trays are most common because they are less costly than other types of trays, but they do not have good air turndown capability.
- Low-profile stripping units use long liquid flow-path trays or aeration tanks with fine-bubble air diffusers.
- Cooling towers can serve as air strippers and are especially economical in situations in which organic vapor emissions do not require treatment.
- In situ stripping can be accomplished in wells or by air sparging the aquifer. Volatiles can be recovered under negative pressure with soil venting techniques.
- Air sparging may be effective in stripping dissolved organics, LNAPL, and residual DNAPL in which volatiles are the contaminants of concern.
- Air sparging may enhance biodegradation, but it may result in partial plugging of the aquifer because of the formation of biomass, ferric oxide, manganic oxide, or calcium carbonate.
- Air sparging that proves to be effective results in faster aquifer remediation than pump-and-treat systems, and it may be used as a direct cleanup technique or as a groundwater plume containment strategy.
- Sparge points generally have a much smaller radius of influence than do extraction wells, and sparge systems are usually installed in conjunction with wells or trenches for soil venting. Therefore, sparging may require a higher capital investment than pump-and-treat systems, especially for deep aquifers for which drilling is expensive.
- The common air blower arrangement for strippers is with forced-draft conventional systems in which air is blown under the packing or trays under pressure.
- Vacuum blowers can be used for negative-pressure strippers, in which the blower is located in the off-gas system and induces air to flow through the stripper.
- With a forced-draft conventional blower arrangement, a mist separator can conveniently be located within the stripper above the groundwater inlet.
- Groundwater feed pumping and off-gas emission abatement can be shutdown automatically in the event of a blower failure. Devices such as a pressure switch in an air duct can be interlocked accordingly.
- The blower should be controlled to automatically shut down shortly after any stoppage of liquid flow to the top of a stripper. Otherwise, internal wetted surfaces will rapidly go dry and become scaled.
- With packed towers, good liquid distribution at the top of the packing is vital, and with large depths of packing, redistribution part-way down the tower is needed.
- Spray nozzles are the least expensive liquid distributors, but they have higher pressure drops than do trough distributors. They also have more potential problems with nozzle erosion and plugging and have poor turndown capability.
- Spray nozzles cannot be used for liquid redistribution part-way down the tower.

- Water-recycled strippers help prevent ice formation in freezing weather, keep tower internals from going dry if groundwater flow stops, and provide constant flow, which has a number of advantages.
- Water recycling consumes extra pumping energy and does not maximize the A/W ratio.
- Heating the groundwater increases H, making stripping more effective or allowing stripping of semivolatiles.
- Steam can be used instead of air as a method of achieving heated stripping and less volume of emissions. Condensation can be used as an economical emission abatement and organics-recovery technique.
- Heating accelerates fouling.
- Effective stripping can be attained with less heating by operating strippers under partial vacuum.
- For steam stripping, the steam can be injected directly, or it can be generated from stripped water in the stripper sump by using a reboiler.
- If vapors from a stripper are not condensed, the most common emission abatement systems use either vapor phase carbon adsorption or oxidizers.
- Optimum performance with vapor phase carbon is achieved with the off-gas relative humidity adjusted to 45% at as cold a temperature as can be reasonably achieved.
- Air strippers have dilute off-gas streams, making high organics removal efficiencies difficult unless multistage adsorption is used.
- Vapor phase carbon has more adsorptive capacity than aqueous phase carbon, thereby making a stripper attractive compared to using aqueous phase carbon alone for groundwater cleanup.
- Used carbon can be shipped off-site for reactivation. It is, however, practical to regenerate vapor phase carbon on-site — usually with steam.
- Oxidizers can be direct thermal types or catalytic types.
- Direct thermal oxidizers are operated at greater than 1,200° F (649° C) or, when air pollution control regulations are stringent, at greater than 1,400° F (760° C). Destruction efficiencies greater than 99% are possible.
- Stripped chlorinated solvents form hydrogen chloride gas in oxidizers. If the HCl concentrations are high, scrubbing with water or caustic may be required.
- Catalytic oxidizers require heating the off-gas to only greater than 600° F (316° C), after which combustion and a further temperature rise occur in the presence of the catalyst.
- Destruction efficiencies with catalytic oxidizers are approximately 95%; higher efficiencies can be achieved with higher preheat temperatures and larger catalyst beds.
- Catalysts are subject to poisoning from certain heavy metals, such as lead, and some catalysts are deactivated by chlorides.
- Catalytic oxidizers use less fuel than direct thermal oxidizers. With the dilute organics concentrations present in air stripper off-gas streams, potential savings from using catalysts are significant.

- The amount of oxidizer auxiliary fuel needed depends on the concentration of organics, because the organic contaminants have fuel value.
- Heat recovery equipment can be added to oxidizers that typically reduce fuel consumption by 50% to 70%. Regenerative designs are available that reduce fuel consumption by more than 90%.
- Pilot testing of a packed stripper aids in determining the HTU and $K_L a$ values needed for full-scale design.
- Computerized programs are available that aid in the preliminary process design and cost estimating of packed air strippers, air sparging systems, and vapor phase carbon adsorption units.

Chapter 6

Aqueous Chemical Oxidation

This chapter deals with ex situ oxidation techniques for the destruction of organic compounds in groundwater, including wet air oxidation, supercritical oxidation, peroxide/Fenton catalysis, UV light with oxidants, and electrochemical oxidation. Also, some discussion is devoted to using UV light without added oxidants for sterilization of bacteria and for organics destruction in the presence of a catalyst.

6.1 Basic Principles

Organic molecules can be destroyed (converted to water and carbon dioxide or to nontoxic substances) by oxidants in the aqueous phase. Oxidants include air or oxygen, ozone, high-valence metals, and certain chemicals, such as hydrogen peroxide. Electrochemical processes can also be used. All of these oxidation schemes are enhanced if UV light is applied. The most widely used chemical oxidation scheme uses hydrogen peroxide with UV light.

6.1.1 Ranking of Oxidants and UV Oxidation Power Consumption

Oxidizing agents can be ranked by their oxidation potential or by an equivalent measure of oxidative power. Fluorine has the highest oxidative potential and is the only oxidant stronger than hydroxyl radical. Fluorine, however, is not commonly used as an oxidant in water; chlorine is. If chlorine is assigned an oxidation power value of 1, oxidants rank relative to chlorine as shown in Table 6-1.

Absolute oxidation potentials in water solution for fluorine and some common oxidants are as given in Table 6-2.

6.1.2 UV Light

UV light can be used at a mild intensity to kill bacteria or at a high intensity to destroy organic compounds. The disinfection mode is used ahead of groundwater treatment equipment, such as air strippers, to prevent biological fouling of internals and of downstream equipment. Groundwater frequently contains bacteria that grow when oxygen is added to the water, forming a biological mass that can plug equipment. Biomass formation is avoided if the groundwater is disinfected before entering a treatment system.

Table 6-1 Relative oxidative power of commonly used oxidants.

Oxidant	Relative Oxidative Power
Fluorine	2.23
Hydroxyl radical	2.06
Atomic oxygen	1.78
Hydrogen peroxide	1.31
Permanganate	1.24
Chlorine dioxide	1.15
Hypochlorous acid	1.07
Chlorine	1.00
Bromine	0.80

Table 6-2 Absolute oxidation potentials of commonly used oxidants.

Oxidant	Absolute oxidation potential (volts)
Fluorine	3.0
Hydroxyl radical	2.8
Ozone	2.1
Hydrogen peroxide	1.8
Potassium permanganate	1.7
Chlorine dioxide	1.5
Hypochlorous acid	1.5
Chlorine	1.4
Oxygen	1.2

With intense UV light, titanium dioxide can be used as a catalyst to speed destruction of organics in groundwater, with or without oxidants such as hydrogen peroxide present. Without an oxidant, most organic compounds, except saturated halogenated hydrocarbons, are destroyed by UV light with titanium dioxide catalyst.

6.1.3 Emerging Technology Using Electrochemical Oxidation

Two emerging technologies that use electrochemical oxidation techniques for treating organic contaminants are mixed oxidants (MIOX, developed by Los Alamos Technology Associates and marketed by MIOX Corp., Albuquerque, New Mexico) and mediated or catalytic electrochemical oxidation (MEO or CEO).

MIOX uses a mixture of chlorine, hypochlorite, chlorine dioxide, peroxide, and free hydroxyl radical formed by electrolyzing sodium chloride solution in a separate cell. The MIOX solution thus produced is used in low concentrations for disinfecting water. In high concentrations, either with or without UV light, the systems can be used for destroying organic compounds. However, without UV light, destruction efficiencies are not high for symmetric molecules, such as benzene.

MIOX claims that the oxidation potential for their mixed oxidant solution is approximately 1.7 times that for hydrogen peroxide solutions or 1.3 times that for hypochlorite solutions at comparable concentrations. The relatively high oxidation potential indicates the probable presence of hydroxyl radical or of higher oxidation states of chlorine. The mixed-oxidant solution shows promise of being a low-cost substitute for hydrogen peroxide solution in UV oxidation systems in which chlorine compounds are permitted for treatment.

Mediated systems apply a low-voltage electric field directly to groundwater flowing through the cell with sulfuric acid or nitric acid electrolyte. The mediating elements (catalytic agents) used are metal ions such as cobalt (3+), silver (2+), and cerium (4+). At the anode, electrolytic action restores the metal mediator from the lower-valence state to the higher-valence state, which supplies oxidative potential for organics destruction. Other reactions at the anode produce gases, such as chlorine or fluorine, from halides. Carbon dioxide is formed from organics destruction, with the oxygen coming from electrolysis of water. Hydrogen is evolved (as another electrolysis product) at the cathode, which is separated from the anode portion of the cell by a diaphragm. MEO units typically operate at 55° to 60° C, with just enough pressure to pump the water through the units. UV light may also be applied to enhance the efficiency of organics destruction.

The US Department of Energy (DOE) is leading the development of this MEO process in the United States. A demonstration unit has been in operation at the DOE laboratory operated by Battelle in Richland, Washington. Another unit is at the DOE laboratory operated by the University of California in Livermore, California. Small-scale units have been operated at the DOE installation at the Savannah River Site, Aiken, South Carolina. Full-scale units are in operation in France. A number of electrochemical processes, including mediated oxidation systems, are described by Parkinson and Ondrey (1996).

6.2 Wet Air and Supercritical Water Oxidation

At elevated pressures, water dissolves more air or oxygen than at atmospheric pressure. The higher the pressure, the more oxygen will dissolve. Also, high temperatures speed the destruction of organic compounds. Carbon dioxide is the main product of aqueous oxidation reactions; however, if the oxidation is not complete, other products form, such as acetic acid.

Wet air oxidation uses air or oxygen dissolved typically at 123 atm and at temperatures up to 578° F (303° C). It has been applied for a number years to destroy sewage treatment sludge. The systems can also be applied to wastewater with chemical oxygen demand (COD) exceeding 15,000 ppm. The systems can handle high suspended solids contents and organic concentrations up to approximately 10% by weight.

An emerging technique is supercritical water oxidation. Supercritical water oxidation uses pressures greater than 218 atm and temperatures greater than 705° F (374° C), the critical point for water. At or beyond this critical point, water exists only as one fluid phase. There is no elevated temperature at which water vapor or steam can be distinguished from liquid water when the pressure is higher than 218 atm. Typical operating conditions for supercritical water oxidation systems are a temperature greater than 450° C at 272 atm. Under such conditions, organics destruction is usually complete in 10 sec. Treatability testing is needed to determine the residence time required for complete complex organic compound destruction. The higher the temperature, the faster and more complete is organics destruction.

At supercritical conditions, oxygen and organic compounds readily dissolve in water. Oxidation of carbon and hydrogen in organic compounds proceeds rapidly. Halogens in organic compounds are converted to acids, so an alkaline agent should be added to feed streams containing halogenated hydrocarbons to neutralize these halogenated acids. Salts that form from acid neutralization and minerals in the groundwater are not soluble in supercritical water and can be removed as a solid phase.

The first commercial supercritical water oxidation system was reported in *Environmental Engineering World* (May/June 1995). It was designed for 19 L/min and operated for cleanup of laboratory wastes.

A system marketed by General Atomics (San Diego, California) pumps the contaminated water at 272 atm, injects oxygen, and heats the mix to 500° to 600° C in a reactor with 0.1 to 0.5 min residence time.

An important component of supercritical systems is the pressure letdown valve. Special erosion-resistant designs must be used. The energy released can be recovered for electric power generation or for heating purposes. As with wet air oxidation, carbon dioxide is a product. To ensure that intermediates such as alcohols and organic acids are destroyed, either the residence time can be extended or the temperature can be increased. Treated water is recycled if the feed-stream organics content is such that the feed-stream heating value is greater than 1,800 Btu/lb (4,600 kJ/kg). Thermal energy in the treated effluent can be used as steam.

With the SRI International (Menlo Park, California) process, sodium carbonate is added to counteract acid formation when treating chlorinated contaminants and to speed up the reaction. Temperatures used in the process can be as low as 380° C.

Because of the very high pressures involved, some supercritical water oxidation units are installed with secondary containment designed to withstand overpressure.

6.3 Fenton's Reagent

Although hydrogen peroxide will rapidly kill bacteria, it is very slow in destroying organic compounds unless a catalyst, ozone, or UV light is also used. The traditional catalyst used is ferrous ion, which forms free hydroxyl radical (OH*) from hydrogen peroxide, according to the reaction shown in Equation 6-1.

$$H_2O_2 + Fe(2+) \rightarrow Fe(3+) + OH^- + OH^* \qquad (6\text{-}1)$$

Free hydroxyl radicals react rapidly to convert organic compounds to carbon dioxide and water, while ferric ion is reduced to ferrous ion. The combination of hydrogen peroxide and ferrous ion is called Fenton's reagent. This reagent is not efficient for destroying saturated chlorinated hydrocarbons, saturated hydrocarbons, methylene chloride, and acetone unless UV light is also used. If UV light is not incorporated into the reaction, the process can be used without the need to pretreat to remove turbidity, oil, and grease. Bigda (1996) indicates Fenton's reagent can also be used to reduce hexavalent chromium and precipitate heavy metals, as well as to oxidize organics.

The reaction with organic compounds is exothermic, so peroxide additions should be temperature controlled. According to Plant and Jeff (1994), the optimal pH for the Fenton reaction is 3 to 4. The pH drops as the reaction proceeds. Residence time is 30 to 60 min. With batch systems, after the reaction proceeds, ferric hydroxide is precipitated by raising the pH to 9 with hydroxide addition. The ferric hydroxide can be separated, acidified, reduced, and reused in the next batch.

Bigda (1996) describes the following operating controls and conditions for a batch Fenton process:

1. Oxidation Reduction Potential (ORP) sensor for controlling hydrogen peroxide additions
2. pH sensor for controlling acid/base addition
3. Temperature monitoring and shut down at 60° C
4. Ferrous sulfate addition at a ratio of one part iron to 10 parts hydrogen peroxide
5. Control of temperature above 18.3° C, to prevent slow initiation of the reaction
6. Addition of peroxide to prevent excessive heating from the exothermic reaction

An in situ application of Fenton's reagent is reported by Bryant and Wilson (1998) for cleanup of DNAPLs derived from chlorinated solvents.

6.4 UV Light with Oxidants

An efficient method of generating hydroxyl radical is to expose peroxide and/or ozone in aqueous solution to UV light. Most UV oxidation systems operate at a wavelength of 254-nm, the wavelength of maximum adsorption of UV light by ozone. The UV light is radiated from lamps (shrouded within quartz tubes) into groundwater within a reactor. Glass tubes are not used because glass absorbs UV light.

Key factors in successful application of any UV remediation system include lamp cleanliness and lamp replacement. Suspended solids in the groundwater will interfere with light transmission through the groundwater and may foul quartz surfaces, as will scaling substances. The groundwater must be prefiltered, and it may be necessary to acidify the feed to the system to inhibit scaling. The quartz tubes must be cleaned intermittently by scraping. Lamp life may be as low as 1,000 hr in some systems. Because a typical UV system operates thousands of hours per year, a lamp replacement program must be incorporated into the standard maintenance activities for these systems.

The chemical reactions involved in a UV/hydrogen peroxide system are typified by the Equations 6-2 and 6-3.

$$H_2O_2 + UV \rightarrow 2\ OH^* \qquad (6\text{-}2)$$
$$R\text{-}Cl + OH^* \rightarrow H_2O + CO_2 + Cl^- \qquad (6\text{-}3)$$

in which R-Cl represents a chlorinated hydrocarbon. With ozone, UV forms atomic oxygen, which reacts with water to form hydroxyl radical.

Hydrogen peroxide solution can be purchased from chemical suppliers and stored in tanks or drums. However, ozone cannot be purchased or shipped because it is too unstable. It is usually generated adjacent to the reactor by passing air or oxygen through an electric corona discharge. Only 1% to 5% of the air or oxygen is converted to ozone in this process. Although ozone is many times more soluble in water than is oxygen, only 20 mg/L can generally be attained in the groundwater being treated because it is generated in such dilute form. Excess ozone is controlled with a reducing agent (e.g., bisulfite) or by catalytic decomposition to oxygen.

Adverse chemical reactions occur when certain metal ions in groundwater are oxidized along with organic compounds. Chromic ion may be converted to the more toxic chromate ion, whereas manganous ion may be converted to insoluble manganic oxide.

Sirabian, Sanford, and Barbour (1994) give pilot data indicating at least 99.7% destruction of tetrachloroethylene with a 0.6-min exposure time; 78% to 83% destruction of tetrachloroethane with a 5.5-min exposure time; and 96% to 98% destruction of methylene chloride with a 5.4-min exposure time. They recommend following the UV/peroxide unit with a polishing air stripper in order to achieve higher cleanup efficiencies for tetrachloroethane and methylene chloride.

Another combination of technologies is the use of UV oxidation plus activated carbon. Calgon Carbon (Pittsburgh, Pennsylvania) markets such a dual-technology system.

Solarchem independently markets UV/peroxide systems that incorporate a reducing agent (e.g., iodide) for dechlorination of halogenated organics. This method improves destruction efficiency and shortens exposure time requirements.

Fletcher (1991) describes a 15-gal/min UV oxidation system using ozone and peroxide with 24 65-W UV lamps. The hydraulic residence time was 40 min. Excess ozone concentrations in the off-gas of up to 20,000 ppm were reduced to less than 0.1 ppm with a nickel catalyst.

A study was conducted by the Research and Development Department of Bechtel based on data obtained from three vendors. A summary of these three vendors' experiences compiled in 1993 is given in Table 6-3. Their reported effectiveness in treating organic contaminants is given in Table 6-4.

6.5 Main System Design Parameters

A method of estimating the UV electrical power requirement for UV/oxidation systems based on information from Solarchem Environmental Systems (since acquired by Calgon Carbon, Pittsburgh, Pennsylvania) and on Bolton et al. (1992) suggests the use of an evaluation factor EE/O, electrical energy per order. EE/O is the electrical energy required to reduce the concentration of a contaminant by one order of magnitude in 1,000 gal, calculated as shown in Equation 6-4.

$$EE/O = (kW/gal/hr) \ (1{,}000 \ gal/\log[c_{in}/c_{eff}]) \tag{6-4}$$

in which c_{in} is the initial contaminant concentration and c_{eff} is the effluent concentration. The term is applied, as given in Equation 6-5.

$$kW = (UV \ dose) \ (gal/hr)/1{,}000 \tag{6-5}$$

in which the UV dose is EE/O $[\log(c_{in}/c_{eff})]$. The units for UV dose are kW-h/1,000 gal.

The EE/Os for some typical groundwater contaminants range up to the following values:

- 3 for vinyl chloride
- 4 for trichloroethylene (TCE)
- 5 for mono-aromatic hydrocarbons, chlorobenzene, phenol, and dichloroethylene (DCE)
- 8 for tetrachloroethylene (PCE)
- 15 for chloroform and dichloroethane (DCA)

For wastes with multiple contaminants, the highest UV dose required for the removal of any one contaminant is used for design purposes. EE/O UV dose values are not added for systems with multiple contaminants.

Table 6-3 Key UV oxidation system vendor experience.

Company	Site	Scale	Capacity (gpm)	Contaminant Source or Waste Type	Influent Contaminant	Effluent Concentration
Ultrox Note 2	Sealed Power Technologies Muskegon, MI	full	200	Process Water	3-4 ppm total VOCs including TCE	2 ppb TCE
	Koppers Corporation Denver, CO	full	10	Wood Treatment Process Water	0.35-1.35 ppm pentachlorophenol	POTW discharge
	Xerox Corporation Rochester, NY	pilot	50	Process Water	3-5 ppm total VOCs	>99% reduction of TCE
	EPA SITE Demonstration Lorentz Barrel & Drum San Jose, CA	pilot	10	Groundwater	Including 120-170 ppb VOCs: TCE; 11,1-DCA; 1,1,1-TCA	12-20 ppb
Peroxidation Systems, Inc. Note 1	Sacramento Army Depot Sacramento, CA	full	100-400	Groundwater	TCE, PCE, DCE, DCA, and chloroform	<5 ppb
	Rocketdyne El Segundo, CA	batch	60	Groundwater	10 ppm VOC	1.2 ppm
	Lawrence Livermore National Laboratory Livermore, CA	pilot	50	Groundwater	<1 ppm VOCs	15 ppb
	ERM West New Jersey	full	50	Groundwater	1-2 ppm TCE	Non-detectable
	Lorentz Barrel & Drum San Jose, CA	full	100	Groundwater	250-1,000 ppb TCE; 1,1-DCA; 1,1,1-TCA	Surface water discharge
Solarchem Note 1	Nestle Foods New York	pilot	10	Process Water	15 ppm TCE, methylene chloride, 1,2-DCE	>99% reduction of TCE
	Waterloo Regional Water District Elmira, Ontario, Canada	full	600	Groundwater	20 ppb dimethyl nitrosamine	0.003 ppb
	Uniroyal Chemical Elmira, Ontario, Canada	pilot	50-150	Process Water	300 ppb dimethyl nitrosamine	<0.25 ppb

Note 1. Peroxidation Systems and Solarchem have been acquired by Calgon Carbon Corporation (Pittsburgh, Pennsylvania).
Note 2. Ultrox has been acquired by US Filter (Santa Clara, California).

Table 6-4 The effectiveness of UV/hydrogen peroxide, ozone oxidation technologies in the treatment of contaminants.

Company	Contaminant	Influent (µg/L)	Effluent (µg/L)
Ultrox Note 2	Total VOCs	170	16.0
	Phenol	6,000	<100
	TCE	65	1.2
	Chlorinated VOCs	6,000	<5
	1,1-DCA	11	5.3
	1,1,1-TCA	4	0.75
	BTEX	12,500	<5
Peroxidation Systems, Inc. Note 1	Naphthalene	891	<2
	Acenaphthene	205	<0.1
	1,2-Dichloroethene	5,500	0.20
	Methylene chloride	1,100	8.0
	Dichloropropane	110	4.0
	Trichloroethylene	200	<1
	Benzene	5,140	0.50
	BTEX	33,500	2.0
Solarchem Note 1	TCE	110,000	<5
	1,4-Dioxane	120,000	<2
	PCBs	300	<.02
	Pentachlorophenol	8,000	<1
	Total petroleum hydrocarbon	15,000	<1
	Total BTEX	5,000	<1
	Trichloroethane	2,000	<5
	Dichloroethane	1,000	<5
	Chloroform	400	<10
	Atrazine	5,000	<20
	Phenol	1,000	<1
	PAHs	2,000	<1
	Freon-112	80	<2
	Acetone	62,000	<5
	Nitrosamines	20	<0.003

Note 1. Peroxidation Systems and Solarchem have been acquired by Calgon Carbon Corporation (Pittsburgh, Pennsylvania).
Note 2. Ultrox has been acquired by US Filter (Santa Clara, California).

The following is an example for 50 gal/min (3,000 gal/hr) of groundwater containing 700 µg/L of TCE, to be remediated to 5 µg/L. The EE/O given for TCE is 4 kW-h/1,000 gal treated. The UV dose is 4 [log(700/5)] = 8.585 kW-h/1,000 gal. Applying Equation 6-5, the UV electric power consumption is calculated as shown in Equation 6-6.

(UV dose) (gal/hr)/1,000 = 8.585 (3,000)/1,000 = 26 kW (6-6)

Besides light intensity, which is affected by turbidity and lamp cleanliness and is a function of electrical power input, key factors important to design and operation include:

- Distance between the UV source and the dissolved organic molecules in groundwater flowing past the lamps
- Peroxide or ozone dosage
- Percentage of organic compounds that are saturated (for example, unsaturates such as TCE oxidize more readily than trichloroethane (TCA))
- pH (raising the pH provides more hydroxyl ions that can become hydroxyl radicals; lowering the pH removes bicarbonate and carbonate, which can consume oxidants)
- Exposure time, a function of groundwater throughput and reactor volume.

The minimum oxidant dosage can be calculated from stoichiometry. In practice, the oxidant dosage is best determined by performing treatability studies on samples of the groundwater, because organic substances besides the contaminants and reducing agents in the groundwater consume oxidant. Also, the approach to stoichiometric equilibrium depends on exposure time. If peroxide is the only added oxidant, the peroxide dose required is twice the COD. It is also good practice to maintain at least 25 ppm of residual peroxide if it is used alone for contaminant oxidation. Treatability studies help determine the optimum pH and the exposure time required with UV/oxidation systems to achieve the desired degree of contaminant reduction for a specific waste stream.

6.6 Treatability Studies for Aqueous Oxidation

When using Fenton's reagent, bench-scale tests are used to determine the amount of acid needed to attain a pH of 3 to 4 and how much caustic is needed to later raise the pH to 9. The tests are also used to check additions of coagulant aid needed for good separation of the ferric hydroxide precipitate. If alkanes or chlorinated saturated hydrocarbons are present, UV light augmentation should be evaluated.

For UV/oxidation systems it is essential that the required exposure time be determined. Sirabian, Sanford, and Barbour (1994) indicate that contaminant reduction is related to exposure time, as shown in Equation 6-7.

$$C_{in}/C_{eff} = e^{-kt} \qquad (6\text{-}7)$$

If t is in minutes, the rate constant k is in 1/min and can be determined from the slope of $\ln(C_{in}/C_{eff})$ plotted against time from pilot data. For testing UV/oxidant treatment, Calgon Carbon (Pittsburgh, Pennsylvania) can run batch tests with drum-size samples in their laboratory. They also sell a skid-mounted flow-through unit for field use. Because UV system performance is highly dependent on clarity of the water, a significant portion of laboratory testing may be devoted to determining pretreatment requirements. A series of tests are run with varying hydrogen peroxide or ozone dosages and UV dosages, both with and without catalyst. If chlorinated hydrocarbons are contaminants of concern, Solarchem can test the use of a reducing agent that will break the chlorine-hydrocarbon bonds to improve the rate of destruction and decrease the amount of electrical power consumption needed for their destruction.

If use of ozone is a consideration, adjustment of pH should be tested also. Ozone will usually work better at a high pH. UV/peroxide treatment is generally more effective at an acidic pH. With some groundwaters, it may be beneficial to use ozone with the pH adjusted to 9 to 12 in a pretreatment stage without UV light.

Tests using UV light, peroxide, and ozone combinations can also be run by Ultrox, which has been acquired by US Filter (Santa Clara, California). They have bench-scale laboratory equipment and a portable unit available for field pilot testing.

If highly volatile organics are present, a fraction of these compounds will change from the dissolved phase to the vapor phase from mixing and handling in a batch reactor. If there is a potential for such volatilization, a control "mixing only" test should be run without UV or chemical addition. This allows the test results to be corrected for the amount of volatilization that occurs during sample handling.

The inorganic chemistry of the groundwater can severely affect the UV and chemical dosages required for efficient contaminant destruction. Chloride, nitrite, sulfite, sulfide, and carbonates all will compete with organic molecules in consuming hydroxyl radicals that are needed for successful treatment. Laboratory batch tests can help in selecting optimum combinations of pretreatment, oxidants, catalyst, and pH conditions.

6.7 Costs for Aqueous Oxidation

Wet air oxidation systems are used for organic sludge destruction and can be applied to the cleanup of groundwater and soil wash water. Based on information from Zimpro Environmental (Rothschild, Wisconsin), investment costs are as given in Table 6-5. Utilities cost $0.01/gal, mainly for compressing air.

Table 6-5 Costs of wet air oxidation systems.

System Capacity (gal/min)	Cost ($ million)
5 to 10	2 to 3.5
50	7
150	15

For batch systems using Fenton's reagent, Bigda (1996) reports basic equipment costs for a 1,000-gal unit at $40,000 to $75,000; add 30% for a 2,000-gal unit. Costs for groundwater pumps, filters, and groundwater storage tanks are extra. Labor and chemical costs for operations for two case histories were $0.14/gal and $0.48/gal, respectively.

A study conducted by the Research and Development Department of Bechtel, based on data obtained from three vendors, indicated capital investment costs for 25-gal/min units are approximately $15,000/gal/min. Present-worth costs for a 20-yr life cycle, including investment, electric power, chemicals, operations, and maintenance, would be in the range of $0.008 to $0.0134/gal treated, at a discount rate of 8%.

UV/peroxide systems are widely used for groundwater remediation, and a method of estimating equipment investment costs and operating costs are outlined, based on information from Solarchem Environmental Systems (Markham, Ontario and Nashville, Tennessee).

- UV electrical power consumption is obtained from Equation 6-5, in kW.
- The equipment cost in 1995 dollars ranges up to the value calculated from Equation 6-8.

$$\text{Equipment cost, 1995\$} = (1{,}160)(kW) + 52{,}000 \tag{6-8}$$

- The yearly UV electrical power cost is the number of kW multiplied by the number of hours per year anticipated for operation times the utility price for power, in $/kW-h.
- Lamp replacement costs typically range up to 50% of the electrical cost.
- Hydrogen peroxide costs range up to $0.008/ppm of peroxide for each 1,000 gal treated. The ppm peroxide concentration required is taken as the greater of 25 ppm, or twice the concentration of COD.

The example begun in Section 6.5 for 50 gal/min (3,000 gal/hr) of groundwater containing 700 μg/L of TCE to be remediated to 5 μg/L is continued below to estimate the cost of such a system.

Given: Flow rate is 3,000 gal/hr; C_{in}/C_{eff} is 700/5, or 140; electric power cost at this location is $0.07/kW-h

Assumption: Equipment is to run 80% of the hours in a year.

Find: Equipment cost, yearly electric power cost for UV lamps, yearly lamp replacement cost, and yearly hydrogen peroxide cost

Step 1: The UV electric power consumption from Equation 6-6 is 26 kW.

Step 2: Applying Equation 6-8, the equipment cost is calculated, as shown in Equation 6-9.

$$1{,}160 \text{ kW} + 52{,}000 = 1{,}160(26) + 52{,}000 = \$82{,}000 \tag{6-9}$$

Step 3: If the equipment is expected to run for 80% of the hours in a year and the utility price for industrial electric power is $0.07/kW-h at this location, using the power consumption determined in Step 1 gives the annual cost for UV power, as shown in Equation 6-10.

26 kW (0.80) (8,760 hr/yr) ($.07/kW-h) = $12,800/yr (6-10)

Step 4: Lamp replacement costs are estimated from item D above at 50% of power cost, or $6,400/yr.

Step 5: The groundwater COD is not given, so the hydrogen peroxide concentration requirement is taken from item E to be 25 ppm, and the cost is therefore 25 x ($0.008) for each 1,000 gal treated. The annual cost for hydrogen peroxide at 3,000 gal/hr is calculated, as shown in Equation 6-11.

25 ($0.008/1,000 gal) (3,000 gal/hr) (0.80) (8,760 hr/yr) = $4,200/yr (6-11)

Note that in this example, the estimated investment cost does not include costs for site preparation, foundations, site access, fencing, extraction wells, groundwater storage tanks, pumps and instrumentation, piping, disposal facilities or injection wells for treated water, electrical service, lighting, structures, enclosures, permits, and engineering. The annual costs do not include costs for maintenance other than lamp replacements, operator labor, supervision and overhead, supplies, pumping and lighting energy costs, disposal costs and fees, monitoring, and taxes and insurance.

If ozone is to be used in conjunction with a UV oxidation system, the cost of an ozone generator adds to the equipment cost. Nelson and Brown (1994) report the cost of ozone generators as given in Table 6-6.

Table 6-6 Costs of ozone generators.

Capacity (lb Ozone/day)	Cost
<10	$10,000 to $20,000
20+	>$75,000

Fletcher (1991) reports treatment costs for a UV/oxidation system using ozone and peroxide as little as $0.10/1,000 gal, with costs being much higher with hard water.

6.8 Summary of Important Points for Aqueous Chemical Oxidation

- A variety of aqueous oxidation schemes have been used with and without UV light. UV light enhances organics destruction with any aqueous oxidation scheme.
- At elevated pressures and temperatures, high concentrations of air or oxygen can be dissolved in water, and organics can be destroyed relatively rapidly.

- At influent organic concentrations greater than approximately 5 wt%, an explosion can occur in a wet air oxidation reactor, so treated effluent should be recycled to dilute the influent.
- At supercritical conditions, organic compounds become miscible with water and can be destroyed more rapidly than at subcritical conditions.
- The letdown valve in a supercritical system is subject to erosion.
- Hydrogen peroxide reacts very slowly to destroy organics unless a catalyst, ozone, or UV light is used.
- Hydroxyl radical is the oxidant that has the highest oxidation potential among oxidants used for waste treatment, but it must be formed from other oxidants or from electrolysis.
- Fenton's reagent uses hydrogen peroxide with an iron catalyst, with or without UV light.
- Electrochemical oxidation systems are being developed that electrolyze brine to form oxidants or directly electrolyze solutions with a metal catalyst present.
- UV light is used at a mild intensity to kill bacteria or at a high intensity to destroy organic compounds. The most effective UV systems for organics destruction use a catalyst (e.g., titanium dioxide) and/or oxidants.
- The most widely used oxidants with UV are hydrogen peroxide and ozone. Either oxidant, or both together, can be used.
- UV systems work best with filtered (nonturbid) water. Lamps must be cleaned of scale, and they need periodic replacement to maintain the treatment efficiency of UV systems.
- UV light is absorbed by oxidants that form hydroxyl radical, a highly reactive oxidant.
- Key factors affecting UV oxidation system effectiveness include the distance between the UV source and dissolved organics, the oxidant dosage, the degree of organics unsaturation, the pH, and the exposure time.
- If ozone is used, it must be generated on-site.
- Costs of equipment, electric power, and lamp replacement in a UV/peroxide system can be roughly estimated as a function of kilowatts of energy consumed.

Chapter 7

Bioremediation Systems

This chapter describes the use of biological systems for the treatment of groundwater and soil contaminated with organics or nitrates. Topics covered in this chapter include fundamentals of microbial growth and metabolism; environmental conditions that enhance bacterial growth and contaminant degradation; in situ versus ex situ treatment; acclimation versus bioaugmentation; aqueous and solid phase treatment systems; system design parameters; treatability studies; and costs. Most of the discussion focuses on the use of aerobic bacteria for the removal of biodegradable organic compounds, as this is the most commonly used biological system for both soil and groundwater treatment. Some discussion is also devoted to anaerobic bacterial degradation, as many impacted natural environments are rapidly depleted of oxygen. In addition, much recent activity has been focused on developing quantitative descriptions of natural processes taking place in contaminated soil and groundwater under the label of *natural attenuation*. The primary mechanism for contaminant removal during natural attenuation at most sites is via anaerobic processes.

7.1. Basic Principles

Bioremediation encompasses all of the biologically mediated processes that result in the destruction or transformation of hazardous organic materials into less complex and more environmentally benign byproducts. These biological treatment processes use the complex enzyme systems of microorganisms in either ex situ or in situ configurations to carry out the desired biodegradation reactions. Attempts are made to maintain these systems under aerobic conditions so that the microorganisms contained within them are encouraged to utilize the hazardous material as a source of energy to grow on. This results in the desired conversion of waste material to carbon dioxide, water, other oxidized inorganic compounds, less complex byproducts, and microbial biomass. These hazardous compounds are used as an energy and carbon source for the microorganisms (electron donor). Oxygen, a source of electron acceptor, is supplied to produce biological degradation reactions that proceed at accelerated rates.

Some microorganisms are also capable of supporting the indirect degradation of hazardous compounds through a process called cometabolism. Cometabolism occurs when nonspecific enzymes produced by an organism to degrade one compound are also capable of degrading other compounds with similar structure, even though the organism derives no energy or carbon from these other compounds. A prime example of such a biological reaction is the oxidation of TCE by methane-oxidizing bacteria.

A variety of engineered biological systems for soil, air, and groundwater treatment have been developed and are summarized in Figure 7-1. Biological systems have advantages over other types of treatment technologies in that if they are applicable to a given site/soil/waste situation.

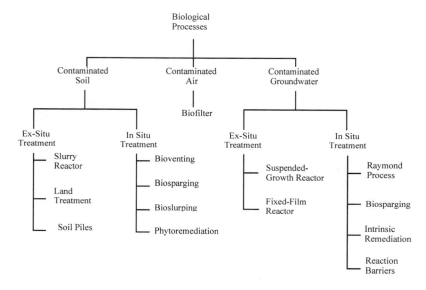

Figure 7-1 Biological treatment processes that can be used for contaminated soil, air, and groundwater remediation (adapted from US EPA, 1989).

- They lead to the destruction of the contaminants of concern under ambient pressure and temperature conditions.
- They are highly cost-effective systems because of the low temperatures and pressure reaction conditions under which they operate.
- They generate no chemical sludges and generally require minimal chemical addition to maintain operating conditions at optimal levels.
- No solid residues are produced if the processes are conducted in situ.
- Advances in process design have extended the applicability of biological reactors to soils, slurries, water, and gas phase treatment.

Because they are biological in nature, however, their applicability is limited to:

- Those sites where the contaminants of concern are biologically degradable, i.e., the contaminants must be able to serve as a source of energy and carbon for the cells to grow or they must be capable of cometabolic degradation
- Those sites where toxicants or chemicals that can inhibit microbial activity are not present, because these inhibitory materials include high concentrations of some organic compounds and moderate to high concentrations of a variety of heavy metals

Biological processes are applicable for a broad spectrum of organic chemicals of environmental concern. Table 7-1 presents some common chemical classes of concern that have been shown to be susceptible to biodegradation. It should be noted that high concentrations (greater than approximately 1 wt%) of otherwise biodegradable compounds may not be amenable to biodegradation because of toxicity and/or inhibition that develops at these high contaminant levels. When free product exists, some inhibition can be expected, and product recovery is highly recommended to enhance the potential performance of bioremediation schemes.

7.1.1 Microbial Metabolism

Biological treatment systems incorporate indigenous microorganisms into the treatment scheme through modifications of the existing groundwater and soil environment to encourage their growth and reproduction. In the process of growth, microorganisms require a source of energy (electron donor) and a means of extracting this energy from the electron donor via an appropriate electron acceptor. The microbial degradation process is conceptually shown in Equation 7-1.

Microbes + electron donor + nutrients + electron acceptor $\rightarrow$
more microbes + oxidized end products   (7-1)

In the application of biological systems for waste treatment, indigenous microorganisms are encouraged to use the waste contaminants of concern as the electron donor, while they are supplied with the electron acceptors and nutrients that they require.

In the general application of biological treatment for waste destruction, the limiting factor in full-scale engineered systems is the rate of transfer of the electron acceptor to the reaction site. Electron acceptors are chemicals that can be utilized by biological systems for the extraction of energy from electron donors so that energy is available for cell replication and growth. The main electron acceptors of interest include the following compounds in order of relative energy yield to the microorganisms: oxygen, nitrate, iron, manganese, sulfate, carbon dioxide, and organic carbon. Oxygen

Table 7-1 Biodegradable chemical classes of concern that have been shown to be susceptible to biodegradation. (Numbers indicate sources of biodegradation information).*

Compound Type	Aerobic	Anaerobic	Cometabolism
Straight-Chain Alkanes	2		3
Branched Alkanes	2		
Halogenated Aliphatics	1^a	1, 3	1 (TCE), 3
Esters, Glycols, Epoxides	4	4	
Alcohols, Aldehydes, Ketones	1	1	3
Carboxylic Acids, Amides, Esters	1	1	
Nitriles, Amines, Pthalate Esters	4	4	3
Nitrosamines		4	
Cyclic Alkanes	2		3
Nonhalogenated Aromatics	1	1, 3	3
Halogenated Aromatics	1	1^a Anoxic, 1^b	3
Simple Aromatic Nitro Compounds	4	4	3
Aromatic Nitro Compounds with Other Functional Groups	4	4	3
Phenols	1	1, 3	3
Fused-Ring Hydroxy Compounds	4		
Nitrophenols		3	3
Halophenols	2	2, 3	3
Phenols-Dihydrides, Polyhydrides	4		4
2-, 3-, 4-, & 5-Ringed Fused Polycyclic Aromatic Hydrocarbons	1		
Biphenyls	4		
Poly Chlorinated Biphenyls	2^a	2^b, 3	1
Organophosphates	4	4	
Pesticides and Herbicides	3	2	3

* Biodegradable does not indicate complete mineralization nor does it infer the rate nor extent of degradation. It indicates the potential for loss of parent compound under the conditions listed.

a - Less Chlorinated
b - More Chlorinated

1. Anderson, W. C. 1995. Innovative site remediation technology, Bioremediation, AAEE.
2. Chapelle, F. H. 1993. Ground-Water Microbiology & Geochemistry. John Wiley & Sons, Inc. NY. 424 pp.
3. Alexander, M. 1994. Biodegradation and Bioremediation. Academic Press, San Diego, CA. 302 pp.
4. US EPA. 1985. Seminar publication. Corrective action: technologies and applications. Center for Environmental Research Information, Cincinnati, OH. EPA/625/4-89/020.

is the preferred electron acceptor, as it results in a maximum energy yield to the microorganism, producing the maximum amount of cell production and organism growth per unit amount of electron donor utilized.

The relative energy release from each of the electron acceptors in order of decreasing energy yield is oxygen > nitrate > iron/manganese >> sulfate > carbon dioxide/organic carbon. In addition to yielding higher energy as compared to the other terminal electron acceptors (TEAs), a wide variety of chemicals are degraded using oxygen as an electron acceptor. In addition, oxygen generally yields a more rapid rate of contaminant degradation, and it produces oxidized end products that can be safely released into the environment. Oxygen-based biological systems are also the most preferred engineered systems because of their inherent stability and efficient process performance.

7.1.2. System Environmental Requirements

7.1.2.1 Microbial Populations

An acclimated indigenous population of microorganisms capable of degrading the compounds of interest must exist at the site if bioremediation is to be successful. A wide variety of organisms have been isolated from the environment with the capability to degrade a wide range of chemicals of concern, as indicated in Table 7-1. It is highly unlikely, then, that the addition of specially cultured organisms will be necessary at most contaminated sites. The delivery and control of microbial amendments are difficult at best, and microbial augmentation should not generally be considered, particularly for in situ treatment options. If an active population of indigenous microorganisms does not exist at a site, inhibitory and/or toxic conditions should be suspected, and alternatives to bioremediation should be considered.

7.1.2.2 Oxygen

As indicated, oxygen is the preferred electron acceptor and is necessary for aerobic biodegradation of organic contaminants. Residual oxygen concentrations of greater than 1.0 mg/L in the aqueous phase and greater than 2 vol% in the gas phase for vapor systems should be maintained to ensure that oxygen is not limiting overall microbial reaction rates. Oxygen transfer rates normally do limit overall contaminant removal rates and remediation system performance, and much of the development of remediation technologies in recent years has focused on improving oxygen supply to contaminated soil and groundwater. Oxygen transfer efficiency is low because of the low aqueous solubility of oxygen in water (approximately 8 mg/L at 20° C) and the short air/water or air/soil contact times that often occur at contaminated sites, particularly where in situ treatment techniques are being applied. In addition to oxygen transfer efficiency limitations, relatively large oxygen demands expressed by contaminants of concern also limit the ultimate efficiency of aerobic biological systems. As shown in Equation 7-2 for the oxidation of benzene to carbon dioxide and water, 7.5 gmol of oxygen are required for the oxidation of 1 gmol of benzene. Expressed on a weight basis this is 7.5 (32)/78, or 3.1 g oxygen/1 g benzene. This oxygen equivalent of 3 g oxygen/g organic contaminant can be used as a rule of thumb when estimating oxygen supply requirements.

$$C_6H_6 + 7.5\ O_2 \rightarrow 6\ CO_2 + 3\ H_2O \tag{7-2}$$

As an example of the magnitude of oxygen transfer requirements at contaminated sites consider the following example:

Given: An aquifer is contaminated with 100 gal of a compound (specific gravity is 0.8) that is known to be biodegradable under aerobic conditions. An in situ bioremediation system (the Raymond System, discussed in Section 7.2.2.1) is being considered, which would consist of extraction wells and an above-ground aeration system to oxygenate and nutrient-amend the extracted groundwater before it is reinjected through an infiltration gallery up-gradient of the source material.

Find: The volume of oxygen and nutrient-amended groundwater that must be applied to the contaminated site to aerobically degrade the contaminant according to the stoichiometry given in Equation 7-2.

Step 1: The volume of oxygen-saturated, air-sparged groundwater required to transfer 1 lb of oxygen is determined from Equation 7-3, realizing that 1 gal of water weighs 8.34 lb and 1 gal/MG is equivalent to 1 ppm, or 1 mg/L.

lb O_2 = (O_2 concentration, mg/L) (volume, MG) [(8.34 lb/MG)/(1 mg/L)] (7-3)

in which MG is million gallons, or:

1 lb oxygen = (8 mg/L) (X MG) [(8.34 lb/MG)/(1 mg/L)] (7-4)
X MG = (1)/[(8) (8.34)] = 0.015 MG = 15,000 gal (7-5)

This indicates that approximately 15,000 gal of oxygen-saturated groundwater (8 mg/L) must be pumped to supply 1 lb of oxygen to the contaminated aquifer.

Step 2: Using the equivalent oxygen demand of 3 lb oxygen/lb organic contaminant required for biological oxidation from Equation 7-2, determine the required aerated groundwater volume that must be pumped to the contaminated aquifer for each pound of contaminant biodegraded.

volume/lb contaminant biodegraded = (3 lb O_2/lb contaminant) (15,000 gal/lb O_2)
= 45,000 gal/lb contaminant biodegraded (7-6)

Step 3: Calculate the aerated groundwater volume requirement for a spill of 100 gal.

lb of contaminant = (100 gal) (8.34 lb water/gal water) (0.8, specific gravity)
= 667 lb contaminant (7-7)
Groundwater volume = (667 lb contaminant) (45,000 gal/lb contaminant)
= 30,015,000 gal of aerated groundwater (7-8)

Note: This value assumes that 100% utilization of the transferred oxygen takes place within the aquifer. This will never be the case, making the required groundwater volume even larger.

These calculations underscore the significance of oxygen demand and resultant treatment volumes that must be managed to stabilize contaminated media when contaminant levels are high.

7.1.2.3 Soil Water

This parameter is important in soil-based systems, as the microorganisms rely on soil water to supply a habitat for their growth and survival. Soil water also provides a medium for transfer of contaminants from the product or solid phases to the microorganisms. Soil water should be in the range of 25% to 90% field capacity (the water content of soil after it freely drains by gravity) to sustain microbial activity. Optimal soil water content is generally found in the range of 55% to 90% field capacity (AAEE, 1998).

7.1.2.4 pH

pH is a measure of a solution hydrogen ion concentration and should be about 7 (neutral pH) for optimal biological treatment performance. pH within the range of 5.5 to 8.5 will support biological activity but should be adjusted to 7, if possible, to improve process performance. The amount of neutralizing agent that must be added to change groundwater pH cannot be predicted without knowing either the complete water chemistry or running laboratory experiments.

Acids that can be used include phosphoric, sulfuric, and hydrochloric. Bases include sodium hydroxide (caustic), lime, and potassium hydroxide (caustic potash). Other compounds that affect the solution pH by their buffering action include sodium carbonate, sodium bicarbonate, sodium dihydrogen phosphate, and disodium biphosphate. It is difficult to modify soil and groundwater pH in situ, so pH levels outside of the acceptable range can be used as a primary indicator of the feasibility of site bioremediation based on initial site assessment data.

7.1.2.5 Nutrients

Nutrients can be classified into macro, minor, and micro element groupings. The macronutrients of concern for bioremediation include fixed nitrogen and phosphorous, while the minor nutrients include sodium, potassium, calcium, magnesium, iron, chloride, and sulfur. The macronutrients are required at order of magnitude higher levels than the minor and micronutrients and subsequently are the nutrients of concern in terms of managed bioremediation. A typical carbon/nitrogen/phosphorus (C/N/P) ratio of approximately 100/10/1 on a weight basis is often utilized to ensure that adequate levels of nitrogen and phosphorus exist for unhindered bioremediation. The nitrogen and phosphorus requirements described by this ratio are approximately half of those found in cell material (generally estimated to be $C_5H_7O_2N$, for which the C/N/P is approximately 50/10/1) because of the assumption that half of the carbon in the contaminant is used to produce cell material, whereas the balance is used for energy production by the cells. One additional note is important. For in situ soil remediation systems (i.e., bioventing, natural attenuation, etc.), nutrient addition has not been shown to improve bioactivity, and it is generally thought that nutrient addition for these systems is unnecessary.

An example of the macronutrient requirements for the ex situ treatment of groundwater is presented below.

Given: A groundwater contains 15,000 μg/L of benzene, 2,500 μg/L of toluene, 0.11 mg/L of nitrogen, and 0.09 mg/L of phosphorus

Find: Nitrogen and phosphorus additions required to bring the C/N/P ratio to 100/10/1

Step 1: Determine the carbon content of the groundwater, with carbon having an atomic weight of 12 g/gmol.

Compound	mg/L	Formula	Molecular Weight	C Portion	C Content, mg/L
Benzene	15.0	C_6H_6	78	(6)(12) = 72	(72/78)(15) = 13.85
Toluene	2.5	C_7H_8	92	(7)(12) = 84	(84/92)(2.5) = 2.28

Total mg C/L = 13.85 mg/L + 2.28 mg/L = 16.1 mg/L   (7-9)

Step 2: Determine the total nitrogen concentration required. The nitrogen added must be in a form other than N_2 as is present in air. If, for example, ammonium nitrate (NH_4NO_3) is added so that a C/N ratio of 100/10 is to be achieved, the ammonium nitrate addition rate per liter of groundwater can be determined as follows, with nitrogen having an atomic weight of 14 g/gmol and ammonium nitrate having a molecular weight of 80 g/gmol:

C/L groundwater, from the above calculation = 16.1 mg/L   (7-10)
N/L groundwater = (10/100)(16.1 mg) = 1.61 mg/L   (7-11)
Nitrogen fraction of ammonium nitrate = 2(14/80) = 0.35   (7-12)
Ammonium nitrate/L groundwater = 1.61/0.35 = 4.6 mg/L   (7-13)

Step 3: Determine the net nitrogen concentration required. The nitrogen requirement should be reduced by the amount of nitrogen that already exists in the groundwater:

Net N/L groundwater = 1.61 mg/L - 0.11 mg/L = 1.5 mg/L   (7-14)
Net ammonium nitrate/L groundwater = 1.5/0.35 = 4.3 mg/L   (7-15)

Step 4: Determine the total phosphorus concentration required. If trisodium phosphate (TSP) (Na_3PO_4) is added so that a C/P ratio of 100/1 is to be attained, the TSP addition per liter of groundwater can be determined as follows, with phosphorus having an atomic weight of 31 g/gmol and TSP having a molecular weight of 164 g/gmol.

C/L from the first calculation = 16.1 mg/L   (7-16)
P/L groundwater = (1/100) (16.1) = 0.16 mg/L   (7-17)
Phosphorus fraction of TSP = (1) (31)/(164) = 0.19   (7-18)
TSP/L groundwater = 0.16/0.19 = 0.85 mg/L   (7-19)

Step 5: Determine the net phosphorus concentration required. As with the nitrogen requirement, the phosphorus requirement should be reduced by the amount of phosphorus that already exists in the groundwater. With phosphate having a molecular weight of 95 g/gmol:

Phosphate portion of TSP applied = (95/164) (0.85 mg/L) = 0.49 mg/L   (7-20)
Net phosphate/L groundwater = 0.49 mg/L - 0.09 mg/L = 0.40 mg/L   (7-21)
Net TSP/L groundwater = (0.40 mg/L) (164/95) = 0.69 mg/L   (7-22)

In actual practice, excess phosphate is added so that some residual phosphate stays in solution, measured analytically as orthophosphate. The excess makes up for phosphate losses, such as from chemical precipitation.

7.1.2.6 Temperature

Biological systems can operate over a wide range of temperature conditions from 5° to 60° C. Three temperature ranges have been identified based on the growth of distinct groups of microorganisms: psychrophilic, less than 15° C; mesophilic, 15° to 45° C; and thermophilic, greater than 45° C. In general, most contaminated site conditions result in mesophilic temperatures, which are adequate to support active microbial growth. If systems are exposed to temperatures below 10° C for extended periods of time during winter months, their performance should be expected to deteriorate until temperatures are raised again. A rule of thumb suggests that reaction rates will increase or decrease by a factor of 2 for each 10° C rise or fall of temperature. It is interesting to note, however, that soil-based treatment systems in northern climates have not shown a significant lag period in system performance when temperatures are raised, indicating that summertime performance should rapidly resume as soil/groundwater temperatures are raised within the mesophilic range.

7.1.2.7 Toxicants in Waste

Because of the biological nature of bioremediation systems, any material that disrupts biochemical processes taking place within the microorganisms used in the treatment system will cause a disruption and, often, eventual failure of that system. A variety of organic and inorganic toxicants can adversely affect the biological treatment system. The microbial consortium existing within the biological treatment system can acclimate to some of these materials, and, by design (i.e., blending of contaminated soil with adjacent uncontaminated soil in a soil pile or land farm system), concentrations of the toxicants can be reduced below inhibitory levels to allow their eventual degradation over time. Because of the site-specific nature of toxicity, procedures for field-scale toxicity assessments will be discussed later in the chapter.

7.1.3 In Situ Versus Ex Situ Treatment

In situ biological treatment systems include those groundwater or soil systems in which remediation and contaminant removal/destruction are conducted in place without the removal of soil and/or water. The addition of exogenous agents takes place within these systems, but contaminated material remains in the location in which it was found during the site assessment process. In situ groundwater treatment systems include (Figure 7-1) the Raymond process (extraction of groundwater and reinjection of a portion of the nutrient- and electron acceptor-amended groundwater via injection wells or infiltration galleries), biosparging (air injection below the water table at low rates to maximize oxygen transfer and minimize air stripping of contaminated groundwater), intrinsic remediation, and biologically active reaction barriers. In situ unsaturated zone treatment systems include (*see* Figure 7-1) bioventing, biosparging, bioslurping, and phytoremediation.

Ex situ biological treatment systems include those groundwater or soil systems in which remediation and contaminant removal/destruction are conducted after the removal of soil and/or water from the location it was found during the site assessment process. Ex situ treatment systems are preceded by excavation of contaminated soil or extraction of groundwater for above-ground treatment. They include contaminated soil treatment in bioreactors, land treatment prepared-bed systems, biomounds and soil piles, and above-ground treatment of extracted groundwater in suspended growth reactors or ponds, or fixed-film reactors.

In situ saturated zone bioremediation systems have the following advantages over ex situ treatment alternatives:

- In situ systems are able to treat contaminants sorbed to aquifer material, as well as those in the dissolved plume, reducing the time required to remediate subsurface contamination as compared to ex situ approaches because of the limited solubility of many contaminants of concern.
- The aerial zone of treatment may be larger than ex situ remedial options, as treatment moves with the plume.
- Costs would be expected to be lower compared to ex situ systems, as no extensive pumping or excavation expenses are incurred with in situ treatment methods.

A number of disadvantages arise using in situ biological treatment process for contaminated groundwater, however. These include:

- Toxicity within the site will limit the applicability and performance of in situ systems. A lack of toxicity at the site is essential for in situ processes, as modification of toxicity in situ is difficult to impossible to achieve.
- Injection well clogging has been a recurring problem because of the abundant growth of microbial mass immediately adjacent to the injection point of the electron acceptor and nutrients.
- Reactants must be moved to the site of contamination in order for the biological reactions to take place. Nutrient and electron acceptor transfer efficiency limitations caused by low soil permeability, fractures, heterogeneities, etc., can severely limit the efficacy of in situ remediation technologies.
- Heavy doses of nutrients can impact groundwater quality adversely. This is a particular concern for nitrate, as it is of human health significance at 10 ppm.

To assess the chance of success for potential in situ bioremediation, three methods are described by Lantz (1991). One of these methods uses Brubaker's screening criteria, as summarized in The Hazardous Waste Consultant (1992). A number of parameters are individually given a score ranging from -2 to +1. If the total is 0 or greater, the site appears to be suitable for in situ treatment. A total of -1 to -2 indicates possible areas of concern; -2 to -4 indicates significant concern; less than -4 indicates that success with in situ bioremediation technologies is unlikely. Scoring of parameters is given in Table 7-2.

In situ biological systems utilized for unsaturated zone treatment have the general positive attributes of in situ biological saturated zone processes, including showing cost advantages, as they do not require soil excavation; and accelerating the time required for remediation, particularly for semivolatile and nonvolatile, biodegradable contaminants. They do overcome a number of the limitations of saturated zone treatment systems in that no injection well clogging nor permeability limitations generally result, as they are applied within the vadose zone. They, as all biological systems, do have limitations, however, in terms of being able to treat only biodegradable contaminants and being affected by toxicants existing throughout the site.

Table 7-2 Test for determining the feasibility of in situ bioremediation (Reprinted from The Hazardous Waste Consultant. *Evaluating the Feasibility of In Situ Bioremediation.* **Jan/Feb 1992; pp. 1.16-1.20. With permission from Elsevier Science).**

Parameter	Score
Contaminant Characteristics	
Structure:	
Simple Hydrocarbon C1 to C12	0
Hydrocarbon C12 to C20	-1
Hydrocarbon greater than C20	-2
Alcohols, phenols, amines	0
Acids, esters, amides	0
Ethers, monochlorinated hydrocarbons	-1
Multichlorinated hydrocarbons	-2
Pesticides	-2
Sources:	
Well-defined point source	1
Undefined multiple sources	-1
Hydrogeology	
Aquifer permeability (cm/sec):	
Greater than 10^{-3}	0
10^{-3} to 10^{-4}	-1
10^{-4} to 10^{-5}	-2
Aquifer thickness:	
6 m (20 ft)	1
3 m (10 ft)	0
1.5 m (5 ft)	-1
Less than 0.6 m (2 ft)	-2
Homogeneity:	
Uniform, well-defined geology	1
Heterogeneous, poorly-defined geology	-1
Depth to Aquifer:	
6 m (20 ft)	1
3 m (10 ft)	0
1.5 m (5 ft)	-1
Less than 0.6 m (2 ft)	-2
Soil and Groundwater Chemistry	
Groundwater pH:	
Greater than 10	-2
8 to 10	-1
6.5 to 8	0
4.5 to 6.5	-1
Less than 4.5	-2
Groundwater chemistry:	
High Fe, S, Ca, Mg, Cu, Ni	-0.5
High NH_4 and Cl	-0.5
Heavy metals (As, Cd, Hg)	-0.5

Interpreting the total score: 0 or greater, site appears suitable for bioremediation; -1 to -2, possible areas of concern; -2 to -4, areas of significant concern; less than -4, success is unlikely.

Ex situ systems are attractive alternatives for soil remediation when site excavation costs can be minimized, i.e., where shallow contamination of a limited aerial extent exists or where site constraints require immediate excavation and removal of contaminated soil. Ex situ systems have advantages over in situ systems as they allow:

- Enhanced control over reactant delivery and contaminant/product recovery by providing an opportunity to construct soil reactors that minimize flow pathways, that have provisions for leachate recovery, etc., or groundwater reactors that maximize electron acceptor and nutrient transfer efficiency and consequently yield optimal reaction rates
- An opportunity to modify existing site soil or groundwater nutrient characteristics through addition of nutrients or the addition of bulking agents during reactor/pile construction to improve nutrient status and air permeability
- Enhanced control over migration pathways away from contaminated soil or groundwater with the addition of a "tent" for volatiles and/or a "tub" for leachables
- Management of the site to optimize biological activity by providing:
  - Aeration management — bulking agents (soil) with forced air via extraction (soil) or injection (soil and water)
  - Moisture management---providing optimal moisture at 75% to 90% field capacity in impacted soils
  - Nutrient management---providing optimal nutrient concentrations and intimate and effective mixing during pile/reactor construction or groundwater treatment that is easier to achieve ex situ
  - Soil texture and toxicity management---providing optimal soil texture (i.e., loam) and toxicity reduction, by blending contaminated soil with desirable uncontaminated soil that is impossible to carry out in situ

7.1.4 Bioaugmentation Versus Bioacclimation

If treatability studies show no degradation (or an extended lag period before significant degradation is achieved), inoculation with strains known to be capable of degrading the contaminant may be helpful. This process of bioaugmentation increases the reactive enzyme concentration within the bioremediation system and subsequently may increase contaminant degradation rates over the nonaugmented rates, at least initially after inoculation. Bioaugmentation has proven successful in a number of laboratory applications, and a few field trials have been documented, including a novel strain of *Pseudomonas cepacia* added to a system to stimulate the degradation of TCE; the addition of *Phanerochaete chrysosporium* (white rot fungus) that biodegrades a wide range of organic compounds with nonspecific extracellular peroxidases; and the addition of active biomass that has been grown on another substrate to accelerate pentachlorophenol degradation in soil bioreactors (AAEE, 1995).

What most of the successful applications have in common is that they are ex situ treatment systems, i.e., those that can be engineered to optimize mixing and mass transfer, and in which reactor conditions can be controlled to optimize the environment for the organism being added. The success of bioaugmentation is particularly difficult with in situ systems because of the limitations to reactant delivery and recovery as discussed in Section 7.1.3, and for the additional reasons listed.

- Most species do not migrate significant distances in soil systems because of the competitive advantage microorganisms receive when they colonize soil surfaces.
- If the augmented organisms are not indigenous to the contaminated site, they have a difficult time competing with organisms that have adapted to the complex soil environment existing at most sites. Stimulation of indigenous organisms outside of the site environment and their reinoculation would be expected to generally prove more successful than the introduction of new species to complex, contaminated environments.
- The augmented organisms and the medium in which they were grown can serve as carbon and energy sources that might be more readily available and degradable than the contaminant of interest, emphasizing the point that simply an increase in activity does not translate to an increase in specific contaminant degradation reactions that might be desired.
- The addition of externally acclimated organisms to a contaminated site may be perceived by the public as the addition of engineered, mutant organisms that may not be publicly acceptable. It may be difficult to obtain permits from groundwater control agencies to use bacterial strains formed with DNA splicing techniques. (In the September 1, 1994 issue of the *Federal Register*, the EPA published a proposed rule defining how it will regulate processes involving genetically modified microorganisms under the Toxic Substances Control Act.)

Bioacclimation is the term used to describe the modification of environmental conditions to encourage the development of enzyme systems within the indigenous population of microorganisms capable of degrading the contaminants of interest. Bioacclimation can include simple actions such as blending and tilling of contaminated soils with uncontaminated media to reduce concentrations of contaminants below levels toxic to the indigenous microorganisms so that their populations can grow and contaminant reaction rates can increase. Bioacclimation can also include the slow increase in liquid reactor contaminant concentrations so that degradative enzymes systems can be induced in the reactor population and total microbial numbers can be increase to stable levels. The relative merits of bioacclimation include:

- Development of populations that are native to the contaminated site environment and that have developed competitive capabilities over time to exist within a niche in this environment

- No exogenous microbial agents are added to the site, making this approach a natural one that will meet with more public support than bioaugmentation schemes

Atlas (1995) reports that significant research is being directed toward searching for microorganisms that have more favorable degradation kinetics; metabolize more compounds; and thrive in solvents, in high alkalinity, and at high temperatures. Much of this work has been stimulated from observations made in describing natural attenuation processes taking place in contaminated environments. These efforts lay the groundwork for improving our ability to stimulate indigenous activity based on an improved understanding of the metabolic requirements and capabilities of microbial communities in the natural environment.

7.2. Aqueous Phase Treatment

Aqueous phase treatment systems are those designed to biologically remove contaminant mass contained within contaminated groundwater. These aqueous phase treatment systems include both those applied ex situ after extraction of contaminated groundwater and in situ groundwater treatment technologies. In situ treatment includes both the conventional, active remediation scheme of the Raymond process and the relatively new approach of plume management through natural attenuation evaluation. Each of these treatment options is discussed below.

7.2.1. Ex Situ Treatment

7.2.1.1. Suspended Growth Systems

Suspended growth systems are designed to provide intimate contact between the hazardous waste/material stream and the microorganisms used for its biodegradation by suspending the microorganisms in and mixing them with this waste material. These systems are capable of treating both liquids and solid slurries. The most commonly used suspended growth bioremediation system is activated sludge. Such treatment systems are commonly used for municipal sewage and industrial wastewaters from petroleum refining and other processes involving organic compounds.

As indicated in Figure 7-2, these activated sludge systems are typically comprised of a reaction or aeration tank, in which oxygen is added and actual biodegradation and growth of microorganisms occurs, plus a liquid/solids separation tank. This liquids/solids separation tank, or clarifier, allows the treated liquid to be separated from the microbial solids so that it can be discharged, and it provides a means for the recycling of necessary microbial solids (termed activated sludge) back into the reactor for use in treating additional hazardous waste/material that enters the process. A dissolved-air flotation (DAF) unit can be used instead of a conventional clarifier (Shin,

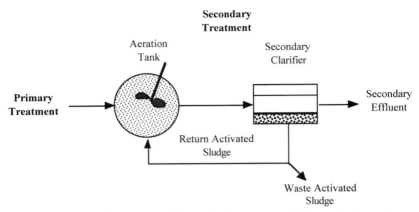

Figure 7-2 A typical process configuration for a suspended growth bioreactor.

1996). With a DAF unit, compressed air is injected into the water being treated at approximately 276 kPa (40 psig). The water is transferred to an atmospheric tank, in which the air comes out of solution as fine bubbles. The bubbles float sludge to the surface where the sludge/froth is skimmed.

Most of the sludge separated from water in the clarifier is recycled to the reactor. A high sludge recycle rate is critical for providing a suitable bacterial residence time, substrate/bacteria (food/microorganism) ratio, and sludge age within the reactor. Excess biomass is processed further (stabilization and drying) before it is disposed of.

With adequate bacterial residence time (at least 48 hr) and an appropriate food/microorganism ratio, 99% destruction of organic compounds can be achieved. For effective separation of sludge and water in the clarifier, a good floc must form, requiring a suitable sludge age and maintenance of favorable, steady environmental conditions. Maintaining fairly constant conditions of oxygen and nutrient content, pH and temperature are essential to forming a good floc.

These suspended growth systems are able to treat waste streams with moderate (1,000 mg/L biochemical oxygen demand, BOD) to high (10,000 mg/L BOD) organic content. They are generally able to resist variable loadings of influent waste materials because of the dilution effect provided in the aeration tank and the continual supply of active microorganisms entering the system with the recycle stream. They can, however, be upset if conditions are not stable or if concentration limits of toxic compounds being treated are exceeded. This can then lead to rapid die-off of the bacteria in the reactor, and several days to weeks may be required to re-establish a viable bacterial sludge mass needed to achieve the organics degradation efficiency required of the system.

Mixing intensity must be adequate to keep the waste material in suspension, which is a particular concern when treating high solids content slurries, and to provide an adequate supply of oxygen to the microorganisms. Finally, if the hazardous waste/material being treated contains high levels of volatile compounds, the aeration tank may have to be covered to capture and further treat these released vapors so they do not pose an unacceptable exposure risk to neighboring receptors.

7.2.1.1.1. Reactor configurations of various types have been shown to be successful in practice. These include the following.

Sequencing batch reactors. In these suspended growth reactors, influent fills a reactor in which the water is held without flow-through while treatment proceeds. During treatment, additional influent is diverted to fill parallel batch reactors or is allowed to accumulate in an influent storage tank or basin. The treatment steps for each batch include:

1. Initial mixing with added nutrients
2. Injection of air and long-term mixing
3. Sludge settling, with the mixer off
4. Decanting/removal of treated water in preparation for next batch

Sludge can be removed from the reactor after several treatment batches, or a portion can be wasted after each treatment batch. Batch reactors are the easiest to control. Conditions favorable to bacterial growth can be readily adjusted for each batch. The pH can be corrected near the beginning of the cycle. Potentially excessive toxic concentrations can be diluted by recycling some treated water. Nutrient additions can be precisely metered and adjusted in these batch systems as well.

An example of a sequencing batch reactor system design for wastewater treatment is described by Shin (1996), in which the BOD_5 is reduced by 99.5%. The aerobic reactor is a 34.5 m (115 ft) diameter concrete tank that is 7.2 m (24 ft) deep and handles 0.017 m^3/sec (0.43 MGD) of mixed-strength wastewater. During peak organic loads, wastewater from a food-processing plant is pretreated in an anaerobic lagoon before being transferred to this batch treatment plant.

Powdered activated carbon treatment (PACT) systems utilize the addition of powdered activated carbon (PAC) to a bioreactor to improve the treatment performance and stability of the system. The carbon provides relatively large surface areas for bacterial reactions and growth. The carbon quickly adsorbs organic contaminants, and then bacteria destroy the biodegradable substances sorbed to the carbon over a period of time. The nonbiodegradable components of the waste stream remain adsorbed and are removed from the reactor with the wasted carbon. A bioreactor such as an activated sludge unit becomes stabilized with carbon addition, even with variable influent concentrations. Also, fewer volatiles escape to the air, and consequently odors are reduced following PAC addition. Advantages over using GAC adsorption alone include:

- The reduction of carbon consumption because bacteria regenerate the sorption capacity of carbon by destroying the organic substances that otherwise would occupy adsorption sites
- An influent that usually does not have to be filtered

Either PAC or GAC can be used. PAC costs less than GAC and is easier to keep in suspension. PAC systems are marketed by Zimpro (Rothschild, Wisconsin), which also markets wet-air oxidation units that can destroy accumulated bacterial sludge and regenerate the carbon. Zimpro has small prefabricated and mobile systems available and can design large-volume systems as well.

Lagoons/ponds are suspended growth reactors without solids recycle. They can be operated aerobically, anaerobically, or both simultaneously depending on system depth and on the organic loading they are subject to. For aerobic operation, lagoons or ponds are generally used at depths of less than 1.5 m (5 ft). Because most bacterial action requires temperatures greater than 10° C (50° F), efficiencies fall off greatly during cold seasons. Because shallow lagoons present such a broad surface to air, most of the oxygen needed for aerobic action is dissolved into the water and continuously replenished by natural mass transfer through the surface of the lagoons. Spray systems or floating mechanical aerators that mix air into the water are sometimes added to increase oxygen transfer rates and increase the acceptable organic loading rate a pond can treat.

High biological degradation efficiency can be achieved only with long residence times, calculated as lagoon volume divided by volumetric water throughput rate. The calculated residence time will not be achieved if a large fraction of the influent travels from a single inlet at one end in the shortest pathway to a single outlet. Either multiple water inlets and outlets should be used, or baffles should be installed that direct the flow through a sinusoidal pathway.

Bacterial sludge will accumulate on the bottom of the pond, but it is generally so slow that it does not prevent effective operation of the lagoon during its entire life in remediation service. If accumulation rates are high, it may be necessary to remove sludge by some means, such as suction dredging or drying a pond section and excavating the sludge.

Constructed wetlands are created by planting marsh-type plants in very shallow ponds. Plants remove some organic compounds by uptake and provide sites in their roots for bacterial growth, where most of the bioconversion of contaminants takes place. Wetlands can also be used for anaerobic biodenitrification. In the denitrification process, nitrate is used as the primary electron acceptor, and some form of organic carbon must be available to serve as the electron donor. One common form of electron donor added to biological denitrification systems is methanol; however, for denitrification using wetlands, less toxic forms of dissolved organic carbon may be desired. These may include acetic acid (vinegar), other organic acids, acetate salts, and various forms of simple sugars. More discussion of biodenitrification is provided in Section 7.2.1.2.2.

With most wetlands, the plants are harvested periodically. Lemna USA (St. Paul, Minnesota) reports duckweed harvesting frequencies of three to 10 times per year when treating municipal sewage, depending on climate and treatment goals. The typical Lemna harvest amounts to 32 m^3/ha (16 yd^3/ac) (drained duckweed at 95% moisture at 500 kg/m^3, or 800 lb net weight/yd^3). Anaerobic conditions are maintained (except for the top 10 cm, or 4 in., of the pond) with pond depths ranging from 1.5 m to 4.8 m (5 ft to 16 ft). Hydraulic residence times for ponds partially covered with a duckweed mat range from 16 days to 22 days. Such systems remain stable during and after short-term upsets in influent pH, temperature, and dissolved oxygen (DO) levels.

7.2.1.1.2. Design parameters for suspended growth bioreactors can be determined from the mathematical model by Monod (1950), but the biokinetic constants must be determined for relating microbe growth rate to contaminant (substrate) concentration. Rozich (1994) indicates that measuring oxygen uptake with respirometers that collect data automatically in laboratory batch tests can be used to determine the biokinetic constants. Such tests are more rapid and have less interferences than conventional shake-flask tests.

Bhattacharya (1992) gives these equations (derived from Monod, 1950) for completely mixed systems:

$S = [K_s(1+t_s)]/[t_c(YK-b)-1]$ (7-23)
$S_{min} = bK_s/(Yk-b)$ (7-24)
$X = (t_s/t_h)Y(S_o-S)t_s/(1+bt_s)$ (7-25)

in which S is substrate concentration, in mass per unit reactor volume; S_o is influent concentration, in mass per unit reactor volume; S_{min} is the lowest S at which bacteria can use the carbon source; X is microbial biomass concentration, in mass per unit reactor volume; t_s is solids retention time (SRT), in days; t_h is hydraulic retention time (HRT), in days; K_s is the Monod half-velocity coefficient, equal to S when dS/dt is 0.5k, in mass per unit reactor volume; k is the maximum rate of substrate per unit weight of microorganisms, 1/time; Y is the growth yield coefficient, in mass per mass; and b is the microorganism decay coefficient, 1/time.

S must be determined because it is the effluent concentration. X must be determined in order to calculate the sludge generation rate. Each microbial community that develops within an activated sludge system in response to a specific contaminant mixture has its own values for K_s, Y, and b that are given in references such as Metcalf and Eddy (1990) or are determined from laboratory treatability tests. The key to optimal performance of suspended growth bioreactors with solids recycle (activated sludge systems) is maintaining the proper SRT by controlling the sludge wasting rate. An optimum design maximizes the SRT and minimizes the HRT. S_{min} is when bacterial growth equals decay. If the regulatory requirement for effluent concentrations of a given contaminant are lower than S_{min} for the microbial consortia in a treatment system, a cometabolite must be added; or microorganisms must be selected that have more favorable values of kinetic constants. One of many available mathematical models (with a graphic interpretation) of the activated sludge process is given by McHarg (1993 and 1994).

7.2.1.2. Fixed-Film Systems

Fixed-film systems are applicable for the treatment of low-organic-content waste material and for complex organic waste streams that require slow-growing, specialized microbial populations for their efficient degradation. The reactor portion of these systems contains inert packing material on which the microorganisms (biomass) grow and are concentrated. Unlike a suspended growth system, biomass in a fixed-film system is retained for long periods of time by the packing material and degrades the contaminants as the waste stream passes through the reactor. Because of the concentrated biomass within the reactor, reactor volume can be small, making the system cost-effective. Figure 7-3 shows a schematic of a typical packed tower fixed-film system, showing the packed bed reactor and clarifier used for liquid solids separation. It should be noted that, as indicated in this figure, fixed-film systems typically recirculate treated water rather than settled solids, as was the case in suspended growth systems. The purpose of the liquid recycle stream in fixed-film systems is to provide: (1) dilution of the influent waste stream to reduce the impact of high incoming concentrations; (2) hydraulic shearing so that excess biomass within the packing material does not clog the media; and (3) increased contact between the waste stream and the biomass on the media by returning the wastewater to the packed bed several times before it leaves the treatment system. Fixed-film systems can be used only for liquid waste streams because of their internal packing material. However, the in situ treatment of contaminated soils is possible using the soil media as a fixed-film reactor as described below.

Fixed-film bioreactors are not generally as sensitive to upsets as are activated sludge units. One exception might occur when the influent concentration is just above the toxic level and is constant. For example, if the phenol concentration in the influent is 600 mg/L and the toxic level is 500 mg/L, bioremediation may still succeed in a completely mixed activated sludge reactor. The average concentration could be in the 300 mg/L range in this completely mixed activated sludge reactor. On entering the

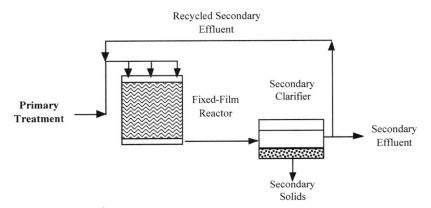

Figure 7-3 Typical configuration of a fixed-film biological reactor.

reactor, the influent is immediately diluted by mixing with water at this lower concentration, and nowhere within the reactor is the phenol concentration greater than 500 mg/L for a long-enough time to kill the bacteria. By contrast, a plug-flow, fixed-film reactor would suffer from bacteria die-off over the entire length of the reactor.

Except for such borderline toxic conditions, fixed-film bioreactors usually work reliably without extreme sensitivity to varying food/microorganism ratios and other environmental conditions that must be held steady in activated sludge systems.

7.2.1.2.1. Reactor configurations. Reactor configurations of various types have been successfully used in practice. These include the following:

Submerged fixed-film reactors have plastic packing material that is completely under the water flowing horizontally through the reactor. Air-distribution piping admits air near the bottom, along the length of the reactor. The air bubbles upward through the packing at a low rate to minimize short circuiting through the reactor.

Rotating biological contactors (RBCs), which are large plastic disks that rotate slowly with a central shaft mounted near the top of an open water trough. As the disk rotates, it is sequentially submerged in the contaminated water and then exposed to the ambient air. Bacteria on the disk surface contact organic contaminants (substrates) when the disk is submerged, whereas oxygen is supplied for aerobic bacterial degradation of substrates when the organisms are exposed to the air.

Trickling filters contain large rocks or plastic, ceramic, or wood media, usually in a large, shallow, circular tank a few feet deep. A horizontal rotating arm with water spray nozzles spreads the influent over the rocks. The rock surfaces are exposed to air, as well as to the water spray, and microbial biomass accumulates on the surfaces. The accumulated biomass sloughs and regrows as contaminants are removed from the water. Another common trickling filter configuration is a vertical downflow tower with plastic or redwood packing material.

Design parameters and design equations for these fixed-film systems can be found in Namkung and Rittman (1987).

7.2.1.2.2. Denitrification reactors utilize bacteria to destroy nitrates, primarily by the conversion of nitrate to nitrogen gas when it is used by the denitrifying bacteria as an electron acceptor. If municipal sewage is not present, phosphorus must be added. The process is anoxic, and some form of dissolved organic carbon must be added as the electron donor to drive the use of the nitrate for metabolism by these organisms. Methanol is generally the least expensive form of dissolved organic carbon available, and it is usually chosen for anaerobic denitrification. If less-toxic forms of dissolved organic carbon are desired, organic acids (e.g., formic acid, acetic acid) or salts formed from organic acids (e.g., sodium acetate) can be used.

Submerged fixed-film reactors, wetlands, or fluidized bed reactors are used in this process. With fluidized bed units, a solid, fluidizing medium, such as sand or GAC, is added to the reactor to provide support for microbial growth. The velocity of the water flowing upward in the reactor must be high enough to maintain these solids in suspension (at least 0.525 m/min, or 1.75 ft/min). The reactor volume should be approximately 0.12 m^3 for each kg/day (2 ft^3 for each lb/day) of nitrogen in the influent. The effluent must be recycled if the groundwater flow rate falls below the design rate.

If the dissolved carbon is in the form of methanol or acetate, the approximate stoichiometric amounts required and the carbon dioxide produced can be determined from the following molar ratios:

- 1 mole methanol/1 mole nitrate produces 1 mole carbon dioxide
- 3 moles acetate/4 moles nitrate produces 6 moles carbon dioxide

It should be noted that these relationships are not exact. Additional methanol or acetate is needed if DO or nitrites are present; less is needed if dissolved organic compounds are present and become degraded. Also, some of the carbon in the methanol or acetate becomes part of the increased biomass that is produced, resulting in approximately a 30% reduction in carbon dioxide production.

If, for simplification, these corrections are not applied, the minimum amount of sodium acetate that must be added to reduce the nitrate concentration from 300 mg/L to 45 mg/L as NO_3^- (federal drinking water standards maximum contaminant level) and the carbon dioxide produced are calculated as given below.

Given: DO and nitrite concentrations are negligible. All carbon becomes carbon dioxide. A starting concentration of 300 mg/L of nitrate is to be reduced to 45 mg/L.

Find: The minimum amount of sodium acetate addition needed per liter of contaminated groundwater treated, and the volume of carbon dioxide generated.

Step 1: Determine the net molar nitrate removal, and apply the molar ratio of 3 gmol acetate/4 gmol nitrate to calculate the sodium acetate requirement.

The nitrate to be removed is 300 mg/L - 45 mg/L, or 255 mg/L of nitrate, with a molecular weight of 62 g/gmol. Then, the moles of nitrate to be removed is:

$255/62 = 4.113$ millimoles of nitrate/L (7-26)

With sodium acetate having a molecular weight 82 g/gmol, the required sodium acetate concentration is:

(3 gmol acetate/4 gmol nitrate) (4.113 millimoles of nitrate/L)
= 3.085 millimoles acetate/L (82 mg/millimole sodium acetate)
= 253 mg sodium acetate/L (7-27)

Step 2: Apply the molar ratio of 6 gmol carbon dioxide/4 gmol nitrate to calculate the amount of carbon dioxide production.

(4.113 millimoles of nitrate/L) (6 gmol carbon dioxide/4 gmol nitrate)
= 6.17 millimoles of carbon dioxide/L groundwater
= 0.00617 gmol carbon dioxide/L groundwater (7-28)

The volume of carbon dioxide produced is calculated as follows, considering that each gmol of a gas at 0° C (273 K) and 1 atm has a volume of 22.4 L:

(0.00617 gmol/L) (22.4 L/gmol) = 0.14 L carbon dioxide/L groundwater @ standard temperature and pressure (7-29)

Step 3: Convert to actual volume, considering that actual volume is proportional to absolute temperature. If the carbon dioxide volume per liter of groundwater is measured at a room temperature of 20° C (293 K), its actual volume it would be:

0.14 L (293/273) = 0.15 L carbon dioxide/L groundwater at 1 atm and 293 K (7-30)

The amount of sludge that will be produced can be determined in a laboratory batch equilibrium test as described in Section 7.4 (Treatability Studies for Bioremediation Systems). With methanol, the sludge mass produced will be approximately 0.1 times the nitrate removed, and with acetate, the sludge mass is approximately 0.157 times the nitrate removed.

7.2.2. In Situ Treatment

7.2.2.1. Raymond Process

The in situ treatment of contamination below the groundwater table is carried out using the soil below the groundwater table as a fixed-film reactor. Infiltration galleries or injection wells are used to supply the required nutrients and oxygen to the subsurface to sweep through the contaminated soils and groundwater toward the groundwater recovery system (Figure 7-4). This process was first field tested in 1972 by Raymond (Raymond et al., 1976; Raymond et al., 1978; Brown et al., 1984) and became commercial in the mid-1980s. A comprehensive introduction to the Raymond process can be found in the monograph "Innovative Site Remediation: Bioremediation" (AAEE, 1995), in Chapter 2 of the *Handbook of Bioremediation* (Norris et al., 1994), in *A Guide for Railroad Industry Use of In Situ Bioremediation* (Brubaker et al., 1994), and in *In Situ Bioremediation: When Does It Work?* (National Research Council, 1993). Detailed design guidance can be found in US EPA (1995a) and Dupont et al. (1998).

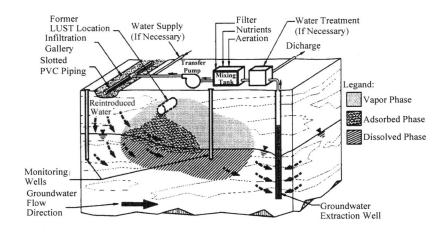

Figure 7-4 Schematic of a typical Raymond process using an infiltration gallery and recovery wells for the treatment of contaminated groundwater (US EPA, 1995a).

The process relies on the injection of electron acceptors and nutrients to stimulate microbial growth and enhance the rate of biodegradation of organic contaminants. The engineering challenge to this technology is to supply the aquifer with electron acceptors and nutrients in a manner such that dispersion occurs throughout the contaminant plume. Depending on the solubility of the contaminants, the hydrogeology, and the system design, varying proportions of overall contaminant mass reduction occur as a result of simple mass removal in the recovered water. This extracted mass must be treated in an above-ground reactor before the groundwater can be reinjected for further aquifer treatment.

The first Raymond systems used oxygen as the electron acceptor by sparging the groundwater in the injection wells with air (Raymond, 1974). Subsequently, other sources of electron acceptors have been used, including hydrogen peroxide (Raymond et al., 1986), nitrate (Hutchins et al., 1991; Bouwer, 1994; Reinhard, 1994), sulfate (Beeman et al., 1993; Bouwer, 1994; Reinhard, 1994), and solid oxygen-releasing compounds or ORC (Bianch-Mosquera et al., 1994; Kao and Bordon, 1994; Marlow et al., 1994; Dupont et al., 1998).

The Raymond process, in its current state of development, is applicable for the remediation of petroleum-based hydrocarbons (including commercial fuel blends), creosote, oxygenated solvents such as alcohols and ketones, some chlorinated compounds, and (to some extent) polynuclear aromatic hydrocarbons (PAHs). Methods are being field tested with some success for treatment of chlorinated solvents such as TCE and TCA. Commercial treatment of pesticides, herbicides, polychlorinated biphenyls (PCBs), high molecular weight hydrocarbons, and most munitions is not likely to be available for some time.

Generally, the process is most suited to relatively permeable soils and is most easily applied to homogeneous aquifers. The aquifer hydraulic conductivity and saturated interval thickness determine the rate at which groundwater can be transported to deliver electron acceptors and nutrients. The minimum acceptable hydraulic conductivity is dependent on the mass loading of the contaminants; however, the flow rate of water injection is typically low, and degradation rates are so rapid that nutrients, particularly oxygen, do not typically travel far from the point of injection before they are completely utilized. In addition, these systems tend to clog rapidly around the injection areas because of either stimulation of microbial growth or oxidation and precipitation of dissolved iron in the groundwater. Care must be taken in the use and application of in situ groundwater treatment because of these limitations.

The Raymond process, as it is currently applied in practice, uses pumping to control the migration of contaminants in the groundwater, with the amendment and filtration of this pumped water ex situ in above-ground equipment, as shown in Figure 7-4. Process modifications have focused on the use of alternatives to air for the supply of

TEAs, as well as alternative injection/recovery systems. These modifications are summarized below.

7.2.2.1.1. The use of alternatives to air to supply electron acceptors has been evaluated because of the limited solubility of oxygen in water and the impact of this low solubility on the rate of oxygen transfer to contaminated aquifer systems. Two approaches have been evaluated: the use of alternative oxygen forms that increase the DO concentration in water, and the use of alternative electron acceptors that can be utilized by microorganisms to degrade the contaminants of interest.

Alternative oxygen forms. The use of pure oxygen in place of air offers the possibility of introducing the electron acceptor at a five-fold increase in rate and, presumably, a similar reduction in the required time for remediation of the contaminated aquifer. Oxygen is generated on-site or brought on-site in the liquid (cryogenic) form (Prosen, 1992). Other aspects of operation are virtually the same as in the conventional Raymond process. Oxygen-generating units, such as those used in remote hospitals, can provide sufficient oxygen for modest systems at reasonable capital investments. If liquid oxygen is used an evaporator, additional equipment are required, making this approach more applicable to larger systems. Fully pressurized systems, in which the well bore is maintained full of water, reduce the hazard of high oxygen levels in the presence of flammable vapors.

Hydrogen peroxide, H_2O_2, was first used in the mid-1980s as an alternative source of oxygen supply because it is miscible in water and decomposes in the aquifer to yield oxygen and water (Brown and Norris, 1988). Two kg of pure hydrogen peroxide produce almost 1 kg of oxygen. Hydrogen peroxide systems use metering pumps to transfer hydrogen peroxide (35% to 70%) from storage tanks for continuous addition to the injection water. Hydrogen peroxide at higher concentrations can serve as a bactericide. The sterilization of bacteria at the well screen and in the sand-pack in the annulus between the screen and the bore-hole wall can be beneficial. Bacterial sludge cannot build up to amounts that would otherwise clog the screen and its immediate surroundings. High concentrations of peroxide can be undesirable for another reason; however, if present at too high a concentration, peroxide decomposition will yield oxygen too rapidly, leading to loss of oxygen to the unsaturated zone or gas blockage in the aquifer. When applicable, hydrogen peroxide can be introduced at concentrations of 100 to 1,000 mg/L (typically, 200 to 500 mg/L). At 500 mg/L, oxygen is theoretically provided at a rate that is 25 times greater than that achieved by the sparging of air and more than 10 times faster than that achieved with pure oxygen. However, hydrogen peroxide stability is frequently a problem and must be evaluated before completing system design (Flathman et al., 1991; Lawes, 1991). If hydrogen peroxide stability and the groundwater flow rate under operating conditions are insufficient to permit hydrogen peroxide, and/or the elevated levels of oxygen that it generates, to penetrate several meters into the aquifer, alternative approaches should be considered.

Solid phase ORC has also been recently developed and commercially marketed by Regenesis™ (San Juan Capistrano, California). This ORC can be applied as 0.3- to 0.6 m (1- to 2-ft) long "socks" that are placed down existing 2.5- or 5.0-cm (1- or 2-in.) monitoring wells, or it may be applied as a slurry in the form of "pencils" (slender slurry rods extending from the depth of groundwater contamination across the water table) or as a continuous ORC slurry wall. The ORC is placed in wells or as a slurry in columns or trenches perpendicular to the direction of groundwater flow. The release of oxygen into the aquifer creates a zone of increased oxygen through which the contaminated groundwater must flow. The distribution of oxygen around the well is controlled by diffusion and dispersion. In order for wells to serve as a barrier to contaminant migration, they must be closely spaced, i.e., from 0.9 to 1.8 m (3 to 6 ft) apart depending on site conditions. ORC systems are passive and, once installed, require no mechanical or electrical equipment. As a result, systems require no maintenance other than periodic replacement of the ORC.

Various derivatives of hydrogen peroxide have been evaluated for use as an effective ORC material for in situ biodegradation processes and soil cells, in which a continuing source of oxygen is required for aerobic biodegradation processes. These peroxide compounds include magnesium peroxide, calcium peroxide, sodium carbonate peroxide, and urea peroxide. Sodium carbonate peroxide and urea peroxide release oxygen much too rapidly to be of practical use, so most bioremediation studies have been conducted with either magnesium peroxide or calcium peroxide. The ORC marketed by Regenesis™ is comprised of magnesium peroxide.

Alternative electron acceptors. The two alternative electron acceptors that have been most widely studied are nitrate and sulfate. These alternatives to oxygen are known to support a variety of biodegradation reactions (*see* Table 7.1) and provide advantages over the use of oxygen because of their much higher solubilities in groundwater. It is essential, however, that the degradation of the contaminant of concern be documented from reported laboratory or field data, as degradation reactions under alternative electron acceptor conditions can be highly organism- and contaminant-specific.

Nitrate systems are similar to the conventional Raymond process, but they use concentrated nitrate solutions (10 to 100 mg/L) to provide the continuous addition of electron acceptor to injected groundwater (Hutchins et al., 1991). Nitrate can serve as a nutrient source for all metabolism; however, organisms that use it as an electron acceptor are able to use only a limited range of organic compounds as a source of electron donor. These compounds include aromatic compounds, except benzene (AAEE, 1995), and some chlorinated solvents, but not aliphatic hydrocarbons. Nitrate is highly soluble in water and is not retarded by soils. Thus, nitrate is, technically, an ideal electron acceptor for some organics. Its use, however, is regulated, and many regulatory agencies will limit the concentration of nitrate in the injection water to 10 mg/L or less as nitrogen (44 mg/L as NO_3^-), the drinking water level for nitrate to prevent metahemaglobanemia ("blue baby" syndrome).

Sulfate systems use concentrated sodium or potassium sulfate solutions to provide the continuous addition of electron acceptor to injected groundwater (Beeman et al., 1993). Sulfate has been used, in combination with a degradable organic substrate/electron donor such as sodium benzoate, at several sites for the reductive dechlorination of PCE and TCE under anaerobic conditions. Sulfate is highly soluble in water and is not retarded by soils so that it is easily distributed through the aquifer. Sulfate can serve as an electron acceptor for a wide range of chemical classes of concern (*see* Table 7-1), as is a primary TEA observed at many natural attenuation sites as described below. The secondary health standard for sulfate is 250 mg/L, and it has a laxative effect on humans at concentrations greater than 750 mg/L. In addition, sulfide is a byproduct that can be produced anaerobically, and that can be a problem because of taste and odor concerns.

7.2.2.1.2. The use of alternative injection systems has been investigated because of the limited volume of influence that generally exists for conventional injection wells. When alternative injection systems are used, the amended water is introduced over a wide area, compared with well systems in which introduced amendments must travel through the aquifer before reaching areas distant from the injection point. Thus, remediation is initiated over a wide area as soon as amended water percolates to the water table. There are drawbacks, however, to alternative systems that introduce amended water above the water table as compared with conventional injection wells, namely some groundwater mounding may occur and most of the nutrients and electron acceptor will be applied preferentially to the upper saturated zone, the capillary fringe, and the lower portion of the vadose zone. Although some mixing within the aquifer occurs a result of normal water table fluctuations, the interval over which the nutrients and electron acceptors are being provided with these infiltration methods is limited to shallow aquifer systems.

Alternative injection systems commonly in use include (Cookson, 1995) the addition of amended water at the surface; the addition beneath the surface but above the groundwater (percolation systems); or the addition via trenches extending beneath the groundwater surface. These systems have been used in various combinations, including combinations with conventional recovery wells.

Surface addition systems. Systems that introduce amended water from the surface are limited to sites in which the unsaturated zone consists of soils with adequate percolation (Cookson, 1995). Sands and gravel are most appropriate. As the fines content of the soil increases, percolation rates decrease, adsorption of nutrients increases, and potential losses of electron acceptors, especially hydrogen peroxide, increase. A number of methods can be used for the addition of amended water to unpaved surfaces, including via sprinklers, irrigation hoses, etc. This allows water to be added over a large area, usually in batch additions. Provisions need to be made to prevent run-off, and the surface is preferably covered with a porous geomembrane to facilitate more homogeneous percolation by minimizing redistribution of fines.

Water addition in the vadose zone. Systems that inject amended water below ground surface but above the water table include injection galleries. These systems have in common a gravel or sand-filled void created by shallow excavations, trenches, or the bottoms of tank pits. These systems have advantages over surface addition in that they introduce amended water closer to the water table surface, have larger water retention capacity, cause less water loss via evaporation and wind drift, are not subject to run-off, and are out of sight.

Water addition in the saturated zone. Systems using trenches extending into the aquifer operate much like conventional well systems except that the radius of influence limits no longer applies as long as the trenches transverse the entire treatment zone. Trenches also provide more surface area and are less prone to plugging than are conventional wells.

7.2.2.1.3. Process limitations negatively impacting the efficiency of in situ treatment of contamination below the groundwater table primarily relate to the low efficiency with which reactants can be delivered to the location of the contaminants of concern at many contaminated sites. Unless the aquifer consists of uniform sand, reactants will not be delivered uniformly, resulting in incomplete remediation in some locations. With heterogeneous soils, the amendment solutions will flow through the most permeable pathways and will diffuse into fine-grained materials very slowly. If significant contamination exists within the fine-grained deposits throughout the site, in situ bioremediation via the Raymond method may not provide the desired remediation efficiencies within an acceptable timeframe.

Nutrient mobility can be a significant challenge in soils with a large clay fraction. Phosphate, for example, will attach to clay particles by ion exchange and may not move a significant distance away from the injection point. Some of this limitation can be overcome by the use of polyphosphate instead of orthophosphate additives, but rapid decomposition of the polyphosphate may still severely limit its overall migration potential. Electron acceptor reactivity is an additional concern. For example, hydrogen peroxide reacts with natural organic matter and decomposes rapidly in clay-rich soils, once again not reaching the location of the contaminants any distance away from the point of application. Other adverse reactions that may occur in situ include conversion of pyrites to sulfuric acid with oxygenated water and systems plugging with dense biofilms because of rapid consumption of nutrients and electron acceptors by soil microorganisms immediately adjacent to injection wells and infiltration galleries.

Aquifer plugging can be assessed by measuring the pressure or head needed in the injection wells to maintain the flow rate. Some plugging can be expected, and the system should be designed accordingly. For example, the injection pump could be specified such that the pressure can ultimately be increased to twice the initial pressure for a given flow rate. The system may be considered successful if the time elapsed before excessive pressures are required to maintain the flow rate is greater than the time required to convert the mass of contaminants targeted for biodegradation.

7.2.2.1.4. In situ biodenitrification is another application of the Raymond process for the removal of high levels of nitrate in the groundwater through biological denitrification reactions. Here, the use of nitrate as a TEA is stimulated through the addition of a simple, biodegradable, soluble carbon source to drive the aquifer nitrate reducing. Substances such as methanol or acetic acid can be injected either continuously at low concentrations or intermittently at higher concentrations. The best scheme for electron donor concentration and injection timing may be determined experimentally in the field by analyzing groundwater samples from monitoring wells surrounding the injection well. If the denitrifying Raymond system is designed with continuous extraction and reinjection wells, the dissolved carbon addition should be metered so no residual organic carbon appears in the extraction wells. The dissolved carbon addition rate should be low at first and can be increased as the carbon metabolism rate increases along with buildup of bacterial cell mass in the aquifer. Field testing should be carried out to determine the radius of biodegradation, the rate of carbon donor and nitrate utilization, and the potential for plugging of wells and of the aquifer. Mathematical modeling (Clement et al., 1998) can aid in evaluating variables affecting field tests, and in determining required well spacing. One such computer code, RT3D, can be downloaded at http://bioprocess.pnl.gov/RT3D.htm.

7.2.2.2 Natural Attenuation

Natural attenuation describes the process of site assessment, data reduction, and data interpretation that is focused on the quantification of the capacity of a given aquifer system to assimilate groundwater contaminants through physical, chemical, and/or biological means. Through the natural attenuation plume management approach, a determination is made of the nature and extent of soil and groundwater contamination and of the extent and rate of natural contaminant degradation at a site. The natural attenuation approach is appropriate for a given site if the plume has not impacted a downgradient receptor and if the rate of contaminant release from the source area is equal to or less than the contaminant degradation rate observed at the site.

7.2.2.2.1. Principles of operation. The natural attenuation plume management approach is described in a number of field sampling protocols available from a variety of sources. These protocols describe approaches for collecting and analyzing data necessary to verify that natural attenuation processes are taking place at a given site (Wiedemeier, 1996; Wilson et al., 1994). The connection of these data with decisions regarding source removal activities or with estimates of source lifetime has not generally been presented in the literature. An approach for implementing natural attenuation concepts from data collection through source removal and source lifetime considerations has been developed for the US EPA and the US Air Force (Dupont et al., 1996, 1998). These concepts and procedures are presented in detail below.

7.2.2.2.2. Process design principles for a natural attenuation assessment are described in Dupont et al. (1996) and involve the seven-step process outlined in Figure 7-5. This process involves: (1) determining whether steady-state plume conditions exist at the site; (2) estimating contaminant degradation rates; (3) estimating the source mass term; (4) estimating the source lifetime; (5) predicting long-term plume behavior with and without source removal; (6) decision making regarding the use of natural attenuation and the impact and desirability of source removal at a given site; and (7) developing a long-term monitoring strategy if natural attenuation is selected for plume management.

7.2.2.2.3. Determination of steady-state plume conditions for a contaminant plume at a given site is critical in establishing that natural attenuation processes are taking place and are likely to provide continued plume containment under current site conditions. Steady-state plume conditions occur when the rate of contaminant release from the source area is equivalent to the rate of contaminant assimilation by biotic and abiotic processes taking place within the aquifer. Steady-state conditions can be identified by observing contaminant concentrations at specific groundwater monitoring locations over time. A more desirable approach is to evaluate contaminant concentration and contaminant mass distribution changes throughout the entire plume over time. This latter approach involves the collection of centerline concentration or total integrated mass and center of mass data within the delineated plume at various time intervals to determine if the entire plume has reached steady-state concentrations.

Plume centerline concentrations that are consistent from one sampling interval to the next are indicative of steady-state plume conditions. Comparison of data between sampling events should be done carefully, however, as the vertical distribution of contaminant at the site can have a significant impact on the mass of dissolved contaminant in the groundwater at a given groundwater table elevation. If large groundwater table fluctuations occur at a site, significantly different volumes of contaminated soil can exist below the groundwater table at any given point in time, producing significantly different dissolved plumes from one sampling time to the next. Although these variations in groundwater plume characteristics are important in understanding the overall risk posed by a given site, they tend to confuse the issue of steady-state plume evaluation. It is thus recommended that if groundwater table fluctuations can be expected to produce highly variable contaminant plume profiles on a seasonal basis, the steady-state evaluation should be based on comparison of data with comparable groundwater elevation values, even though these data sets may be 6 mo. to 1 yr apart in time.

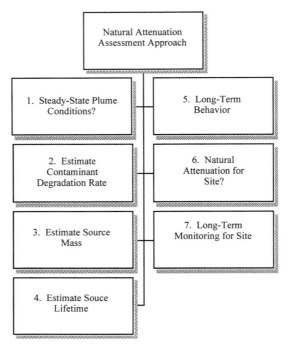

Figure 7-5 Components of the natural attenuation assessment approach.

If steady-state or receding plume conditions are indicated based on three to four sets of comparable monitoring data, the plume can be considered to be stable under existing aquifer conditions, and the natural attenuation plume management option should be considered for the site. If the plume is observed to be growing, i.e., plume centerline concentrations are increasing over time, either monitoring should be continued or aggressive containment and source removal activities should be carried out if a sensitive receptor has already been impacted by groundwater contamination.

The evaluation of the total dissolved mass and movement of that mass in a contaminant plume provide a more rigorous evaluation of plume steady-state conditions than simple centerline analysis. In order to develop an estimate of the dissolved plume mass, M_D, an aquifer volume associated with each monitoring point must first be determined. Once an aquifer volume is associated with each monitoring point, the product of contaminant concentration, C_i, and the aquifer volume for each monitoring point are summed to yield a total dissolved mass for the plume.

Aquifer volume is determined from the product of the aquifer porosity, θ; the average aquifer thickness, h, (generally the length of the largest sampling interval used within the monitoring network at a given sampling time); and a plume surface area associated with each sampling point, A_j, as indicated in Equation 7-31.

$$M_D = \sum mass_i = \sum C_i (\theta) (h) (A_j) \qquad (7\text{-}31)$$

in which $mass_i$ is the dissolved contaminant mass associated with sampling location i.

One procedure that can be used to obtain an estimate of area associated with each sampling point is the Thiessen polygon method (Chow et al., 1988). This method was developed in the field of hydrology for estimating areas associated with point rainfall measurements within rain gage networks. The Thiessen method assumes that the concentration measured at a given sampling point is constant out to a distance halfway to the sampling points located next to it in all directions. The relative weights (areas) represented by each sampling point are determined by the construction of a Thiessen polygon network, the boundaries of which are formed by the perpendicular bisectors of lines connecting adjacent points. The construction of an example polygon network is shown in Figure 7-6.

The outer boundary of the Thiessen polygon network is estimated based on the outermost well locations. It is important to be consistent with boundary definition if mass calculations are to be comparable among different sampling events. It is also

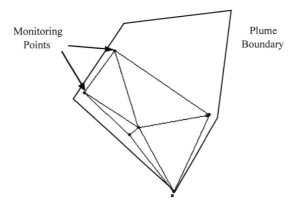

A. The outer boundary of the sampling network is identified based on logical, physical boundaries of the problem. Each sampling location is then connected to all adjacent points to form a series of polygons with the sampling points as their corners.

Figure 7-6 An example of Thiessen polygon network construction.

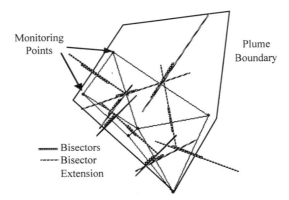

B. The lines between these sampling points are bisected, and perpendicular lines are drawn at the bisection points. These perpendicular lines are then extended so that they intersect one another.

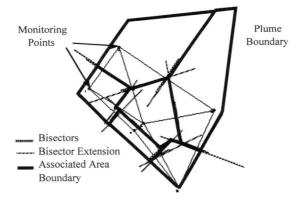

C. The intersecting lines are connected to form polygons associated with each original sampling location to yield unbiased and consistently generated areas. These areas can then be used to generate associated groundwater and soil volumes that allow the determination of the mass of contaminant within the assigned plume boundary and the changes in that mass over time.

Figure 7-6 An example of Thiessen polygon network construction (continued).

important to note that this method can be used for estimating mass within a monitoring network consisting of as few as three monitoring wells. An increase in sampling point density throughout the plume will improve the accuracy of the plume mass calculations, however, as interpolation among data points will be improved the shorter the interpolation distance. Ideally, from 10 to 20 monitoring points spread throughout the site, both inside and outside the contaminant plume, can be used to provide reasonable accuracy in plume mass estimates and plume delineation for a reasonable cost.

In addition to estimating the total mass of a compound within the dissolved plume at a given point in time, the representative center point of the combined plume mass can also be calculated. This representative mass center, termed the centroid of the mass (CoM), is calculated by taking the first moment of inertia of the mass at each sampling location within the contaminant plume about specified X and Y axes. Mathematically this can be expressed as shown in Equations 7-32 and 7-33, respectively.

$$X = \sum_{i=1}^{n} x_i (mass_i) / \sum_{i=1}^{n} (mass_i) \quad (7\text{-}32)$$

$$Y = \sum_{i=1}^{n} y_i (mass_i) / \sum_{i=1}^{n} (mass_i) \quad (7\text{-}33)$$

in which x_i and y_i are the x and y coordinates of each sampling location within the Thiessen area network.

Table 7-3 provides an example of calculations for total mass and CoM results generated from a monitoring network for which contaminant concentration and Thiessen area values were obtained. Results from these calculations with a uniform contaminated thickness of 4.1 m (13.67 ft) indicate a total dissolved benzene plume mass of 180.6 g, with CoM X and Y coordinates of +3,713.9 m and +91,105.3 m, respectively.

These CoM calculations are useful for tracking and interpreting the movement of contaminants, reactants, and products within the contaminant plume over time. They can also aid in assessing the status of the plume and in interpreting its migration pattern over time.

If plume centerline analysis and CoM calculations suggest that the plume is growing over time, steady-state conditions have not been reached. Under these conditions, either ongoing monitoring should take place to ensure that attenuation of the plume occurs in the future or active source removal and/or site remediation should occur if a sensitive receptor is or will be impacted in the near term. If the contaminant plume is shown to have reached steady-state conditions, further quantitation of the nature and extent of plume attenuation taking place under site conditions is warranted.

Table 7-3 Total mass and CoM calculations for benzene from sample monitoring network for which Thiessen areas have been determined.

Monitoring Location Designation	Well Depth (m)	Easting (m)	Northing (m)	Bottom of Well Depth (m BToC)	Depth to Water (m BToC)	Water Depth (m)	Associated Area (m²)	Pore Volume (m³)	Benzene (µg/L)	Dissolved Benzene Mass (g)	X Moment (g-m)	Y Moment (g-m)
U1-078	7.64	2881.0	90179.00	8.10	4.56	3.54	59.7	73.4	20.5	1.5	4336	135715
U1-089	10.38	2725.3	89634.39	6.79	4.35	2.44	18.7	23.0	5.8	0.1	364	11973
U1-097	5.80	5072.0	91345.50	8.11	4.13	3.98	106	130	3.2	0.4	2115	38095
U1-098	4.40	3813.0	91898.20	5.30	1.62	3.68	62.4	76.7	12.1	0.9	3540	85326
U1-099	4.46	3267.8	92852.70	5.55	2.30	3.25	45.2	55.6	27.9	1.6	5068	143998
U1-103	12.22	2681.6	90002.60	10.58	8.73	1.84	32.8	40.3	335	13.5	36207	1215220
U1-104	10.08	2948.1	90606.19	10.52	6.41	4.10	86.6	107	59.0	6.3	18528	569428
U1-105	10.50	3127.7	91482.00	11.09	8.93	2.16	158	195	345	67.3	210405	6154144
U1-108	10.08	3807.9	91268.64	7.76	5.63	2.14	138	169	379	64.2	244335	5856280
U1-109	7.64	5533.9	91457.00	8.00	5.73	2.27	7.1	8.7	28.6	0.3	1384	22880
U1-110	7.02	5251.2	91924.55	7.61	4.56	3.04	13.0	15.9	36.5	0.6	3056	53498
U1-111	5.19	4839.9	92115.01	5.50	3.05	2.45	47.0	57.8	0.5	0.0	140	2662
U1-112	4.58	3896.6	92422.03	5.19	2.31	2.88	36.5	44.8	5.9	0.3	1031	24453
U1-115	10.72	1746.2	89175.93	11.12	7.29	3.83	23.2	28.6	7.1	0.2	354	18088
U1-116	9.93	1802.6	90242.74	9.75	7.11	2.64	144	177	133	23.5	42448	2125034
				Max Depth =	4.10		978			Σ = 180.6	573312	16456793
										CoM (m,m) =	3173.9	91105.3

7.2.2.2.4. Estimation of contaminant degradation rates can be carried out using dissolved plume contaminant mass data if a declining mass of contaminant is observed over time or using contaminant groundwater concentration data if the source is found to produce steady-state dissolved mass in the plume over time. If steady-state mass is observed, degradation rates for the contaminants can be estimated directly from centerline concentration data or through the calibration of a contaminant fate and transport model to field groundwater data. If aquifer flow data are available, the use of a fate and transport model accounting for advection, dispersion, sorption, and degradation, is preferred over the use of plume centerline concentration data alone. In addition, the use of less-degradable plume-resident tracer compounds in the calibration process is desirable for the calibration of the transport component of the fate and transport model if data for these tracer compounds are available. The use of the less-degradable tracers simplifies the model calibration process, as the transport components of a model can be calibrated without having to consider degradation reactions.

Dissolved plume mass changes over time can be used as an indicator of the type of plume existing at a site, as summarized in Table 7-4. If a large source of contaminant produces the dissolved plume, then the mass of contaminant in this dissolved plume will appear constant over time.

If the source area is finite in size, or if the source material generating the dissolved plume is highly weathered, the flux of contaminant out of the source area and into the dissolved plume will decrease to zero over time. This decrease in contaminant flux from the source area will cause the total mass of contaminant in the dissolved plume to also decrease over time. The position and concentration profile within a dissolved plume produced by this finite source will not be at steady-state. To estimate the degradation rate of contaminants within the plume resulting from this finite source, the changes over time of total contaminant dissolved plume mass should be analyzed.

A classical approach to the evaluation of contaminant degradation rates in biological systems is to analyze the time course of changes in contaminant concentration or mass to investigate the relationship between concentration or mass versus reaction time using zero- or first-order reaction rate laws. Zero-order reactions are described by a contaminant reaction rate independent of contaminant mass, i.e., a constant degradation rate over time, or:

$$dM/dt = -k_o \tag{7-34}$$

in which k_o is the zero degradation rate, in units of mass per time. The integrated form of this equation is shown in Equation 7-35.

$$M = M_o - k_o t \tag{7-35}$$

Table 7-4 Possible changes in contaminant mass and mass center coordinates for a contaminant plume and the corresponding interpretation of these changes relative to plume mobility and persistence.

Contaminant Mass	Centroid of Mass	Interpretation
Increasing	Moving Downgradient	Continuous source; unstable plume; contaminant migration
Constant	Moving Downgradient	Finite source; plume migration; minimal natural attenuation
Constant	Stable	Continuous source; stable plume; contaminant attenuation
Decreasing	Moving Downgradient	Finite source; plume migration; contaminant attenuation
Decreasing	Moving Upgradient	Finite source; plume attenuation; rapid contaminant attenuation; optimal intrinsic bioremediation

in which M is the contaminant mass at time t; and M_o is the initial contaminant mass at time t equals zero. If the reaction taking place is governed by a zero-order degradation rate law, a plot of contaminant mass versus time produces a linear relationship, the slope of which equals k_o, and whose intercept value should equal M_o.

First-order reactions are described by a contaminant reaction rate that is dependent on contaminant concentration or mass, i.e., a degradation rate changing over time, or:

$$dM/dt = -k_1 M \qquad (7\text{-}36)$$

in which k_1 is the first-order degradation rate, 1/time. The integrated form of this equation is shown in Equation 7-37.

$$M = M_o e^{-k_1 t} \qquad (7\text{-}37)$$

A plot of contaminant mass versus time produces a nonlinear relationship that can be linearized by plotting the natural log of contaminant mass versus time. The slope of this linearized relationship is equal to k_1.

Plume centerline concentration data can be used to quantify contaminant degradation rates if the dissolved plume mass does not change significantly over time, i.e., if a continuous source is indicated (*see* Table 7-4). Using the data reduction approach described in Equations 7-34 through 7-37 for dissolved plume mass, contaminant concentration data can be analyzed using zero-order reactions with Equation 7-38.

$$dC/dt = -k_o \quad (7\text{-}38)$$

in which k_o is the zero-order degradation rate, in units of mass per volume per time. The integrated form of this equation is shown in Equation 7-39.

$$C = C_o - k_o t \quad (7\text{-}39)$$

in which C is the contaminant concentration at time t, in units of mass per volume; and C_o is the initial contaminant concentration at time t equals zero, in units of mass per volume. A plot of contaminant concentration versus time produces a linear relationship, the slope of which equals k_o, and whose intercept value should equal C_o.

First-order reactions using contaminant concentration data are written as:

$$dC/dt = -k_1 C \quad (7\text{-}40)$$

in which k_1 is the first-order degradation rate, 1/time. The integrated form of this equation is shown in Equation 7-41.

$$C = C_o e^{-k_1 t} \quad (7\text{-}41)$$

A plot of the natural log of contaminant concentration versus time is linear when first-order degradation is taking place, with the slope of this linearized relationship equal to k_1. This data analysis approach has been incorporated into a natural attenuation protocol developed by Chevron Research and Technology Company (Buscheck et al., 1993).

Calibration of analytical fate and transport groundwater models. The use of a contaminant fate and transport model provides the best estimate of contaminant degradation rates when a continuous source scenario is observed at a site, as these models integrate transport, retardation, and degradation using site-specific contaminant and aquifer properties. An analytical, one-dimensional flow, three-dimensional dispersion model developed by Domenico (1987) is one such model that can be incorporated into this natural attenuation methodology. Use of this model accounting for flow and contaminant sorption characteristics in addition to degradation yields a dilution-corrected degradation rate. Used with nondegradable tracer compounds (i.e., dimethylpentane or trimethylbenzene isomers) existing in the source area, flow calibration can be accomplished leading to improved degradation rate estimates for the more reactive compounds (i.e., BTEX) of health significance.

Equation 7-42 is the form of the advection-dispersion equation (ADE) presented by Domenico (1987) that describes contaminant transport in three dimensions. The first three terms of this equation describe contaminant dispersion in the x, y, and z directions. The fourth term describes contaminant advection with the moving groundwater, whereas the last term on the left side of Equation 7-42 is a generic kinetic term used to simulate processes that result in the degradation of the contaminant during migration.

$$\alpha_x v \frac{\partial^2 C}{\partial x^2} + \alpha_y v \frac{\partial^2 C}{\partial y^2} + \alpha_z v \frac{\partial^2 C}{\partial z^2} + v_x \frac{\partial C}{\partial x} - \lambda C = \frac{\partial C}{\partial t} \qquad (7\text{-}42)$$

The analytical solution for the ADE given in Equation 7-42 for a continuous source is provided in Equation 7-43 (Domenico, 1987),

$$C(x,y,z,t) = \left(\frac{C_o}{8}\right)\left(\exp\left\{\left(\frac{x}{2\alpha_L}\right)\left[1-\left(1+\frac{4\lambda\alpha_L}{v_r}\right)^{1/2}\right]\right\}\right)\left(\mathrm{erfc}\left[\frac{x - v_r t \left(1 + \frac{4\lambda\alpha_L}{v_r}\right)^{1/2}}{2(\alpha_L v_r t)^{1/2}}\right]\right)$$

$$\left(\mathrm{erf}\left[\frac{\left(y+\frac{Y}{2}\right)}{2(\alpha_T x)^{1/2}}\right] - \mathrm{erf}\left[\frac{\left(y-\frac{Y}{2}\right)}{2(\alpha_T x)^{1/2}}\right]\right)\left(\mathrm{erf}\left[\frac{(z+Z)}{2(\alpha_z x)^{1/2}}\right] - \mathrm{erf}\left[\frac{(z-Z)}{2(\alpha_z x)^{1/2}}\right]\right) \qquad (7\text{-}43)$$

in which C(x,y,z,t) is the concentration at point x, y, and z at time t of a given contaminant C, with units of mass per volume; Co is the initial concentration of contaminant C adjacent to the source area, with units of mass per volume; v_r is the retarded groundwater velocity equal to v/R, with units of length per time; v is the aquifer pore water velocity, with units of length per time; R is the contaminant retardation factor, which is unitless; λ is the contaminant first-order decay constant, 1/time; α_L is the dispersivity in the x direction (longitudinal), with units of length; α_T is the dispersivity in the y direction (transverse), with units of length; α_Z is the dispersivity in the z direction (vertical), with units of length; Y is the source dimension in the y direction, with units of length ; and Z is the source dimension in the z direction, with units of length. This solution assumes a constant plane source perpendicular to the direction of ground-water flow, and that ground-water velocity is one-dimensional, i.e., no vertical flow occurs within the flow field. Hydraulic and chemical properties affecting the transport of contaminants within the subsurface, and incorporated into the multidimensional transport model given in Equation 7-43, include aquifer pore water velocity and dispersivity and contaminant retardation. The following describes methods used in the determination of these parameters for input into the modeling effort at study sites.

Pore water velocity. Aquifer pore water velocities can be calculated based on measured values of hydraulic gradient and hydraulic conductivity and estimated values of total aquifer porosity using Darcy's law (Equation 7-44).

$$v = K\ (\partial H/\partial L)/\theta \qquad (7\text{-}44)$$

in which K is hydraulic conductivity, with units of length per time; ∂H is the change in groundwater table elevation, with units of length; ∂L is the corresponding horizontal distance between head measurements, with units of length; and θ is the porosity, which is unitless.

Sorption coefficient/retardation factor. The term retardation factor, R, defines the reduction in contaminant velocity in an aquifer because of its sorption to aquifer solids. It is the factor by which pore water velocity is reduced to estimate contaminant velocity in groundwater systems. The retardation factor is a function of the soil/water partition coefficient of the compound, bulk density, and porosity of the aquifer, as defined by Equation 7-45.

$$R = 1 + (\rho_b\ K_d)/\theta \qquad (7\text{-}45)$$

in which R is the retardation factor, which is unitless; ρ_b is the soil bulk density, with units of mass per volume; and K_d is the soil/water partition coefficient, with units of volume per mass.

Compound soil organic carbon normalized distribution coefficients, K_{oc}, available from the literature, can be used to provide estimates of compound K_d values using the relationship between K_d and K_{oc} as follows:

$$K_d = K_{oc}\ (f_{oc}) \qquad (7\text{-}46)$$

in which f_{oc} is the weight percent organic carbon in the aquifer material.

Dispersivity. Based on current practice in the field, a longitudinal dispersivity of 0.1 times the plume length is generally used. Transverse dispersivity can be assumed to be four to 20 times less than longitudinal dispersivity. Unless a significant vertical component to flow is known to occur at a site because of vertical velocity gradients or vertical density-driven flow, vertical dispersivity is generally assumed to be negligible (0.0003 m). The vertical flow component can be evaluated based on a measurement of the vertical distribution of contamination indicated by multilevel sampling probes. The reader is referred to ASTM (1995) and Newell and McLeod (1996) for more details regarding the estimation of dispersivity values for specific site conditions.

The following is provided as an example of the application of Equation 7-43 for the prediction of contaminant concentrations produced downgradient from a source area.

Given and assumptions: A continuous source of groundwater contamination has been identified at the site of a former underground storage tank. The source area is approximately 30 m (100 ft) wide, with a smear zone of 1.8 m (6 ft) thick below the water table. The plume appears to extend 37.5 m (125 ft) downgradient from the source area and shows very little vertical migration. The aquifer material has a bulk density of 1500 kg/m³, a total porosity of 0.3, and a soil organic carbon content of 0.2%. The groundwater velocity is 0.045 m/day (0.15 ft/day) as determined from hydraulic gradient and conductivity measurements. The contaminant's K_{oc} is 100 mL/g, its groundwater concentration at the source area is 30 mg/L, and its estimated first-order degradation rate, λ, in the aquifer estimated from Equation 7-41 is 0.005/d.

Find: Using Equation 7-43, predict the contaminant plume as it develops over time and space from the source material, and estimate the time that it takes for the plume to reach steady-state conditions. Carry out calculations using Equation 7-43 for groundwater concentrations 0.3 m (1 ft) below the water table.

Step 1: The dispersivity values are estimated based on the observed characteristics of the contaminant plume using the relationships described.

$\alpha_L = 0.1$ (plume length) = 0.1 (37.5 m) = 3.75 m (7-47)

$\alpha_T = \alpha_L/4 = (3.75$ ft$)/4 = 0.94$ m (7-48)

$\alpha_Z = 0.0003$ m if vertical dispersivity is stated to be insignificant (7-49)

Step 2: A K_d for the compound is calculated using Equation 7-46.

$K_d = K_{oc} (f_{oc}) = (100$ mL/g$) (0.002) = 0.2$ mL/g (7-50)

Step 3: A retardation factor is determined for the compound using Equation 7-45, realizing that 1,500 kg/m³ equals 1.5 kg/L equals 1.5 g/mL.

$R = 1 + (\rho_b K_d)/\theta = 1 + (1.5$ g/mL$) (0.2$ mL/g$)/0.30 = 1 + 1 = 2$ (7-51)

Step 4: Summarize the input data to be used for calculations using Equation 7-43 as summarized below.

Input Variable	Value Used for Calculations
α_L	3.75 m (12.5 ft)
α_T	0.94 m (3.12 ft)
α_Z	0.0003 m (0.001 ft)
λ	0.005/day
R	2
v	0.045 m/day (0.15 ft/day)

Input Variable	Value Used for Calculations
Co	30 mg/L
Y	30 m (100 ft)
Z	1.8 m (6 ft)
z	0.3 m (1 ft)

Step 5: Solve Equation 7-43 for various time increments and at various positions within the aquifer using the input data presented in Step 4. An example of the intermediate calculations involved in each concentration estimate are provided for 1 yr or 365 d after the release at a downgradient distance, x, of 3 m (10 ft) and a cross-plume distance, y, of 6 m (20 ft).

Term $Co/8 = (30 \text{ mg/L})/8 = 3.75 \text{ mg/L} = 3,750 \text{ µg/L}$ (7-52)

Term $x/(2 \alpha_L) = (3 \text{ m})/[2 (3.75 \text{ m})] = 0.4$ (7-53)

Term $v_r = v/R = (0.045 \text{ m/day})/2 = 0.0225 \text{ m/day}$ (7-54)

Term $(4 \lambda \alpha_L)/v_r = [4(0.005/\text{day})(3.75 \text{ m})]/(0.0225 \text{ m/day}) = 3.33$ (7-55)

Term $2(\alpha_L v_r t)^{1/2} = 2[(3.75 \text{ m})(0.0225 \text{ m/day})(365 \text{ days})]^{1/2}$
$= 2(30.8 \text{ m}^2)^{1/2} = 2(5.55 \text{ m}) = 11.1 \text{ m}$ (7-56)

Term $2(\alpha_T x)^{1/2} = 2 [(0.94 \text{ m}) (3 \text{ m})]^{1/2} = 2 (2.82 \text{ m}^2)^{1/2}$
$= 2 (1.68 \text{ m}) = 3.36 \text{ m}$ (7-57)

Term $2(\alpha_Z x)^{1/2} = 2[(0.0003 \text{ m})(3 \text{ m})]^{1/2} = 2 (0.0009)^{1/2}$
$= 2 (0.03 \text{ m}) = 0.06 \text{ m}$ (7-58)

Term
$$1 - \left(1 + \frac{4 \lambda \alpha_L}{v_r}\right)^{1/2} = 1 - (1 + 3.33)^{1/2} = 1 - (4.33)^{1/2}$$
$= 1 - 2.08 = -1.08$ (7-59)

Term
$$\exp\left\{\left(\frac{x}{2\alpha_L}\right)\left[1 - \left(1 + \frac{4 \lambda \alpha_L}{v_r}\right)^{1/2}\right]\right\} = \exp[(0.4)(-1.08)]$$
$= \exp(-0.432) = 0.649$ (7-60)

Term
$$\text{erfc}\left[\frac{x - v_r t \left(1 + \frac{4 \lambda \alpha_L}{v_r}\right)^{1/2}}{2 (\alpha_L v_r t)^{1/2}}\right] = \text{erfc}\left(\frac{3 \text{ m} - (0.0225 \text{ m/d}) (365 \text{ d}) (1 + 3.33)^{1/2}}{(11.1 \text{ m})}\right)$$

$$\text{erfc}\left(\frac{3 \text{ m} - (8.21 \text{ m}) (4.33)^{1/2}}{11.1 \text{ m}}\right) = \text{erfc}\left(\frac{3 \text{ m} - (8.21 \text{ m}) (2.08)}{11.1 \text{ m}}\right)$$

$$\text{erfc}\left(\frac{3 \text{ m} - 17.08 \text{ m}}{11.1 \text{ m}}\right) = \text{erfc}\left(\frac{-14.08 \text{ m}}{11.1 \text{ m}}\right) = \text{erfc}(-1.27) \tag{7-61}$$

and by definition: $\text{erfc}(-1.27) = 1 + \text{erf}(1.27) = 1 + 0.9275 = 1.9275$ (7-62)

Term

$$\text{erf}\left[\frac{\left(y + \frac{Y}{2}\right)}{2(\alpha_T x)^{1/2}}\right] - \text{erf}\left[\frac{\left(y - \frac{Y}{2}\right)}{2(\alpha_T x)^{1/2}}\right]$$

$$= \text{erf}\left(\frac{6 \text{ m} + (30 \text{ m}/2)}{3.36 \text{ m}}\right) - \text{erf}\left(\frac{6 \text{ m} - (30 \text{ m}/2)}{3.36 \text{ m}}\right)$$

$$= \text{erf}(21/3.36) - \text{erf}(-9/3.36) = \text{erf}(6.25) - \text{erf}(-2.68)$$
$$= 1.0 - (-0.99985) = 1.99985 \tag{7-63}$$

Term

$$\text{erf}\left[\frac{(z + Z)}{2(\alpha_z x)^{1/2}}\right] - \text{erf}\left[\frac{(z - Z)}{2(\alpha_z x)^{1/2}}\right] = \text{erf}\left(\frac{(0.3 \text{ m} + 1.8 \text{ m})}{0.06 \text{ m}}\right) - \text{erf}\left(\frac{(0.3 \text{ m} - 1.8 \text{ m})}{0.06 \text{ m}}\right)$$

$$= \text{erf}(2.1/0.06) - \text{erf}(-1.5/0.06) = \text{erf}(35) - \text{erf}(-25) = 1 - (-1) = 2 \tag{7-64}$$

The concentration at a t of 1 yr, using an x of 3 m, a y of 6 m, and a z of 0.3 m is then:

$$C(3,6,0.3,1) = (3,750 \text{ µg/L})(0.649)(1.9275)(1.99985)(2) = 18,762 \text{ µg/L} \tag{7-65}$$

The results of additional calculations carried out for a time of 0.25 to 5 yr, using an x of 0.003 to 75 m and a y of zero to 30 m using an Excel spreadsheet are shown in Table 7-5.

Step 6: Based on the results presented in Table 7-5, it appears that the plume reaches approximately steady-state conditions by year 5 after the release of the contaminant. From that time on, the concentrations within the dissolved plume will remain constant until the source material weathers to the point that the flux of contaminant into the aquifer is reduced.

Model calibration. Both statistical and visual methods should be used to select model input values that best match contaminant groundwater concentrations observed over time at field sites. Model parameters that are varied to fit the measured data include elapsed time since contaminant release and various contaminant degradation rates. Model parameters that are held constant during calibration and model simulation runs include contaminant sorption coefficients, aquifer pore water velocities and dispersivity values, source area dimensions, and the initial source strength. With the hydraulic parameters and source dimensions set for a site, only the time since contaminant release and each contaminant degradation rate are changed to produce the best model fit to observed field data for plume centerline concentrations observed at a site.

The best-fit model results are selected based on the mean square error (MSE). The MSE is used to determine goodness of fit of the model along the centerline transect of the plume. The MSE represents the sum of the square of the difference between the actual data points and model estimates, normalized by the number of data points available for model fit evaluation. The lower the value of the MSE, the better the model predictions fit the observed data. The MSE is determined from the output of the fate-and-transport model versus the observed centerline concentrations for each contaminant, for a range of time and length values. The best-fit time and λ values are selected that yield the smallest MSE for each of the contaminants of concern at the site. Equation 7-66 gives the equation for the MSE term at a given time.

$$MSE = \frac{\sum_{i=1}^{n} \left(C_{i(observed)} - C_{i(predicted)} \right)^2}{n} \qquad (7\text{-}66)$$

in which $C_{i(observed)}$ and $C_{i(predicted)}$ are the observed (measured) and predicted concentrations, respectively, at point i, with units of mass per volume; and n is the number of observations. Once trends in MSE values are identified, continual refinement of the modeling effort can be carried out by visual data fitting to further minimize the calculated MSE.

The following is a summary of the step-wise procedures carried out during model calibration efforts:

- Hydraulic properties for the aquifer at a field site are set constant at values determined from field measurements or selected as representative of site aquifer conditions.
- The source vertical dimension and the simulated plume elevation are set constant at values based on the thickness of the contaminated water column observed at each site.
- The source lateral dimension and contaminant site characteristics are set constant to the values appropriate for each contaminant of concern at the site.
- A range of values for the time since contaminant release in the source area (based on known site history), is determined and is used for all contaminants of interest at the site. A given value of t is used for all contaminants along with contaminant-specific λ values to determine MSE values generated for the model estimate of the centerline profile for each contaminant. This MSE is generated by comparing

Table 7-5 Summary results for contaminant concentrations using Equation 7-43 with input data as presented in the example problem.

| | 0.25 Years | | | | 1 Year | | | |
| | Y Direction | | | | Y Direction | | | |
X Direction (m)	0 m	6 m, -6 m	15 m, -15 m	30 m, -30 m	0 m	6 m, -6 m	15 m, -15 m	30 m, -30 m
0.003	25,842	25,842	12,921	0.0	29,546	29,546	14,773	0.0
3	12,209	12,208	6,104	0.00	18,763	18,762	9,379	0.0
7.5	2,090	2,073	1,045	0.07	9,044	8,971	4,522	0.3
15	10.8	10.3	5.4	0.03	2,077	1,993	1,044	4.9
30	0.0	0.0	0.0	0.0	18.9	17.5	9.9	0.5
45	0.0	0.0	0.0	0.0	0.0	0.0	0.0	0.0
75	0.0	0.0	0.0	0.0	0.0	0.0	0.0	0.0
	2 Years				3 Years			
	Y Direction				Y Direction			
X Direction (m)	0 m	6 m, -6 m	15 m, -15 m	30 m, -30 m	0 m	6 m, -6 m	15 m, -15 m	30 m, -30 m
0.003	29,956	29,956	14,978	0	29,985	29,985	14,992	0.0
3	19,415	19,414	9,708	0	19,460	19,458	9,730	0.0
7.5	10,088	10,006	5,044	0.3	10,164	10,081	5,082	0.3
15	3,288	3,156	1,652	7.7	3,419	3,281	1,718	8.0
30	245	226	128	5.8	356	329	187	8.5
45	6.7	6.2	3.8	0.4	27.7	25.5	15.4	1.6
75	0.0	0.0	0.0	0.0	0.0	0.0	0.0	0.0
	4 Years				5 Years			
	Y Direction				Y Direction			
X Direction (m)	0 m	6 m, -6 m	15 m, -15 m	30 m, -30 m	0 m	6 m, -6 m	15 m, -15 m	30 m, -30 m
0.003	29,987	29,987	14,993	0.0	29,987	29,987	14,994	0.0
3	19,463	19,462	9,732	0.0	19,463	19,462	9,732	0.0
7.5	10,170	10,087	5,085	0.3	10,170	10,087	5,085	0.3
15	3,431	3,292	1,724	8.1	3,432	3,293	1,724	8.1
30	376	347	197	9.0	378	349	198	9.0
45	38.1	35.0	21.2	2.2	40.5	37.2	22.5	2.3
75	0.2	0.1	0.1	0.0	0.3	0.3	0.2	0.0

model estimates to measured contaminant concentrations at various distances downgradient from the source area along the plume centerline.
- The λ values are then varied over ranges applicable for each contaminant to provide minimum MSE values for each contaminant. A new value of t is selected to evaluate its impact on contaminant MSEs, and λ values are once again varied for each contaminant until minimum MSEs are generated. This procedure is repeated until a combination of t and λ produces the smallest MSE values for all contaminants.
- Based on λ equals zero versus calibrated λ degradation rates, a determination is made regarding evidence for biologically mediated contaminant degradation based on significant differences observed between these two simulation runs.

- Finally, the effects of source removal on the lifetime of the plume and the maximum plume travel distance are assessed using the site-specific, field-data calibrated model.

Details of the application of Domenico's fate-and-transport model to a number of field sites using this stepwise procedure are provided in Dupont et al., (1996, 1998a, 1998b) and Gorder et al. (1996).

Use of modeling in intrinsic remediation assessment. As indicated above, the use of groundwater fate-and-transport modeling is essential for the integration of contaminant- and site-specific parameters that control the overall fate of hazardous chemicals at a given site. Models can be used to assess existing monitoring data and to make determinations of both short- and long-term behavior of a contaminant plume under existing site conditions. They are also essential in providing quantitative estimates of the impact of source removal or source control on the ultimate size and duration of a contaminant plume. Models can be used effectively to evaluate the desirability and cost-effectiveness of implementing source removal at a given site, but they are only as good as the data put into them. The Domenico model will not provide meaningful results if the required input data cannot be reliably determined, or if dynamic flow and/or source release conditions at the site do not justify the use of this constant, plane source, one-dimensional groundwater velocity model.

7.2.2.2.5. Estimation of source mass and source lifetime. The next step in the protocol is the estimation of source area mass and the lifetime of this source mass caused by natural weathering processes. With an estimate of the rate of contaminant degradation taking place at a site, management decisions regarding the appropriateness of source removal actions and the effect of such actions on the projected lifetime of contamination at the site can be made.

If a finite source is observed at a site based on declining dissolved contaminant mass over time, the mass of contaminant in the source has been weathered to the point that contaminant flux from the source is decreasing rapidly. The lifetime of the source and plume can be estimated based on the degradation rate of the dissolved plume mass determined from Equations 7-34 through 7-37. Equations 7-35 or 7-37 would be used to determine the time to reach a final dissolved mass, M, based on the dissolved plume mass, Mo, that currently exists at the site. The final mass is determined from the endpoint that must be reached in the aquifer at the site in order for the site to be considered for closure. This may be the maximum contaminant level (MCL), or some alternative contaminant level based on a risk assessment for the site. The following example illustrates the determination of the lifetime of a finite, pulse source.

Given: The mass of a dissolved benzene plume produced from a source area has been found to be decreasing over time at a zero-order rate of 775 kg/yr using Equation 7-35. The mass of benzene that was found in the dissolved plume at the last monitoring event was 6,313 kg, with a maximum contaminant concentration observed in the plume of 379 μg/L, as indicated in Table 7-3. Based on a risk assessment conducted at the site, it was determined that a 75% reduction in the maximum contaminant concentration will be required to allow risk-based closure of the site.

Find: The time required for the dissolved plume to reach this risked-based contaminant level is unknown. The regulatory agency is also interested in the time required for 90% mass reduction in the dissolved plume by natural attenuation processes.

Step 1: Because the degradation of mass within the dissolved plume has been observed to be zero-order with respect to contaminant mass, the lifetime of the dissolved plume can be calculated by rearrangement of Equation 7-35, solving for the time for 75% reduction in the current mass of benzene existing at the site.

$$t = -(M - M_o)/k_o = -(0.25\ M_o - M_o)/k_o = (0.75\ M_o)/k_o \tag{7-67}$$

Step 2: Substituting for M_o and k_o allows the determination of the time to 75% mass removal as follows:

$$t = [0.75\ (6{,}313\ kg)]/(775\ kg/yr) = 6.1\ yr \tag{7-68}$$

Step 3: The time to reach 90% mass removal in the dissolved plume can be determined in a similar manner with the following approach:

$$t = -(M - M_o)/k_o = -(0.10\ M_o - M_o)/k_o = (0.90\ M_o)/k_o$$
$$= [0.90\ (6{,}313\ kg)]/(775\ kg/yr) = 7.3\ yr \tag{7-69}$$

It should be noted that the plume lifetime calculations summarized above indicate relatively short time periods and would be expected once the source material has been depleted of the contaminants of concern caused by natural weathering processes or by active removal or treatment of the source area. The predicted lifetimes for unweathered source material are projected to be much longer, as indicated later in this section.

If a continuous source is found at a given site, contaminant mass within the source area continues to contribute mass to the groundwater, maintaining the contaminant plume that has developed over time. To estimate the potential lifetime of this plume, an estimate must be made of contaminant existing as residual saturation both above and below the groundwater table, along with that residing in any free product that might be found at a site. These estimates should ideally be based on soil core and free product samples collected within the source area throughout the site. This total mass estimate requires that the soil and free product volumes associated with each soil core or monitoring location be defined using a procedure such as the Thiessen Polygon Method that was described above.

Figure 7-7 indicates the configuration of soil cores and associated geometry used in the following equations to determine average bore hole concentration, C_{ave}, and total contaminant mass, M_T, estimates in a source area.

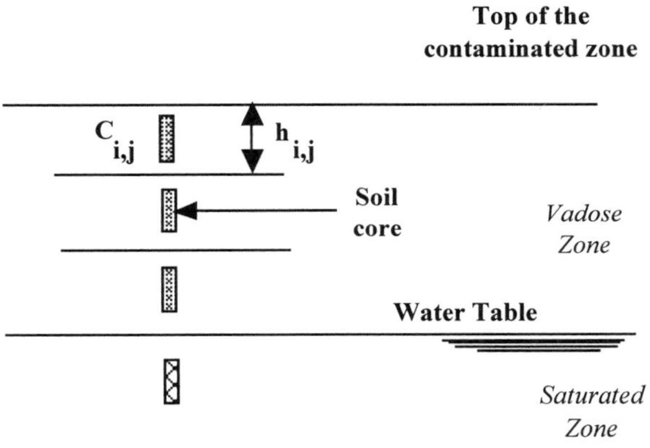

Figure 7-7 Configuration of soil cores and associated geometry used for average bore hole contaminant concentrations and total mass calculations.

$$C_{ave,j} = \frac{\sum_{i=1}^{n} C_{i,j} h_{i,j}}{\sum_{i=1}^{n} h_{i,j}} \tag{7-70}$$

in which $C_{i,j}$ is soil contaminant concentration in core j at depth i in the core, with units of mass contaminant per mass soil; $h_{i,j}$ is the core j interval thickness at depth i, with units of length; and n is the total number of soil cores collected at the site. Total source area mass is calculated as:

$$M_T = \sum_{j=1}^{n} A_j \left(\sum_{i=1}^{n} C_{i,j} h_{i,j} \right) \tag{7-71}$$

in which A_j is the Thiessen area associated with core j, with units of length squared.

The denominator in Equation 7-70 has two values depending on whether average concentrations are being described for above or below the water table. It is the thickness of vadose zone contamination for mass above the groundwater table, whereas it is the thickness of contaminated soil below the groundwater table for saturated zone mass determinations. Generally, total mass calculations provided in Equation 7-71 are carried out separately for mass above and below the groundwater table so that a picture of the vertical distribution of contaminant mass can be developed.

If soil core data are not available, then contaminant mass within a source area can be estimated based on a determination of the extent and composition of free product or residual saturation that exists at a site. The extent of free product can be estimated from observation well and monitoring point locations that are found to contain free product during sampling. Recognizing that the thickness of product observed in a monitoring well can be substantially greater than that actually existing within the formation, care must be taken in interpreting monitoring well product thickness results. Attempts should be made, however, to estimate free product volume both above and below the water table so that source term lifetime predictions can be attempted. Knowing the composition of contaminant within the free product at a site, estimating the lateral and vertical extent of free product distribution, and estimating the formation total porosity allow the estimation of the mass of contaminant existing within the free product at the site, M_{fp}, as follows:

$$M_{fp} = (Area)(Thickness)\,\theta\,\rho_{fp}\,C_i \qquad (7\text{-}72)$$

in which Area is the estimated aerial extent of free product, with units of length squared; Thickness is the estimated vertical extent of free product, with units of length; θ is the formation total porosity, expressed as a decimal; ρ_{fp} is the free product density, with units of mass per length cubed; and C_i is the contaminant concentration within the free product, with units of mass per mass.

Product may exist at a site as residual saturation if there are high dissolved contaminant concentrations but no free product is observed within the sampling network. Estimates of the maximum amount of residual phase product existing at a site can be made based on the characteristics of the soil at a site using the quantitative relationships presented by Parker et al. (1987) and Mobil Oil Corporation (1995). These relationships describe the typical residual hydrocarbon saturation within the smear zone at and below the groundwater table as a function of soil texture. For coarse sands, a residual saturation of 25% of the total pore volume is suggested, while this value drops to 15% for sandy silts and fine sands and to only 5% for silty clays.

With residual phase product, the composition of the product can be inferred from groundwater concentration data. This can be done by assuming that equilibrium exists between the residual phase and the groundwater and by using the following relationships based on an analogy to Raoult's law, along with an assumed molecular weight of the residual product, MW_{fp}, and the known aqueous solubility of the individual compounds of interest.

Equilibrium Concentration = (Mole Fraction) (Aqueous Solubility) (7-73)
Mole fraction = (equilibrium concentration)/(aqueous solubility) (7-74)
Moles in product = (mole fraction) $(M_{fp})/MW_{fp}$ (7-75)
Mass in product = (moles in product) $(MW_{compound})$ (7-76)

The use of this procedure for estimating residual phase product and contaminant mass in the source area is shown in the field site example below.

Given: A site at which former gasoline tanks had released product to the soil and groundwater is the location of an intrinsic remediation study. No free product, only a light sheen, has been observed in soil core and ongoing groundwater monitoring samples. However, TPH and BTEX groundwater concentrations approaching levels in equilibrium with free product have been observed throughout much of the site. These elevated groundwater concentrations were used to delineate the apparent aerial extent of residual phase material, representing approximately 752.4 m^2 (8,360 ft^2). The vertical extent of potentially contaminated soil consists of 3 m (10 ft) of measured groundwater contamination and 1.05 m (3.5 ft) of capillary fringe and smear zone. The soil is composed of fine sands and clays. The specific gravity and molecular weight of the product are 0.8 and 120 g/gmol, respectively. Contaminant concentrations in the groundwater in equilibrium with the product were analyzed by gas chromatography, as shown in Table 7-6.

Table 7-6 Summary of estimated total residual contaminant mass based on residual product volume estimates and dissolved plume mass measured at a field site.

Compound	MW (g/gmol)	Aqueous Solubility (mg/L)	Measured Groundwater Concentration (mg/L)	Mole Fraction	gmol in Product	Mass in Product (g)
Benzene	78	1,780	4.9	0.00275	2,512	195,936
Toluene	92	759	3.2	0.00422	3,854	354,568
Ethylbenzene	106	135	1.9	0.01407	12,851	1,362,206
p-Xylene	106	221	6.3	0.02851	26,039	2,760,134
Naphthalene	128	30.6	0.79	0.02582	23,582	3,018,496
TPH	120		86.1			109,600,000

Find: The volume of contaminated soil existing below the site that contains this residual saturation, the volume of residual product estimated to be contained within this soil volume, and the mass of residual product and each contaminant of interest that must be assimilated before natural attenuation processes allow site closure.

Step 1: The total volume of contaminated soil is estimated based on the area of the apparent residual phase and the thickness of contaminated soil below the site. Using the data given, the volume of contaminated soil is:

Soil Volume = (Area) (Thickness) = (752.4 m^2) (4.05 m) = 3,047 m^3 (7-77)

Step 2: The total pore volume of contaminated soil is then estimated based on the assumed porosity for the contaminated soil. For fine sands and clays, a total porosity is estimated to be 0.3.

Pore Volume = (Soil Volume) (Porosity) = (3,047 m^3) (0.3) = 914 m^3 (7-78)

Step 3: The volume of residual phase contamination that is estimated to exist at the site is then determined from the predicted volume of pore space that is occupied by hydrocarbon liquids at residual saturation in the soil type existing at the site. Using Parker et al. (1987), it is estimated that 15% of the pore volume of the fine sand and clay soil at the site contains this residual product.

Product Volume = (Pore Volume) (% Residual Saturation, fraction)
= (914 m^3) (0.15) = 137 m^3 (7-79)

Step 4: The mass of product is estimated based on this product volume and the specific gravity of the product.

Mass of Product = (Product Volume) (specific gravity) (unit weight of water)
= (137 m^3) (0.8) (1000 kg/m^3) = 109,600 kg (7-80)

Step 5: From this estimate of product mass and the measured concentration of specific components in equilibrium with this product in the groundwater, the mole fraction of each contaminant of interest within the source area residual product can be estimated using Equation 7-28. This calculation for benzene is shown below, and results of this calculation for all other compounds of interest are summarized in Column 5 of Table 7-6.

Mole fraction = (groundwater concentration)/(aqueous solubility)
Benzene mole fraction = (4.9 mg/L)/(1,780 mg/L) = 0.00275 (7-81)

Step 6: The number of moles of each contaminant in the source area is then calculated using Equation 7-29. The calculations shown are for benzene, with all other results summarized in Table 7-6.

Moles of Contaminant in Product = (Mole Fraction) (M_{fp})/(MW_{fp}) (7-82)
Moles of Benzene = (0.00275) (109,600,000 g)/(120 g/gmol) = 2,512 gmol (7-83)

Step 7: The mass of each constituent in the residual product within the source area is then estimated using Equation 7-30. The calculations shown are for benzene, with all other results summarized in Table 7-6.

Mass of Contaminant in Product = (Moles in Product) ($MW_{compound}$) (7-84)
Mass of Benzene = (2,512 gmol) (78 g/gmol) = 195,936 g (7-85)

Once the total mass of contamination is estimated above and below the groundwater table, estimates for the total lifetime of the plume can be made based on the approach described above for a plume generated from a continuous source. The total mass disappearance rate is equated to the contaminant degradation rate determined from Equation 7-14 through 7-17 to yield an estimate of total plume lifetime, $T_{continuous}$, of:

$T_{continuous}$, zero = ($Mass_V$ + $Mass_{SZ}$ + M_T)/k_o (7-86)
$T_{continuous}$, first =- ln[M/($Mass_V$ + $Mass_{SZ}$ + M_T)]/k_1 (7-87)

for zero- and first-order degradation rate relationships, respectively, in which $Mass_V$ is the contaminant mass located above the groundwater table in the source area; $Mass_{SZ}$ is the contaminant mass located below the groundwater table in the source area; and $Mass_T$ is the contaminant mass located in the dissolved groundwater plume.

7.2.2.2.6. Predicting the long-term behavior of the dissolved plume. The long-term behavior of a contaminant plume is impacted both by the characteristics of the source, affecting the duration of the release of contaminant into the aquifer, and by the characteristics of the aquifer itself, affecting the transport and degradation of contaminant once it is released from the source area. The procedure for evaluating long-term source behavior is based on whether the plume is generated from a finite or continuous source.

If the plume can be considered generated from a finite source, no residual source area exists, and the long-term behavior of the plume is related to the projected lifetime of the plume (Equations 7-35 or 7-37). If the site is shown to contain a significant source area, producing a "continuous source" plume, long-term plume behavior can be evaluated based on various source removal scenarios. If no source removal is to be carried out, a worst-case scenario exists in terms of the length of time the plume will persist, as the plume lifetime calculations for the sum of vadose zone, saturated zone, and dissolved plume masses estimated above apply (Equations 7-86 or 7-87). If contaminant source removal is being considered, analysis of the effect on plume lifetime of mass removal from various locations at the site and at various levels of removal efficiency can be carried out using the expressions shown below.

$T_{continuous}$, zero = [(1-r) ($Mass_V$ + $Mass_{SZ}$ + M_T)]/k_o (7-88)
$T_{continuous}$, first =- ln[M/{(1-r) ($Mass_V$ + $Mass_{SZ}$ + M_T)}]/k_1 (7-89)

in which r is the removal of contaminant mass expressed as a decimal fraction. If 100% vadose and saturated zone source removal is assumed, the continuous source plume lifetime equation reduces to that of a finite source, as shown in Equations 7-35 and 7-37.

Once source removal strategies are investigated, the complete long-term behavior of the contaminant plume can be predicted using a groundwater fate-and-transport model. As indicated in a paper by Gorder et al. (1996), modeling of source removal activities can be carried out by superposition of a negative continuous source plume on top of the existing steady-state plume concentration profile. This negative plume is generated using a source concentration equal to the negative of the initial source concentration, at a point in time corresponding to the time of source removal. What this superposition accomplishes is the modeling of the movement of the steady-state plume away from a source area that has been eliminated following source removal activities. It is important to note that the steady-state contaminant plume profile represents the highest downgradient concentration profile that would be expected at a given site. The concentrations at a given point in space will decrease over time following source removal activities if the plume is truly at steady-state, and if all other site conditions remain the same over time.

7.2.2.2.7. Decision-making regarding natural attenuation. The decision to proceed with natural attenuation at a given site should be based on the impact the plume has on downgradient receptors and on the potential success of intrinsic attenuation reactions providing plume containment and control. This decision is made from a documentation of the TEA pool available to aquifer microorganisms for contaminant assimilation.

The final questions that must be answered regarding application of a natural attenuation management approach at a site are: (1) whether or not a sensitive receptor is being impacted now or will be in the future when the plume is projected to reach steady-state conditions; and (2) whether or not the projected lifetime of the plume is acceptable to the owner/operator, regulatory agencies, and other interested parties. In general, if an existing or projected receptor impact exists, active source removal and plume control/remediation will be required unless institutional controls (i.e., deed restrictions, etc.) can be put into place to restrict the long-term use of contaminated soil and/or groundwater. The issue of plume lifetime tends to be a more complicated one. If significant contaminant mass remains in the source area of a site, the plume that results may persist for decades. If remediation goals are established with shorter timeframes (i.e., for property transfer reasons, etc.), this assimilation time will likely not be acceptable, and active remediation may be required.

The focus of the discussion above has been on quantifying the transport and degradation of contaminants taking place under actual site conditions. Once a projected source lifetime is deemed acceptable and natural attenuation is considered viable at a site, final supporting evidence for verification that degradation reactions are biologically mediated must be provided through an analysis of the changes in background TEA mass compared to that within the plume itself. If contaminant biodegradation is taking place, indigenous organisms will utilize TEAs (dissolved oxygen, nitrate, dissolved iron and managenese, sulfate, carbon dioxide) at a rate and to an extent that should correspond to contaminant loss observed at the site. The stoichiometry associated with microbial metabolism known to occur under various TEA conditions allows a determination of the potential contaminant assimilative capacity of background groundwater moving into the source area and that available within the plume itself. If this theoretical assimilative capacity is equal to or greater than the level of contamination observed at the site, expressed both on a maximum concentration and total mass of contaminant basis, biological natural attenuation processes can be expected to play a major role in contaminant attenuation. If assimilative capacity is limited, some source removal and/or active site remediation action is likely warranted. The assumed stoichiometric relationships for various TEA conditions are indicated in Table 7-7 for alkane and aromatic compounds normally found at petroleum contaminated sites.

Table 7-7 Potential aromatic and aliphatic hydrocarbon assimilative capacity relationships for TEAs of importance at contaminated sites.

TEA Indicator	Compound Degraded	Molar Relationship (mol/mol HC Degraded)	Mass Relationship (wt/wt HC Degraded)
Oxygen	Aromatic	- 7.5	- 3.1
	Alkane	- 9.5	- 3.5
Nitrate	Aromatic	- 6	- 1.07
Fe^{3+} to Fe^{2+}	Aromatic	+ 30	+ 21.5
	Alkane	+ 38	+ 24.7
Mn^{4+} to Mn^{2+}	Aromatic	+ 15	+ 10.6
	Alkane	+ 19	+ 12.1
Sulfate	Aromatic	- 3.75	- 4.6
	Alkane	- 4.75	- 5.3
Organic to CH_4	Aromatic	+ 3.75	+ 0.77
	Alkane	+ 4.75	+ 0.88

It should be noted that negative values indicate TEA use in a reaction, whereas positive values indicate product generation in a given reaction. For solid phase reactants, i.e., iron and manganese, quantification of product generation will normally underestimate the total assimilative capacity associated with these TEAs.

Evaluation of potential site assimilative capacity is provided in the following example.

Given: A dissolved plume at an abandoned underground storage tank (UST) site has the maximum observed BTEX concentrations and upgradient and downgradient electron acceptor as reported below.

Find: The mass of each electron acceptor utilized within the contaminant plume is unknown, as is the corresponding assimilative capacity of the aquifer for BTEX and hydrocarbon contaminants. Also comment on the feasibility of long-term plume management using a natural attenuation approach.

Assumptions: The stoichiometry for TPH reactions can be approximated by the alkane relationships summarized in Table 7-7. In addition, no significant source of inorganic reducing agents exists at the site, so that all TEA utilization can be attributed to biodegradation processes.

Compound	Maximum Concentration (mg/L)
BTEX	4.6
TPH	10.0

TEA	Upgradient Concentration	Downgradient Concentration
Oxygen	6.5	0.5
Nitrate	9.5	0.5
Fe^{2+}	0.5	12.5
Mn^{2+}	0.5	0.5
Sulfate	45.5	2.5
CH_4	0.0	1.5

Step 1: Calculate the change in TEA concentration across the plume from upgradient and downgradient TEA concentration data presented in the problem statement. The results of these calculations are as follows:

TEA	Δ TEA (mg/L)	TEA Stoich. (g/g alkane)	TEA Stoich. (g/g BTEX)
Dissolved Oxygen	6	3.5	3.1
Nitrate	9		1.07
Iron	12	24.7	21.5
Manganese	0	12.1	10.6
Sulfate	43	5.3	4.6
Methane	1.5	0.88	0.77

Step 2: Calculate the hydrocarbon equivalent of the TEA utilization observed across the plume using the results of Step 1 and the stoichiometric relationships indicated in Table 7-7. These calculations for dissolved oxygen (DO) are detailed below, and they are summarized for all TEAs in tabular form below.

HC equivalent of DO used = (ΔDO, mg/L)/(g DO/g TPH used)
= (6.0 mg DO/L)/(3.5 g DO/g alkane used) = 1.71 mg/L HC   (7-90)

258

BTEX equivalent of DO used = $(\Delta DO, mg/L)/(g\ DO/g\ BTEX\ used)$
= (6.0 mg DO/L)/(3.1 g DO/g alkane used) = 1.93 mg/L BTEX   (7-91)

Step 3: Based on the estimated assimilative capacity of 12 to 22 mg/L hydrocarbon and the maximum observed contaminant concentration within the dissolved plume of only 5 to 10 mg/L, it appears that adequate assimilative capacity exists at the site to ensure long-term stability of the contaminant plume. Based on these calculations and observations of plume stability, a strong case could be made for implementation of natural attenuation for plume management at this site.

TEA	Δ TEA (mg/L)	TEA Stoich. (g/g alkane)	TEA Stoich. (g/g BTEX)	TPH Equiv. (mg/L)	BTEX Equiv. (mg/L)
DO	6	3.5	3.1	1.7	1.9
Nitrate	9		1.07		8.4
Iron	12	24.7	21.5	0.49	0.56
Manganese	0	12.1	10.6	0	0
Sulfate	43	5.3	4.6	8.1	9.3
Methane	1.5	0.88	0.77	1.7	1.9
Total Assimilative Capacity				12.0	22.2
Maximum Observed Concentration				10.0	4.6

7.2.2.2.8. Long-term monitoring. A long-term monitoring plan must be developed if a natural attenuation plume management approach is selected for a given site. It should be noted that the requirements of the monitoring strategy are twofold, namely for compliance purposes, as well as for natural attenuation process monitoring. A compliance monitoring program must be established to confirm that plume containment and risk management continue to take place at the site. Compliance monitoring normally involves the use of an upgradient, background monitoring well, one to two monitoring wells within the contaminant plume, and one to two downgradient compliance wells used to detect contaminant migration toward potential receptors. Groundwater elevation, contaminant concentrations, and minimal groundwater quality data (pH, temperature, total dissolved solids) are generally required to be reported from these monitoring wells.

The information generated for compliance monitoring purposes is necessary but not sufficient for natural attenuation process monitoring, and additional monitoring locations and analyte data should be collected for process monitoring purposes. Figure 7-8 shows a monitoring well network that is appropriate for initial natural attenuation evaluation during the site assessment phase, as well as for long-term compliance and natural attenuation process monitoring. The analytes that should be collected for natural attenuation process monitoring in addition to those collected for compliance monitoring include: TEAs that are utilized (DO, nitrate, sulfate); products that are formed (dissolved iron and manganese, methane) during contaminant biodegradation; and water quality characteristics of the contaminated aquifer, such as

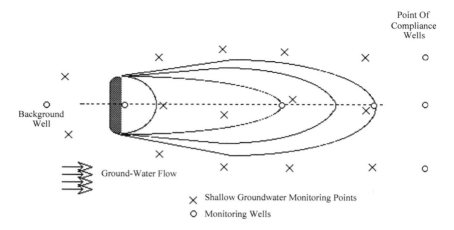

Figure 7-8 Example groundwater monitoring network applied at a natural attenuation site for both compliance and natural attenuation process monitoring.

alkalinity (an indicator parameter for the effect of microbial carbon dioxide production on a system's buffering capacity) and oxidation/reduction potential (an indicator parameter describing the relative oxidation state of a system in response to changes in TEA concentrations resulting from microbial metabolism).

With this network established and additional process monitoring data collected, the conceptual model of the site (including land use assumptions) and model calibration results can be periodically updated. These updates should provide ongoing refinements to source lifetime predictions and to risk assessment considerations for the site.

Finally, the frequency of groundwater monitoring must be established as part of the long-term monitoring plan. Compliance monitoring schedules generally require quarterly to annual sampling. Under most circumstances, however, annual sampling will be the shortest time interval necessary for natural attenuation process monitoring, as low groundwater velocities observed at most sites do not warrant more frequent sampling intervals. At a site with a groundwater velocity of 0.01 m/day (0.033 ft/day), unretarded groundwater moves less than 3.9 m (13 ft) in a year. With a retarded velocity one-third to one-sixth that of the groundwater velocity (appropriate for benzene and xylene, respectively), contaminant movement of less than 0.6 to 1.5 m (2 to 5 ft) would be expected over a 1-yr period. With a monitoring grid spaced at 10-m (approximately 30-ft) intervals, a 1-yr change in plume position cannot be detected. Again, the sampling interval should be assessed on a site-specific basis, but generally an annual to biennial sampling schedule should be sufficient to ensure that adequate data are collected while minimizing the sampling and analysis burden at natural attenuation sites.

7.3. Solid Phase Biological Treatment

Solid phase biological treatment systems describe those treatment processes designed to remove contaminant mass contained within contaminated soil using biochemical reactions. These solid phase treatment systems include both those applied ex situ following excavation of contaminated soil and in situ soil treatment technologies. In situ soil treatment systems include bioventing, biosparging, bioslurping, and the emerging area of phytoremediation (*see* Dupont et al. (1998a) for a recent review of phytoremediation principles as they apply to contaminated site remediation). The ex situ technologies of land treatment, soil piles, and slurry-phase reactors and the in situ technologies of bioventing and bioslurping are described in detail below.

7.3.1 Ex Situ Treatment

7.3.1.1 Land Treatment Systems

Land treatment systems are biological processes utilizing the upper soil zone for the treatment and ultimate disposal of waste materials utilizing the natural site/soil assimilative capacity for a given set of waste contaminants of concern. These systems rely on degradation, transformation, and immobilization of contaminants of concern within a treatment zone to provide cost-effective treatment of soils and waste materials contaminated primarily with biodegradable organic compounds. These systems are governed by the same fundamentals as the other bioremediation options discussed above, including requiring optimal soil water (55% to 90% field capacity), optimal soil pH (6 to 9), adequate oxygen (greater than 2 vol% in soil air pore space), and moderate soil temperatures ($10°$ to $30°$ C).

Land treatment units are constructed soil systems configured as shallow soil reactors, typically placed below grade. Historically these systems have been used for the treatment and disposal of industrial waste sludges from the petroleum refining, wood preserving, and food processing industries. Use of land treatment systems for industrial waste treatment and disposal has been drastically curtailed in response to RCRA Land Disposal Restrictions. However, the technology remains highly technically and cost-effective for many wastes, particularly for hydrocarbon-contaminated soils, and its use has actually been promoted for CERCLA remedial actions when appropriate.

As indicated in Section 7.1.3, the ex situ nature of the land treatment process has a number of distinct advantages over in situ treatment options. These include:

- Enhanced control over reactant delivery and contaminant/product recovery
- The opportunity to modify existing site soil and/or nutrient characteristics
- Enhanced control over migration pathways, when modified, to include:
  - "Tent" for volatiles, using enclosures
  - "Tub" for leachables, using liners and leachate collection in a prepared bed mode
- Management of the site to optimize bioactivity, including:
  - Aeration management — till for soil aeration
  - Moisture management — adjust to 55% to 90% field capacity
  - Nutrient management — optimize by improved mixing ex situ
  - Soil texture management — create a loam soil by blending
  - Toxicity management — blend to reduce contaminant concentrations below those toxic to the microorganisms

The philosophy of the land treatment approach is grounded in the concept of utilizing the natural assimilative capacity of the soil environment for the responsible management and disposal of wastes so that the soil used for treatment is not irretrievably removed from future beneficial use. This philosophy leads to the consideration of the characteristics of the contaminants of a waste to be managed and the interaction of these contaminants with the soil environment within the land treatment unit. Based on these contaminant/soil interactions, three distinct contaminant loading rate criteria can be defined, any of which can control overall system design. These three contaminant loading rate criteria are: application-limiting, rate-limiting, and capacity-limiting loading rates.

7.3.1.1.1. Application-limiting contaminant loading defines a waste loading rate limited by the release of a contaminant of concern from the treatment zone as a result of volatilization or leaching as affected by soil and micrometeorlogical conditions. The basis for design using this loading criterion is the mobility of contaminants from the treatment zone. A contaminant's mobility is compared to its degradability so a system can be designed to degrade/transform/immobilize the contaminant before it leaves the treatment zone. Typically, a laboratory evaluation of contaminant degradability and mobility is conducted to determine the effect of loading rate on the fate of contaminants in a land treatment system. One approach described in the literature is a volatilization-corrected degradation rate (VCDR) laboratory procedure (Dupont, 1986; Park et al., 1990), which utilizes a mass balance approach to quantitate contaminant loss via biotic, abiotic, and volatilization pathways that occur under simulated waste application and tilling conditions. The results of this laboratory evaluation are then used in the development of design constraints for a larger-scale field demonstration.

If a waste is highly volatile or contains a significant portion of highly soluble constituents that can leach readily from the land treatment unit, the rate of waste loading must be controlled to limit contaminant migration. Pretreatment to remove volatile constituents via vacuum stripping, or water soluble constituents by dewatering, can be very effective in removing the contaminants. This pretreatment step may effectively remove the constituent limiting the waste loading rate on an application-limiting basis and may be cost-effective in the long-term because, with higher loading rates, land areas for treatment are reduced.

7.3.1.1.2. Rate-limiting contaminant loading defines a waste loading rate limited by the migration of a contaminant of concern from the treatment zone as a result of limited degradation, transformation, and/or immobilization during its residence time in the treatment zone. This loading rate is defined by those contaminants that do not readily degrade, transform, or sorb to the solid matrix in the land treatment unit before they move out of the treatment zone. Even constituents that have high retardation factors can represent rate-limiting constituents if their degradation rates are also very low. The VCDR laboratory approach (Dupont, 1986; Park et al., 1990) has been used to assess the rate-limiting effects of loading rate on contaminant degradation through the assessment of the toxicity and mutagenicity of wastes in the soil environment. The assessment of waste toxicity to soil microorganisms is carried out using the Microtox™ toxicity assay, while the mutagenicity of waste/soil mixtures to possible receptors is determined using the Ames mutagenicity assay.

To overcome rate-limiting loading constraints, efforts must be focused on increasing contaminant reaction rates, while reducing their mobility within the land treatment unit. One common practice is to amend the land treatment unit with external carbon sources (straw, manure, etc.) to increase the bioactivity of the treatment zone (TZ), provide possible co-substrates, and increase the organic carbon level in the treatment soil to retard the migration of the contaminant.

7.3.1.1.3. Capacity-limiting contaminant loading describes a waste loading rate limited by the accumulation of nondegradable contaminants to levels within the land treatment unit that limit or prevent the future beneficial use of the waste-amended soil. Inorganic contaminants (i.e., heavy metals) that have plant and animal toxicity thresholds are typically the contaminants of concern for the capacity-limiting criterion. Highly chlorinated, high molecular weight compounds (i.e., dioxins, PCBs, etc.) can also accumulate to unacceptable levels within the land treatment unit if they are not monitored. The maximum contaminant concentration for nonbiodegradable constituents is the basis for design using capacity-limiting loading criteria.

Avoidance of capacity limitations is best done through a thorough analysis of a waste stream prior to selection of land treatment as a remediation alternative. For example, segregation of highly degradable oil-water separator sludges and waste oils from primarily inorganic waste streams from boiler clean-outs (alkaline cleaning operations, etc.) not amenable to biological treatment was easily accomplished at a midwestern refinery (Dupont, 1991). This allowed the large volume of biodegradable wastes to be cost-effectively treated in a land treatment system while separately disposing of the low volume, toxic inorganic waste without impacting the land treatment unit.

7.3.1.1.4. Process configuration. A typical land treatment system is conceptually shown in Figure 7-9. This land treatment unit consists of a biological system composed of soil with active, acclimated microorganisms; an adequate nutrient pool; an irrigation system for moisture control as needed; and adequate levels of appropriate electron acceptors in a constructed soil treatment bed. The soil treatment bed consists of microbially active, fertile soil of moderate to high organic carbon content, with good water holding capacity and cation exchange capacity, preferably of agricultural origin.

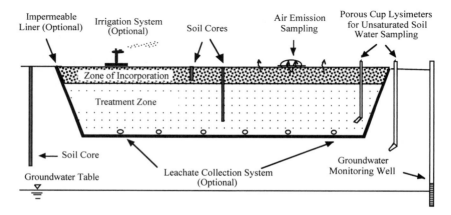

Figure 7-9 Schematic of a typical land treatment system with optional liner and leachate collection system, indicating the typical sampling type and locations.

An upper 0.15- to 0.45-m (0.5- to 1.5-ft) thick zone of incorporation (ZOI) is used to intimately mix the waste with soil in the land treatment unit. If the contaminated soil can support active microbial populations and does not contain toxic or inhibitory levels of contaminants of concern, then it may be placed directly in the land treatment unit in 6- to 18-in. layers without being blended with treatment soil. It is in the ZOI or upper soil zone that the primary degradation/transformation reactions take place. It is also this ZOI that provides oxygen transfer to the soil below it by surface tilling to encourage and maintain aerobic degradation reactions within the land treatment system.

A lower treatment zone extends from below the ZOI a total of 1.5 m (5 ft) below ground surface. The treatment zone is the soil zone in which additional contaminant degradation/transformation takes place, and where long-term contaminant immobilization capacity is provided through sorption of organics and precipitation surface exchange of metals.

Monitoring to ensure adequate and ongoing assimilation of contaminants within land treatment units is an important aspect of land treatment system operation. An extensive process monitoring system should be utilized at land treatment sites. An ideal monitoring system consists of:

- Unsaturated zone sampling to monitor soil pore liquid for the detection of contaminant migration via leachate
- Groundwater sampling if no liner is utilized to monitor for contaminant migration to the groundwater via vapor and/or leachate migration
- Surface emission/ambient air sampling, if necessary, to monitor for the release of volatile contaminants during waste application and tilling
- Soil core sampling to provide definitive verification of waste treatment via contaminant mass removal within the land treatment unit

If process monitoring indicates that sufficient contaminant degradation has taken place in the ZOI, additional contaminant loading can then take place, in essence reusing the assimilative capacity of the ZOI until reaching the capacity limitations for the soil. A first-cut determination of sufficient contaminant degradation should be made based on measurements of parent compound concentrations below regulatory cleanup levels within the ZOI. In addition, reductions in toxicity or mutagenicity of the land treatment unit soil should also be assessed to ensure that no toxic or mutagenic daughter products are accumulating in the ZOI soil over time.

7.3.1.1.5. Process design for land treatment systems is based on a general hierarchy that has been developed in the "Permit Guidance Manual on Hazardous Waste Land Treatment Demonstrations" (Sims et al., 1986). This hierarchy includes evaluation of the land treatability of a given waste using first laboratory-scale treatability studies, then a field demonstration, followed by modeling of long-term performance at full-scale.

Laboratory treatability studies are designed to evaluate the potential for the treatability of a given waste/soil mixture using land treatment technology. This stage of land treatment design is used to determine the general acceptability of the technology and to identify potential limitations to its efficacy based on incompatibilities between the waste and the soil being used for the treatment medium. A mass balance approach is used for fate and transport assessment via the VCDR method described above and results in the quantification of contaminant degradation/transformation and volatilization rates as a function of loading rate, loading method (i.e., surface versus subsurface application), loading frequency, tilling frequency, application of amendment, etc. Conducting an initial assessment of waste treatability provides a rapid and cost-effective way to screen a range of waste treatment options prior to their assessment at a limited field scale.

For treatment of wastes or contaminated soils that have been shown in the literature to be amenable to a land treatment in a wide range of soils (i.e., fuel contaminated soils, petroleum production wastes, wood preserving wastes) using land treatment technology, the phased design process can be greatly simplified. What must be determined, however, on a case-by-case basis is:

- Whether there are any constituents within an otherwise land treatable waste that may prove toxic to soil microorganisms in the land treatment unit
- Whether the site soil can be utilized as a treatment medium (i.e., whether the site soil has desirable properties leading to contaminant degradation and attenuation)
- Whether there is adequate soil volume available for on-site treatment based on acceptable application-, rate-, and capacity-limiting loading rates

These questions may be answered by conducting limited laboratory treatability studies (i.e., using the site soil and a readily available alternative soil at two loading rates, with stoichiometric nutrient requirements and soil moisture at 85% field capacity) to assess the volatilization and toxicity potential of the contaminated soil and to determine relative degradation rates expected for contaminants of concern. These studies would be carried out using the steps outlined in Section 7.4.

Field demonstrations are designed to test the results of the laboratory-scale study for "fatal flaws" at the field scale. A field demonstration also uses a mass balance approach for the assessment of contaminant fate and transport at the field scale, focusing on a subset of loading rates, management techniques, etc., shown to provide optimal performance in laboratory-scale studies. With a successful field demonstration, proof of concept for land treatment of a given waste material is provided prior to full-scale implementation of the technology. A typical field demonstration takes 9 to 18 mo. and is best conducted over at least a 12-mo. period to assess the impact of seasonal moisture and temperature fluctuations on contaminant degradation rates observed in the land treatment unit.

Modeling of long-term performance is used to integrate soil/site/waste data from laboratory- and field-scale studies to predict the long-term fate and transport of contaminants at a given land treatment facility. A variety of unsaturated zone fate and transport models are available to describe degradation and transport of contaminants in soil above the water table, and readers are directed to the US EPA laboratory in Ada, Oklahoma for information regarding availability of these models. The Vadose Zone Interactive Process model is specified for use in the *Permit*

Guidance Manual on Hazardous Waste Land Treatment Demonstrations (Sims et al., 1986), and details of this modeling approach can be found in the guidance manual.

7.3.1.1.6. Process operation. Land treatment system performance is affected by four main operating variables, which include:

- *Loading rate and frequency* which is managed to control waste toxicity to soil microorganisms, to allow acclamation of soil microorganisms to waste materials, and to control the mobility of contaminants from a land treatment unit before they are degraded.
- *Tilling frequency* which is managed to control soil aeration and oxygen, provide soil/contaminant mixing to ensure intimate contact with active soil microbial population, and to limit the potential for volatile emissions of waste constituents.
- *Waste application method* which is managed to control waste volatilization potential (i.e., via surface versus subsurface application) and to manage contaminant vapor retention in land treatment unit.
- *Moisture and nutrient control* which are managed to provide stimulation of microbial growth and contaminant degradation and to limit the mobility of contaminants from a land treatment unit before they are degraded.

These variables should be assessed in the treatment demonstration stage of land treatment design to determine which if any of these characteristics can be managed to optimize land treatment system performance.

Operational and performance monitoring approaches for field-scale land treatment systems include both contaminant degradation rate verification and quantification of contaminant mobility at full-scale. Soil core measurements provide the best data for the definitive determination of progress in contaminant degradation and soil remediation. Contaminant degradation rates are determined from soil core data collected from representative locations throughout the land treatment unit. Such samples are taken over time during a relevant treatment cycle (i.e., weekly between monthly waste application and tilling events, to monthly between 6-month application cycles, etc). The proportion of contaminant loss associated with volatile contaminant emissions determined from surface emission sampling and with contaminant leachate production determined from leachate collection system sampling should also be determined over time as necessary to quantify the fraction of apparent contaminant loss that can truly be attributed to biodegradation. Soil/waste detoxication rates are also generally determined based on analyses of aqueous extracts with the Microtox™ and from analyses of solvent extracts of soil core samples using the Ames test.

7.3.1.1.7. Process design calculations commonly carried out in the design of land treatment systems are demonstrated in the following example.

Given: A land treatment system is being evaluated for a soil/waste mixture that will result from the excavation of contaminated soil from a RCRA Corrective Action. A 0.3-m (1-ft) ZOI is to be used for land treatment systems design. A lab treatability study was conducted to evaluate the feasibility of using land treatment for the remediation of this excavated soil. Uncontaminated, native soil adjacent to the site is a sandy loam and is ideal for use as a treatment medium. Initial analysis of the contaminated soil/waste mixture and the uncontaminated background soil yielded the following results:

Parameter	Soil/Waste Mixture	Background Soil
Oil & Grease	100,000 mg/kg	100 mg/kg
Microtox™ EC_{50}	25%	> 100%
Ames assay	Nonmutagenic	Nonmutagenic
Water content	0.5 wt%	50% Field capacity
Field capacity*	Not applicable	5 wt%
pH	8.5	7.1
Specific gravity	1.8	1.3

* Field capacity is defined as the soil water content after a saturated soil is allowed to freely drain by gravity and is determined by a soil's textural characteristics.

A VCDR experiment was conducted to assess the relative fraction of volatile loss to degradation loss associated with the treatment of this soil/waste mixture at a variety of waste loading rates and to determine the impact of waste loading method on volatile compound emission rates. Two application methods were evaluated in this study. Surface application was used to evaluate losses following surface waste spreading and tilling, while a subsurface application experiment was used to simulate volatilization following injection of the waste below the soil surface. A summary of these findings is presented below:

Find: Provide an interpretation of the initial soil/waste mixture and background soil analyses. Provide an interpretation of the VCDR experimental results. Make a recommendation regarding the loading rate and waste application method to use for this soil/waste mixture. Determine the uncontaminated soil requirement for this recommended loading rate, and determine the surface area required for the treatment of 240 m³ (300 yd³) of contaminated waste/soil mixture.

	% Total Apparent Loss Attributed to Volatilization	
Loading Rate	Surface Application	Subsurface Application
10% Oil & Grease	100	90
5% Oil & Grease	85	75
2.5% Oil & Grease	60	30
1% Oil & Grease	40	2

Step 1: Based on the initial analyses of the background soil and the contaminated soil/waste mixture given above, the following interpretation can be given:

Contaminated soil/waste mixture — The Oil & Grease content of the contaminated soil is excessive (10 wt%) and is potentially inhibitory/toxic to the soil microorganisms, as indicated by the 25% EC50 in the Microtox™ test. This 25% result indicates that a 25% dilution of a water extract of the contaminated soil is concentrated enough to reduce the light output of the Microtox™ assay organisms by 50%, a response that has been strongly correlated to soil and aquatic organism toxicity in natural environments. These results indicate that the contaminated soil must be diluted by at least a factor of 4 in order for water soluble constituent concentrations to be below inhibitory levels within the soil matrix. Although the soil/waste mixture is toxic to soil organisms, it does not appear to be mutagenic, as determined in the Ames test. Part of the inhibition may be the result of contaminants producing the alkaline waste pH, but further testing is necessary to confirm this.

Background soil — This soil is representative of uncontaminated, background soil typically found at field sites. It is low in Oil & Grease, its water soluble extract is not toxic, as indicated by the Microtox™; and a solvent extract of this soil is not mutagenic, as indicated by the Ames test. Its field capacity is indicative of a loamy textured soil, and it has a neutral pH supportive of biological activity. It would serve as an effective treatment medium within a land treatment context.

Step 2: The VCDR test results confirm the inhibitory nature of the soil/waste mixture. The 10% Oil & Grease loading represents a 100% contaminated soil/waste mixture, and subsequent loadings represent 1/2, 1/4, and 1/10 dilutions of the contaminated soil with the uncontaminated background soil, respectively. VCDR results indicate significant volatilization when surface applying the waste mixture, with progressively lower emission rates as the waste toxicity is reduced by blending with uncontaminated soil. Subsurface application significantly reduces the emission rate of volatiles, even with the application of the undiluted soil/waste mixture.

Step 3: Based on the outcome of this study, the subsurface application of a 1% Oil & Grease mixture of the contaminated soil should be used to optimize treatment and minimize volatile emissions from the land treatment unit.

Step 4: At a 1% Oil & Grease loading, a 1/10 blending of contaminated soil to total soil mixture would be required. This loading is given on a weight basis, so the following calculations are used to determine the uncontaminated soil requirement for the treatment of 240 m^3 (300 yd^3) of contaminated waste/soil mixture:

Soil/waste mixture unit weight = 1.8 (1,000 kg/m^3) = 1,800 kg/m^3 (7-92)

With this unit weight, the 240 m³ has a weight of:

Weight of soil/waste mixture = (240 m³) (1,800 kg/m³) = 432,000 kg (7-93)

At a 1/10 blend, the total weight of the contaminated and uncontaminated soil is:

Total weight = (10) (432,000 kg) = 4,320,000 kg (7-94)

The weight of the uncontaminated background soil needed is:

Background soil weight = (0.9) (4,320,000 kg) = 3,888,000 kg (7-95)

The volume of uncontaminated background soil needed is determined as follows:

Background soil unit weight = 1.3 (1,000 kg/m³) = 1,300 kg/m³ (7-96)
Volume of background soil needed = (3,888,000 kg)/(1,300 kg/m³) = 2,991 m³ (7-97)

Step 5: With a 0.3-m thick ZOI utilized in this land treatment unit, the volume of soil contained in 1 ha of ZOI 0.3-m thick is:

(0.3 m) (10,000 m²/ha) = 3,000 m³/ha (7-98)

The total volume of soil/waste mixture plus background treatment soil needed for this land treatment system is:

Total soil volume = (240 m³) + (2,991 m³) = 3,231 m³ (7-99)

So the total area required for the land treatment system with a 0.3-m thick ZOI is:

Total area required = (3,231 m³)/(3,600 m³/ha) = 0.9 ha (7-100)

7.3.1.2 Soil Piles

Soil piles or biomounds are prepared-bed systems in which biological process are utilized for the treatment of excavated contaminated soils. These systems are governed by the same fundamentals as other bioremediation options, including requirements of minimal soil water (50% to 85% field capacity), optimal soil pH (6 to 9), adequate DO (greater than 2 vol% in bed air pore space), and moderate soil bed temperatures (10° to 30° C) for optimal, unhindered biodegradation to take place within them. As the name implies, biomounds provide biological ex situ treatment of contaminated soil in controlled mound environments.

Biomound systems are attractive alternatives for soil remediation when site excavation costs can be minimized, i.e., where shallow contamination of a large aerial extent exists or where site constraints require immediate excavation and removal of contaminated soil. Biomound systems have advantages over in situ systems, as they can provide enhanced control over reactant delivery and contaminant/product recovery. They provide an advantage over land treatment systems, as they require less land area and need no tilling. They also provide an opportunity to modify existing site soil nutrient status through addition of nutrients, and can improve native air permeability through addition of bulking agents during mound construction.

If contaminant mobility and off-site migration are of concern, soil piles are also advantageous because they offer a chance to control migration pathways if they are modified to include a "tent" for volatiles or a liner to act as a "tub" for leachable constituents in the contaminated soil. Management of soil piles to optimize biological activity is easily accomplished through a variety of different operational strategies. Aeration management can be carried out by physically turning the piles in a composting mode or by using forced air via injection or extraction of air through the pile. Moisture management can be provided by adjustment of pile soil moisture content to between 50% and 90% field capacity during construction and through routine moisture addition during pile treatment. Nutrient and soil texture management can be provided in a similar fashion through the incorporation of slow-release nutrient blends and bulking agents into the pile as it is constructed.

7.3.1.2.1. Process configuration. A typical biomound system includes a biological component and a structural/physical component. The biological component must have active, acclimated microorganisms; an adequate nutrient pool consisting of requisite levels of nitrogen, phosphorus, potassium, and trace elements; and appropriate levels of electron acceptor, namely oxygen.

The structural/physical component of a biomound system generally includes liners for leachate collection and control; a delivery/recovery system that may include an irrigation system for nutrient and water management, a leachate collection and recycle system, and an air injection or extraction system; leachate and gas treatment systems; and an impervious cover or gas containment structure for off-gas minimization and control. A simple scheme includes plastic sheeting underlayment draped over hay bales or 0.3 m (1 ft) plastic pipe placed around the perimeter to form a surrounding berm. The average biomound system ranges in volume from 200 to 1,250 m^3 (260 to 1,600 yd^3) and typically has height, width, and length ranges of 1.5 to 2.5 m (5 to 8 ft), 9 to 20 m (30 to 65 ft), and 30 to 50 m (100 to 165 ft), respectively. Air piping of 10 to 15 cm diameter (4 to 6 in.) is used for oxygen supply, while 2.5- to 5-cm (1- to 2-in.) diameter pipe is utilized for vapor monitoring and measurement of bioactivity within a mound. A schematic of a typical biomound system, using a blower to aerate the piled soil, drip irrigation for soil water management, and an impervious cover to minimize volatile contaminant releases, is shown in Figure 7-10.

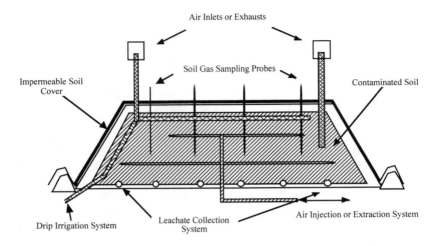

Figure 7-10 Schematic of a typical biomound soil treatment system using air extraction or injection to aerate the piled soil, drip irrigation for soil water management, and an impervious cover to minimize volatile contaminant release.

7.3.1.2.2. Process design of a biomound system primarily involves the optimization of microbial activity within the mound to enhance the efficiency of cleanup performance at the site. Physical characteristics of the mound can be carefully controlled during construction. Construction of the mounds to ensure adequate air phase permeability requires minimizing soil compaction and/or adding soil bulking agents (straw, wood chips, etc.) during mound placement. The pile should generally be kept irrigated during construction and continued operations to maintain the soil moisture content within 50% to 90% of its field capacity. Nutrients (fixed nitrogen and phosphate) can be dissolved in the spray water, or nutrient pellets can be dispersed on to the pile and gradually dissolved by the irrigation water.

Mound heterogeneity can be minimized through effective blending and mixing of the contaminated soil to produce desirable soil textural characteristics. The effectiveness of these construction techniques can be determined through soil gas measurements of VOC and respiration gas composition, through mound soil gas pressure measurements, and through soil core contaminant concentrations measured at various locations within the mound.

As with the in situ processes described in Section 7.3.2, the characterization of viable indigenous populations via respiration process monitoring during site assessment activities is essential to determine the applicability of a biological approach to contaminated soil remediation. Nutrient analysis of contaminated soils indicates the need for nutrient amendment, and soil moisture measurements can be used to assess the need for soil water management in the mound configuration.

Once the existence of biological activity is verified, and a biomound system is chosen for soil remediation, an operating scheme for the mound system should be developed through the quantitative evaluation of biodegradation potential via respiration rate monitoring within the mound. In situ respiration tests are conducted utilizing soil gas probes and oxygen soil gas analyzers (Hinchee et al., 1992). These procedures are detailed in Section 7.3.2.1 on bioventing. One slight modification is suggested in the biomound application. That is, oxygen concentrations should be monitored at several locations throughout the mound during mound aeration to determine the duration of the air injection/extraction cycle required to adequately oxygenate the contaminated soil. This measurement establishes the duration of the operating phase, while the data generated from the standard respiration test establish the duration of the incubation phase. The example case study presented below illustrates the use of respiration and oxygen transfer data in a biomound system for operating protocol development.

Oxygen transfer rate determinations can be made from the measurement of oxygen concentrations in soil vapor probes located throughout the biomound following air injection/extraction at a given operating flow rate. The recovery of oxygen concentrations during air input can be quantitatively analyzed, as described, using simple zero- and first-order rate relationships (Equations 7-38 through 7-41). From data collected in limited field studies, oxygen recovery generally follows a zero-order relationship (McGinnis et al., 1994). These data allow the determination of the required operating time at a given mound air flow rate to provide a desired level of oxygen within the biomound soil gas.

Once the physical oxygen transfer rate is determined, the oxygen requirement for contaminant degradation must be determined from in situ respiration measurements carried out as per Hinchee et al. (1992) with the air injection/extraction system shut off following initial biomound aeration. Soil gas oxygen and carbon dioxide data are collected over time from biomound vapor probes and are regressed using zero- and first-order rate relationships to generate quantitative oxygen uptake and carbon dioxide production rates for operating system design. These respiration rate measurements indicate oxygen consumption rates during biomound incubation and allow the selection of design flows and a system operating mode based on oxygen transfer requirements and oxygen uptake rates observed under operating field conditions.

7.3.1.2.3. Process operation of biomounds is designed to minimize contaminant volatilization and optimize biodegradation because of their reliance on biological mechanisms for contaminant degradation. Volatilization is minimized by using low (4.6 to 23 actual L/sec, or 2 to 10 acfm) or intermittent (230 to 920 actual L/sec, or

100 to 400 acfm) air flow rates and by pulling (25 to 50 cm [10 to 20 in.] of water) or injecting air into the mound to maximize vapor retention within the mound. Biodegradation is optimized through the operation of the mound at optimal soil moisture contents (i.e., 50% to 90% field capacity) by providing nutrient amendment (an ideal carbon/nitrogen/phosphorus ratio of approximately 100/10/1 on a weight basis is desirable) when necessary and by modifying soil contaminant concentrations (blending background soil with contaminated soil) to obtain concentrations below inhibitory/toxic contaminant levels.

Table 7-8 summarizes general physical and operational characteristics of typical biomound systems. Operational and performance monitoring approaches for biomound treatment systems are similar to those for soil vapor extraction (SVE) and particularly bioventing systems (Section 7.3.2.1). Biomounds utilize the following general measurements for the monitoring of system performance associated with air flow and contaminant volatilization considerations: vacuum pressure and contaminant gas composition at soil gas monitoring probes to ensure uniform and effective flow in the mound; system air flow rates; contaminant concentrations in the vent gas; vent well and blower discharge contaminant distribution; and soil core measurements from within the biomound for definitive treatment progress determinations.

In addition, the rate and extent of microbial activity at a biomound site under field conditions is monitored utilizing mound vapor monitoring probe oxygen and carbon dioxide composition data. These data are collected during operation and during shutdown "incubation" periods and are compared to background rates to quantify microbial activity. These data may be collected following any addition of amendments used for process stimulation to ascertain the true impact the amendment had on baseline bioactivity. Finally, oxygen uptake and carbon dioxide production data are used to evaluate biomound performance and to determine when biomound treatment is complete when mound rates reach uncontaminated soil background respiration rate levels determined at the site.

7.3.1.2.4 Composting is a biopile process option that is used widely for the remediation of a number of waste types, such as wood preserving and munitions and explosives wastes. It is characterized by high water content (50 to 60 wt%), relatively low nitrogen content, and high operating temperatures (50° to 60° C).

In the composting process, the soil is generally mixed with bulking agents and placed into piles 1.8 to 3 m (6 to 10 ft) high and 4.5 to 7.5 m (15 to 25 ft) wide. Bulking agents that can be used include sewage treatment plant sludge, manure, straw, or wood chips, which naturally contain or have been enriched with nutrients to yield an optimum carbon/nitrogen ratio of 20/1 to 30/1. Nutrient enrichment can be synthetic — (e.g., applying fixed nitrogen and phosphate fertilizer to wood chips) or natural — (e.g., manure with bedding straw). The addition of a bulking agent makes possible the bioremediation of soils with high clay content, as well as coarser-grained soils.

Table 7-8 General physical and operational characteristics of typical biomound systems.

Parameter	Typical Range of Values
Compound Type	Biodegradable
Soil Concentration	Below Inhibitory/Toxic Levels (< 10 wt% TPH)
Operating Mode	Maximum Retention Time & Aerobic Conditions
Air Flow Rates	4.6 to 23 actual L/s (2 to 10 acfm) Continuous 230 to 920 actual L/s (100 to 400 acfm) Intermittent
Operating Vacuum	25 to 50 cm (10 to 20 in.) H_2O
Optimal Soil Moisture	≈ 75% to 90% Field Capacity
Nutrient Requirement	C:N:P ≈ 100:10:1†
Soil Gas O2 Levels	> 2 vol%
Pile Configuration	Height: 1.5 to 2.5 m (5 to 8 ft) Width: 9 to 20 m (30 to 65 ft) Length: 30 to 50 m (100 to 165 ft)
Average Volume/Pile	200 to 1,250 m^3 (260 to 1,600 yd^3)
Air Piping Diameter	10 to 15 cm (4 to 6 in.)
Monitoring Well Diameter	2.5 to 5 cm (1 to 2 in.)

†This ratio represents a maximum theoretical requirement that may or may not be needed at a given site.

Oxygenation of the compost piles is provided through mechanical turning of the pile in a windrow configuration, through air injection via perforated pipe in a static pile configuration, or through compost mixing and air injection and extraction in a mechanical in-vessel composting configuration. Windrow pile mixing and air flow rate control in the static pile and in-vessel composting systems are also used for the control of pile temperatures during composting. Pile temperatures increase during composting in response to microbial activity, and temperature measurements are a primary method for monitoring of composting process performance. Pile temperatures rise from initial levels of 50° to 55° C to the 60° C range during active composting.

The time required for waste stabilization in a compost unit is directly proportional to the level of mixing and oxygenation that is provided. Treatment times for simple windrow piles typically range from 6 to 15 wk for composting and curing. With forced air injection in a static pile mode, this treatment time can be reduced to 8 wk. In-vessel composting systems reduce the required treatment time even further, to only 1 to 6 wk, because of the high levels of control provided in a reactor environment.

Composting was applied in conjunction with ex situ soil venting in a case study reported by Davis and Russell (1993). The soil was contaminated with TCE, PCE,

TCA, bis(2-ethylhexyl) phthalate, fluoranthene, aromatics, and other petroleum hydrocarbons. A pile was formed in an enclosure with embedded perforated pipes within a 15- to 30-cm (6- to 12-in.) layer of wood chips on a double liner system with leachate collection. The soil was mixed with municipal sludge and manure in 12 to 16 m³ (15 to 20 yd³) batches in an agricultural-type mixer. Each batch was mixed for 3 to 6 min and then spread evenly over the wood chips. Thermocouples were embedded and used to control air blown into the perforated pipes such that the pile temperature was maintained at 55° C (131° F). The pile was 30 m (100 ft) long, 18 m (60 ft) wide, and 2.4 m (8 ft) high, forming a volume of almost 1,440 m³ (1,800 yd³). A sprinkler system was suspended just under the roof to keep the pile moist. The system was first operated in the vapor extraction mode to remove VOCs that could be toxic to the bacteria. During this mode, the air blower was operated continuously, with the air discharging through two activated carbon canisters for 90 days. VOC reduction was greater than 80%. Then the sprinkler system was activated to initiate the composting mode. The air blower was on an average of 10 min each hour for an additional 180 days. Semivolatile organic compound and total petroleum hydrocarbon reductions of greater than 90% should be attainable with such a system.

7.3.1.2.5. Process design case study. Typical procedures for the design of a biomound system and management for performance enhancement is demonstrated through the following case study for a pentachlorophenol- (PCP-) and hydrocarbon-contaminated site (McGinnis et al., 1994).

This site was located at an abandoned wood treatment facility in the southeastern United States and contained more than 10,400 m³ (13,000 yd³) of surface and near-surface soils contaminated with PCP and petroleum waste oils resulting from more than 50 yr of wood preserving activities. The system had the following characteristics:

- Seven soil piles utilized with the following dimensions: 49.5 m (165 ft) long, 19.5 m (65 ft) wide, 2.4 m (8 ft) deep
- Ductwork placed in the piles so no point in the soil pile was more than 1 m from an air source
    - Perforated, flexible, 10 cm (4 in.) in diameter plastic piping was used for ductwork.
    - Ductwork was sealed within the pile and connected to 1.5-kW (2-hp) blowers.
    - Two blowers delivered 12 actual m³/min (400 acfm) at 25 cm w.c. pressure to each pile.
    - Sealed sampling ports were installed in the ductwork for monitoring of pressure and oxygen and carbon dioxide concentrations inside the ductwork.
    - Atmospheric oxygen concentrations were maintained inside the ductwork during blower operation.

- One to three monitoring wells were utilized per pile for soil pile monitoring.
  - The monitoring wells consisted of 1.8-m (6-ft) lengths of 5 cm (2 in.) diameter slotted PVC connected to 1.5 m (5 ft) of solid 5 cm (2 in.) diameter PVC.
  - The wells extended from natural ground to 0.9 m (3 ft) above the surface of the piles.
  - The sampling ports consisting of a single resealable opening were installed through the cap of each well.
  - Flashing was used to prevent short circuiting of air from the surface of the piles.
- Contaminated soils ranged from sandy top soils to semipermeable clays.
- Pile PCP and TPH concentration and soil type varied from pile to pile.
  - Three piles were of sandy clay topsoils with PCP concentrations at 30 to 100 mg/kg of soil and TPH concentrations at 2,500 to 3,300 mg/kg of soil.
  - Two piles were of clay soils from deeper excavations with PCP concentrations at 200 to 300 mg/kg of soil and TPH levels at 4,500 to 5,000 mg/kg of soil.
  - Two piles were of sandy, loamy mixtures from intermediate depths with PCP concentrations from 60 to 180 mg/kg of soil and TPH concentrations at 300 to 3,000 mg/kg of soil.

Bulking agents were added to increase the water retention and pore space in the fine textured clay and sandy clay soils at the site. In addition to deficits of nutrients such as nitrogen and phosphorus, these soils contained little organic carbon. Alternate carbon sources have been shown to enhance degradation rates of chlorinated hydrocarbons (Davis and Madsen, 1991), and chicken manure was added to these contaminated soils, providing nitrogen, phosphorus, and organic carbon in an effort to enhance microbial activity.

System treatment performance evaluation was based on a comparison of treatment piles with an unaerated control pile; electron acceptor utilization (oxygen) and respiration product (carbon dioxide) generation rates using the method of Hinchee et al. (1992); and quantitative assessment of site contamination before and after treatment via soil core data. The system operating protocol was based on the relationship between the oxygen transfer rate determined through oxygen recovery studies and the oxygen uptake rates determined from the respiration tests described above.

Field soil respiration rates and corresponding hydrocarbon and PCP degradation rates for all of the aerated piles were significantly greater than for the unaerated control pile (McGinnis et al., 1992). These rates were also shown to be generally unrelated to soil texture and generally followed a zero-order reaction rate law, as well as a first-order (with respect to oxygen concentration) reaction rate law. Multiple sampling points from a given soil pile indicated the uniformity of each pile's microbial activity.

These data suggest that biomound systems designed to stimulate aerobic biodegradation significantly enhance microbial activity when managed for oxygen status, moisture, and nutrient levels. With the addition of the bulking agent to improve soil texture and air permeability that is possible using ex situ treatment, starting soil texture has little to no impact on resulting bioactivity. This characteristic makes biomound systems attractive for the remediation of fine-grained soils that are not easily remediated in place. Multiple sampling locations can be utilized to indicate the success of pile homogenization and the uniformity of oxygenation and nutrient supply throughout a given biomound. Once pile uniformity is proven, continuing performance checks can be provided by a single, representative sampling point.

A stepwise regression analysis was conducted (McGinnis et al., 1992).between the significant zero- and first-order respiration rates versus amended nutrient levels and soil organic carbon (TOC), PCP, and TPH (Oil & Grease) concentrations to determine what combination of management variables could be used to describe observed respiration rates within the field piles. The results of this stepwise regression indicated that although the average TOC, nitrogen, and phosphorus ratio for the soil piles monitored in this study was 5,230/1/10, and although an apparent significant deficiency of nitrogen was evident (a ratio of 100/10/1 for TOC, nitrogen, and phosphorus, respectively, has been recommended for optimal soil remediation), this deficiency was not found to affect microbial activity.

Stepwise regression results suggest that zero-order respiration rates (and corresponding contaminant degradation rates) can be enhanced with the addition of phosphorus at the rate of approximately 0.02 (vol% O_2/h)/(mg/L added available P) greater than that currently being provided with the chicken manure amendment. No correlation was found between PCP concentration, and microbial respiration rates when PCP was the lone predictor variable. When PCP concentration was combined with available phosphorus concentrations, it was consistently identified as a significant variable in the prediction of microbial respiration rates. This result indicates that PCP levels in the piles of 30 to 300 mg/kg dry soil were not inhibitory to the soil microorganisms. The slight dependence of respiration rate on soil TOC levels shown from the zero-order respiration rate stepwise regression results suggests that the addition of complex soil organic carbon (i.e., chicken manure) provides additional substrate to enhance microbial activity. This carbon addition should result in improved TPH and PCP degradation in these field soil piles at the rate of 0.09 (vol%O_2/h)/(% TOC added).

Finally, air blower schedules were established based on respiration and measured oxygen transfer rates into the piles.

An example of observed changes in soil pile oxygen concentrations over time is illustrated in Figure 7-11. The decrease in oxygen concentration follows zero-order behavior (zero-order oxygen uptake rate calculated to be -0.47 vol%/hr) to Point A, using approximately 2 vol% oxygen, after which time the rate of oxygen depletion (i.e., microbial respiration rate) goes to zero. These results were consistent among piles, indicating that aerobic respiration rates are inhibited at soil gas oxygen concentrations less that 2 vol% or at equivalent soil water concentrations of approximately 1 mg/L using Henry's law to predict air/water oxygen distribution. To maintain a maximum respiration rate, the soil gas oxygen concentrations should not be allowed to fall below 2 vol% at any point within the piles.

Air blower schedules were then established from the oxygen transfer and microbial respiration rates observed in the field soil biomound systems. The operating protocol was determined as follows:

- Based on results shown in Figure 7-11, the blowers should be restarted when soil gas oxygen concentrations approach 2 vol%.
- Blowers should be operated for approximately 16 hr (13 hr x 1.4 vol%/hr = 18 vol%) to raise soil oxygen concentrations to ambient levels.
- Blower should then be shut off for approximately 36 hr (36 hr x -0.47 vol%/hr = 18 vol%) to reduce soil pile oxygen concentrations to just above inhibitory levels.

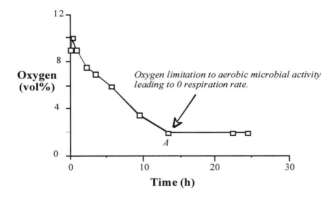

Figure 7-11 Oxygen utilization over time in soil pile 5 following shutdown of the air injection system at the PCP/TPH-contaminated soil biomound field site.

This intermittent blower operation optimizes soil pile remediation with respect to energy utilization and system maintenance costs and can be particularly cost-effective in situations in which soil pile VOC emissions must be controlled. In addition, ongoing monitoring of oxygen utilization and oxygen recovery over the course of soil remediation allows continual optimization of blower operation as bioactivity and contaminant levels decrease over time.

7.3.1.3 Slurry Reactors

Slurry reactors are a type of suspended growth system, discussed in Section 7.2.1.1, modified to mix and transfer oxygen to highly concentrated solids waste streams and highly contaminated soils (Dupont et al., 1998). Slurry-phase reactors should be particularly considered for highly contaminated soils and sludges that have contaminant concentrations ranging from 2,500 to 250,000 mg/kg. Slurry reactors are used to relieve the environmental factors typically encountered with treating relatively recalcitrant constituents in soil or sediments. By suspending the soil or sediments in an aqueous system, the availability of carbon sources, inorganic nutrients, and an electron acceptor (typically oxygen) are greatly improved as a result of maximizing mass transfer rates and contact between the contaminants and the microorganisms. A typical slurry treatment system requires significant pre- and post-treatment handling of the waste material (Figure 7-12), and it is nearly always considerably more expensive than treatment using a soil pile or land treatment. The use of this technology is generally limited to treatment of more recalcitrant compounds or highly contaminated soils or sediments. Typically, oil refinery wastes, principally sludges from storage and treatment lagoons, and wood preserving wastes, such as impoundment sludges and the surrounding soils contaminated with creosote and/or PCP, are treated using slurry reactors.

Like many solids treatment processes, the first step in the application of a slurry reactor is screening of the contaminated soil to remove material greater than a specific size, in general about 5 cm (2 in.), that is not effectively treated in the reactor. This soil screening equipment is typically capable of processing 72.6 to 90.7 metric tons/hr (80 to 100 tons/hr) of excavated material and can also reduce the size of large soil clumps. The soil screening equipment is typically operated to screen all of the soil independent of the downstream processing schedule. The removal of large objects and the reduction in size of others enhances soil and sediment mixing and enables the contaminated material to be suspended and efficiently mixed within the slurry reactor. Further size reduction will occur as a result of mixing, thus exposing more soil/sediment surfaces to the aqueous phase. In soils that contain a range of particle sizes (e.g., sand through clay), a soil washing system should be considered as organic contaminants tend to be concentrated in the humic and fine grain size fractions of the soil (i.e., organic matter, silt, and clay), while the contaminants associated with the coarse grain size fraction (i.e., sand and gravel) are primarily surficial.

A slurry preparation/soil washing system processes the screened material at an average rate of 7.3 to 10.9 metric tons/hr (8 to 12 tons/hr). The process separates the

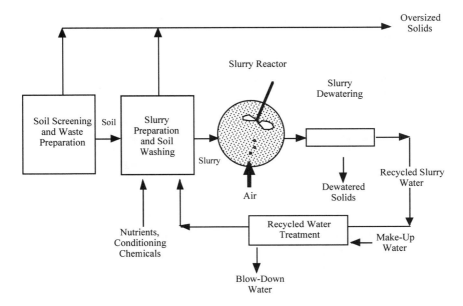

Figure 7-12 Typical schematic of a slurry reactor for contaminated soil remediation.

stockpiled material into several size fractions utilizing several separation techniques. The equipment may include wet and dry screens, hydrocyclones, sand screws, and hydroclassifiers. The slurry is treated with conditioning chemicals, and the larger size fractions are washed and separated from the slurry. The remaining slurry is generally amended with nutrients and pH adjusted prior to transfer to the bioreactors.

In some cases, chemical pretreatment has been used on contaminated soils prior to slurry reactor treatment to partially degrade relatively recalcitrant chemicals. Hydrogen peroxide has been used in conjunction with slurry treatment to pre-oxidize biologically recalcitrant targeted contaminants, forming oxidized intermediate compounds that are more susceptible to biological treatment than the parent compounds. Applications of this strategy have been predominantly targeted at polynuclear aromatic hydrocarbon contaminants found in municipal town-gas site pollution. There are some distinct concerns with the use of hydrogen peroxide. Doses of hydrogen peroxide in high concentration can effectively disinfect the contaminated solids likely requiring culture addition following pretreatment. The presence of iron in the soil also limits the effectiveness of this pretreatment process because trace amounts of iron are capable of autocatalytically decomposing hydrogen peroxide. The oxygen formed by decomposition will be lost from the liquid phase and will have little effect on the treatment. Peroxide mixtures are available that avoid rapid decomposition, but these mixtures are less reactive than pure hydrogen peroxide.

In the actual slurry reactor, soils or sediments are mixed with water to form as high a solid content as can be managed, up to 35% to 40% solids. The limiting factors are usually the density of the solids and the ability of the mixing equipment to maintain a well-mixed system. Sediments are generally less dense than soils and thus can be maintained in suspension more easily and at lower energy requirements, which range from 0.019 to 0.19 kW/m^3 (0.1 to 1 hp/1,000 gal). Mixing is provided by mechanical means, by aeration, or by a mixture of mechanical means and aeration. Although mixing is generally considered necessary, Retec (US EPA, 1990) has reported a case in which degradation rates were actually faster in an unmixed system than in a mixed system.

At the completion of treatment in a slurry bioreactor, the soil and water making up the slurry must be separated for further handling. A number of options exist for liquid-solid separation, including drying beds, gravity filtration, filter presses, centrifuges, and thermal dryers. Depending on project requirements, the aqueous or solid fractions may require further treatment (i.e., stabilization of metals) prior to discharge or disposal. Polishing processes for the aqueous effluent are the same as for the groundwater treatment systems discussed above. The treated aqueous phase is typically transferred to a storage tank for reuse in the slurry preparation process, or it can be discharged provided it meets permit limits for release into an appropriate receptor. The treated solids can be placed back into the ground or disposed of at an appropriate landfill provided the requirements of the Land Disposal Restrictions (LDRs) have been met. On-site disposal will require that moisture be adjusted to allow compaction requirements to be met. Transport may require some drying to prevent leakage of water during transportation.

Off-gas treatment may also be necessary, especially for soils impacted with a mixture of chlorinated solvents that degrade poorly or not at all under aerobic conditions or with highly volatile petroleum hydrocarbons. Depending on the concentrations of volatile species; the reactor temperature; the mode of oxygen supply; applicable local, state, and federal regulations; as well as nuisance considerations; off-gas treatment may be a significant factor in the design and cost of above-ground reactors. Reactors may be designed as enclosed vessels to facilitate the collection and treatment of emissions and may even be operated under reduced pressure conditions.

7.3.1.3.1. Process design requirements for slurry reactors are best determined by means of treatability studies (Section 7.4). Slurry reactors can be operated in single or sequencing batch modes or in either continuous or semicontinuous mode. In all systems, the two main design criteria are mixing and aeration. Although mixing can, in some cases, be provided by aeration, it is typically achieved through mechanical means, especially for treatment of soils, which require more energy to maintain suspension than do sediments.

Stripping is a concern with many slurry reactor designs, as most use aeration to provide all or part of the energy required for solids suspension. Use of oxygen instead of air reduces the volatile emissions but also minimizes contributions toward suspension of solids, thus requiring more substantial mechanical mixing. Recovery of an aqueous stream, with subsequent saturation with oxygen prior to recycling of the aqueous stream to the reactor, almost totally eliminates emissions. In cases in which heavy PAHs or other relatively nonvolatile compounds comprise most of the contaminant mass, volatilization may not be an issue. In cases in which solvents or low molecular weight hydrocarbons make up a substantial fraction of the organic loading, system design will normally include provisions for minimizing air emissions. This could include providing a reactor cover fitted with an air capture and treatment system.

Nutrients are added, and chemicals are added for pH adjustment, to the initial slurry reactor feed. During treatment, further nutrient addition and pH adjustment are typically automated as needed.

Although not a primary removal mechanism for most organic compounds, sorption on biological solids does occur, can contribute to organics removal, and may also affect waste sludge characteristics. For slurry reactors, desorption rather than sorption generally controls the applicability and success of solid-phase treatment. The rate of biodegradation in many cases is limited by bioavailability, especially for heavier, less soluble compounds, such as five- and six-ring PAHs. The longer organic contaminants have been in contact with solids, especially silts and clays with high organic carbon content, the less available the compounds are for biodegradation, and the less successful biological treatment systems will be in yielding treated solids with low residual contaminant concentrations. Some slurry reactor designs have incorporated the addition of surfactants into a slurry preparation phase to improve contaminant desorption and bioavailability.

Slurry reactors have been constructed in lined lagoons, unlined lagoons (for treatment of lagoon solids), or constructed reactors.

7.3.1.3.2. Process operation considerations for slurry reactors includes an important requirement for the careful management of soil during pre- and postprocessing. Stockpile locations as well as soil pre- and post-treatment areas, need to be located to minimize soil handling requirements. Equipment selection, both type and size, is important to the efficiency of the operation. If rain is heavy during pre- and postprocessing, provisions need to be made to cover the contaminated soils within the pretreatment, stockpile, and post-treatment areas. All soil processing and stockpiling areas should be located so that transportation requirements to the slurry reactor are minimized, and provisions should be made to minimize suspension and transport of contaminated soils away from the treatment area during process and transport.

During system start-up, prepared soils are diluted to solids concentrations below design levels, and the slurry is transferred from the slurry preparation system to one of normally multiple reactors (operating volumes are typically 380 to 680 m^3, or 100,000 to 180,000 gal, each), initially in small increments. Influent and effluent slurry and effluent water phase concentrations are monitored as the reactor solids content is increased to design levels to ensure that system performance is achieved during start-up. Slurry reactors are typically operated in batch mode by filling the reactor, treating for a fixed length of time, and discharging the slurry at the completion of treatment. Continuous/semicontinuous mode reactors have a constant influent and effluent stream, with the slurry remaining within the reactor based on the design hydraulic residence time.

When operated in a batch mode, the more readily degraded compounds are removed relatively rapidly during the initial period of treatment. The more recalcitrant compounds subsequently degrade at a slower rate. Cell growth rates behave similarly, and as early formed cells die, they lyse, releasing internal stores of cellular material and recycling any nutrients they contain. Thus the nutrient addition rates during the initial phase of operation are relatively large and then taper off over time during treatment. Typical cell densities may be 10^9 cells/mL, or 1 g of carbon/L. This equates to approximately a 100 mg/L initial nitrogen requirement.

As indicated in Section 7.1, biological reaction rates increase with increasing temperature, following the Arrhenius relationship of an approximate doubling of rate constants for each 10° C temperature increase up to a maximum tolerable temperature of approximately 35° C. Heating full-scale reactors is particularly important during cold-season operation when biological degradation rates would otherwise fall. The decrease in treatment times achieved at higher reactor temperatures has obvious benefits but requires additional capital and operating costs to insulate reaction vessels and heat incoming waste flows. One additional drawback to reactor heating is that elevated temperatures may increase the amount of volatiles emitted from the reactor, requiring additional off-gas treatment that may not be needed at lower temperatures.

Reactor heating can be advantageous, however, as indicated by Dupont et al. (1998), in the slurry reactor treatment of soils containing approximately 10,000 mg/kg of PAHs. In this example, 9 days were required to reach cleanup levels at 25° to 27° C versus only 6 days at 35° to 37° C. Overall treatment costs were 10% less at the elevated temperature than at lower temperatures despite increased operating costs for reactor heating because of the increase in reaction rate and decrease in treatment time required at the elevated operating temperature.

Because of the nature of the contaminated materials being treated in a slurry-phase reactor, in order to effectively and accurately monitor their performance, care must be taken to collect representative samples at various points in the process. These samples must be analyzed for the appropriate chemical, physical, and biological parameters. Sample collection procedures must be developed for both the slurry reactors and the process equipment and piping. In order to collect representative samples, a statistically-based plan must account for the variability in the feed characteristics and for the effect of particle size distribution and solids content on the contaminant distribution within the solids. Statistically significant samples must be collected and must be tested for particle size density, total solids, and slurry density to determine how representative the sample is of the entire batch. Analytical results will be biased if samples are not representative, i.e., higher fraction of fines in sample, increased total solids, etc. In general, composite samples are preferable to discrete, grab samples. Once the samples are collected, specific, repeatable procedures are required for sample handling, preparation, and extraction.

7.3.2. In Situ Treatment

7.3.2.1. Bioventing

Bioventing describes the process in which air is utilized to deliver oxygen to the vadose zone to stimulate the in situ biodegradation of organic contaminants. Air is an extremely efficient oxygen transfer medium because of its high oxygen content (20.9 vol%, i.e., 209,000 ppmv) and low viscosity compared with saturated water. Bioventing represents a hybrid physical/biological process utilizing SVE systems for oxygen transfer, which focuses not on contaminant stripping but rather on in situ aerobic contaminant biodegradation for the remediation of a contaminated site. Consideration of soil vacuum extraction for oxygen transfer to the subsurface was proposed in 1988 by Wilson and Ward, who noted that systems designed for the removal of volatiles from soil could also be used to transport oxygen. A number of other authors have discussed the potential improvement of in situ, aerobic, subsurface bioremediation using SVE for oxygen transfer (Bennedsen, 1987; Riser, 1988; Ely and Heffner, 1988; Stapps, 1989), and there has been ample recent evidence demonstrating field-scale bioventing system effectiveness for fuel-contaminated site remediation (Dupont et al., 1991; Miller et al., 1991; Hinchee et al., 1991; Ong et al., 1994; van Eyk, 1994; Leeson et al., 1995; Leeson and Hinchee, 1995).

Bioventing systems are composed of hardware identical to that of conventional SVE systems, with vertical wells and/or lateral trenches, piping networks, and a blower for gas extraction or injection. They differ significantly from conventional systems, however, in their configuration and philosophy of design and operation. The primary purpose of a bioventing system is to use moving soil gas to transfer oxygen to the subsurface where indigenous organisms can utilize it as an electron acceptor to carry out aerobic metabolism of soil contaminants. As such, bioventing system extraction wells are not placed in the center of the contamination, as in conventional SVE systems, but on the periphery of the site. In addition, low flow rates (0.3 to 1.5 actual m^3/min, or 10 to 50 acfm, versus 3 to more than 45 actual m^3/min, or 100 to more than 1,500 acfm) for conventional SVE systems) are used to maximize the residence time of vent gas in the soil to enhance in situ biodegradation and minimize contaminant volatilization.

Because it is a biological treatment approach, however, bioventing does require the management of environmental conditions to ensure maintenance of bioactivity at the site. Management of soil moisture and soil nutrient levels to avoid inhibition of microbial respiration within the vadose zone can be accomplished fairly easily, and it has been used to optimize contaminant biodegradation at field sites when other variables (i.e., toxicity) do not limit microbial activity (Dupont et al., 1991; Miller et al., 1991).

Oxygen transfer to the subsurface via SVE systems is generally more rapid than oxygen uptake rates observed under field conditions. This results in the oxygenation of soil gas to near ambient levels if vent system blowers are operated on a continuous basis. To minimize system operating costs, and more importantly to reduce or even perhaps eliminate off-gas treatment requirements entirely, cyclic or surge pumping of vent systems in bioventing operations is recommended. Surge pumping in a bioventing mode entails operating the blower system until soil gas oxygen levels reach near-ambient conditions throughout the site being remediated. The system can then be shut off for some period of time during which soil gas oxygen concentrations are routinely monitored until they reach a level that inhibits aerobic microbial activity. Once this limiting soil gas concentration is reached, the vent system is restarted, and the on-off cycle continues. Based on the H for oxygen, a limitation would be expected to occur at a soil gas concentration of approximately 2.0 vol%, corresponding to soil water oxygen concentrations of approximately 1 mg/L. An inhibition of soil respiration has been reported at the 2.0 vol% soil oxygen level in venting systems treating JP-4 contaminated soils (Dupont et al., 1991) and in vented soil piles contaminated with PCP waste (McGinnis et al., 1994), suggesting that this value represents a good operating number for field-scale applications.

Based on observed field respiration data from various JP-4 jet fuel-contaminated sites (Dupont et al., 1991; Hinchee and Ong, 1992; Ong et al., 1994) and bioventing of PCP-contaminated soil piles (McGinnis et al., 1994), field oxygen uptake rates of 0.03 to 1.4 vol%/hr (0.8 to 39.7 g O$_2$/m^3 of soil-d at air-filled porosity, of 40 vol%) can be expected. These rates can be nearly an order of magnitude lower as remediation progresses to near de minimus soil hydrocarbon levels (Dupont et al., 1991). This allows typical bioventing systems to be operated on schedules of 8-hr on, 16-hr off at the initiation of remediation, to 8-hr on, 7-days off near the end of the field effort, while still maintaining aerobic conditions within the contaminated soil during nonventing periods. Table 7-9 summarizes the general design, operational, and

Table 7-9 General design and application considerations appropriate for conventional versus bioventing SVE systems.

Parameter	Conventional SVE	Bioventing
Compound Type	Volatile @ Room Temperature	Biodegradable
Vapor Pressure	> 100 mm Hg	NA
Hc (dimensionless)	> 0.01	NA
Aqueous Solubility	< 100 mg/L	NA
Soil Concentration	> 1 mg/kg	< 1%
Depth to Groundwater	> 20 ft	NA
Air Phase Permeability	> 1 x 10^{-4} cm/s	
Subsurface Conditions	Little or No Stratification	
NAPL Phase	Little or None	Biodegradable
Extraction Well Placement	Within Contamination	Outside Contamination
Injection Well Placement	Outside Contamination	Within Contamination
Operating Mode	Maximum Soil Gas Exchange Rate	Maximum Retention Time & Aerobic Conditions
Operating Air Flow Rates	3 to 45+ actual m^3/min (100 to 1,500+ acfm)	0.3 to 1.5 actual m^3/min (10 to 50 acfm)
Pore Volumes/d	1 to 15	0.1 to 0.5
Optimal Soil Moisture	≈ 25% Field Capacity	≈ 75% Field Capacity
Nutrient Requirement	NA	C:N:P ≈ 100:10:1†
Soil Gas O_2 Levels	NA	> 2 vol%
Toxicants	NA	Little or None

† Caution should be used in considering a nutrient requirement as field-scale bioventing research has shown mixed results in performance with nutrient addition. This ratio represents a maximum theoretical requirement that may or may not be needed at a given site.

application considerations appropriate for conventional SVE systems versus those utilized in a bioventing operating mode.

7.3.2.1.1. Process design. Bioventing system design has only recently been codified despite the large number of bioventing systems implemented in both the public and private sectors. The US Air Force has been a leader in the development and implementation of bioventing systems for remediation of many of their fuel release sites, and they have developed, through the Air Force Center for Environmental Excellence (AFCEE), a field treatability protocol for bioventing system design (Hinchee et al., 1992). In addition, an addendum to this protocol document detailing the integration of soil gas survey results into bioventing system evaluation was published by AFCEE in 1994 (Downey and Hall, 1994). These AFCEE field bioventing protocol documents were written as a guide for the field-scale evaluation of the potential application of bioventing for remediation of Air Force sites, and they focus heavily on field methods for in situ respiration/degradation rate determinations using procedures adapted from Hinchee and Ong (1992).

These field methods were used to document the applicability of bioventing for petroleum-contaminated sites through a comprehensive, 145-site, field bioventing study conducted by Battelle for the US Air Force. The results of this study were published by Leeson et al. (1995) and Leeson and Hinchee (1995), and they form the basis for the "Bioventing Design" text published by CRC Press in 1997 (Leeson and Hinchee, 1996). The findings of this study suggest that bioventing has almost universal application as an enhancement to the natural biodegradation of petroleum hydrocarbons in the unsaturated zone and that no single factor (nitrogen, phosphorous, moisture, or pH) had a dominant influence on the rate of in situ biodegradation observed at these field sites.

Based on the importance of the in situ respiration rate in reflecting biological activity at a site and on the lack of correlation between it and other environmental variables that can impact biodegradation rates, the direct measurement of in situ respiration rates is recommended to provide a site-specific assessment of the applicability of bioventing at a given site. In the following system design and performance evaluation approach, in situ respiration measurements form the basis for preliminary screening of bioventing as a remedial alternative, and they are a key element in the monitoring and evaluation of the progress of remediation at a site. This design approach relies heavily on the methodology described in the AFCEE protocol (Hinchee et al., 1992; Downey and Hall, 1994), but it attempts to improve on the methodology by more completely integrating field bioventing system evaluation and design from initial site characterization activities through the system design, process monitoring, and performance evaluation steps. Schematically, this bioventing system design and performance evaluation protocol is summarized in Figure 7-13, describing the objectives, activities, and outcome/interpretation of each of its five phases of bioventing system design.

Phase I — assessment of the potential for contaminant biodegradation under field conditions. This phase is the first step in the evaluation of the potential application of bioventing, or more generally, of any biologically based remediation system at a given field site under existing site conditions. To determine the potential for in situ biodegradation of vadose zone contaminants via bioventing, existing soil microbial activity should be quantified during site assessment investigations. This can be readily accomplished through the analysis of soil gas oxygen and carbon dioxide composition, in addition to the more routinely measured total hydrocarbon concentrations, prior to venting activity at the site. TPH concentrations, as well as oxygen and carbon dioxide concentrations, can be measured during standard soil gas surveys using a variety of measurement techniques. Although both respiration gases can be easily measured, oxygen concentrations are considered a better indicator of microbial activity in soil systems because there are rarely abiotic sinks for oxygen in these environments. Carbon dioxide is produced through anaerobic microbial activity, as well as aerobic microbial activity, and can also be affected by precipitation or dissolution of carbonate rock.

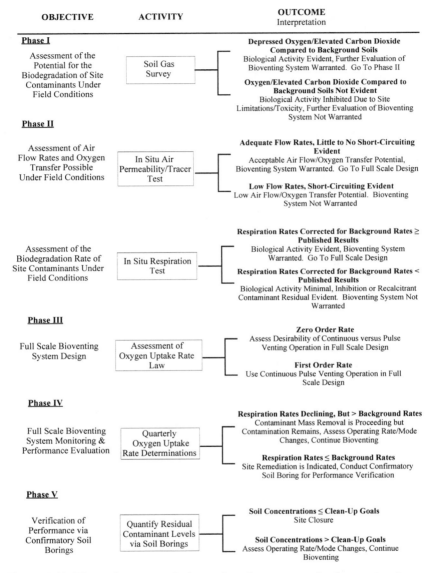

Figure 7-13 Bioventing system design and performance evaluation protocol.

The key to the evaluation of soil bioactivity using these methods is the determination of the extent of oxygen depletion and carbon dioxide enrichment in soil gas at a site with respect to background, uncontaminated soil levels. It cannot be overemphasized that these determinations must be based on a comparison to uncontaminated soil conditions, as only levels of oxygen depletion and carbon dioxide enrichment in excess of background are indicative of increased microbial activity compared to normal, basal respiration levels seen in uncontaminated soils at a site. It is also important to note that despite the common practice of collecting hydrocarbon concentration data during initial site investigations using soil gas surveys, respiration gas (oxygen/carbon dioxide) measurements are still not routinely made, even though they can be collected using the same soil gas probes and virtually at the same time as hydrocarbon measurements are taken. These respiration gas readings are unequivocal indicators of microbial activity at the site under actual field conditions, and they are critical in evaluating the next step in bioventing feasibility assessment at a given site.

If soil gas organic vapor and soil core data show contamination but microbial respiration has not yielded oxygen and carbon dioxide soil gas concentrations above background levels, conditions within the contaminated soil have resulted in soil microbial toxicity and/or severe inhibition, or significant nutrient or moisture limitations exist at the site. Unless soil moisture is the cause of this limitation, bioremediation has limited application, and alternative remediation schemes should be considered.

If soil contamination exists and microbial activity above background levels is evident from soil gas measurements, quantification of maximum respiration rates under field conditions can be carried out utilizing the in situ respiration measurement techniques described below. (See the addendum to the AFCEE bioventing test protocol by Downey and Hall, 1994, for additional examples of soil gas data interpretation related to the feasibility of the application of bioventing for site remediation).

Phase II — assessment of air flow and in situ respiration rates under field conditions. Once biological activity has been verified at the site, quantitation of the rate of air/oxygen supply, as well as the rate of in situ oxygen utilization, must be determined. As described above, the remediation of most contaminated sites is limited by the supply of electron acceptor, namely oxygen, and rational engineering design of bioventing systems requires a focus on supplying the oxygen needed to meet the in situ oxygen demand.

Air flow and tracer tests are first conducted, as described by Hinchee et al., (1992), the US EPA (1992c), Leeson and Hinchee (1995), and Leeson et al. (1997), to provide data regarding the existence of short-circuit pathways, stagnant zones, and the general conditions of vapor flow and oxygen transport in the subsurface. These data are essential, as efficient oxygen supply to the subsurface is key to optimal bioventing system design. Once subsurface air velocities are estimated from these air flow/tracer tests, oxygen transfer rates and transfer efficiencies can be estimated for various points throughout the area of contamination.

As stressed in these air flow/tracer test protocol, data from multiple lateral and vertical points throughout the contaminated soil should be collected to provide information regarding the spatial distribution and heterogeneity of air flow and oxygen transfer throughout a site. A minimum of three radial distances and three vertical locations (a minimum of nine total sampling points) should be used to provide the air flow and permeability data necessary to assess gas transport conditions at a typical site. Potential oxygen transfer rates can be estimated knowing that 0.03 standard m^3/min (1 scfm) of air equals 0.0063 standard m^3/min (0.21 scfm) of oxygen, which, from the ideal gas law, is equivalent to 7,700 mg of oxygen/min at 1 atm and 25° C. Table 7-10 provides an estimate of oxygen transfer rates for various soil types under a uniform SVE system operating condition of 76 cm (30 in.) w.c. for a 10-cm (4-in.) diameter extraction well, a radius of influence of 9 m (30 ft), and well slotting of 3 m (10 ft), assuming simple one-dimensional, radial flow into the well.

Table 7-10 Potential oxygen transfer rates in various soils.

Soil Type	Air Flow Rate (m^3/min)	Oxygen Transfer Rate (mg oxygen/min)
Medium Sand	1.07	275,000
Fine Sand	0.107	27,500
Silty Sand	0.011	2,750
Clayey Sand	0.001	275

This table indicates that even in clayey soils in which operating flow rates are low, significant oxygen transfer rates (275 mg of oxygen/min, or 396,000 mg of oxygen/day) are possible in bioventing systems.

These potential oxygen transfer rates, as calculated from the air flow/tracer data and subsequent air velocity/flow rate determinations, are representative of actual field conditions, and they are directly indicative of system performance that can be expected under full-scale conditions. The final data necessary in this phase of the bioventing treatability assessment are for the in situ oxygen demand or for the in situ oxygen respiration rate produced by the site microbial population in the degradation of the contaminants.

In situ respiration tests are conducted following the air flow/tracer tests to quantify the rate of oxygen demand expressed under actual field-limiting conditions. In order to expedite and minimize the efforts and costs of bioventing treatability assessment and design, the in situ respiration tests are optimally coupled to the air flow/tracer test efforts. Using this approach, in situ respiration tests would be initiated following completion of the air permeability/tracer tests when the entire area of influence of the permeability/tracer test extraction well is oxygenated. In this way, all tracer injection and soil gas monitoring points installed as part of the air permeability/tracer test can be utilized for soil respiration rate determinations. Use of these identical monitoring points provides air flow and respiration data that correspond directly to one another, in addition to the laterally and vertically distributed data necessary to ascertain the spatial variability and distribution of microbial activity throughout the site. Finally, the background, uncontaminated site location used as a baseline for the soil gas survey should also be incorporated into the in situ respiration test effort. This allows for a quantitative determination of the significance of measured respiration rates within the contaminated area with respect to background oxygen uptake rates in uncontaminated soil. This approach requires that the background point be oxygenated separately via air injection for a 16- to 24-hr period if it does not fall within the area of influence of the air permeability/tracer test extraction well.

With the entire flow field oxygenated, the in situ respiration test is initiated by first stopping air flow to the contaminated soil (as would be done at the completion of the air flow/tracer test), followed by the measurement of oxygen uptake and carbon dioxide production at the soil gas probes over time. Sampling and analysis of soil respiration gas composition over time are carried out to yield data similar to that shown in Figure 7-14. Selection of an appropriate sampling interval should be flexible based on actual site conditions, and it should be adjusted based on initial readings collected at 3- to 4-hr intervals following blower shutdown. With this as a guide, a typical respiration gas sample collection schedule would be as follows: at 0, 3, 6, 12, 18, 24, 30, 42, 56, 68, 80, 104, 128, 176, and 224 hr.

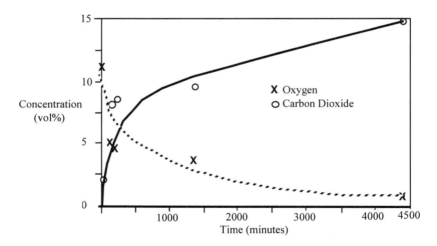

Figure 7-14 Typical soil respiration gas data collected during a field in situ respiration test.

Respiration rate data reduction is carried out using either a zero- or first-order reaction rate model to generate either zero- or first-order respiration rate values (vol%/hr or 1/h, respectively) from the slope of these linear regression relationships. Section 7.2.2.2.4 discusses biodegradation rate laws and governing rate equations.

A zero-order relationship is linear when measured respiration gas concentrations are plotted versus time. The slope of this linear relationship is the zero-order degradation rate constant, k_o.

A first-order reaction is described as one in which the change in the dependent variable (in this case respiration gases) over time is directly related to the variable's concentration. This relationship is nonlinear when the measured respiration gas concentration is plotted versus time, but it can be linearized by plotting the natural log of gas concentration versus time. The slope of this linear plot of log concentration versus time is the first-order degradation rate constant, k_1.

These regression relationships are generated from linear least-squares analysis of the field respiration data. The least-squares regression calculations can be carried out using standard statistical packages available on microcomputers and many handheld calculators. The first-order regression analysis and plot presented in Figure 7-15 were generated using StatViewII on a Macintosh computer. These figures show the measured data, the regression line of best fit, the 95% confidence bands (curved lines) of the slope of the best-fit regression line for the data, the resultant linear regression equation, and the r^2 value for the relationship. Quantification of the observed respiration reactions using this statistical approach provides a description of microbial activity observed at each monitoring point, and it allows for quantitative comparison of respiration rates spatially at a given point in time and temporally at a given location in the contaminated site. These comparisons can be made using data shown in Figure 7-16, which is for the respiration data shown in Figure 7-15, that include: the F-test and t-test statistics, the p value or probability of a significant regression, and the confidence interval of the slope of the regression relationship

The first question that must be answered regarding the regression data should be whether the relationship is significant (i.e., whether the measured oxygen uptake rate is significantly greater than zero using a zero- or first-order regression model). If the slope of the regression line is statistically greater than zero, the p-value of the regression will be less than 0.05, and the 95% confidence intervals will not include zero. As shown in Figure 7-16, with a p-value of 0.0038, the slope of this regression is significantly different from zero and is represented by a mean zero-order oxygen uptake rate of -0.145 vol%/hr. The 95% confidence interval for the slope of this regression data (the range of the zero-order oxygen uptake rate described by the data accurate within 5% of the true value) is -0.089 to -0.201 vol%/hr.

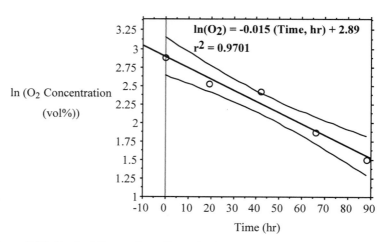

Figure 7-15 Typical first-order soil respiration gas data collected during a field in situ respiration test. The first-order respiration rate is less than 0.015/hr. The curved lines are 95% confidence intervals for the slope of the relationship.

Simple Regression X1: Time(hr) Y1: O2 Concentration (vol%)

Count:	R:	R-squared:	Adj. R-squared:	RMS Residual:
5	.9785	.9575	.9434	1.2443

Analysis of Variance Table

Source	DF:	Sum Squares:	Mean Square:	F-test:
REGRESSION	1	104.7431	104.7431	67.6497
RESIDUAL	3	4.6449	1.5483	p = .0038
TOTAL	4	109.388		

Variable:	Coefficient:	Std. Err.:	Std. Coeff.:	t-Value:	Probability:
INTERCEPT	16.7158				
SLOPE	-.145	.0176	-.9785	8.2249	.0038

Confidence Intervals Table

Variable:	95% Lower:	95% Upper:	90% Lower:	90% Upper:
MEAN (X,Y)	8.7091	12.2509	9.1704	11.7896
SLOPE	-.2011	-.0889	-.1865	-.1035

Figure 7-16 Typical regression results for linear regression analysis of field respiration data for bioventing systems.

Once the respiration rates are evaluated for statistical significance, background soil respiration rate values should be used to correct contaminated soil values for basal soil respiration taking place at the site. An inert gas tracer may be injected during soil aeration so that respiration rate measurements can also be corrected for diffusion of oxygen away from and carbon dioxide diffusion to the sampling probe during respiration rate determinations (Hinchee and Ong, 1992). However, the determination of background respiration rates in uncontaminated soils accounts for physical diffusion, as well as biological reaction mechanisms; tracer use during respiration rate determinations at any time other than immediately following air flow/tracer tests is not necessary. If background respiration rates are significantly greater than zero, they should be subtracted from respiration rates determined at locations throughout thecontaminated soil to yield background-corrected respiration rates. If background rates are not significantly different from zero, no correction to rates measured in the contaminated soil is necessary.

Finally, background-corrected respiration rates can be compared to rates published in the literature for field-scale bioventing systems to assess the relative biological activity measured at the field site. Table 7-11 provides an example of reported treatability test and field demonstration respiration rate data from various sources that can be used for this comparison. If background-corrected field respiration rates compare favorably with these reported data, significant biological activity is evident, and full-scale bioventing system design and implementation are warranted. If background-corrected field respiration rates are significantly lower (based on 95% confidence interval values) than these reported data, less than optimal conditions are evident because of moisture or nutrient limitations and/or the presence of inhibitory materials. The application of a bioventing system may not be practical under these limiting field site conditions. Consideration should then be given for the use of a laboratory treatability study to attempt to identify the cause of this limited microbial activity, or an alternative, nonbiological remediation scheme should be evaluated for use at the site.

Determination of the governing rate law can be made by investigating the nature of the regression residuals generated during the regression analysis. A residual is defined as the difference between the actual data point and the value of the dependent variable on the regression line; it can have either a positive or a negative value. A standardized residual is the residual divided by the value of the dependent variable at the point where the residual is calculated. If a given rate expression describes a data set, not only should the p-value be less than 0.05 and the 95% confidence of the regression slope not include zero, but the standardized residuals should also be randomly distributed over the range of the independent variable used in the regression. If a pattern is observed in the standardized residuals plot, the assumption that a particular linear model fits the data is not valid, and an alternative model should be selected for use to describe the data. If the residuals plots for a number of models are similar,

Table 7-11 Example treatability study and full-scale bioventing system respiration rates reported from various sources.

Site Location	Oxygen Utilization Rate(vol%/d)	Temperature (°C)	Source of Data*	Reference
Alaska	13.2	4 to 5	Treat.	Ong et al. (1994)
	6.9 ± 0.0	16	Treat.	Ong et al. (1994)
	4.2 ±2.6	8	Treat.	Ong et al. (1994)
	7.7	4 to 5	Field	Ong et al. (1994)
Florida	10 ± 0.5	25	Treat.	Hinchee et al. (1992)
Maryland	3.0 ± 0.2	21	Treat.	Hinchee et al. (1992)
Nevada	6.0 ± 0.2	21	Treat.	Hinchee et al. (1992)
Oklahoma	4.0 ± 0.5	17	Treat.	Hinchee et al. (1992)
Utah†	0.19 to 7.7	15	Field	Dupont et al. (1991)
	0.10 to 3.6	15	Field (+H2O)	Dupont et al. (1991)
	0.06 to 1.3	15	Field (+Nutr.)	Dupont et al. (1991)
	0.0 to 0.02	15	Field (Back.)	Dupont et al. (1991)

* - Treat. = field in situ respiration treatability test results, Field = field bioventing system performance data, $+H_2O$ = with moisture addition, +Nutr. = with inorganic nutrient addition, and Back. = background soil respiration rates.
† - These data were described by the first order relationship, $C = C_o\, e^{-k_1 t}$, with values in the table representing maximum rate values calculated from k_1 (21 vol%).

showing no particular pattern, then as a matter of practice, the simpler model form is selected.

Figure 7-17 shows the zero-order linear regression for a set of hypothetical in situ respiration rate data indicating that the regression coefficient is high (0.8964) and the p-value is well below the 0.05 criterion point. The residual plot for this data set, Figure 7-18, clearly indicates, however, that the linear model is not an adequate descriptor of these data because of the obvious pattern of the residual values. The first-order model, Figure 7-19, is much improved, particularly when one inspects the residuals plot for this set of data, as shown in Figure 7-20. Based on these results, a first-order oxygen uptake rate constant of -0.01/hr should be reported for these data.

Hydrocarbon degradation rate determinations can be made from field-determined in situ respiration data, assuming the 3.5/1 oxygen/hydrocarbon mass stoichiometry presented in Equation 7-101, and from known or estimated properties of the site soil using the following expression (Leeson et al., 1997):

Hydrocarbon degradation rate (mg/kg of soil/d)
$= (10\, k_o/3.5)\, (\theta_a/BD)\, (32\, P)/[0.08205\, (273 + T)]$ (7-101)

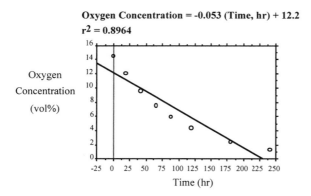

Figure 7-17 Zero-order linear regression results for hypothetical in situ respiration rate data.

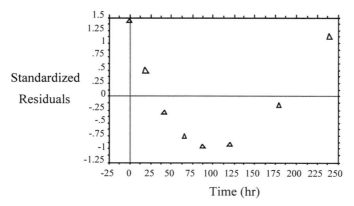

Figure 7-18 Residuals plot for zero-order linear regression results for hypothetical in situ respiration rate data.

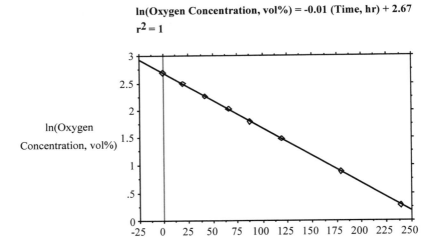

Figure 7-19 First-order linear regression results for hypothetical in situ respiration rate data.

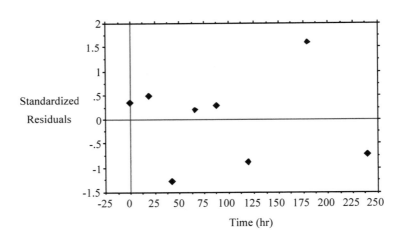

Figure 7-20 Residuals plot for first-order linear regression results for hypothetical in situ respiration rate data.

in which θ_a is air-filled porosity, which is unitless; BD is soil bulk density, kg/L soil; P is pressure, atm; and T is temperature, °C. Assuming average values for these parameters, θ_a is 0.3, BD is 1.4 kg/L of soil, P is 1 atm, and temperature is 10° C, the following relationship between measured zero-order degradation rates and equivalent hydrocarbon removal rates can be developed:

Hydrocarbon degradation rate (mg/kg of soil/d)
= (10 k_o/3.5) (0.3/1.4) (32) (1)/[0.08205 (283)]   (7-102)

These calculations can be made for sites where first-order oxygen utilization rates are observed by multiplying the first-order decay constant, k_1 (1/time) by the concentration of oxygen occurring within the soil at the site (typically 21 vol%). This yields an equivalent oxygen utilization rate with units identical to that of k_o. However, the oxygen utilization rate changes directly with oxygen concentration if removal follows a first-order relationship, so predicted hydrocarbon degradation rates will be dependent on oxygen concentrations maintained within the contaminated site under these conditions.

Remediation time can be estimated from these hydrocarbon degradation rates knowing the initial concentration of contaminant existing at the site. Respiration rates decrease linearly with decreasing contaminant concentrations below approximately 3,000 mg/kg TPH (Ravipaty, 1996), so the time for site remediation provided by Equation 7-103 is the minimum time to site cleanup. The actual time to reach soil closure levels will likely be two to three times longer; however, Equation 7-103 provides an initial estimate of the best-case time for preliminary evaluation of the feasibility of a bioventing system applied at a given site.

Minimum time
= (soil concentration, mg/kg TPH)/(degradation rate, mg/kg of soil/d)   (7-103)
Expected time to remediation
= approximately 2 to 3 x (minimum time to remediation)   (7-104)

Phase III — bioventing system design – and the interpretation of utilization rate data. Once in situ oxygen respiration rate values have been estimated, corrected for background respiration at a site, and determined to reflect biological activity that warrants application of a bioventing system at the site, full-scale system design should be carried out.

Air flow considerations are critical to optimal bioventing system design. The air flow rates that must be maintained can be determined directly from oxygen uptake rate measurements, as described above. It is imperative that the full-scale system effectively deliver the required oxygen to the contaminated locations and at a rate needed to maintain optimal aerobic activity, while minimizing air flow to reduce or even eliminate volatile emissions from the site. This requires careful consideration of tracer test/air permeability test results so that design components can be incorporated into the full-scale system that overcomes air flow limitations resulting from dead zones, low permeability lenses, short-circuit pathways, etc., which limit vapor flow rates through the zone of contamination. These design components may include passive/active air injection wells strategically placed to minimize dead zones, multiple air injection/extraction wells used to treat distinct soil layers existing at the site, etc. *See* Hinchee et al., (1992), the US EPA (1992c), Leeson and Hinchee (1995), and Leeson et al. (1997) for further details regarding the conduct of tracer tests and test data interpretation.

System operating conditions for full-scale design are determined in large part by the governing rate law for oxygen utilization observed throughout the field site. As indicated above, if the oxygen uptake data are governed by a zero-order rate law, oxygen utilization and concomitant contaminant degradation rates are independent of soil gas oxygen concentrations until oxygen limitations (approximately 2 vol% oxygen) occur. From a practical standpoint, this means that a constant inflow of oxygen to the subsurface is not required to maintain optimal oxygen uptake rates. Under these conditions, a pulse pumping system is possible, in which a blower would be operated for short periods of time for soil oxygenation followed by longer periods of with no air flow during which time soil incubation and contaminant removal takes place without air movement and possible air emissions. This pulsed system could take advantage of existing schedules of facility operations and maintenance personnel or existing blower equipment, or it might provide a cost-effective approach for satisfying stringent requirements on mass emission rates, limitations on operating hours caused by noise considerations, etc. Regardless of any specific project needs, zero-order uptake rates allow a much wider range of operating modes than does a system exhibiting first-order in situ oxygen uptake rates.

Field systems governed by first-order oxygen uptake relationships offer no flexibility in their operating mode. Performance of these systems is enhanced with increased soil oxygen concentrations, and they must be operated in a continuous mode to maximize removal of contaminants. One option that is currently being assessed by the US Air Force to maximize contaminant removal while still minimizing vapor emissions at sites dominated by first-order rates is the use of pure oxygen in place of atmospheric air as the gaseous oxygen source. Although this option has increased costs associated with pure oxygen generation, a reduction in air flows by a factor of approximately five significantly reduces air emissions and reduces many overall system costs associated with these reduced flow rates. The overall cost-effectiveness of these pure oxygen systems has yet to be proven, but it may be the only way to significantly enhance bioventing systems in which performance is found to be sensitive to soil gas oxygen levels.

7.3.2.1.2. Process operation. System operation is initiated in *Phase IV — system monitoring and performance evaluation* of this bioventing design protocol. Routine system monitoring is essential to the optimal operation and control of a field-scale bioventing system. The air extraction/injection well and vapor monitoring probes installed for conducting the initial air flow/permeability and in situ respiration testing should be incorporated into the full-scale field system as much as is practical. Additional wells should be placed as described above to overcome air flow limitations evident from tracer test results. In addition, multilevel vapor probes should be added as necessary to provide a representative, three-dimensional picture of contamination existing at the site prior to initiating the full-scale system. The logical places for these monitoring points are at the locations from which soil core samples are collected for initial contaminant quantitation. Multilevel, nested probes, such as those described in the AFCEE protocol by Hinchee et al. (1992), minimize the effort and expense of probe placement, as well as field sample collection and analyses, by utilizing common bore holes for multiple vapor probe depths. Use of these multilevel probes should be considered strongly at new bioventing field sites. As a rule of thumb, a minimum of nine vapor monitoring points (three spatial, radial locations at three depths each) should be installed per air extraction or injection well to provide the data necessary to adequately monitor vapor flow, respiration, and contaminant removal within a field site. The reader is referred to Hinchee et al., (1992) and Leeson et al. (1997) for additional discussion related to vapor probe design and placement.

Following complete bioventing field system installation, the soil vapor probes should initially be monitored daily to verify that the specified system design and operating mode is providing air flow to the site that was anticipated. System operating flow configuration and/or flow rate changes may be necessary to adapt the bioventing system to actual full-scale conditions encountered at the field site. This shakedown period is expected to last 1 to 2 wk, with operation after that time being fairly stable, requiring only minimal adjustment and maintenance. It is recommended that routine system monitoring be conducted monthly for the first 6 mo. of operation, and then quarterly thereafter, to verify proper system operation and allow system fine-tuning as remediation takes place throughout the site.

Routine system monitoring should include at the minimum the following parameters: system air flow rate (preferably flow rate to each injection/extraction well in the system); extraction system gas characteristics (oxygen/carbon dioxide, TPH, temperature, relative humidity, vacuum) if an air extraction system is utilized; soil gas monitoring point characteristics (oxygen/carbon dioxide, TPH, temperature, relative humidity, vacuum/pressure); and blower vacuum/pressure.

In addition to the collection of routine system monitoring data to ensure system operating effectiveness, quarterly to semi-annual system shutdown tests should be conducted to assess the progress of remediation taking place throughout the site. These shutdown tests are conducted in a manner identical to those described above, in which the air injection/extraction system is shut off and oxygen uptake is allowed to proceed without oxygen replacement. As was done with the Phase II in situ respiration data, data collected from these routine shutdown tests are statistically evaluated for their significance. Comparisons of overlapping confidence intervals of the slope of the oxygen uptake relationships, measured at given sampling locations but at previous time periods, are made to evaluate whether a significant reduction in respiration rates is occurring (inference that a significant reduction in contaminant levels is occurring as well). In addition, contaminated site respiration rate confidence intervals are compared to background respiration rates to determine if microbial activity at the site is reaching background activity, suggesting that cleanup levels are being reached at the site.

Respiration rates can decrease because of limited nutrient availability and/or low soil water contents (less than 25% field capacity) in addition to reduced soil hydrocarbon levels. With ongoing soil vapor relative humidity measurement, soil drying should be evident over time. Drying is minimized in bioventing systems because of low air flow movement through the soil. If drying does become an obvious limitation to system performance, however, controlled surface irrigation in coarse-grained soils, or injection of water-saturated vapor into fine-grained soils, can aide in modifying soil water content to within acceptable levels for improved microbial activity. Nutrient limitation is a more difficult matter, as nutrient supply and profusion into soils are limited by the high sorption capacity of soils for typical inorganic nutrients used in remediation systems. Evidence from field-scale bioventing systems treating fuel-contaminated soils indicates that nutrient addition in the field does little to nothing to improve the performance of these systems (Dupont et al., 1991; Miller, 1991). Nutrient limitation should not be considered a major cause of significant respiration rate reductions observed at a field site.

The primary cause of significant decreases found in in situ respiration rates, if soil water is not limiting, is a significant reduction in degradable contaminant concentrations in the soil. This respiration rate/contaminant concentration relationship has been suggested from field data and is well documented in the wastewater literature, but it has been verified only recently in laboratory-scale microcosm studies conducted at Utah State University (Ravipaty, 1996). Some of the results of this work for JP-4 contaminated soil from Hill AFB, Utah is presented in Figure 7-21 for 10 JP-4 concentration levels from 0 to 10,000 ppm at two soil water contents of 50% and 75% field capacity. As indicated in Figure 7-21, respiration rate appears to vary linearly with contaminant concentration to a soil level of approximately 1,000 ppm TPH, beyond which respiration rate reaches a pseudo steady-state value. The actual steady-state value reached varies with soil moisture content. This variance with soil water content is postulated to be related to an increased mass of contaminant available to the soil organisms resulting from an increase in the volume of soil water as soil water content increases. More critical to the use of field-determined respiration rates for the evaluation of the progress of

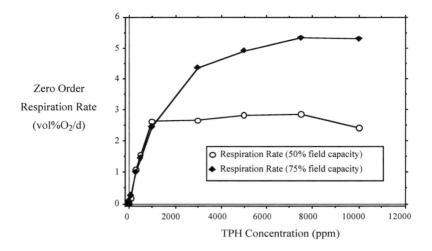

Figure 7-21 Respiration rate/contaminant concentration relationships generated in laboratory-scale microcosm studies conducted with JP-4 contaminated soil from Hill AFB, Utah (from Ravipaty, 1996).

remediation, however, is the fact that respiration rate approaches zero as a function of soil contaminant level independent of soil water conditions at which this respiration is taking place. This makes the use of declining respiration rates, particularly as they approach zero, or background respiration rate levels a powerful indicator of the extent to which contaminant mass removal is taking place at the site. With this being the case, the possible outcomes of this phase of the bioventing protocol are as follows: (1) respiration rates are shown to remain statistically greater than background respiration rate levels and within the range reported in the literature (Table 7-10), indicating that contaminant concentrations remain high (greater than 2,000 ppm TPH) and bioventing should continue; (2) respiration rates are shown to be greater than background levels but are decreasing over time, indicating that contaminant mass removal is continuing, contaminant concentrations in the soil are approaching approximately 1,000 ppm TPH, and bioventing should continue; or (3) respiration rates are shown to be equal to background levels, indicating that contaminant removal and site remediation are indicated, and confirmatory soil borings (Phase V) should be collected to verify that remedial soil concentration goals have been met.

Phase V — verification of system performance. The final step in the bioventing design protocol, that of verification of bioventing system performance using confirmatory soil core results, is reached through a positive outcome from Phase IV. Again, if respiration throughout the site approaches or is statistically equivalent to background levels, low residual contaminant levels are indicated, and verification of this result should be provided from soil concentration values. Soil core samples should be collected in a manner identical to that used in preliminary site assessment activities to allow direct comparison of results between sampling time intervals. In addition because of the large variability inherent in soil sampling and contaminant distribution, confirmatory samples should be collected as close as possible to the locations of the original soil cores if valid comparisons of contaminant levels are to be made over time.

If contaminant mass removal has occurred and low respiration rate results are indicative of low residual contaminant concentrations, these confirmatory soil core results should show low levels of both volatile and semivolatile constituents remaining at the site. If measured soil concentrations are below regulated site soil cleanup levels, the site would be considered for a closure action and the bioventing system would no longer have to operate at this site. If soil concentrations remain above regulatory action levels, the rate and mode of operation of the bioventing system should be evaluated and system modifications should be made to enhance the removal of remaining contaminant so that closure can be accomplished in the future. In the latter case, a modified bioventing system would go back into operation, and respiration rates during shutdown periods would continue to be monitored on a quarterly basis until once again background oxygen uptake rates are observed, initiating the collection of a new round of confirmatory soil core samples.

7.3.2.2. Bioslurping

Bioslurping, as coined by the US Air Force (1995), is a relatively new technology that is designed to simultaneously recover free-phase product and remediate the vadose zone utilizing vacuum-assisted free-phase product recovery. Bioslurping systems are used to extract free product from above the water table and to transfer oxygen and aerate vadose and smear-zone soils with the use of soil venting. The bioslurper system withdraws groundwater, free product, and soil gas through the same extraction well using a single, liquid-ring vacuum blower and can be operated to achieve hydraulic control of a contaminant plume, as is done with conventional pump-and-treat technology. Vacuum-enhanced pumping minimizes liquid drawdown near the extraction well, maximizing fluid recovery as liquid moves horizontally along high-transmissivity flow paths and into the well (*see* Figure 7-22).

Application of a vacuum (up to 0.51 m, or 20 in. Hg) moves liquid (water and product), as well as soil gas. In moving the latter, oxygen is transferred to the vadose and smear zones, adding a bioventing component to the overall product recovery scheme. It is because of the bioventing component that the Air Force termed this process bioslurping.

7.3.2.2.1. Process design. Bioslurper system design involves conducting a series of pilot tests to evaluate free product recovery and contaminant mass removal under a

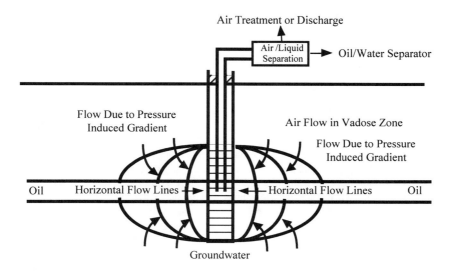

Figure 7-22 Theoretical air, water, and product flow occurring from the operation of a bioslurping system. Modified from the US Air Force (1995).

variety of extraction well operating modes, including conventional product skimming, vacuum-enhanced product recovery, and a dual-pump drawdown mode. Pilot test procedures for the evaluation of bioslurper systems have been published by AFCEE (US Air Force, 1995) and form the basis for the treatability tests described below. Whereas the initial components of these treatability tests are primarily focused on the recovery of product from the subsurface, the latter stages of the testing methodology are used to describe the biodegradation of contaminant in the smear zone and capillary fringe and are therefore relevant to the main topic of this chapter.

The basic pilot test procedure is conducted by placing a drop tube down into an extraction well to remove liquid and product off of the water table with a liquid-ring vacuum blower in a conventional skimming simulation test. Next, the extraction well is sealed (so that a vacuum can be maintained on it) to evaluate the impact of vacuum enhancement on overall free-phase product recovery. Finally, the drop tube is lowered, and with the same vacuum blower, a drawdown product recovery test is conducted. More specific procedures and monitoring requirements for each stage of the bioslurper treatability study are summarized below.

Conventional skimmer simulation. The US Air Force (1995) bioslurper treatability test protocol begins with the assessment of the performance of conventional product skimming using a bioslurping well configured to allow only atmospheric pressure to develop within the well (valve open in Figure 7-23). The slurper tube is set at the NAPL elevation, and total NAPL-plus-groundwater recovery volumes are recorded periodically over a 48-hr test period, as are the LNAPL and groundwater depths below ground surface.

Bioslurper system simulation. Following the evaluation of conventional skimming, the valve at the extraction well head is closed, allowing vacuum to build up around the well (valve closed in Figure 7-23). Operation of the well in this bioslurper mode continues for a 96-hr period, during which time the following process variables are monitored: extraction well vapor discharge (minimum of two samples, one after startup and one at the end of the 96-hr operating period); aqueous well discharge samples (minimum of two samples, one after startup and one at the end of the 96-hr operating period); LNAPL recovery volume (every 30 min for 2 hr, every 2 hr to hour 12, then every 12 hr till the completion of the test); continuous vapor and groundwater discharge volumes; data for soil gas permeability determinations (soil gas pressure readings every 5 min for 20 min, every 20 min to hour 1, every hour to hour 6, every 6 hr to hour 24, and every 24 hr until the end of the bioslurping test) measured in multilevel vapor probes installed around the extraction well; and biodegradation monitoring using procedures outlined in Section 7.3.2.1.1. Soil gas permeability and biodegradation data are collected for bioventing system performance assessment during the bioslurping phase of this treatability protocol, as described in the bioventing section of this chapter.

Following the 96-hr bioslurping test, an additional conventional skimmer test is recommended by the US Air Force (1995) to provide "...a more accurate basis for comparing sustainable LNAPL recovery rates with conventional technology and bioslurping." This second conventional skimmer test is conducted in a manner identical to that described above, and leads into the final operating mode of the bioslurping test protocol, that being a dual-pump/drawdown simulation mode.

Dual-pump/drawdown simulation. A 48-hr drawdown simulation recovery test is conducted following the second conventional skimmer test by operating the atmospheric recovery well with the slurper tube dropped below the water table to a depth that corresponds to the vacuum observed during the bioslurping portion of the test. The recovery well operating characteristics during this phase of the treatability test are shown in Figure 7-24.

Results of the treatability assessment. Based on the results of this field-scale bioslurping treatability test, the following system design parameters can be quantitatively determined:

- The rate of product and groundwater recovery under conventional skimming, vacuum-assisted skimming (bioslurper), and dual-pump drawdown operating conditions

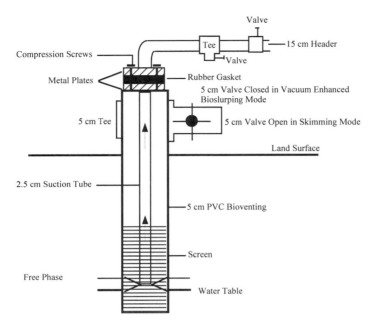

Figure 7-23 Configuration and product flow occurring in a bioslurping well operated in a conventional skimming mode and in a bioslurping (vacuum enhanced) mode. From the US Air Force (1995).

- The radius of influence, soil pneumatic permeability, and vacuum/air flow relationships for operation of a bioslurper system
- Soil oxygen mass transfer rate during bioslurper operation based on soil gas oxygen and carbon dioxide measurements
- In situ soil respiration rate and equivalent hydrocarbon degradation rate based on soil gas oxygen and carbon dioxide data collected during in situ respiration measurements

From these findings, the design engineer can determine if vacuum-assisted LNAPL recovery provides significant benefits to site remediation compared to conventional methods. It could also be determined that although free product recovery can be provided to some extent with the use of a bioslurper system, the main advantage of its use may be the biodegradation of residual-phase material in the smear zone and at the water table. Some sites have actually operated a bioslurping system in a drawdown mode, where a liquid-ring vacuum blower system provides both groundwater depression and bioventing of the capillary fringe and dewatered smear zone.

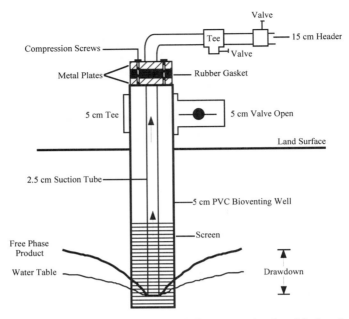

Figure 7-24 Configuration and product/air flow occurring in a bioslurping well operated in a drawdown simulation mode. From the US Air Force (1995).

7.3.2.2.2. Process operation. Bioslurper well operation is carried out in a fashion similar to that described for bioventing systems in Section 7.3.2.1.2, with additional requirements for the free-phase product and groundwater recovery, and groundwater and vapor treatment systems that may be required for legal operation of these systems. Routine system monitoring is essential to the optimal operation and control of a field-scale bioslurping system. The extraction well and vapor monitoring probes installed for conducting the initial treatability testing should be incorporated into the full-scale field system as much as is practical. Additional wells should be placed based on knowledge of the distribution of free-phase product and residual-phase material throughout the site and on the pneumatic permeability test results. In addition, multilevel vapor probes should be added as necessary for full-scale system performance monitoring. Multilevel, nested probes (*see* Figure 7-25) minimize the effort and expense of probe placement by utilizing common bore holes for multiple vapor probe depths. The reader is referred to Hinchee et al. (1992), the US Air Force (1995), and Leeson et al. (1997) for additional discussion related to vapor probe design and placement.

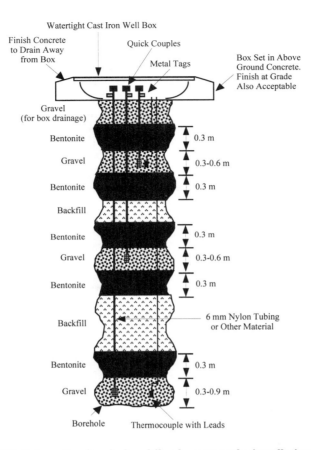

Figure 7-25 Schematic of typical multilevel vapor probe installation in a single bore hole. From the US Air Force (1995).

Following complete bioslurping field system installation, extraction wells and soil vapor probes should initially be monitored daily to verify that the specified system design and operating mode are providing product recovery and air flow that was anticipated from the treatability study results. System operating flow configuration and/or flow rate changes may be necessary to adapt the bioslurping system to actual full-scale conditions encountered throughout the entire field site. This "shake-down" period is expected to last 1 to 2 wk, with operation after that time being fairly stable, requiring only minimal adjustment and maintenance. It is recommended that routine system monitoring be conducted monthly for the first 6 mo. of operation, and then quarterly thereafter, to verify proper system operation and allow system "fine-tuning" as remediation takes place throughout the site.

Routine system monitoring should include at the minimum the following parameters: system air extraction rate; blower vacuum; system groundwater and product recovery rates; extraction system gas characteristics (oxygen/carbon dioxide, TPH, temperature, relative humidity, vacuum); soil gas monitoring point characteristics (oxygen/carbon dioxide, TPH, vacuum); and appropriate vapor and groundwater treatment system operating parameters (i.e., flow rates, inlet and outlet contaminant loading rates, operating temperatures, removal efficiency, etc.).

In addition to the collection of routine system monitoring data to ensure system operating effectiveness, quarterly to semiannual system shutdown tests, as described in the bioventing section of this chapter, should be conducted to assess the progress of remediation taking place throughout the site. As with bioventing systems, the collection of soil core data would be used to verify the level of contaminant removal indicated from these background respiration data.

7.4 Treatability Studies for Bioremediation Systems

The use of bench- and/or pilot-scale studies may be necessary to avoid technology misapplication at a full-scale site once candidate treatment technologies have been selected. When first evaluating the potential feasibility of using biological treatment, it is necessary to understand the biodegradability of the contaminants present. The reported biodegradability of typical groundwater and soil contaminants is provided in Table 7-1. Site characterization data should be used to estimate the nature and projected concentrations of contaminants that will exist in the extracted groundwater and excavated contaminated soil that must be treated. If the selected technology has a long performance history for wastes and under site conditions that are found at the contaminated site being considered, treatability studies will not be necessary, and full-scale remedial system design can begin. However, if the technologies selected for consideration are new or unproven, or if performance of an old technology has not been documented at the field-scale for the specific waste type or under site and/or soil conditions that exist, bench- and/or pilot-scale treatability studies may be necessary.

After it is determined that the constituents or the majority of the constituents present are biodegradable, it is necessary to identify candidate technologies for further consideration. The technologies to be selected for further evaluation will depend on the characteristics of the contaminated media, most notably the overall organic content as measured by BOD or COD, and on whether the media to be treated is groundwater or soil. Generally, for groundwater treatment applications, an aerobic fixed growth process is preferred because of the relatively low organic content of the expected wastewater stream. For treatment of soil with high organic content (up to

250,000 mg/kg concentrations), suspended growth reactors are required in a slurry-phase application.

Bench-scale treatability studies are designed to screen a large selection of possible technologies to determine under ideal laboratory conditions whether soil and waste conditions existing at the site may limit the efficiency of contaminant reduction. A large selection of technologies can be screened for $10,000 to $50,000 within a 4- to 6-wk period (US EPA, 1989b). From this bench-scale screening effort, those technologies that prove incapable of providing the contaminant reduction necessary at the site are eliminated from further investigation. Once bench-scale screening has been completed, pilot-scale testing at the actual field site may be necessary to collect further site-specific data necessary for the final design of the full-scale remedial technology system. The decision to carry out pilot-scale studies, costing from $25,000 to $1,000,000 or more and lasting from 12 wk to 12 mo., is driven by the cost of replacement of the technology if it fails at the field scale. Again, pilot-scale studies will generally not be necessary if the technology is not new, or if the short-term risk of the release is low so that design and implementation of other technologies could take place without compromising the health and safety of the public if an initial technology does not meet minimum performance requirements.

7.4.1 Treatability Studies Applicable to Aqueous Phase Treatment

Treatability tests are generally recommended for ex situ groundwater bioremediation systems. The tests are conducted to determine a number of design variables, including:

- Micronutrient addition requirements
- Determination of reaction rate constants
- Identification of the presence of toxins, or formation of toxic byproducts as a consequence of microbial metabolism
- Neutralizing chemical requirements
- Potential use of cometabolites to aid contaminant degradation

Most field pilot tests require more extensive preparations. However, a field bioreactor can be set up rapidly and economically as follows: an open-top cloth filter bag or sock is placed in a plastic bucket or pail, and an air diffusion stone (obtained from a aquarium supply store or pet store) is placed in the bottom. Air is fed to the stone through plastic tubing attached to the discharge of a very small aquarium air pump. Groundwater, nutrients, and bacterial innoculum are added to the interior of the cloth bag. Starting after 48 hr, samples of treated groundwater are withdrawn from the annular space between the cloth bag and the vessel wall and analyzed for contaminant removal. Usually, soil at or above the contaminated groundwater contains indigenous bacteria acclimated to the groundwater. The best source of bacterial innoculum is derived by placing some of this soil into the cloth bag. The second-best chance of obtaining acclimated bacteria is to use sludge from the nearest municipal sewage treatment plant. Bacterial sludge that builds up on the interior surface of the cloth bag can be cultured for seeding further tests or for seeding full-scale operation of a bioreactor.

For laboratory testing of denitrification, a batch equilibrium test can be conducted as follows: a sample of groundwater with added methanol or acetate and some bacterial sludge from a municipal sewage denitrification plant are magnetically stirred in the absence of air for a few weeks or until gas production ceases. The sludge produced is collected and analyzed for MLVSS (mixed liquor volatile suspended solids) and quantified in comparison to the sludge added originally. The same type of laboratory test can be used to determine if excess dissolved carbon addition is needed to achieve the final nitrate concentration goal. Based on an analysis of the stoichiometry of denitrification, 3 millimoles (246 mg) of sodium acetate is the minimum amount needed to reduce 4 millimoles (248 mg) of nitrate. One experiment should be run with this stoichiometric amount of sodium acetate. Parallel experiments should be run with 15% and 30% additional sodium acetate. If degradable organic compounds are in the groundwater, an experiment should be run with reduced acetate concentration.

If continuous-flow testing of denitrification is attempted, the reactor solid-support surfaces must first build up a film of denitrifying bacteria. The treated effluent is recycled for a period of time until stability is achieved for the nitrate and total organic carbon removal rates, as indicated by the amounts of nitrate and dissolved carbon additions needed to maintain steady-state recycle solids concentrations. Also, if a fluidized bed test reactor is being used, the bacterial film on the sand or on the carbon granules is deemed suitable when the settled bed height remains at steady-state.

7.4.2. Treatability Studies Applicable to Solid Phase Systems

7.4.2.1. Land Treatment Systems

The first step in land treatment system design and treatability assessment is the selection of an appropriate medium to support biological reactions that take place in the unit. Ideally, the contaminated soil will serve this purpose. However, if the contaminated soil is of unacceptable texture, or if it contains concentrations of contaminants of concern that are toxic or inhibitory to the soil microorganisms, dilution of this toxicity must be carried out using additional soil media. Uncontaminated on-site soil is the best alternative for this purpose, as a major expense in land treatment systems can be the cost of the purchase and transport of off-site soil. Whether using on-site or off-site soil, the soil medium should be selected for its ability to support active microbial metabolism (i.e., loamy texture, good water holding capacity, good aeration, etc.).

Next, waste loading rates to be used in the laboratory treatability studies should be selected from literature data for the waste of concern. For fuel-related wastes for example, application rates for successful treatment have ranged from 1 wt% to 6 wt% Oil & Grease. The loading rate is based on the assumption that the waste is mixed intimately with soil in the ZOI of the land treatment unit. If the fuel/soil mixture at the site has Oil & Grease concentrations less than 1%, the land treatment system can be designed without the need for additional treatment media. With this level of Oil & Grease, the excavated soil forms the entire volume of the land treatment unit. Management of this system requires nutrient and moisture management, as well as tilling for soil aeration.

Water extracts of the soil/waste mixture are then obtained for toxicity evaluation (Sims et al., 1986). Water extracts represent a simulated leachate and provide some indication of the composition of soil solution available to the soil microorganisms. A Microtox™ analysis is conducted to evaluate the potential toxicity of a given waste loading rate. Toxicity in the Microtox™ suggests a loading rate that would be inhibitory to soil microorganisms within the land treatment unit, requiring a lower waste loading rate to eliminate this observed toxicity.

VCDR procedures would then be conducted to evaluate the degradation and volatilization of contaminants of concern at those loading rates shown to be nontoxic in the Microtox™ assay. Parent compound volatilization rates are quantified by monitoring reactor head space and off-gas concentrations over the course of the laboratory studies. Parent compound degradation rates are quantified by monitoring their concentration in reactor soils at five to six sampling intervals over time.

Design loading and management techniques are then chosen, from the results obtained above, to minimize contaminant volatilization and maximize degradation rates. These loading rates are then translated into treatment bed requirements for field-scale system design based on the total mass of contaminated soil or waste requiring treatment.

7.4.2.2. Soil Piles

After it is determined that the constituents or the majority of the constituents present are biodegradable, the evaluation of the effectiveness of nutrient augmentation and/or soil matrix manipulation (i.e., organic matter addition) can be assessed using laboratory-scale treatability tests. These tests are similar to those described above for land treatment systems and should be conducted using samples of actual contaminated media to be treated. They should also be conducted under representative soil moisture and aeration conditions so that laboratory results provide meaningful data that can be effectively scaled up to actual field-scale conditions. These treatability studies normally focus on the evaluation of enhanced treatment performance as a function of: soil moisture content (50%, 75%, 90% field capacity, for example); nutrient addition (one and two times stoichiometric nitrogen and phosphorus concentrations, for example); and organic carbon augmentation (10 wt% and 25 wt% manure addition, for example). Soil pile conditions shown to be optimal from these laboratory studies are then used to construct full-scale piles. Soil pile operating conditions can then be optimized at the field scale by carrying out field oxygen transfer and in situ respiration tests in full-scale field piles, as described in Section 7.3.1.2.5.

7.4.2.3. Slurry-Phase Reactors

Treatability testing for slurry-phase reactors will generally always be necessary, as only a limited data set currently exists that defines their performance in the treatment of highly contaminated, complex soil matrices. The first step in determining if biological treatment is feasible for the specific contaminant and matrix is to perform a biotreatability evaluation using representative samples of the material. Pilot-studies will be critical in demonstrating the technical and economic feasibility of slurry reactors before field implementation.

For slurry-phase applications, bench-scale reactor studies can identify potential mixing problems with the contaminated solids and can aid in determining optimal solids concentrations for field-scale reactor operation. Testing should consist of operation for at least 8 wk of bench- or pilot-scale systems such that multiple sludge ages are achieved. Data collection during the treatability phase should focus on influent and effluent quality, as well as operational parameters such as oxygen uptake rate, sludge yield, sludge settleability, etc., that will be important for design development. It is generally advisable to evaluate at least two operating conditions (e.g., organic loading rate, hydraulic residence time, etc.) for each process under consideration. Temperature effects, bioaugmentation options, foaming control options, and surfactant enhancements can all be evaluated effectively at the bench-scale phase prior to initiating full-scale bioslurry remediation.

7.4.2.4. Bioventing

Treatability testing for bioventing systems is generally carried out at the field scale using the step-wise protocol detailed in Section 7.3.2.1.1 and presented graphically in Figure 7-13. A determination of the potential for the use of in situ bioprocesses is made based on soil gas oxygen and carbon dioxide measurements made during the site investigation phase. Further development of air flow system design and process operation to meet oxygen uptake requirements is carried out using a field-scale, one-well pneumatic pump test once evidence of active biodegradation has been collected from the site.

7.4.2.5. Bioslurping

As with bioventing systems, field-scale treatability testing is an integral part of the design process for bioslurping systems. Air flow, groundwater flow, product recovery rates, and in situ respiration rate determinations are made from a single-well extraction study conducted before the final selection of bioslurping as the remedial approach for a site. The results from this field treatability assessment indicate the feasibility of vacuum-assisted product removal and the enhancement to conventional free-phase product recovery that can be expected from the use of a bioslurper system under a fixed set of field site constraints. These data also indicate the optimal operating strategy for a site based on both free-phase product recovery and contaminant removal from the site via residual-phase biodegradation in the capillary fringe and smear zone. Details of procedures for field bioslurping treatability evaluation can be found in the US Air Force (1995) and are summarized in Section 7.3.2.2.

7.5. Cost-Estimating for Bioremediation Systems

Costs and performance of a variety of field applications of bioremediation technologies are included in the the EPA Office of Research and Development (phone 513-569-7562) database "Bioremediation in the Field Search System." The database is available on disk for personal computer applications.

Site assessment costs associated with soil core and groundwater sampling can vary dramatically based on the characteristics of the site, depth to groundwater, soil texture, surface structures, etc. These costs are similar, however, for both in situ and ex situ systems. The extent of contamination must be accurately determined either to establish the extent of excavation necessary for ex situ systems or to design an in situ treatment network that will provide treatment for the entire area of contamination at a given site. Site assessment costs would be expected to range from $5,000 to $25,000 per typical site.

Soil excavation costs for ex situ systems can also vary dramatically based on the target cleanup level, extent of contamination, depth to contamination, etc. Depth of contamination is a particularly important variable as excavation costs increase exponentially below 15 m (50 ft). At a depth of 15 m (50 ft) or greater, excavation becomes prohibitively costly, and in situ treatment systems should be considered. The US EPA (1985) provides data for backhoe and crane/dragline use that show excavation costs of $2,000 to $4,000 in 1992 dollars for a typical site requiring the removal of 800 m^3 (1000 yd^3) of contaminated soil from depths ranging from 3 to 15 m (10 to 50 ft).

7.5.1 Costs for Aqueous Phase Treatment

7.5.1.1 Costs from Case Histories

Costs for in situ and ex situ groundwater biotreatment were reported by DuTeaux (1996), as summarized in Table 7-12. Costs for an ex situ 2.7-L/sec (40-gal/min) RBC system with four parallel 0.67-L/sec (10-gal/min) contactors were given in the

Table 7-12 Costs for in situ and ex situ groundwater biotreatment (DuTeaux, 1996).

Total Cost	Treatment	Contaminants
In Situ		
$7.81/m³ ($31.25/1,000 gal)	Lagoon with surface aerators	Oil & Grease
$40 to $57.5/L jet fuel ($160 to $230/gal jet fuel)	Nitrate-enhanced anaerobic treatment	Jet Fuel
$21/L jet fuel ($84/gal jet fuel)	Infiltration gallery/nitrate and phosphate addition	Jet Fuel
Ex Situ		
$0.30/m³ ($1.20/1,000 gal)	Activated Sludge	Phenols, ketones, toluene benzene, vinyl chloride, chloroform
$0.075/m³ ($0.50/1,000 gal)	Methanotrophic fluidized bed and trickling filter	TCE, PCE

California Department of Health (1988) as follows: capital investment costs were $1,244,000; and operations and maintenance costs were $3.59/m³ ($13.60/1,000 gal). A breakdown of the capital investment costs for this RBC system is given in Table 7-13, and a breakdown of its estimated operating and maintenance costs is given in Table 7-14.

Costs for an ex situ 5-L/sec (75-gal/min) activated sludge treatment system combined with PACT were given by the California Department of Health (1988) as follows: capital investment costs were $1,570,000; and operations and maintenance costs at 4.3 L/sec (65 gal/min) were $9.10/m³ ($34.46/1,000 gal).

A breakdown of the capital investment costs for this PACT system is given in Table 7-15, and its estimated operating and maintenance costs are summarized in Table 7-16.

Unit costs for in situ air sparging systems are given by Shrauf, Sheehan, and Pennington (1993, 1994) at $37.50/m³ ($30/yd³) for monitoring and maintenance, whereas other in situ costs range from $82.50/m³ to $154/m³ ($66/yd³ to $123/yd³). These costs are almost doubled if hydrogen peroxide is used (Geselbracht, Donovan, and Greenwood, 1986).

7.5.1.2 Computer Programs for Process Design and Cost Estimating

The computer models Composer Gold/ECHOS marketed by Building Systems Design (Atlanta, Georgia) include ex situ and in situ bioremediation modules. The RACER/ENVEST computer models marketed by Talisman Partners Ltd. (Englewood, Colorado) include process design and cost estimating for in situ bioremediation. This

Table 7-13 Estimated capital costs of an RBC system (California Department of Health Services, 1988).

Description	Quantity	Unit	Unit Price ($)	Total Cost (Rounded $)	Assumption	Reference
RBC	4	each	100,000	400,000	Four contactors @ 9,000 m² (100,000 ft²) surface area	Treatability study Cost from vendor
Hood	4	each	10,000	40,000	Fiberglass covers for contactors	Cost from vendor
Clarifier	1	lump sum	30,000	30,000	5.4 m (18 ft) diameter coated steel, 10 m/d (225 gal/d/ft²) surface overflow rate	Cost from vendor
Sand filter	1	lump sum	50,000	50,000	0.22 m/min (5 gpm/ft²) surface loading rate, 0.9 m (3 ft) diameter	Cost from vendor
Biomass controls	4	each	6,000	24,000	Chemical and nutrient feed systems and appurtenances	Cost from vendor
Freight	4	each	15,000	60,000	Delivery to site of all equipment	Cost from vendor
Concrete Basin	4	each	80,000	320,000	Foundation and walls to house 100 m³ (25,000 gal) each installed	Treatability study Cost from vendor
Installation	4	each	50,000	200,000	For all equipment	Cost from vendor
Sludge handling equipment	1	lump sum	120,000	120,000	Sludge storage tank/aerobic digester, plate and frame filter press, piping	Cost from vendor
Total				1,244,000		

Table 7-14 Estimated annual operation and maintenance costs of an RBC system (California Department of Health Services, 1988).

Description	Quantity	Unit	Unit Price ($)	Total Cost (Rounded $/yr)	Assumption	Reference
Operation and maintenance labor	3,000	hr	50	150,000	8 man-hr/day, 365 day/yr for a biological system. Includes process control monitoring.	Professional judgment
Power costs	120,000	kW-hr	0.1	12,000	Unit price based on 1988 rate	KW-hr based on engineering judgement
Parts and supplies	1	lump sum	30,000	30,000	Estimated at 3% of capital equipment costs	Allowance based on professional judgement
Nutrient addition	1	lump sum	50,000	50,000	Estimate based on anticipated Stream A flow rate of 0.16 m³/min (40 gpm)	Based on engineering judgment
Sludge disposal	44	metric ton	1,000	44,000	Sludge production of 44 metric ton/yr (88 ton/yr) estimated from treatability study results. Includes filter cake analysis.	Cost from Stringfellow Pretreatment Plant cost data
Total				286,000		

Table 7-15 Estimated capital costs for PACT treatment of 5 L/sec (75 gal/min) of extracted groundwater (California Department of Health Services, 1988).

Description	Quantity	Unit	Unit Price ($)	Total Cost (Rounded $)	Assumption	Reference
PACT treatment system including basin, blowers, clarifier, sludge storage/aeration, PAC and chemical feed systems, instrumentation.	1	lump sum	600,000	600,000	Sized based on anticipated maximum flow rate for cost comparison to existing activated carbon system. Volume = 320 m3 (80,000 gal) based on 18 hr hydraulic retention time from treatability study.	Cost from vendor. Design based on Zimpro treatability study
Installation and start up	1	lump sum	240,000	240,000	Estimated at 40% of treatment system cost.	Cost from vendor
Piping and electrical, installed	1	lump sum	300,000	300,000	Estimated at 50% of system cost	Allowance based on professional judgment
Solids filter press and pumps, installed	2	each	75,000	150,000	Two presses sized for 175 kg/d (350 lb/d) of dry solids, ≈ 0.3 m^3 (10 ft^3) each	Cost and design from CH2M-Hill; Zimpro
Dual-media filtration, installed	2	each	140,000	280,000	Two 1.8 m (6 ft) diameter gravity flow filters @ 0.09 to 0.44 m/min (2 to 10 gpm/ft^2) each	Design and cost extrapolated from CH2M-Hill
Total				1,570,000		

Table 7-16 Estimated annual operation and maintenance costs for PACT treatment of 5 L/sec (75 gal/min) of extracted groundwater (California Department of Health Services, 1988).

Description	Quantity	Unit	Unit Price ($)	Total Cost (Rounded $/yr)	Assumption	Reference
Carbon	136,500	kg	1.5	205,000	Based on dosage of 953 mg/L carbon or 0.95 kg/m^3 (8 lb/1,000 gal) treated	PACT treatability study; cost/lb from vendor
Nutrients	21,500	kg	2	43,000	49 kg/d (98 lb/d) N, 10 kg/d (20 lb/d) P	Zimpro
Polymer	285	kg	3.6	1,000	0.8 kg/d (1.6 lb/d) at a dose of 2 mg/L	PACT treatability study; cost/lb from vendor
Power costs	1,186,250	kW-hr	0.1	119,000	Based on 3,250 kW-hr/d power usage for PACT system and solids handling. Unit price based on 1988 rate.	kW-hr/d from Zimpro, professional judgment
Sludge disposal	585	metric ton	502	294,000	8.5 kg/m^3 (68 lb/1000 gal) wet solids @ 40% solids content	PACT treatability study
Process lab analysis	365	each	70	26,000	$70/analysis, 7 analyses/wk, 52 wk/yr	CH2M-Hill
Labor	6,500	hr	50	325,000	18 hr/day, 365 day/yr	Labor hours based on vendor; professional judgment; CH2M-Hill
Maintenance materials	1	lump sum	130,000	130,000	Estimated at 10% of PACT system equipment capital cost	Allowance based on professional judgement
Total				1,143,000		

bioremediation model is based on a system with injection wells. No extraction or circulation of groundwater is included. Required input parameters include injection rate per well; number of wells; depth to the groundwater table; and aquifer thickness.

The ENVEST model report suggests that the injection rate for a confined aquifer in gal/min can be approximated by:

Injection rate (gal/min) = $2Kb(h-H)/[1055\{\log(r/R)\}]$ (7-105)

and for an unconfined aquifer by:

Injection rate (gal/min) = $K(h^2 - H^2)/[1055\{\log(r/R)\}]$ (7-106)

in which h is the feet of head above the bottom of the aquifer while recharging, H is the head without injection ongoing, r/R is the dimensionless ratio of well radius of influence to well radius, b is the feet of aquifer thickness, and K is the hydraulic conductivity (gpd/ft^2). K values can be approximated by 10,000 for gravel, 100 for sand, and 1 for silt. If wells vary in depth or diameter, they should be grouped and the model run for each group of similar wells.

The model assumes for a drilling method that a hollow-stem auger is used for depths to 45 m (150 ft) and a water/mud rotary rig for depths greater than 45 m (150 ft). Air rotary drilling is an option, or water/mud rotary can be chosen for any depth. The model assumes use of stainless steel screens and casings, but schedule 40 or 80 PVC can be chosen. The well diameter can be chosen in the 5- to 20-cm (2- to 8-in.) range, or the model will assume the following values:

- Up to 3.33 L/sec (50 gal/min)   5-cm (2-in.)
- 3.33 to 10.0 L/sec (50 to 150 gal/min)   10-cm (4-in.)
- 10.0 to 20.0 L/sec (150 to 300 gal/min) 15-cm (6-in.)

Labor hours for drilling supervision by a staff hydrologist are included in the cost estimate, assuming a drilling rate of 6 m/hr (20 ft/hr), plus 1 hr per 15 m (50 ft) for well construction plus 2 hr per well for completion. Costs are included for decontamination of well structures, drilling tools, and samplers if chosen.

Other options available include whether soil sampling, drumming of cuttings, and installation of monitoring wells are to be done. The influent piping is assumed to be 30 m (100 ft) per well, using 5-cm (2-in.) stainless steel, unless other lengths or PVC is chosen. Nutrient feed tanks and monitoring well purge water tanks — stainless, polyethylene, or steel — can be specified in sizes ranging up to 130 m^3 (35,000 gal). Microorganism cultures purchased from laboratories can be included with the initial installation as an injection option. If the option for adding nutrients is chosen, the model assumes that pulverized fertilizer is used, in 25-kg (50-lb) bags, containing 20%

each of nitrogen, phosphorus, and potassium. Alternatively, if the user does not specify the quantity of nutrients, the model estimates nutrient cost to be 30% of the hydrogen peroxide cost.

The model assumes that hydrogen peroxide will be used as the oxidizing agent (electron acceptor). Pricing is based on using 227-kg (500-lb) drums of 50% peroxide solution that is mixed with water to form a 300 mg/L solution for injection from a tank.

The model can estimate the cost of operations and maintenance. If such an option is chosen, estimated costs include a maintenance crew that comes to the site once a week for the first month and once a month thereafter. Ongoing sampling and analysis costs can be estimated using a separate ENVEST model, and this model suggests the following protocol be used each month: measurement of DO, total heterotrophic plate count for bacteria, count for hydrocarbon degraders, count for fluorescent pseudomonas, purgeable halocarbons, and purgeable aromatics.

7.5.2 Costs for Solid Phase Treatment

DuTeaux (1996) has tabulated case histories with costs for ex situ soil biotreatment. His findings are summarized in Table 7-17.

Table 7-17 Cost of ex situ soil treatment based on case histories (DuTeaux, 1996).

Total Cost	Treatment	Contaminants
$563/m^3 ($450/yd^3)	Windrow Composting	TNT, RDX, and HMX explosives
$33.75/m^3 (27/yd^3)	Heap Pile	TPH
$88/metric ton ($80/ton)	Aerated Pile	TPH
$250/m^3/($200/yd^3)	Tilled to 0.45 m (18 in.) depth	BTEX and gasoline
$250/m^3/($200/yd^3)	Slurry in Sequencing Batch Reactors	Oil and PAHs
$439/m^3/($351/yd^3)	Tilled	TPH
$201/m^3/($161/yd^3)	Prepared-Bed Land Treatment	PCP, PAHs, and dioxins/furans
$398/m^3/($318/yd^3)	Slurry	PCP
$238 to $250/m^3/($190 to 200/yd^3)	Slurry	PAHs

Costs for this last slurry-phase bioreactor system are broken down as follows for remediation of 15,510 metric tons/8,400 m^3 (14,100 tons/10,500 yd^3) of creosote-contaminated soil: the treatability study costs $200,000; design engineering costs

$100,000; soil preparation (screening, add water) costs $800,000; treatment operation and maintenance cost $700,000; dewatering costs $400,000; site preparation and closure cost $500,000; and administration costs $200,000.

The RACER/ENVEST^{TM} computer software marketed by Talisman Partners Ltd. (Englewood, Colorado) estimates costs for landfarming and for bioventing, and the use of this model for cost estimating is discussed below along with actual unit treatment costs reported in the literature.

7.5.2.1 Land Treatment Systems

Land treatment system costs can be quite low if the site soil provides an effective treatment medium for waste treatment and if adequate land is available for development of a land treatment unit on site. The main costs of land treatment systems are associated with the contaminated soil excavation and land treatment unit development/construction. Requirements for a liner and/or surface cover for volatiles control can increase the cost of a land treatment system by approximately 25% to 50%.

System construction costs associated with physical makeup of the land treatment system include preparation of liners and leachate collection systems for water management, monitoring well construction, and construction of an optional surface enclosure for volatile control if required. These costs can be expected to range from $5,000 to $15,000 for a typical site with 800 m^3 (1,000 yd^3) of contaminated soil.

System operation and maintenance costs associated with tiller operation and maintenance are modest, ranging from $20 to $50 per month, while system monitoring and performance validation can be expected to represent 10% to 30% of the entire remediation cost. If leachate control and off-gas treatment are required, system operation and maintenance costs will increase by approximately 100%.

Overall system costs would be expected to range from $25/m^3 to $62.50/m^3 ($20/yd^3 to $50/yd^3) of contaminated soil without leachate and gas control and $62.50/m^3 to $125/m^3 ($50/yd^3 to $100/yd^3) with these liquid and gas controls in place. Graves and Leavitt (1991) report typical land treatment costs at $37.50/m^3 to $87.50/m^3 ($30/yd^3 to $70/yd^3).

For landfarming, the ENVEST model assumes that the soil layer is 0.45-m (18-in.) thick. The model does not include the cost of a liner, water source, or excavation. Input to the computer must include values or choices for the following parameters: area (ft^2), depth (up to 0.45 m, or 18 in.), and duration (in weeks). Secondary parameters that the user may choose include tilling frequency, watering frequency, fertilizer frequency, and use of cultured microorganisms.

The default value for watering frequency is once per week. A watering truck is assumed. The user may specify a sprinkler system with subsurface piping (not recommended with tilling). The default value for fertilizer addition is 800 lb/acre, once per month. Use of a fertilizer with 20% each of nitrogen, phosphorus, and potassium is assumed. This dosage equates to approximately 0.31 kg/m^3 (0.5 lb/yd^3) of soil if the fertilizer is mixed with a soil layer 0.3-m (12-in.) thick.

7.5.2.2 Soil Piles

Biomound treatment system costs can vary significantly from those of in situ treatment options if excavation requirements at a UST site, caused by deep and widespread contamination, are significant.

System construction costs associated with physical makeup of the biomound system include preparation of liners and leachate collection systems for water management, mound and air injection/extraction system construction, monitoring well construction, mound surface treatment for erosion and/or run-off/run-on control, and impermeable mound cover construction if required. These costs can be expected to range from $6,000 to $10,000 for a typical site with 800 m^3 (1,000 yd^3) of contaminated soil.

System operation and maintenance costs are associated with blower and piping system operation and maintenance, leachate handling and disposal, monitoring, and off-gas treatment if required. As indicated for bioventing systems, when these systems are operated for biodegradation enhancement, often no off-gas treatment is required. If off-gas treatment is required, system operation and maintenance costs will increase by approximately 50%. Operating costs for small blower systems are modest, ranging from $20 to $50 per month, while system monitoring and performance validation can be expected to represent 10% to 30% of the entire remediation cost.

Overall system costs would be expected to range from $62.50/m^3 to $125/m^3 ($50/yd^3 to $100/yd^3) of contaminated soil for petroleum related site cleanups. Once again, these costs will be dependent on controlling site/waste conditions, but they are felt to represent a realistic range of costs for biomound systems over the spectrum of site conditions and monitoring requirements expected in the field.

7.5.2.3 Slurry-Phase Reactors

Cost estimates for slurry-phase reactors are not as well developed as other bioremediation systems. Dupont et al. (1998) provide a detailed cost estimate for a hypothetical slurry reactor treating 8,000 m^3 (10,000 yd^3) of contaminated soil. The costs were generated based on data collected from full-scale operating treatment systems and are based on the following assumptions:

- 4,000 m³ (5,000 yd³) of contaminated soil are removed in the soil preparation and soil washing process.
- Four slurry-phase reactors with an operating volume of 680-m³ (180,000-gal) are operated at a 25% solids concentration.
- The reactors are operated in a batch mode, with each batch requiring 30 to 35 days for biological treatment.
- The treated slurry is dewatered in a filter press, and the recovered water is recycled to the slurry preparation/soil washing process.
- The treatment system is operated 7 days a week for the duration of the project.

The system costs and capacity are reflective of actual full-scale systems. Engineering design, procurement, and project administration include direct labor costs, benefits, overhead, and per diem for office and field personnel and amounted to $450,000. Equipment costs include direct costs, depreciation, and operation and maintenance requirements. Materials, supplies, and utilities include direct costs plus a markup. These additional equipment, materials, and operation and maintenance charges totaled $1,550,000 in treatment costs. The total estimated cost then for a commercial slurry-phase biological process, operating under these conditions, is approximately $250/m³ ($200/yd³) of material.

Three process variables have a significant impact on the total project cost of a slurry-phase treatment system: reactor solids concentration, residence time in the reactor, and the percentage of material removed in the slurry preparation/soil washing system. In general, for a given reactor configuration, the greater the slurry solids concentration, the lower the unit cost for the contaminated material. The upper solids concentration that can be effectively handled in a conventional slurry reactor configuration is limited to approximately 30% to 35% solids, resulting in a minimum cost of approximately $55/metric ton ($50/ton) of solids. An increase in reactor residence time reduces the throughput of the system, requiring additional labor and equipment costs to treat the same amount of material per batch, resulting in an increase in the per-ton cost for slurry reactor treatment. Finally, the greater the quantity of material removed in the soil washing process, the less material required for treatment and dewatering in the actual slurry-phase reactor, potentially significantly reducing the overall cost of slurry-phase treatment.

Modifications of operating temperatures can also affect slurry reactor economics. Using a full-scale slurry reactor treating PAH-contaminated soil as an example, the total capital and operating costs for a steam boiler system were approximately $150/day/reactor. Using an average solids loading of approximately 170 tons/reactor (Liu et al., 1994), the cost of heating the reactors equates to approximately $0.95/metric ton/day ($0.86/ton/day). Based on the data collected from the field reactors, heating the reactors from approximately 25° to 35° C increased the kinetics for PAH removal by a factor of 1.6. For an initial concentration of 10,000 mg/kg PAHs, slurry-phase biological treatment at 25° C requires approximately 9 days to achieve the treatment criteria of 950 mg/kg, whereas biodegradation at 35° C requires only approximately 6 days to reach the same level of treatment. The cost to operate the unheated treatment system, adjusted for a 4-mo. shutdown period during the winter, is approximately $225.50/metric ton ($205/ton) for a 9-day batch time (Liu et al., 1994). Operation of the treatment system without a winter shutdown at a 6-day batch time yields a total cost of approximately $198/metric ton ($180/ton). Heating the reactors for a period of 6 days requires an additional $5.50/metric ton ($5/ton) in operating costs, raising the total costs to approximately $230.50/metric ton ($185/ton), which is still lower than the $225.50/metric ton ($205/ton) for the extended, unheated operation. This impact would be even greater as the ambient temperature and the slurry temperature decrease to 15° C, resulting in even longer batch operating times to achieve treatment criteria. Heating the reactor allows for continuous operation of the treatment system, reducing required operating times, increasing equipment utilization, and lowering overall life-cycle treatment costs.

7.5.2.4 Bioventing

System construction costs associated with physical make-up of a bioventing systems can vary dramatically based on the extent of contamination, depth to contamination, etc., and they would be expected to range from $10,000 to $50,000 per typical site. System operation and maintenance costs are associated with blower and piping system operation and maintenance, recovered liquid handling and disposal, and off-gas treatment. Operating costs for small blower systems are modest, ranging from $20/mo. to $50/mo. Off-gas treatment, when required, represents approximately 50% of total system cost and can be expensive for carbon replacement, incinerator operating costs, etc. These costs may range from $1,000/mo. to $2,000/mo. This fact highlights the primary advantage to bioventing systems over conventional SVE systems if a bioventing system can be operated to eliminate the off-gas treatment requirement.

Overall system costs have been reported to range from $12.50/m^3 to $62.50/m^3 ($10/yd^3 to $50/yd^3) (Chowdhury and Fouhy, 1994) and to $125/m^3 ($100/yd^3) (Dupont et al., 1991) of contaminated soil for petroleum-related site cleanups. These costs will be dependent on controlling site/waste conditions, but they are felt to represent a realistic range of costs for bioventing technologies over the spectrum of site conditions and monitoring requirements expected in the field.

The ENVEST model assumes the following operating characteristics when carrying out a cost estimate for bioventing systems:

- Contaminants of concern are petroleum hydrocarbons.
- pH is already in the 7 to 8.5 range, and pH is not adjusted.
- A vadose zone well open to the atmosphere provides passive air injection to the subsurface, and a vacuum blower is connected to a number of air extraction wells.

Nutrient solution is infiltrated into the soil starting just below the surface or through the air ventilation well and/or the air extraction well. (In actual practice, nutrient additions are needed only at a few percentage of sites).
- A surface sprinkler system is included that irrigates the area of concern and provides water that infiltrates into the contaminated zone. The program user can instead use other ENVEST models if an infiltration gallery or injection wells are to be used.

Input to the computer must include values or choices for the following parameters:

- Whether vertical wells or horizontal wells are used — The allowable range for depth to the top of the trench for a horizontal well is 9 m, or 30 ft.
- The aerial extent of contamination
- Soil type — The allowable choices and corresponding hydraulic conductivities are as follows:
    - Silty clay, clay — 10^6 to 10^3 cm/sec
    - Mixed sandy, silty, clayey soils — 0.0001 to 0.1 cm/sec
    - Fine to medium sands — 0.01 to 1.0 cm/sec
    - Sand and gravel — 0.1 to 10 cm/sec
- Average depth to the top of the screened interval for vertical wells, or to the top of the trench for horizontal wells
- Well screen length

Computer input values or choices may be stated by the program user for secondary parameters, including air extraction system, drilling of vertical wells, trenching of horizontal wells, and soil additives. If the user opts not to give these values or choices, the program has built-in default values and choices as follows:

- The air extraction system includes well spacing and corresponding number of wells for the aerial extent of contamination, air flow rate, horsepower of vacuum blowers, and number of blowers. The well spacing and air flow rate are given for each soil type, with the air flow rate related to the linear feet of well screen as follows:
    - Silty clay, clay — 4.5 m (15 ft) spacing, 0.06 m^3/min/m (0.6 ft^3/min/lf)
    - Mixed sandy/silty/clayey soils — 10.5 (35 ft) spacing, 0.15 m^3/min/m (1.5 ft^3/min/lf)
    - Fine and medium sands — 15 m (50 ft) spacing, 0.33 m^3/min/m (3.55 ft^3/min/lf)
    - Sand and gravel — 45 m (150 ft) spacing, 1.44 m^3/min/m (15.5 ft^3/min/lf)
- The total standard air flow rate is taken as the product of these rates times screen length per well times the number of wells. The corresponding required blower power and number of blowers are based on the total flow rate Q:

- Up to 2.8 m³/min (98 scfm) — one 0.75-kW (1-hp) blower rated at 2.8 m³/min (98 scfm)
- 2.8 to 3.6 m³/min (99 to 127 scfm) — one 1.0-kW (1.5-hp) blower rated at 3.6 m3/min (127 scfm)
- 3.6 to 4.5 m³/min (128 to 160 scfm) — one 1.5-kW (2-hp) blower rated at 4.5 m³/min (160 scfm)
- Q over 4.5 m³/min (160 scfm) — F 3.7 kW (5-hp) blowers, each rated at 7.9m³/min (280 scfm), in which F is the number of blowers, or Q/7.9 m³/min (280 scfm) rounded up to the next whole number

For drilling of vertical wells, the model defaults to 5-cm (2-in.) diameter casing; the user can specify 10-cm (4-in.). The model defaults to the hollow-stem auger drilling method for wells less than 45 m (150 ft) deep and to water/mud rotary for deeper wells. The user can specify air rotary drilling. For each drilling method and casing diameter, the model has a specific bore hole diameter. If the user does not specify stainless steel, the model defaults to schedule 40 PVC casing for wells less than 25.5 m (85 ft) deep and schedule 80 PVC for deeper wells. All connecting piping is assumed to be above-ground schedule 40 PVC with the same diameter as the well casing, manifolded to a 10-cm (4-in.) diameter schedule 40 PVC header, with a vacuum gauge and appropriate piping appurtenances. The program user can specify whether drilling will include soil sample collection with a split-spoon sampler, used at 1.5-m (5-ft) intervals. Samples are screened in the field with an organic vapor analyzer (OVA). (This model does not estimate laboratory analysis costs; there is another ENVEST model available for soil analyses.) The user can specify whether drill cuttings are to be placed in drums. The model estimates the professional field labor hours spent supervising the drilling, related to whether samples are taken. The model estimates costs for decontaminating well internals and augers prior to and between bore hole/well installations with steam cleaning and detergent and for decontaminating samplers, bailers, and hand augers.

For trenching horizontal wells, the model defaults to these types of equipment:

- Up to 1.2 m (4 ft) deep — train trencher
- 1.2 to 6 m (4 to 20 ft) — hydraulic excavator
- 6.3 to 9 m (21 to 30 ft) — Horizontal Dewatering Systems, Inc. proprietary method

The model defaults to 5-cm (2-in.) diameter schedule 40 PVC perforated or screened casing for depths down to 3 m (10 ft) and to 10-cm (4-in.) for more than 3 m (10 ft). Connecting piping is similar to that for vertical wells connecting piping and manifolding. The user has the option of specifying drumming of trench cuttings, but there is no option for soil sampling.

For soil additives, the parameters are watering, nutrients, and cultured microorganisms. If the program user opts for watering, a surface spray irrigation system is included. If nutrients are selected, nitrogen/phosphorus/potassium pulverized fertilizer is assumed to be applied at 1,000 kg/ha (800 lb/ac). If microorganisms are selected, a monthly application of 12.5 kg (25 lb) of bacteria per 800 m^3 (1,000 yd^3) of contaminated soil is included. (Inoculating cultured microorganisms is rarely practiced and is not recommended with bioventing.)

Other ENVEST models are available for site work (site preparation, overhead electric power service, clearing, grubbing, fencing, and signs), water distribution, and groundwater monitoring wells. The user can choose to have the model estimate operating and maintenance costs. The model defaults to a maintenance/sampling crew to be on-site once per week the first month and once per month thereafter.

7.5.2.5 *Bioslurping*

Only limited data are available describing the cost of bioslurping systems for the recovery of free product and contaminated groundwater and bioventing of vadose zone contamination under full-scale field conditions. The only study reporting the cost for field-scale bioslurping systems is that of Connolly et al. (1995), which described the recovery of gasoline and diesel fuel from a site with fractured rock with a water table depth of approximately 4.8 m (16 ft). The system utilized 11 5-cm (2-in.) bioslurper wells, with both off-gas and groundwater treatment provided by a biofilter. With this system, 3.9 m^3 (1,030 gal) of product have been recovered for a total design and installation cost of $80,000 and annual operating costs of $40,000. This product volume does not include vapor phase product recovery or product destruction provided by in situ biodegradation in the vadose zone, so this is a conservative estimate of bioslurper performance at this site. With a reported liquid volume recovery of 15% to 25% of the estimated total release volume, the unit costs for this bioslurper system were $20.52/L ($77.67/gal) design and capital costs and $10.26/L ($38.83/gal) in annual operating and maintenance costs. Because of the general lack of cost data for this technology, more data are required to improve these cost figures beyond these preliminary values presented here.

A US Air Force field study initiated in 1994 (Leeson et al., 1995), which involved more than 35 bioslurping sites across the United States, Hawaii, and the Johnston Atoll, was designed to evaluate the applicability, cost, and performance of bioslurping systems operated in a wide range of site/soil settings. Results of this study should provide a valuable database of bioslurper costs for the projection of installation and operating and maintenance costs for new systems. The AFCEE at Brooks Air Force Base, Texas, should be contacted for the status of findings for this project.

7.6. Summary of Important Points for Bioremediations

Biological processes are applicable to waste existing as a liquid, a slurry, or as contaminated soil, as long as the contaminants are biodegradable and are not toxic to the microorganisms being used to carry out the biodegradation process. Table 7-1 summarized the range of compounds of concern that are known to be degradable and/or transformable using biological systems.

Once the biodegradability of the waste is determined, then the treatment system variables listed in Table 7-18 must be controlled to ensure biodegradation can be carried out without inhibition. Aerobic processes, those which operate in the presence of oxygen, are the most common biological treatment system used because they can degrade a wide variety of organic compounds, they are the most operationally stable, and they result in the conversion of complex organic compounds to simple oxidized end products. Anaerobic systems are used for removing nitrates and in the dechlorination of alkyl halides and some nitrogen-substituted organics that will not degrade aerobically, but because of their instability and low reaction rates, they are not yet widely used in practice except for denitrification.

Table 7-18 Biological treatment system variables that must be controlled to ensure uninhibited biodegradation rates.

Variable	Effect of Variable	Desirable Range
Toxicity	Results in death of microorganisms, resulting in no viability for the biodegradation process	Non-toxic
Oxygen	For aerobic processes oxygen is the primary reactant necessary for the destruction of organic compounds. When oxygen falls below these levels the system turns anaerobic and efficiency drops.	> 2 mg/L in Water > 2 vol% in Soil Gas
	For anaerobic processes necessary for some hazardous compounds (see Table 7-1). Many anaerobic microorganisms cannot survive in the presence of oxygen and efficiency drops.	< 1 mg/L in Water < 2 vol% in Soil Gas
pH	Optimal treatment usually occurs at neutral pH. Extreme pH values (<4 or >10) tend to be toxic to microorganisms and efficiency drops.	6 to 8
Nutrients	Primary nutrients required for microorganism growth are nitrogen and phosphorous. They are required on a C:N:P mass basis of 100:10:1 If N or P are inadequate relative to the carbon in the waste stream, efficiency drops.	C:N:P $\approx$ 100:10:1
Temperature	Biodegradation rate varies by a factor of 2 for every 10°C temperature change, i.e., goes up by 2 with 10°C temperature increase and down by 2 with 10°C temperature drop.	15 to 45°C

Table 7-19 summarizes the general criteria that affect in situ versus ex situ process selection. The primary criteria are the extent of the release, i.e., whether cost-effective overexcavation and treatment above-ground is possible and whether the in situ delivery and recovery of material are possible at the site. If the release is widespread, if it involves both soil and groundwater contamination, if surface disturbances cannot be allowed, and if the soil beneath the site is uniform and moderately permeable in its structure and texture, in situ treatment systems are generally applicable. If the release is small in extent, if the subsurface is very non-

Table 7-19 Applicability of in situ versus ex situ technologies for contaminated site remediation.

Technology	Site Characteristic	Applicability of Technology
Ex Situ	Extent of Contamination	Small aerial and vertical extent (i.e., shallow) favoring excavation
	Subsurface Soil	Highly variable making migration pathways difficult to determine
	Surface Structures	No structures so that surface excavation is possible
	Proximity of Receptors	No close receptors so releases during excavation pose little risk
	Remedial Objectives	Rapid cleanup required, forcing excavation and ex situ treatment
	Cost Constraints	No cost constraints so costly excavation is possible
In Situ	Extent of Contamination	Large aerial and/or vertical extent which makes excavation costly
	Subsurface Soil	Uniform, moderately permeable, making migration pathways definable
	Surface Structures	Structures which cannot be disturbed, making excavation impossible
	Proximity of Receptors	Close receptors so releases during excavation pose high risk
	Remedial Objectives	Rapid cleanup not required
	Cost Constraints	Cost constraints so costly excavation is traded for longer-term remediation and monitoring costs

uniform, and if excavation from the surface is possible, then ex situ processes should be strongly considered because of their predicted higher efficiency and reliability as compared to in situ process options.

The following is a summary of the main points of the chapter:

- Aerobic biodegradation requires appropriate amounts of oxygen, fixed nitrogen, phosphorus, and micronutrients.
- In situ aerobic biodegradation may proceed naturally or with enhancement from oxygen and sometimes nutrient addition.
- The Raymond method includes groundwater extraction, oxygen and nutrient addition ex situ, and reinjection back into the aquifer where bioreactions take place.
- For anaerobic biodegradation to proceed, high concentrations of dissolved organic carbon are necessary. Sometimes, additions of methanol, sugar, an organic acid, or an organic salt is practical.
- Aerobic bioreactions, if complete, produce water, carbon dioxide, and increased bacterial cell mass.
- Bacterial augmentation can be achieved with indigenous bacteria, with cultured strains, or with engineered DNA-manipulated bacterial strains.
- Reactors designed for optimum bacterial environment are often limited by ineffective oxygen transfer rates.
- Reactors with extra-long residence time are often limited by the amount of effective mixing that they can provide.
- The bacterial environment is affected by levels of nutrients (nitrogen and phosphorus), micronutrients, pH, temperature, and toxins.
- For most situations, nitrogen and phosphorus levels in relation to biodegradable carbon should be such that the carbon/nitrogen/phosphorus ratios are 100/10/1 on a weight basis.

- The two main designs used for bioreactors are the completely mixed type and the fixed-film type with plug flow.
- Bioremediation combined with PACT has advantages over either bioremediation or carbon adsorption alone.
- For bioremediation of nitrates, either fixed-film or fluidized bed reactors can be used anaerobically, with additions of dissolved organic carbon.
- Aerobic ponds are generally less than 1.5 m (5 ft) deep. Large surface areas provide for natural mass transfer of oxygen from air into the water.
- Ponds need long residence times and mixing, which is achievable with either multiple water inlets and outlets or with baffles.
- Wetlands marsh plants provide uptake of some contaminants and sites in their rhisosphere for bacteriological growth.
- With some wetlands, especially when anaerobic denitrification is desired, the plants are not harvested but allowed to decay, thereby providing needed phosphate and additional carbon loading to the system.
- In situ groundwater bioremediation has a potential for success in homogeneous soils of relatively high permeability.
- Peroxide addition and air sparging may aid in situ bioremediation.
- Depending on its concentration and soil characteristics, peroxide will sterilize bacteria near an injection well and then promote bacterial growth as it disperses.
- Injection wells are used for in situ bioremediation for the addition of oxygenated groundwater and nutrients, usually intermittently.
- Incomplete remediation may occur in situ because of nonuniform dispersion, phosphate uptake by soil particles via ion exchange, peroxide decomposition by natural organic matter, adverse chemical reactions, and plugging by bacterial mass buildup within the soil.
- Tools used to assess and/or track in situ bioremediation include Brubaker's screening criteria, Battelle's mathematical models, and, for aquifer plugging, injection pressure and time for tracer transport.
- Moisture and oxygen addition are usually employed in soil phase biological treatment systems.
- Ex situ soil remediation techniques include land treatment, soil piles, and slurry bioremediation.
- If a high concentration of moisture is maintained, organics in excavated soil will biodegrade when air contact is made by spreading in layers up to 20 cm (8 in.) thick, by periodic tilling, by turning or mixing compost, or by distributing air with a mechanical blower via embedded perforated piping.
- Macronutrients (i.e., fixed nitrogen and phosphate) can be added in forms supplied by plant nurseries.
- Spreading and tilling (landfarming) techniques can be accomplished in multiple lifts.
- Runoff and leachate from soil piles should be collected and recycled.
- In composting, an organic bulking agent is used to improve soil water holding capacity and aeration when soils contain high clay content.

- Excavated soil can be slurried with water and treated in mixed, aerated, bioslurry reactors.
- Slurry bioremediation has proven effective for a wide range of organics, with the notable exception of dioxins/furans.
- In situ soil remediation techniques include bioventing, biosparging in the smear zone, and bioslurping.
- Bioventing is an in situ technique by which air is induced to flow in the soil to provide oxygen for aerobic degradation of organics.
- Bioventing degrades VOCs, as well as semivolatiles that cannot be stripped by soil venting at ambient temperatures.
- Bioslurping is a vacuum-assisted free product recovery technique that enhances the in situ aerobic degradation of residual phase contaminants by moving oxygen to the vadose and smear zones as in bioventing.
- Treatability tests address a number of concerns, including micronutrient needs, toxicity, pH adjustment, and cometabolites.
- Field-scale treatability testing is an integral part of the design of bioventing, land treatment, and bioslurping systems.
- Computerized models are available for process design and cost estimating for aqueous phase ex situ (Composer Gold/ECHOS) and in situ (RACER/ENVEST) bioremediation systems.
- Computerized models are available for the soil phase treatment processes of landfarming and bioventing that aid in process design and cost estimating.

Chapter 8

Soil Venting

Soil venting (soil vapor extraction, SVE) applies to removal of volatile compounds, such as gasoline and solvents, that can be removed from contaminated soil with gases, usually air. Removal efficiencies often exceed 90% of gasoline hydrocarbons and 99% of halogenated solvents, generally within a period of months. It can be accomplished ex situ by imbedding excavated piles of soil with perforated piping. The piping is manifolded to a blower. If air pollution control regulations apply, a vacuum blower is used, discharging usually through vapor phase activated carbon or an oxidizer. The pile is covered with plastic sheeting, with ventilation holes cut out to permit air to enter the pile at a number of locations. Ex situ soil venting can also be done with steam instead of air (Meekins, 1997). Hrubetz Environmental, Inc. (Dallas, Texas) claims a patent for using heated air. M.D. Ikenberry has patents on an ex situ heated air system marketed as "The HAVE System^{TM}" by Global Remedial Technologies (Duvall, Washington).

This chapter, however, deals primarily with contaminated soil that has not been excavated, using air at ambient temperature as a stripping medium for removal of VOC from the soil. Geological factors are thereby involved, as well as the volatility of the contaminants. This chapter focuses on the following topics, which include geological factors, as well as process and mechanical design considerations:

- Installing trenches and wells used for extraction of soil vapor
- Using above-ground vacuum-producing equipment
- Vapor treatment systems
- Treating vapors
- Utilities consumption
- Determining extraction well radius of influence(ROI), soil air permeability, and contaminant mass removal rate
- Using ventilation wells to direct underground air flow patterns
- Laboratory testing and field pilot testing
- Estimating costs

Heat-enhanced SVE and in situ soil heating are mentioned but not dealt with in detail. These remediation techniques involve soil venting and can speed the remediation process or extract semivolatile contaminants, as well as VOC, whereas air at ambient temperature can extract only VOC from unheated soil.

8.1 Basic Principles of Soil Venting

Excavation is often used for highly contaminated soils within several meters of the surface. The US EPA (1991d) indicates that excavation costs begin to exceed in situ soil venting costs at a soil volume of 382 m^3 (500 yd^3). In most in situ soil venting applications, horizontal air flow is established through strata in the vadose zone, vaporizing VOCs and stripping them from the soil. Either a natural impermeable stratum above the contaminated soil or an artificial surface seal is needed to impede vertical air flow. In most installations, air is admitted outside of the contaminated zone via natural leakage from the atmosphere, with vacuum induced at trenches or wells. Initially, the air in the main soil channels that is saturated with vapor is swept out, resulting in temporary high vapor concentrations in the extracted air. If volatile liquid organics are present in the main channels, high concentrations may be sustained until vapor flow becomes completely controlled by relatively slow diffusion from soil micropores. During this diffusion-controlled stage of operations, the VOC mass extraction rate is independent of the air flow rate (Bohn, 1997).

Soil moisture content affects soil venting processes. At a given soil permeability and air flow, the amount of vacuum developed increases with soil moisture content.

For in situ soil venting, the extraction of contaminants is via trenches or wells. Generally, the trenches or wells should not extend below the water table. Where the water table is shallow (within 10 ft of ground surface) trenches are more economical than wells. Short trenches may simply be backfilled with gravel, sealed near the top with bentonite and/or grout layers, and fitted with a vertical pipe connected to piping feeding the inlet of a vacuum blower. Most installations include a horizontal perforated or screened pipe near the bottom of the trench in gravel or in a coarse sand filter pack. The pipe is fitted with an elbow to the vertical pipe. If extraction from only one side of the a trench is desired, the other side can be sealed or lined with heavy plastic sheeting prior to backfilling with gravel.

Most commonly, extraction is done with conventional wells, including vertical pipe that is screened for some length within a contaminated stratum and surrounded with a sand pack, as shown in Figure 8-1.

Licensing of SVE technology patented by J. Malot using such conventional wells is done by Terra Vac Corporation (Windsor, New Jersey). (The Hazardous Waste Consultant (1995) reported that a federal District Court ruled against Terra Vac in a court order filed May 16, 1994, in a patent infringement suit brought by Terra Vac, and that Terra Vac appealed the ruling. Subsequently, a federal Appeals Court ruled that J. Malot has certain patent rights that are valid).

Roy F. Weston, Inc. (West Chester, Pennsylvania) has patented a simplified well design. The well boring is filled with gravel for the vertical distance desired, sealed with bentonite and grout, fitted with a vertical pipe through the seal into the top portion of the gravel, and connected to piping feeding the inlet of a device producing negative pressure.

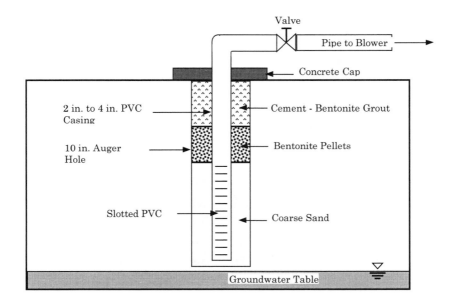

Figure 8-1 Typical extraction well schematic for SVE (US EPA, 1991d).

If a vapor extraction well is screened in the saturated zone, groundwater may need to be extracted simultaneously with vapor. This concept of simultaneous groundwater extraction is licensed by Xerox Corporation (Stamford, Connecticut). Systems using simultaneous air and water extraction combined with bioremediation air sparging are described by Lamarre et al. (1997).

Another combination treatment scheme for contaminated groundwater is to sparge the groundwater with air injected below the water table and to apply vapor extraction in the vadose zone. VOCs are thereby stripped from the groundwater and biodegraded in situ, and the air and contaminants (both from the saturated and vadose zones) are captured by the soil venting system. In-well air sparging can also be conducted in conjunction with soil venting, as described in Section 5.3.4. Reports of success with such applications may affect the way in which future soil venting systems will be designed when contaminated groundwater is present.

Other advances include extraction of heavier hydrocarbons and semivolatile organic compounds through the use of steam heating of the soil or through simultaneous steam and electric heating. Without heating, the range of extractable compounds is quite limited. Bennedsen (1987) suggests that soil venting at ambient soil temperatures applies to compounds having vapor pressures greater than 0.5 mm Hg at $20^\circ C$, with the dimensionless Henry's Law constant being greater than 0.01. Steam, electric heating, or injection of heated air can be used to hasten the extraction of volatile contaminants, as well as to extend the volatility range that is applicable.

Koltuniak (1986) describes an in situ system using heated air. Noonan et al. (1993) describe steam injection applications to SVE. Kent Udell at the University of California at Berkeley and the Lawrence Livermore National Laboratory have developed systems that combine steam heating and electric heating (Bader, 1997).

Four methods of electrically heating soil in situ include the following:

- *Six-phase heating*, developed by the Battelle Pacific Northwest National Laboratory (Richland, Washington), is a process in which six steel-pipe electrodes are driven into the soil in a circular array and a voltage of 200 to 2,000 volts AC at 60 Hz is applied, with a seventh neutral electrode placed in the center of the array.
- *Conventional three-phase heating with imbedded electrodes.*
- *Thermal blankets* made of 8-ft by 20-ft steel boxes from which are suspended stainless steel webs with electrically heated rods threaded through the webbing (Bader, 1997). This system is marketed by TerraTherm, an affiliate of Shell Oil Company (Houston, Texas), and is used for contamination within 3 ft of the ground surface. (Thermal wells are used for deeper contamination.) A silicone rubber vapor barrier over the thermal blanket acts as a hood and collects steam and soil gas that evolve. Air ducting connected to a thermal oxidizer, an air cooler, an activated carbon adsorber, and a vacuum blower in series are usually used.
- *Radiofrequency heating*, as demonstrated by KAI Technologies (Portsmouth, New Hampshire).

The presence of a stratum of impermeable soil above the contaminated zone can greatly extend the zone of influence (ZOI) of each extraction well. Otherwise, the ground surface could be paved or plastic-sheeted to beyond the periphery of the contamination to increase the ZOI. This helps induce horizontal air flow through the contaminated zone toward each extraction trench or well screen.

An essential requirement is for wells to be screened within only one major stratum. If the contamination extends vertically into multiple strata with significantly different permeabilities, then screened intervals for each well should be only in strata in which the soil has the same permeability. Other wells should be screened at each major stratum that has a different permeability. Otherwise, the horizontal air flow path will bypass the less-permeable strata. Zones that are bypassed will not be remediated unless the less-permeable layers are relatively thin. (With thin layers, forces that enhance diffusion will aid in the volatilization and movement of molecules into the nearby horizontal air flow path.)

Within each well, vertical piping conducts the soil gas (air and organic vapors) up to the ground surface. These piping runs can be manifolded into one header that feeds into the suction of a vacuum-producing device. Valves that allow admitting dilution air should be included if there is a chance of explosive levels of flammable vapors being present in the soil gas. Throttling valves and flow indicators should be included so that air flow through more permeable strata is balanced with air flow through less permeable strata.

One well with a relatively large diameter can serve multiple strata, as with a well cluster. The well is sealed above each screened interval, and a separate vertical pipe conducts the soil gas (mainly air) from each interval. Throttling valves above the ground surface can be adjusted to balance air flows.

8.2 Inducing Vacuum

8.2.1 Vacuum Blowers

The vacuum blowers most commonly used are centrifugal types, constructed of aluminum. Nonsparking materials of construction, such as aluminum, stainless steel, and plastic, are preferred, because the vapors may be explosive. For example, the lower explosive limit (LEL) for gasoline vapors is approximately 13,000 to 14,000 ppmv; frequently, soil gas may contain more than 20,000 ppmv of hydrocarbons. LEL values are given in NFPA (1994). (*See* Section 9.1.2.) Shelton (1995) tabulates heats of combustion, autoignition temperatures, and LEL values. The percent LEL in a vapor-air mixture can be estimated from Shelton for up to n number of compounds using Equation 8-1.

$$\% \text{ of LEL} = [(HV_1)(C_1)(d_1) + (HV_2)(C_2)(d_2) + \ldots + (HV_n)(C_n)(d_n)]/48 + 0.04(T-60) \tag{8-1}$$

in which C_i is the concentration of an organic compound in the mix, volume %; d_i is the vapor density at 60° F and 1 atm pressure and is equal to molecular weight divided by 379, lb/scf; T is the temperature of the air if it has been preheated, °F; and HV_i is the heat of combustion, Btu/lb.

Among 46 organic compounds tabulated by Shelton, except for acetylene and the inorganic gases carbon monoxide and hydrogen, this equation predicts LEL values less than 10% above test-derived LEL data. If the heat of combustion of a compound is not known, but its molecular formula is, Shelton gives Equation 8-2.

$$HV = 15{,}410 + 323.5 \text{ (wt\% H)} - 115 \text{ (wt\% S)} - 200.1 \text{ (wt\% O)} - 162 \text{ (wt\% Cl)} - 120.5 \text{ (wt\% N)} \tag{8-2}$$

Clark and Sylvester (1996) estimate the LEL of typical organic fuels as 0.55 times the stoichiometric concentration for complete combustion, and the LEL of a mixture expressed in volume %, with Equation 8-3.

$$\text{LEL\%} = 100\%/(a_1/LEL_1 + a_2/LEL_2 + \ldots + a_n/LEL_n) \tag{8-3}$$

Here, a_i is the volume percentage of fuel i in the total fuel, and n is the total number of fuels.

The amount of vacuum produced with centrifugal blowers applied to soil venting ranges up to 8 in. Hg. (Full vacuum at sea level is 29.92 in. Hg). The most widely used centrifugal blowers are regenerative types. These types can produce a medium-high vacuum with a single blower wheel; however, they are not as energy efficient as other centrifugal or positive-displacement designs.

Sometimes positive-displacement blowers are used for tight soils. Screw types (or rotary-lobe types at lower flow rates) are used to produce a vacuum up to 15 in. Hg. Each such blower has a pair of screws or lobes that mesh together with a small clearance and that rotate in opposite directions. The timing of rotation between the two elements must be precise in order to avoid rapid wear and to avoid sparking when steel construction is used. Maintenance costs and power consumption are generally much higher for positive-displacement blowers than for centrifugal blowers.

Liquid-ring compressors can be used to provide a very high vacuum. Applications to soil venting are described by Hansen, Gates, and Sittler (1998). Figure 8-2 shows a flow schematic for a liquid-ring application.

Any blower or compressor generates heat of compression. In soil venting applications, this results in raising the soil gas temperature. In a liquid-ring compressor, water used to seal the space around the rotor also gets heated. Soil gas discharged from the compressor has entrained water that is separated in vessel V-1. From the bottom of V-1, the water is pumped through a cooler back into the liquid-ring compressor.

Soil gas from V-1 is passed through vessel V-2, which has a mist eliminator near the top outlet. The mist eliminator, usually a mesh pad or baffles, coalesces fine water droplets that then fall to the V-2 bottom outlet. Soil gas from the top of V-2 is either discharged to the atmosphere or passed through an air pollution abatement system.

Aglitz et al. (1995) give details on installing, maintaining, and troubleshooting liquid-ring systems. They refer to "Performance Standards for Liquid Ring Vacuum Pumps" (Heat Exchange Institute, Cleveland, Ohio) for complete information on operating principles and design specifications.

Unless a liquid-ring compressor is used, it is important to include a water knockout pot or mist separator to first remove condensate from extracted air. If the nonvertical extraction piping is not sloped back toward the wells, it should be sloped so that condensate runs into liquid traps or into the knockout pot. Liquid water can damage nonliquid-ring blowers and interfere with vapor treatment. Condensate that is collected should either be automatically pumped out of the system or carefully monitored and manually drained. If groundwater is also being treated ex situ at the same site, the condensate might be combined with the groundwater for treatment. Otherwise, it must be disposed of at a permitted disposal facility and handled properly.

In some climates, vapor extraction piping runs that are not buried below the frost line may need heat tracing and insulation. This prevents ice formation in the lines and retards condensation. If water knockout pots have heaters for freeze protection, the

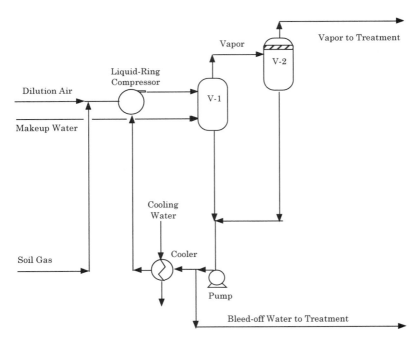

Figure 8-2 Flow schematic for liquid-ring compressor application (Courtesy of the Nash Engineering Company).

heat input should be controlled to prevent water vaporization if there is a chance that condensation will occur downstream of the pots.

Extracted air is sometimes saturated with water vapor as it exits from below ground, and the temperature is usually at 55° to 60° F. If cold ambient conditions cause the flowing air temperature to drop to the mid-30° F level, 65 gal/wk of condensate will form from each 100 scfm of air. This example condensing rate is derived from the values given for saturated water content on a psychrometric chart for air at 60° F and 35° F, as follows:

Assumptions: Standard temperature is 60° F, so that the density of air is 0.076 lb/scf at standard temperature and pressure. The soil venting system operates 24 hr/day every day.

Given: Volumetric air flow rate is 100 scfm; air temperature drops from 60° F to 35° F; absolute humidity of saturated air at 60° F is 0.0114 lb of water/lb of air; and absolute humidity of saturated air at 35° F is 0.0043 lb of water/lb of air.

Find: Amount of condensate that forms per week.

Step 1: Determine air mass flow rate, as shown in Equation 8-4.

100 scfm air (0.076 lb/scf) (10,080 min/wk) = 76,608 lb air/wk (8-4)

Step 2: Determine the mass of water condensed/lb of air from the change in humidity corresponding to the temperature dropping from 60° F to 35° F, as shown in Equation 8-5.

Humidity difference = (0.0114 - 0.0043) lb of water/lb of air
= 0.0071 lb of water/lb of air (8-5)

Step 3: Calculate the amount of water condensed, as shown in Equations 8-6 and 8-7.

0.0071 lb of water/lb of air (76,608 lb of air/wk) = 544 lb of water/wk (8-6)
(544 lb of water/wk)/(8.33 lb of water/gal) = 65 gal/wk (8-7)

Even more liquid water can be drawn into the above-ground extraction equipment if a strong vacuum is applied to a well screened into or near a shallow water table.

8.2.2 Internal Combustion Engines (ICEs)

Automobile engines can generate a very high vacuum; fuel and air are aspirated into the intake manifold when an intake valve is open to a piston that is moving away from the fuel intake port. The vacuum that is generated is used to extract soil gas in certain soil venting systems. Such engines can also be used to combust flammable vapors present in soil gas. Thus, automobile engines or similar reciprocating engines can serve to induce vacuum and also to combust the extracted vapors prior to discharge.

Two types of configurations are used with reciprocating ICEs. The engine itself can induce vacuum, as well as serve as the main air pollution control device, in assemblies marketed by VR Systems (Anaheim, California). Or, the engine shaft can be used to drive a blower that discharges into the engine intake. With either configuration, combustible contaminants in the extracted soil gas serve as fuel for the engine in which the bulk of the combustion takes place. Final combustion of unburned or partially burned organic compounds and carbon monoxide in the exhaust is accomplished in a small catalytic converter (oxidizer), as is done with automobiles.

Soil venting wells with high initial hydrocarbon concentrations will have decreasing concentrations as time progresses and venting continues. When concentrations fall below the level at which combustion is self-sustaining, auxiliary fuel (such as propane) must be consumed. Costs for auxiliary fuel can rise rapidly as concentrations of extracted hydrocarbons continue to decline. Also, engine maintenance costs may rise dramatically if extraction operations run 24 hr every day over a long period of time.

The optimum soil venting system with air pollution control would use an ICE (probably rented) or a thermal oxidizer initially while hydrocarbon concentrations are high and very little auxiliary fuel must be purchased. Then, a catalytic oxidizer would be used when hydrocarbon concentrations are in an intermediate range, with activated carbon adsorption used when concentrations are low. If control agencies will allow it, the initial air pollution control permit application should include provisions for changing over to catalytic oxidation and carbon adsorption when it is economical to do so.

When organic concentrations become low, the most cost-effective use of vacuum-producing equipment and of oxidizers or activated carbon is not possible. A strategy often used is to stop extracting from some or all of the wells for a few weeks, allowing concentrations to rebound. Another strategy is to convert the system to a bioventing mode, which uses low air flow rates that minimize the stripping of volatiles but provides oxygen for biodegradation of organic contaminants.

8.2.3 Passive Soil Venting

Another useful strategy when concentrations fall toward the end of remediation is to use passive soil venting. Rather than inducing vacuum to extract air with contaminant vapors, advantage is taken of natural fluctuations in barometric pressure. During times when atmospheric barometric pressures are falling, soil gas is expelled naturally from vadose zone wells. Air pressure changes in soil pores lag changes in atmospheric barometric pressure. Because of this time lag, expulsion of soil vapor continues for a period of time, even after the barometric pressure starts rising.

The well-head piping can be fitted with a check valve and ducted to a vapor phase carbon adsorption canister or drum if air pollution control is required. Costs for installing, operating, and maintaining above-ground mechanical equipment are eliminated when passive systems are used. This is a slow process applicable after concentrations have decreased to the level at which mechanically enhanced removal is not cost-effective. The concept applies to bioventing (as described in Chapter 7), as well as to stripping of volatile compounds from soil via soil venting.

8.3 Vapor Treatment and Discharge

The treated air stream can be discharged through wells back into the soil or directly to the atmosphere. Soil venting systems can be designed that include adsorption of contaminants with a resin, automatic on-site regeneration of the resin under vacuum with recycled electrically heated nitrogen, condensation of contaminants and water, and return of the treated air stream to the soil. Wells used for injection of treated air should be strategically placed so that the underground air flow path for treated air and uncontaminated air sweeps through the contaminant plume toward extraction wells.

Conventional systems include treating the extracted air stream with an emission abatement system that discharges to the atmosphere after treatment. Some systems have been installed with the soil vapor treated by combustion in an existing boiler used to generate process steam or by adsorption in existing on-site vapor phase activated carbon system used for process solvent recovery. When an ICE is not used, new adsorption systems and new or leased oxidizers are most frequently used for abatement. For high contaminant concentrations, a condenser may be used to collect liquified product ahead of a final abatement device. The applicability of alternative vapor treatment methods is summarized in Table 8-1. In this table, the concentrations apply at optimum efficiency. The techniques will control emissions at lower concentrations, although usually at reduced efficiencies.

Table 8-1 Applicability of soil vapor treatment methods (From the US EPA, 1989, "Soil Vapor Extraction, VOC Control Technology Assessment," EPA/450/4-89/017).

Vapor Treatment System	Emission Stream Characteristics				VOC Characteristics
	Concentration	Flowrate Moisture	Temperature	Variation	
Carbon Adsorption	>0.1 ppmv	<50%	Insensitive	<150°F	Molecular weight of VOC should range from 50 to 150 g/mole for best performance
Thermal Destruction	>100 ppmv	——	Sensitive	——	Can control most VOCs without operational difficulty
Catalytic Oxidation	>100 ppmv	——	Sensitive	——	Phosphorus, bismuth, lead, mercury, arsenic, iron oxide, tin, zinc and halogenated compounds may foul catalyst
Condensation	>5,000 ppmv	——	Sensitive	<200°F	Removal efficiency limited by vapor pressure-temperature characteristics of VOCs present

Innovative vapor treatment methods being applied and emerging technologies being demonstrated include:

- Biofilters---Soil gas is distributed into soil and/or compost in which microbiological action destroys the biodegradable organic compounds.
- The VaporMate™ system — This system, marketed by North East Environmental Products (West Lebanon, New Hampshire), oxidizes the organics with ozone in the presence of a catalyst at 200° F.
- Plasma destruction — Plasma destruction, using high-energy electron beams, is marketed as Zapit by Advanced Solutions for Environmental Treatment (Palo Alto, California), and it has been independently demonstrated by the Massachusetts Institute of Technology Plasma Fusion Center (Cambridge, Massachusetts).

This discussion focuses on the most frequently used, proven technologies:

- Vapor phase adsorption, in which activated carbon or a resin is used to remove organic vapors from the gas stream
- Oxidation, both direct thermal and catalytic, for combustion of the organic vapors

8.3.1 Adsorption

Activated carbon produced specifically for vapor phase adsorption and certain resins are used for treating soil vapor. The particle size distribution of the granular adsorbent is selected by the carbon vendor such that resultant pressure drops during operation are minimal. Although this is important for controlling emissions from most processes (e.g., off-gas from air stripping of groundwater), pressure drop is not a critical concern in most SVE systems using activated carbon, as the vacuum blower used may develop several inches of mercury air compression. Pressure drop through an abatement system is usually only several inches of water. (One in. of mercury is equivalent to more than 13 in. of water.) The cross-sectional area of the bed and bed depth are chosen by the vendor so that the superficial air velocity is in a range that corresponds to a reasonably small pressure drop.

Two stages of adsorption beds in series are usually installed, with the first-stage adsorption bed being removed at a predetermined amount of contaminant breakthrough being detected in the air exiting from the first unit. Two stages in series may be needed to achieve more than 90% overall reduction in contaminant concentrations. For the most effective use of adsorbent media, the second-stage adsorbent is placed in the lead position when the first-stage adsorbent is removed. Fresh adsorbent is used in the last unit. The capacity of vapor phase activated carbon to adsorb various VOCs at $25°$ C and 1 atm is given by Yaws, Bu, and Nijhawan (1995).

If a centrifugal or positive-displacement blower is used with an adsorbent resin or carbon, the adsorbent beds should precede the blower. Adsorption is exothermic, causing an increase in flowing air temperature. Heat of compression at the blower causes an increase in flowing air temperature. High temperatures degrade the amount of adsorption, and, unless the blower discharge is cooled, the combined temperature increases may result in temperatures exceeding acceptable plastic pipe operating limits after the first adsorption bed. High temperatures can also result in no adsorption in the second bed. However, as discussed below, adjusting the flowing air relative humidity to below 50% will result in good adsorption, and relative humidity depends on temperature.

Adsorbents used for vapor-phase emission control can be readily regenerated on the site or sent out for reactivation. On-site regeneration is usually accomplished with steam. Regeneration of activated carbon or other adsorbents is also possible by pulling a vacuum on the used carbon adsorbent bed. Vacuum regeneration does not drive off all of the adsorbed contaminants; a heel remains in the micropores. The larger the heel, the less contaminant quantity can be adsorbed during the next adsorption cycle. In order to minimize the amount of heel, vacuum regeneration is sometimes used in combination with some form of heating (e.g., electric) of the adsorption bed, accompanied with an inert sweep gas.

An alternative to on-site regeneration is periodic replacement of carbon in permanent on-site canisters or use of the canisters as transport vessels for taking carbon to and from an off-site regeneration facility.

When GAC is used to adsorb gasoline hydrocarbons from the vapor phase, it can hold more than 20% (the amount depends on vapor concentration) of its weight as sorbed hydrocarbons. For practical applications, designs should be based on 15% of the carbon weight as being the adsorptive capacity of the first stage. For light aromatics (i.e., benzene, toluene, and xylenes) and chlorinated solvents, the working carbon loading is 25% to 30% (Stenzel and Perryman, 1997).

Note that some commonly encountered VOCs (e.g., vinyl chloride) essentially do not sorb onto activated carbon. Use of a resin or other air emission control options must be assessed when these compounds are present in elevated concentrations in the airstream.

Vapor phase adsorption works best at low temperatures and at relative humidities less than 50%. Blowers raise the soil vapor air temperature caused by the heat of compression. If the temperature exceeds 110° to 150° F, consideration should be given to installing an air cooler or to placing the adsorption unit upstream of the blower. In order to achieve the desired low relative humidity, installing a steam or electric heater ahead of the adsorption unit may be necessary. A combination of equipment that achieves a relative humidity of 45% at temperatures less than 90° F entering the adsorption bed would be the optimum design. Soil vapor often is saturated with moisture (100% relative humidity) when it leaves the soil. The ideal installation handling high contaminant concentrations would include a mist separator (water knockout vessel) placed first in line, a chiller or a refrigerated condenser recovering liquid product, followed by a heating device and then the adsorption beds. Note that the condensate that is recovered as liquid product may have to be managed as a hazardous waste.

8.3.2 Oxidizers

Either direct thermal oxidizers or catalytic oxidizers are used to combust emissions from soil venting vacuum blowers when an ICE is not used. Auxiliary fuel, such as natural gas or propane, is used when the concentration of combustible vapors in the soil gas is less than the concentration that will sustain combustion. Auxiliary fuel or electric heating is always used with catalytic oxidizers for preheating the soil gas. With direct thermal units, auxiliary fuel is often not needed near the beginning of soil venting ; that is when contaminant concentrations are highest.

Direct thermal oxidizers are usually employed used at temperatures greater than 1,200° F with a residence time of at least 0.3 sec. Some air quality agencies require 1,400° F and residence times in excess of 0.5 sec. Hydrocarbon destruction efficiencies can thereby exceed 99%. System designers should confirm regulatory requirements for destruction efficiency, temperature, and residence time. Air quality control agencies frequently require that the operating temperature be continuously recorded, as they can readily inspect the recorder charts as an aid to ensuring maintenance of the required destruction efficiency.

If chlorinated solvents are included in the extracted air stream, air pollution control authorities may require a wet scrubber on the oxidizer exhaust to abate emissions of hydrogen chloride. Also, if a catalytic unit is used, a special catalyst that is resistant to chloride degradation may be needed, depending on chloride concentrations and mass flow rates. Certain heavy metals can poison catalysts; e.g., tetramethyl lead (present in some gasolines) can rapidly degrade platinum catalyst if not first converted to lead oxide in a preheating burner flame. Windmueller and Sykes (1995) describe an economical method for detecting volatile lead compounds in soil gas.

With catalytic oxidizers, both the catalyst inlet and outlet temperatures are monitored. The temperature rise across the catalyst is somewhat proportional to the soil gas hydrocarbon concentration. Hydrocarbon destruction efficiencies of greater than 95% can be achieved depending on catalyst inlet temperature, amount of catalyst, and activity of the catalyst. Auxiliary fuel or electric heating must be used to raise the soil vapor (which is mainly air with contaminant vapors) to the 600° to 700° F temperature range at the catalyst inlet. As long as the catalyst is active, combustible vapors will be oxidized, and the temperature rise through the catalyst will range up to a few hundred degrees Fahrenheit, depending on inlet concentration.

For most applications, the catalyst is impregnated on the pore surfaces of an inert, porous block. Details of catalyst design, deactivation, and restoration are given by Chu and Windawi (1996). Fluidized bed systems are also in use, in which the upward velocity of the soil gas is high enough to keep small beads of catalyst in suspension. The catalyst most often used is platinum (as in automobile exhaust systems), with which inlet hydrocarbon concentrations should be less than approximately 3,000 ppmv. With concentrations higher than 3,500 ppmv, the temperature rise across the catalyst results in outlet temperatures high enough to damage the catalyst.

If the vapor concentration coming out of the soil is too high for the catalytic unit, two alternatives are commonly practiced:

- Oxidation is performed in the direct thermal mode by removing the catalyst block (in oxidizers designed for either direct or catalytic operation) or by switching to a separate direct thermal unit.
- Dilution air is introduced into the vacuum blower suction.

If dilution air is introduced, either a manual control damper can be set such that the oxidizer outlet temperature is below the allowable design limit or an automatic temperature control system can be installed that modulates the position of the dilution air control damper. Figure 2-6 shows a modulating dilution air control system with controller TIC-1 and control valve TCV-2. In addition, a separate automatic high-temperature shutdown system should be included in the design that will shut off the vacuum blower and auxiliary fuel or electric heating if excessive outlet temperatures are approached. In Figure 2-6, this separate shutdown is accomplished with temperature switch TSHH-1B, blower switch HOA-1, and auxiliary fuel control valve TCV-1. Normal auxiliary fuel input or electric heating is accomplished with automatic control of the soil gas temperature entering the catalyst, as shown in Figure 2-6 with TRC-1.

When vapor concentrations are low, the extra cost of catalyst often has an attractive payback when compared to the higher fuel costs with direct thermal applications. The extracted soil vapor is preheated to some extent by the vacuum blower. Then, natural gas or propane fuel is used to heat the soil vapor from blower discharge temperatures of 600° to 700° F for catalytic units versus 1,200° to 1,400°F for direct thermal units. A typical heating duty for direct thermal units will correspond to raising the temperature of soil gas (frequently containing more than 99% air) by 1,100° F, whereas for catalytic units, the temperature rise required is only 500° F. Hydrocarbons and other combustible vapors in the soil gas have fuel value that meets part of the heating duty needs. For example, at a concentration of 1,200 ppmv (0.12%) hydrocarbons in soil vapor at 400 scfm, the fuel value is approximately 35 therms/day (3,500,000 Btu/day). Without that fuel value, the auxiliary fuel requirement of a well insulated, direct thermal oxidizer would be as follows, based on the specific heat of air in the temperature range involved being approximately 0.24 Btu/(lb-°F):

In the direct thermal example, soil vapor temperature is raised by 1,100° F. (*See* Equation 8-8).

1,100° F (0.24 Btu/(lb-°F) (400 scfm) (0.076 lb/scf) (1,440 min/day)
= 11,557,000 Btu/day (= 116 therms/day) (8-8)

Accounting for the 35 therms/day fuel value of the combustible vapors, the net auxiliary fuel demand for the direct thermal example is as shown in Equation 8-9.

(116 - 35) therms/day = 81 therms/day (8-9)

In the catalytic oxizider example, no credit is given for the fuel value of the combustible vapors, and the soil vapor temperature is raised by 500° F instead of 1,100° F. (*See* Equation 8-10).

(500° F/1,100° F) (11,557,000 Btu/day) = 5,253,000 Btu/day (53 therms/day) (8-10)

Auxiliary fuel savings can be realized by including a recuperative heat exchanger in the design. With the high flammable vapor concentrations encountered in many soil venting systems, an automatic heat-exchanger bypass or dilution air system should be included to prevent combustion within the heat exchanger.

In some localities, the public considers direct thermal oxidation to be incineration, which should be rejected. If catalytic operation is shown to be analogous to automobile emission control, public approval may be easier to obtain.

8.4 Main System Design Parameters

8.4.1 Pneumatic Testing

Field pilot tests or laboratory bench-scale tests are needed prior to the design of most soil venting systems. Some determinations that can be made with field pilot testing are as follows:

- Contaminant concentrations in extracted soil gas, and how the concentrations initially vary with time
- Initial rate of contaminant mass removal (for sizing off-gas treatment units)
- Identification of constituents that may affect vapor treatment, e.g., tetramethyl lead or chlorinated solvents may be present that can potentially cause poisoning or deactivation of a catalyst
- Amount of vacuum that is necessary to induce a reasonable volumetric flow rate of air from each well
- Air permeability of the soil at each contaminated stratum
- ROI of each well

The 1991 EPA publications "Soil Vapor Extraction Technology" (EPA/540/2-91/003) and "Guide for Conducting Treatability Studies Under CERCLA: Soil Vapor Extraction" (EPA 540/2-91/019A) give detailed information on soil venting testing. The best preliminary test is a field pilot test, usually done with an ICE or a centrifugal vacuum blower rated at approximately 50 ft^3/min following a pair of drums containing activated carbon for emission control.

Details of laboratory column tests and field pilot tests are given in Section 8.5. A primary objective of field tests is to determine the ROI for extraction wells screened at each stratum. Then the well pattern can be designed to fit the aerial extent of the contaminant distribution that has been determined during site investigations. Test results give information for ft^3/min to vacuum and for calculating soil air permeability. Contaminant concentration measurements are needed to estimate the initial mass extraction rate and the emission abatement system operating requirements.

8.4.2 Radius of Influence of Extraction Wells and Soil Air Permeability

The ROI is that distance from an extraction point at which the observed vacuum reading is not significant. Setting a significant vacuum level is arbitrary, although sometimes it is related to the sensitivity of the vacuum gauges selected for the pilot test. Any amount of vacuum increase noted at an observation well or probe when the blower is on indicates that soil vapor is flowing from the observation point toward the extraction well. Sometimes vacuum levels in the range of 0.05 to 0.10 in. w.c. are selected to indicate the limit of the ROI. (This practice is disputed by Bohn (1997), who contends that the ROI is where the amount of induced vacuum is zero.)

In practice, it is usually found that the selected vacuum level is not exactly observed at the remote wells and probes. A curve is plotted showing the logarithm of observed vacuum readings versus radial distance from the extraction well at a given air flow rate, and extrapolated to an arbitrarily selected small vacuum. Or a linear regression of the absolute pressure (P) values equivalent to observed vacuum readings at each distance r from the extraction well can be fitted to Equation 8-11.

$$P^2 = a + b \, (\log r) \tag{8-11}$$

in which r is the distance from the extraction well to the observation point, and a and b are empirical constants determined from the linear regression. From this equation (or from a plot thereof), the distance r that corresponds to the selected vacuum equivalent to P is the ROI. If all the data points fit well with this equation, the extraction well is probably extracting air from a homogeneous stratum.

Another method of arriving at the ROI is to plot observed vacuum versus distance from the extraction well at a given flow rate, and to select that distance to be the point just before the curve starts flattening out.

It should be noted that after the vacuum blower is started, it takes some time for vacuum readings and soil gas concentrations to become steady, at which time readings should be noted for use in the calculations.

The relationship among quantity of air flow through the soil, radial distance to the observation point, extraction vacuum, and observed vacuum depends on the extraction well screen length, air temperature, molecular weight and viscosity, well radius, and the air permeability of the stratum. The air permeability for successful soil venting should be at least 10^{-8} cm^2. Air permabilty can be calculated from Equation 8-12, adapted from Johnson et al. (1990).

$$K = [Q \mu \ln(r/R_w)]/[\pi b P_w (P_{atm}/P^w)^2 - 1] \tag{8-12}$$

in which k is the soil air permeability, cm^2; Q is the air flow rate, cm^3/sec; μ is the viscosity of air, 1.8×10^{-4} g/(cm-sec); b is the well screen length or sand-pack length, cm; P_w is the absolute pressure at the extraction well, g/(cm-sec^2); P_{atm} is 1 atm pressure, or 1.01×10^6 g/(cm-sec^2); r is the distance from extraction well to observation point, cm; and R_w is the radius of the extraction well bore hole, cm.

Examination of this equation indicates that for a given permeability value, k, the larger the radius of an extraction well bore hole, R_w, the higher is the air flow rate. This radius can be somewhat effectively increased by enlarging the sand packing around the well (Gomez-Lahoz et al., 1991).

The method of Johnson et al. (1990) for designing extraction well systems has been programmed for computer application; it is distributed by the EPA under the name HyperVentilate. The Macintosh computer version costs $17; the PC version costs $22. Both are available from the Superintendent of Documents, Box 371954, Pittsburgh, PA 15250-7954 (phone 202-783-3238). The Macintosh document order number is S/N 055-000-00403-0. The program includes these steps:

- The user may enter the relative amounts of each contaminant compound to be extracted from the soil or enter "fresh gasoline" or "weathered gasoline."
- The user enters the soil texture (i.e., medium sand, fine sand, silty sand, clayey silts) or the soil permeability.
- The user enters the well radius, ROI, length of the screened interval in the wells, and inches of water column vacuum. For the ROI, Johnson et al. (1990) suggest

40 ft (12 m) as a default value if the actual value is not known. Johnson et al. give Equation 8-13, which relates observed test pressure readings and ROI.

$$P = P_w \{1 + [1-(P_{atm}/P_w)^2] \ln(r/R_w)/\ln(R^w/R_1)\}^{1/2} \qquad (8\text{-}13)$$

in which, using consistent units, P is the remotely observed absolute pressure at distance r from the extraction well; P_w is the absolute pressure applied at the extraction well; R_w is the radius of extraction well bore hole; and R_1 is the ROI.

- The program calculates the air flow rate range per well.
- The user enters the soil temperature, and the program calculates the contaminant extraction rates (in lb/day or kg/day) at the chosen amount of vacuum and at lesser amounts of vacuum.
- The user enters the estimated total mass of contaminant to be removed and the desired number of days of remediation, and the program calculates the maximum extraction rate range. (Note that the program assumes ideal circumstances, and the actual extraction rate over a period of time will be less than the predicted rate.)
- For each group of compounds within certain boiling point ranges, the program calculates the corresponding vapor concentrations and the residual concentrations in the soil.
- The program calculates the minimum amount of air that must be drawn through the soil per gram of initial contaminant to achieve at least a 90% reduction. The program calculates the number of wells needed and expresses this result as a range.

The program manual notes that calculated values are predictions that are intended to serve as guidelines. The program helps determine whether soil venting is an appropriate technology to apply at the described site.

The computer program Bioventing Plus™, marketed by Environmental Systems & Technologies (Blacksburg, Virginia) uses equations based on Johnson et al. (1990) and estimates hydrocarbon recovery rate versus time for multiwell systems. It also calculates the air permeability from field measurements of vacuum and air flow rate. Other models of soil venting have been developed by Wilson (Eckenfelder Inc., Nashville, Tennessee).

Figure 8-3, adapted from Appendix E of the US EPA (1991d), gives an order-of-magnitude relationship between soil air permeability and type of soil. The permeability scale units are Darcy's; multiply by 9.87×10^{-9} (approximately 1×10^{-8}) for equivalent permeability k in units of cm^2.

Other methods of estimating the soil air permeability are described as follows:

- Based on a correlation published by the U.S. Department of Commerce (1991), k is $125(D_{15})^2 \times 10^{-5}$ cm^2, in which D_{15} is the 15% particle size diameter in centimeters, passing by weight, as determined from a sieve analysis of the soil.
- Ratio the air permeability to hydraulic conductivity of the soil when saturated. If k is in cm^2 and hydraulic conductivity, K, is in cm/sec, k/K (which decreases with

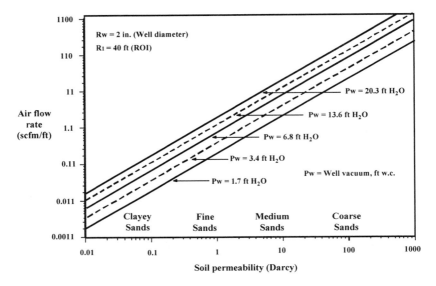

Figure 8-3 Air flowrate versus soil air permeability and applied vacuum (US EPA, 1991d, Appendix E).

temperature) has been estimated to be of the order of 10^{-3} at 50° F from data in the US EPA (1991c), or of the order of 10^{-5} from correlations in the US Department of Commerce (1991).
- Johnson et al. (1990) give Equation 8-14.

$$k = Q\mu/(4A\pi m) \tag{8-14}$$

in which Q is the volumetric vapor flow rate from an extraction well, cm^3/sec; μ is the viscosity of air, 1.8×10^{-4} g/(cm-sec); m is the stratum thickness, cm; and A is the slope of the straight-line curve developed by plotting the pressure noted during testing at an observation well or probe versus the natural log of time. (Note that these pressure readings should be taken frequently during the first minutes when vacuum is applied.) The slope A can alternatively be used in Equation 8-15 from Johnson et al. (1990).

$$k = 10^{-8} r^2 e\mu/(4\,P_{atm}) \exp(B/A + 0.5772) \tag{8-15}$$

in which k is the air permeability, cm^2; e is the air-filled soil porosity (void fraction); B is the y-intercept of the straight-line curve, g/(cm-sec); A is the slope, g/(cm-sec^2); r is the radial distance from extraction well, m; μ is the viscosity of air, 1.8×10^{-4} g/(cm-sec); and P is the ambient atmospheric pressure, 1 atm = 1.013×10^6 g/(cm-sec^2). Note that if the soil moisture content increases, the permeability decreases.

Johnson et al. (1990), reprinted by the US EPA (1991d), provide multiplying factors for predicting the volumetric air flow rate from Figure 8-3 for variations in extraction well radius R_w and in the ROI, R_1, as given in Table 8-2.

Table 8-2 Multiplying factor applied to air rate for extraction wells (US EPA, 1991d, Appendix E).

R_w (in)	R_1 (ft)	Air Flow Rate Multiplier
2	25	1.09
2	75	0.90
3	40	1.08
4	40	1.15
4	25	1.27

Johnson et al. (1990) note that the predicted air flow rate is not sensitive to changes in estimated radius of influence. This is true for the smaller well radii, as indicated in this tabulation for the multipliers corresponding to the 2-in. and 4-in. radius well sizes. An example (adapted from Appendix E of the US EPA, 1991d) using a 4-in. radius in a medium sand with a 5-ft screened interval for wells placed approximately 42 ft apart (25-ft ROI, so the multiplier is 1.27) is given in Table 8-3.

Table 8-3 Example of air flow rates (US EPA, 1991d, Appendix E).

Vacuum = 3.4 ft w.c. (3 in. Hg column)	Flow Rate at Indicated Permeabilty	
scfm/ft, Figure 8-3	0.41 at 1 Darcy	4.1 at 10 Darcy
scfm with 5-ft screen	2.1 at 1 Darcy	20.5 at 10 Darcy
scfm/well, corrected	x 1.27 = 2.6 at 1 Darcy	26 at 10 Darcy
Vacuum = 13.6 ft w.c. (12 in. Hg column)	Flow Rate at Indicated Permeabilty	
scfm/ft, Figure 8-3	1.3 at 1 Darcy	13 at 10 Darcy
scfm with 5-ft screen	6.5 at 1 Darcy	65 at 10 Darcy
scfm/well, corrected	x 1.27 = 8.3 at 1 Darcy	83 at 10 Darcy

In this example, a four-well extraction system with 42-ft well spacing would cover an area greater than 3,500 ft^2 (2 x 42 ft x 42 ft) with overlapping circles (25-ft radius) of influence. A centrifugal blower capable of extracting air with a vacuum of 3 in. Hg (plus the capability to overcome pressure drops in piping and emission abatement equipment) at each well should have a rating of approximately 100 scfm (four wells at 26 scfm per well). This would be a suitable blower selection for a conceptual design. High vacuum applications, such as the 12 in. Hg in this tabulation, would apply to clayey soils better than to the sandy soil given for this example. Final design and blower selection should be based on test results that would give a narrower range of permeability values.

The ROI at a given permeability is somewhat proportional to vacuum at the extraction well and is best determined from field pilot tests. Johnson et al. (1990) determine an air flow range that will result in extracting 90% of the contaminant mass during a desired time period.

8.4.3 Volumetric Air Flow and Contaminant Mass Removal Rate

Bohn (1997) suggests a flow rate of 16 to 50 m^3/hr (10 to 30 ft^3/min) per well. After an initial extraction period, higher air flow rates do not increase contaminant mass removal rates (during the diffusion-controlled period). Excessive air rates waste energy, require oversized vacuum-inducing and vapor treatment equipment, and dry the soil.

The initial mass removal rate of contaminants, and the basis for emission abatement equipment design, is the product of air flow rate, contaminant concentration, and contaminant vapor density, as shown in Equation 8-16.

Rate (lb/min) = Q (concentration, ppmv/10^6) d   (8-16)

in which Q is the air flow rate, scfm; and d is the contaminant vapor density, lb/scf.

Under soil venting conditions, the ideal gas law holds and the air flow rate in scfm is equal to the actual measured air flow in ft^3/min multiplied by (1 atm divided by the actual absolute pressure) times (the actual absolute temperature divided by the standard temperature). If vacuum is measured in inches of mercury, 1 atm = 29.92 in. Hg. The absolute pressure is derived from the vacuum reading at which the actual air flow rate is measured. If the standard temperature is chosen to be 520° R (60° F), the density (in units of lb/scf) of any vapor component at this standard temperature and 1 atm is its molecular weight divided by 379. The effective molecular weight of gasoline is usually in the 95 to 111 range, with the high end of the range corresponding to gasoline that has had time to weather. During this time, the lighter (lower molecular weight) compounds, such as butanes and isopentane, volatilize, and the effective molecular weight gradually increases. An example of a mass extraction rate calculation is as follows:

Given: The air velocity measured in a 2-in. diameter pipe (pipe cross-sectional area is 3.1 in^2, 0.02153 ft^2) is 40 ft/sec; the vacuum is measured at 3.8 in. Hg; the temperature is measured at 55° F; the benzene concentration is 4,000 ppmv (benzene's molecular weight is 78 lb/lb mole); the total petroleum hydrocarbon (TPH) concentration is 50,000 ppmv.

Assumptions: Gasoline has weathered somewhat; the effective molecular weight is 105 lb/lb mole; and the standard temperature is 520° R.

Find: Mass extraction rate for benzene and for TPH.

Step 1: Determine the measured air flow rate in acfm and corresponding scfm units using Equations 8-19 and 8-20.

Vacuum = 3.8 in. of Hg (absolute pressure = 29.92 in. - 3.8 in. = 26.1 in. Hg) (8-17)
Temperature = 55° F (absolute temperature = 460° R + 55° = 515° R) (8-18)
Actual ft^3/min = 40 ft/sec (60 sec/min) (0.02153 ft^2) = 51.7 acfm (8-19)
Standard ft^3/min = 51.7 acfm (29.92/26.1) (520/515) = 59.8 scfm (8-20)

Step 2: Determine the density of the two contaminants of concern in this example using Equations 8-21 and 8-22.

Benzene vapor density = 78/(379 lb/scf) = 0.206 lb/scf (8-21)

TPH (weathered gasoline in this example) vapor density, molecular weight of 105, = 105/(379 lb/scf) = (0.277 lb/scf) (8-22)

Step 3: Calculate the mass rate, scfm (ppmv/10^6), and the density using Equations 8-23 and 8-24.

Benzene rate = 59.8 scfm (4,000/10^6) (0.206 lb/scf) = 0.049 lb/min (8-23)
TPH rate = 59.8 scfm (50,000/10^6) (0.277 lb/scf) = 0.83 lb/min (8-24)

Note that TPH includes benzene and that the rates are not additive.

At a given air flow rate, the contaminant mass removal rate will decay markedly after the easily removed volatiles in the interstices of the soil particles or that are loosely bound on soil particle surfaces have been stripped.

Soil gas contaminant concentration levels may be constant for a period of time when soil venting is initiated. After soil gas concentration levels begin to decline, the decay in mass removal rate can be predicted from Equation 8-25.

$M(t) = (M_o - M_f)e^{-kt} + M_f$ (8-25)

in which $M(t)$ is the mass removal rate at any time t; M_o is the mass removal rate at the start of the decline in concentration from a steady level; M_f is the mass removal rate at the end of remediation; t is the time after the start of the decline in concentration from a steady level; and k is the decay constant.

Because M_o is usually much larger than M_f, the M_f terms can be dropped from the equation. The value of k is chosen so that the area of integration under the curve $M(t)$ (plotted against time) is equal to the total mass to be extracted after the concentration starts declining. Or k can be evaluated at a known time t from the measured value of M_o and of $M(t)$, as given by Equation 8-26.

$k = \log_e[M_o/M(t)]/t$ (8-26)

In many applications, the concentration sometimes falls so low that it is best to stop the extraction for a few weeks and let the concentration build up. The extraction system is operated intermittently for a given set of extraction wells. Thus, the average concentration while operating the vacuum-inducing device is maximized.

For a given target mass removal, this mode of operation uses the least amount of energy and optimizes the use of any emission abatement devices. Bohn (1997) notes that a high rebound in concentration indicates that air flow rates per well have been excessive.

8.4.4 Ventilation Wells

Some of the wells in a soil venting field may be used for ventilation instead of SVE. There are situations in which some wells are installed for the purpose of admitting air into the soil. Three scenarios in which this is advantageous are as follows:

- When the air flows without ventilation wells are inadequate to achieve cleanup in a reasonable time using reasonable operating conditions.
- When ventilation wells are placed on one side of the contaminated soil and extraction takes place on the other side — this arrangement helps ensure maximum use of the extracted air for the purpose of vaporizing contaminants.
- When ventilation wells are placed in a row perpendicular to a line between extraction wells and another source of contamination that is off the property being remediated.

This arrangement accomplishes the same goal as in the first scenario and also helps guard against cleaning up someone else's property or cross-contaminating the main property of concern with different contaminants.

8.5 Treatability Studies for Soil Venting

Field pilot tests are a must for most in situ soil venting projects. Laboratory experiments done by drawing air through a column of undisturbed soil obtained by a geologist as a core sample have some marginal utility. The laboratory tests can determine the initial organic vapor concentration in the soil gas under dynamic air flow conditions; the partitioning that might be achieved — (the ratio of organics that vaporize to organics that stay adsorbed on the soil particles); and the order of magnitude of the soil air permeability. Appendix B of the EPA treatability studies guidance (US EPA, 1991c) recommends an air flow rate of 0.5 to 1.0 L/min through a 2.5-in. diameter soil column. Plots are made of vapor concentrations versus time and versus number of pore volumes of air passed through the column. The pressures recorded are plotted as a straight line against the $\log_e$ of time, and the slope of the line can be used for estimating the soil air permeability, as discussed in Section 8.4.2.

Field pilot extraction tests are best for determining permeability, initial contaminant extraction rate, and well ROI. Some extraction wells in the contaminated vadose zone are needed for field pilot tests. Results of the testing are used to determine the best locations for additional extraction wells. If the site has shallow contamination and is not paved, heavy plastic sheeting could be spread on the ground surface, extending aerially beyond the boundaries of the contaminated area and weighted down at the edges. Field pilot tests are best conducted by extracting from one well at a time, while monitoring vacuum and vapor concentrations versus time. At the same time, vacuum readings are taken at surrounding wells or soil probes. Extraction wells should be screened within only one soil stratum at a time. If remediation of multiple strata is needed, then multiple extraction wells must be tested separately during the pilot test runs. Each extraction well is tested at three different air flow rates, say in the range of 16 to 50 m^3/hr (10 to 30 ft^3/min).

A simple setup for a field pilot test has been used with a regenerative centrifugal blower creating negative pressure in each extraction test well. The blower discharges soil vapor, which is usually more than 90% air, through two 55-gal vapor phase activated carbon drums in series. It is important to monitor the temperature of the air between the two drums. If the air temperature rises above approximately 135° F, adsorption in the second drum will be severely impaired. In fact, the cooler the air, the better the adsorption, as long as the relative humidity is less than 50%. Two factors tend to cause increased air temperatures: heat of compression in the blower raises the temperature, and heat of adsorption on carbon raises the temperature further.

Soil vapor can have very high flammable vapor concentrations. The higher the concentration, the more heating occurs within the carbon bed. High concentrations can be extremely dangerous, and levels greater than 50,000 ppmv have been observed during pilot tests. The LEL for many hydrocarbons is in the range of 11,000 to 14,000 ppmv range. The centrifugal blower should be constructed of nonsparking materials and should accommodate dilution air so that concentrations are below the LEL in the extraction system.

An improved soil venting field pilot setup would have the soil vapor is drawn through a water knockout vessel (mist separator) and carbon vessels ahead of the blower. If the vapor concentration is monitored between the carbon and the blower and if operations are discontinued when breakthrough is attained, so that the LEL is not exceeded, a steel positive displacement blower can be used. The vapor extraction wells are usually screened only in the vadose zone, which is appropriate unless multiphase extraction with a shallow water table is desired. In that event, a liquid-ring vacuum pump would be a suitable vacuum-inducing device. Another method of conducting field pilot tests (for vadose zone extraction only) is to use an ICE.

8.6 Cost Estimating for Soil Venting

The main capital costs for soil venting systems are for the blower, inlet mist separator and filter, vapor treatment system, controls, shop assembly, sound-proofing, and field installation (which typically might include a slab, fencing, soil vapor piping connections, and utilities). As an example, a complete shop-assembled, skid- or trailer-mounted unit with all the main components including a catalytic oxidizer and controls costs $50,000 to $70,000 for units rated at 200 to 600 ft^3/min. Some units can be leased.

Trailer-mounted ICE units are available for purchase or lease. Units rated at less than 100 ft^3/min cost, in 1995, approximately $40,000 or rent for $3,500/mo. Larger units range in cost up to $100,000 or $9,000/mo. Extra costs are entailed to adapt the units for use of natural gas as auxiliary fuel, computerized monitoring accessories, noise muffler, and air pollution control permits.

Two 1991 EPA guidance documents (EPA /540/2-91/003 and EPA/2-91/019A) give capital cost ranges for extraction wells and equipment and costs of preliminary testing. Table 8-4 summarizes the reported 1991 costs for wells and equipment (excluding emission abatement equipment).

Table 8-5 gives the reported capital costs and alternative rental costs as of 1989 for carbon adsorption and catalytic oxidization systems. In this table, the capital cost for the ORS Model 1282008 catalytic oxidizer includes additional option features. The term "NA" is used where there is no available cost information. Table 8-3 also gives electric power requirements for operations. Preliminary laboratory tests in 1991 cost $30,000 to $50,000; field pilot tests cost $10,000 to $50,000 for determining soil air permeability and more than $100,000 for complete pilot tests.

The US EPA (1992d) gives estimated costs for operating vapor treatment equipment. Use of activated carbon costs approximately $45/kg ($130/gal) of gasoline removed. Fuel cost for catalytic oxidizers is given as approximately $400/mo. for each 100 ft^3/min of air treated using propane fuel priced at $1/gal.

Baker and Moore (2000) give a comparison of costs for soil venting, thermal desorption, land farming, and bioventing for projects ranging from 500 to 20,000 yd^3 of soil remediated.

For thermally enhanced in situ soil venting, DuTeaux (1996) tabulates case histories as follows:

- $252/$yd^3$ to $317/$yd^3$ using in situ steam and air stripping of soil via hollow-stem augers for removal of VOCs and semivolatile organic compounds.
- $15/ton to $30/ton using resistive heating and radio frequency heating combined for removing fire training and chemical production wastes.
- $63/$yd^3$ using heated vapor reinjection with vapor phase carbon adsorption for removing PCE, TCE, chloroform, and methylene chloride.

An important parameter in any remediation cost analysis is the life cycle of the operations. The life cycle of soil venting systems cannot be predicted accurately. Changes in the groundwater table elevation and infiltration from rain water can affect concentrations. Except when dealing with water-soluble volatile organics, such as acetone, some soil moisture helps in the desirable desorption of contaminants into the vapor phase. Too much moisture can undesirably decrease soil air permeability, however. Optimum usage of blower energy is usually achieved by ceasing extraction from a particular group of wells for a few weeks at a time so that depleted concentrations are allowed to be restored naturally. Factors such as changing soil water content and on-off operations affect the overall length of time to remediate a given site.

Table 8-4 Typical unit costs of SVE components (US EPA, 1991a).

Component	Type	Size	Capital Cost	Notes
Extraction Well Construction				
Casing	PVC	2 in.	$20 - $40/ft	Sch. 40 PVC
	PVC	4 in.	$2 - $3/ft	Sch. 40 PVC
		6 in.	$3 - $5/ft	Sch. 40 PVC
Screen	PVC	2 in.	$7 - $12/ft	Sch. 40 PVC
		4 in.	$2 - $4/ft	Sch. 40 PVC
		6 in.	$5 - $7/ft	Sch. 40 PVC
Sand Pack			$10 - $15/ft	Any Slot Size
Gravel Pack			$15 - $20/yd^3	
Piping	PVC	2 in.	$20/yd^3	
		4 in.	$1/ft	Sch. 40 PVC
		6 in.	$3/ft	Sch. 40 PVC
Valves (Ball)	PVC	2 in.	$5.25/ft	Sch. 40 PVC
	single union	4 in.	$65	
		6 in.	$300	2 in. & 4 in. threaded socket
Joint (Elbow)	PVC	2 in.	$700	6 in. flange end connection
	90 degrees - slip	4 in.	$2.50	Sch. 40 PVC
		6 in.	$16	Threaded, Socket
			$51	End Connections
Water Table Depression Pumps			$3,700	R. E. Wright Associates w/Explosion Proof Pump Motor Control System
			$3,000	
Surface Seals	Bentonite	6 in.	$3.30/yd^2	
	Bentonite	4 in.	$2.20/yd^2	
	Polyethylene	10 mil	$2.30/yd^2	
	HDPE	40 mil	$5.00/yd^2	
	Asphalt	2 in.	$9.30/yd^2	

Table 8-4 Typical unit costs of SVE components (US EPA, 1991a) (continued).

Component	Type	Size	Capital Cost	Notes
Blower	Fan	1 hp	$1,700	Environ Instruments
		1.5 hp	$2,000	
		2 hp	$2,200	
		3 hp	$2,700	
		5 hp	$3,300	
		10 hp	$5,000	
		30 hp	$6,000	
	Centrifugal	2.5 hp	$600	
		25 hp	$12,000	Includes Installation
		50 hp	$42,000	
Air/Water Separator	Water Resources knockout pots	Medium	$1,500 - $2,400	Vendor - Water Resources
		800 gal	$11,600	Installation 33%
		20 gal	$1,500	of Capital Costs
		35 gal	$1,600	
		65 gal	$1,800	
		105 gal	$2,200	
		130 gal	$2,400	
Instrumentation	Vacuum Gauge (Magnehelic)		$50 - $75	
	Flow (Annubar)		$300	
Sampling Port	brass T		$20 - $30	
Concrete Pad			$450/yd^3	
Heat Exchanger	& housing unit		$1,400	
Fiberglass Shed	8 ft x 10 ft		$8,500	
Flame Arrestor	without SS element		$670	Vendor - Stafford Tech.
	with SS element		$740 - $930	
Air Relief Valve			$230	Vendor - Stafford Tech.
Soil Gas Probe			$30 - $50	Vendor - K. V. Assoc.
Engineering/Design			8 - 15% of system cost	
Diffuser Stacks	Carbon Steel	4 in.	$8/ft	Add 40% for Installation
		6 in.	$10/ft	
	Stainless Steel	4 in.	$30/ft	
		6 in.	$40/ft	

Note: PVC = Polyvinyl Chloride
SS = Stainless Steel

Table 8-5 Typical unit costs for SVE vapor treatment (US EPA, 1991a).

Treatment	Vendor	Model	Maximum Flow (scfm)	Capital Cost ($)	Rental ($/Month)	Lease Period	Operations
Carbon Adsorption	Carbtrol	SVX	105		$1,540	1 yr. + deposit	
			250		$1,830	1 yr. + deposit	
			500		$1,900	1 yr. + deposit	
	Continental Recovery Systems	Manual Beds					
		1 Bed		$20,000	$2,400		5 hp
		2 Beds		$26,000	$2,900		
		3 Beds		$32,000	$3,300		
		4 Beds		$3,800	$3,700		
		5 Beds		$44,000	$3,900		
		6 Beds		$50,000	$4,000		
		Automatic Regenerable		$149,000	$7,500	6 mo.	
Catalytic Oxidation	Dedert Corporation	CATOX	1,000	$85,000			
			5,000	$200,000			fuel
	CSM Systems	Model 28 TORVEX	200	$37,500			
		Model 5A + heat ex.	500	$60,000			
		Model 5B TORVEX	500	$50,000			
		Model 10B TORVEX	1,000	$70,000			
	ORS	Catalytic Scavenger					
		1282001	200	$63,000	$7,000	12 mo.	20 kw
		1282002	500	$78,000			35 kw
		1282008**	300**	$90,000**	$15,000	1 mo.	20 kw
					$10,000	12 mo.	

Table 8-5 Typical unit costs for SVE vapor treatment (US EPA, 1991a) (continued).

Treatment	Vendor	Model	Maximum Flow (scfm)	Capital Cost ($)	Rental ($/Month)	Lease Period	Operations
Catalytic Oxidation	Water Resources Associates	AB 15-5-SVS	100	$11,200			
		AB19-10-SVS	210	$1,530			
		AB22-10-SVS	320	$18,400			
		AB24-10-SVS	420	$20,100			
		AB22-15-SVS	570	$22,900			
		AB15-5-SVS	100	$23,000	$3,900		fuel
		AB19-10-SVS	210	$28,000	$4,700		1.5 hp
		AB22-10-SVS	320	$32,000	$5,400		
		AB24-10-SVS	420	$36,000	$6,000		
		AB22-15-SVS	570	$40,000	$6,700		
Combination Systems	Haz Tech	MMC-5	100	$60,000			fuel
		MMC-2 w/trailer	30	NA			
		MMC-3	1,000	NA			
	CEMI	Ranger	200	$55,000			main contract $7,600/yr
Carbon Cannisters	Carbtrol G-Series Cannisters	G-1	100	$600			
		G-2	300	$800			
		G-3	500	$800			
		G-5	600	$7,700			
	Calgon	Vapor-Pac 1,800 lb	1,000	$5,600			
		Ventsorb Cannister	100	$764			
	High Flow Ventsorb Cannisters	28 in.	400	$1,700			
		36 in.	600	$4,000			
		48 in.	1,100	$6,400			
	Tiggnixtox Series	N50ODB	500	$7,050			
		N75ODB	750	$11,900			
		N150ODB	1,500	$19,200			
		Boxsorber 6 X 6	2,200	$13,800			
		Boxsorber 8 X 8	4,000	$20,500			

* 1989 Estimates, Contact Vendor For Actual Prices. **NA: No available cost information.

8.6.1 Utilities Costs

The main utility costs of a typical soil venting system with an oxidizer are for auxiliary fuel and electric power for the vacuum blower. (For oxidizers energized by electric power, there would be no auxiliary fuel cost, but electric power costs would increase.) For the direct thermal example in Section 8.3.2 using 400-ft^3/min at 81 therms/day, if fuel were priced at $0.50/therm, the fuel cost per month would be approximately $1,200; or for the catalytic example at 53 therms/day, $800. These examples are for systems without heat exchange and do not take into account the fuel value of flammable vapors in air being treated.

If motor efficiency and power factor are taken into account, the kilowatt demand of a blower motor is approximately equal to the actual horsepower (not the motor nameplate horsepower). If the vacuum blower consumed 5 kW of power priced at $0.10/kW-h, the power cost would be $0.50/hr or approximately $360/mo. of continuous operation.

Fuel costs for oxidizers can be reduced by purchasing units with heat exchangers that preheat incoming air with hot exhaust gases. A typical conventional air preheater with metal heat exchange surfaces will reduce the heating duty by 50%; designs for 60% to 70% are also available. However, in applying these designs to direct thermal oxidizers, care must be taken so that ignition does not occur in the heat exchanger.

Regenerative oxidizer designs will reduce the heating duty by 90% to 98%. A regenerative unit has at least two heat exchange chambers in addition to the main oxidation chamber. The heat exchange chambers are filled with solid inert material that has a high heat capacity, such as clean gravel or ceramic shapes. During one part of the cycle, this solid material is heated by direct contact with hot exhaust. Then valve positions are reversed, and incoming soil gas is passed through this heated chamber in direct contact with the solid material. Regenerative units are usually applied only to very large systems with an air flow rates of several thousand ft^3/min and low flammable vapor concentrations.

8.6.2 Carbon Adsorption Costs

The cost of carbon adsorption systems with off-site regeneration is given by Stenzel and Perryman (1997) as follows:

- With purchased permanent on-site canisters, the canisters are priced at $5,800 with 450 kg of carbon for 1,000 ft^3/min and at $8,000 with 910 kg for 2,000 ft^3/min; carbon changeout costs $1/kg to $1.60/kg for 450-kg batches and 10% less for 910-kg batches. With canisters rented and used as carbon transport vessels, replacement with reactivated carbon costs $3,500 to $5,000 for 815-kg canisters, and canister monthly rental is $150 to $725 per unit.
- For adsorption of light aromatic and chlorinated solvent vapors, case studies are presented based on carbon loadings of 25% to 30% of carbon weight as adsorbed vapor. For 2 yr of operation with purchased canisters, estimates for 500 ft^3/min (removing 4.5 kg/day of low concentrations of light aromatics) are $49,000 to

$75,000 with permanent canisters and $28,000 to $79,000 with rented units used for transport. For removing 22.6 kg/day of high concentrations, costs are $104,000 to $164,000 and $80,000 to $154,000, respectively. At 2,000 ft^3/min, removing 90.7 kg/day of chlorinated solvents costs $184,000 to $295,000 with permanent canisters and $161,000 to $269,000 with rented transport units.

8.6.3 Software for Soil Venting Process Design and Cost Estimating

COMPOSER GOLD (Building Systems Design, Atlanta, Georgia) computer software includes cost-estimating models for ex situ vapor extraction, air sparging, and in situ vapor extraction.

RACER/ENVEST™ (Talisman Partners, Ltd., Englewood, Colorado) computer software includes cost-estimating models for ex situ vapor extraction, air sparging, heat-enhanced vapor extraction, and ambient-temperature in situ vapor extraction. This last model can be used to estimate capital costs of wells or trenches, blowers, and connecting piping with budget accuracy with the following parameter values or types given:

- Vertical wells or horizontal trenches up to 30 ft deep
- Area of contaminated soil
- Type of soil
- Average depth to top of screen or trench depth
- Screen length
- Location

Better cost-estimating accuracy can be achieved if values for these secondary parameters are given:

- Spacing between wells or between trenches
- Number of wells or trenches
- Gas flow rate
- Blower capacity in ft^3/min and horsepower, and the number of blowers

If the well spacing, L, is not known, the cost-estimating model assumes a spacing ranging from 15 ft for silty clay to 50 ft for primarily sandy soil or to 100 ft for sand/gravel. The number of wells cannot be directly specified but is derived from the spacing, as shown in Equation 8-27.

Number of wells = area/$(\pi L^2/4)$ (8-27)

For each extraction well or trench, the gas flow rate per foot of screen length is assumed for each type of soil, ranging from 0.6 ft^3/min/ft for silty clay to 3.55 ft^3/min/ft for primarily sandy soil or 15.5 ft^3/min/ft for sand/gravel. Use of one blower is assumed if the total for all extraction points is 280 scfm or less. At larger total flow rates, the number of blowers is taken to be the total air flow in units of scfm divided by 280, rounded up to the next whole number. One of four types of blowers can be specified:

- 98 scfm, 1 hp
- 127 scfm, 1.5 hp
- 160 scfm, 2 hp
- >160 scfm, 5 hp

If the drilling technique is not specified, the model assumes that a hollow-stem auger will be used for depths up to 150 ft or that the water/mud rotary technique will be used for greater depths. Water/mud rotary and air rotary techniques can be specified. Costs for supervision by a hydrogeologist; safety monitoring; containment of cuttings; and decontamination of samplers, bailers, and drilling tools are included. Costs of collecting soil samples may be included at the option of the user of the cost-estimating model. Costs of installing peripheral monitor wells are not included; a separate model is available for this activity.

Separate ENVEST models are also available for estimating the costs of carbon adsorption, overhead electrical power service, fencing and signs, site clearing, demolition and hauling of old pavement, and installing new pavement.

With the basic model, the user can specify a well diameter of 2 in. or 4 in., with either PVC or stainless steel casing.

For estimating the costs of connecting piping, above ground installation is assumed. (Another model is available for estimating the costs of underground piping.) If connecting piping lengths are not specified, the model uses one-half the well spacing multiplied by the number of wells. The diameter of the connecting piping is taken as 2 in. for 2-in. diameter wells and 4 in. for 4-in. diameter wells. If manifold piping length is not specified, the model uses half the length of the connecting piping and a diameter of 4 in. Costs of vacuum gauges and other piping appurtenances are included.

For estimating maintenance and monitoring costs, the model assumes a crew will be on-site once per week for the first month and once per month thereafter to adjust blower flow rate and valves, to perform general maintenance, to use a rented organic vapor analyzer, and to remove a drum of collected condensate. Another model is available for estimating the costs of using GAC for emission abatement.

8.7 Summary of Important Points for Soil Venting

- Soil venting can be applied ex situ to excavated piles imbedded with perforated air piping.
- Either vadose zone wells or trenches are used to extract soil gas from unexcavated soil. Trenches are generally used where the water table is within 10 ft of the ground surface.
- Extraction wells and trenches should be screened above the water table, unless a system designed for two-phase extraction from shallow depths is intended.

- At ambient temperatures, soil venting is used to volatize and remove volatile organics, such as gasoline hydrocarbons and chlorinated hydrocarbon solvents.
- Soil venting can be applied to removing heavier hydrocarbons and semivolatile organic compounds if heat is used.
- Air is induced to flow horizontally through each soil stratum.
- The ground surface could be paved or plastic film should be spread if there is no confining layer of soil above the contaminated zone.
- Extracted soil gas is usually more than 95% air.
- Vacuum can be induced with centrifugal blowers, positive-displacement blowers, liquid-ring compressors, or ICEs.
- Unless a liquid-ring compressor is used, a water knockout pot (mist separator) should be installed upstream of other equipment. Provisions should be made for preventing condensate formation or for removing condensate from extraction lines and the knockout pot.
- Soil vapor concentrations may attain a steady level at some sites near the beginning of extraction and then start decaying.
- Soil gas can be extracted without inducing vacuum by applying passive vapor extraction, relying on natural changes in barometric pressure. This is a slow process applicable after concentrations have decreased to the level at which mechanically-enhanced removal is not cost-effective. The concept applies to both soil venting and bioventing.
- If emission abatement is required, vapor phase carbon adsorption and direct thermal or catalytic oxidation are commonly used.
- Treated air emissions can be discharged into wells or to the atmosphere. Regulatory limitations must be determined for either scenario.
- During early phases of vapor extraction, ICEs (which thermally oxidize contaminant vapors) and direct thermal oxidizers are economical.
- When vapor concentrations decrease to low levels, catalytic oxidation, carbon adsorption, and passive vapor extraction are progressively more economical.
- Operating costs with oxidizers can be reduced by installing heat recovery equipment that exchanges the heat from the hot exhaust to the extracted soil gas.
- Field pilot tests are needed to determine initial soil vapor concentrations and soil air permeability.
- Field pilot tests also help determine the ROI of the extraction wells and volumetric air flow rates versus vacuum developed.
- Computer software is available for applying field pilot test data to extraction wells system design.
- The mass rate of contaminant removal is the product of air flow rate, vapor concentration, and contaminant density.
- Contaminant vapor density is proportional to its molecular weight.
- Ideal gas law relationships are accurate enough for describing air flow conditions for use in design of soil venting systems.
- After vapor concentrations in the soil gas start declining, the contaminant mass removal rate decays to a tailing profile that is driven by diffusion/desorption until all the contaminant mass is removed from the soil.
- During the diffusion-controlled period of extraction, increasing the air flow rate per well often does not increase the contaminant mass removal rate and will waste energy.

- Sometimes it is efficient to stop extracting from certain wells for a few weeks and then resume extracting when concentrations have built up again (i.e., have rebounded).
- Ventilation wells can be strategically placed to help attain optimum underground air flow patterns.
- Skid-mounted and trailer-mounted blower/vapor treatment assemblies are available for purchase or lease.
- Trailer-mounted ICE units are available for purchase or lease.
- A computerized model is available for budget estimating of costs for wells or trenches, blowers, connecting piping, maintenance, and monitoring. The model aids in preliminary configuration of trench systems and well spacing.
- Companion models are available for estimating costs of ex situ vapor extraction, heat-enhanced vapor extraction, air sparging, carbon adsorption, electric power service, fencing and signs, site clearing, and pavement.

Chapter 9

Thermal Treatment for Soils and Sludges

This chapter deals with ex situ thermal treatment of soils and sludges containing organic contaminants such as semivolatile and volatile hydrocarbons and chlorinated solvents, PCBs, dioxins/furans, cyanides, and pesticides. Two treatment technologies will be described: incineration and thermal desorption. With incineration, organics are combusted directly by subjecting the soil or sludge to high temperatures in the presence of oxygen. Thermal desorption describes the vaporization of VOCs and semivolatile organic contaminants at temperatures below the ignition point of the contaminants. The vapors that are released, along with soil particulate matter, are further treated in a variety of ways, including recovery as liquids using condensation or adsorption/regeneration/condensation, combustion in an afterburner, baghouse filtration, and wet scrubbing. (In situ thermal desorption or soil heating is generally associated with soil venting and is discussed in Chapter 8.)

Because some incinerators are fitted with liquid injection nozzles, destruction of organic liquids and slurries can be considered within the scope of this chapter. The most common soil incinerator used is the rotary kiln type. Fluidized circulating bed combustors (CBCs) and infrared furnaces are also used for soil treatment. Table 9-1 summarizes the advantages and disadvantages of several types of incinerators, based on data in Bonner et al. (1981) and the US EPA (1986).

If nonvolatile metal contaminants are also contained in the contaminated media, incineration might be followed by fixation, discussed in Chapter 11.

9.1 Basic Principles

9.1.1 Incineration Basics

As in any combustion process, with soil incineration, the organic destruction efficiency depends on temperature, residence time, and mixing with adequate combustion air. To some extent, for a given destruction efficiency (usually greater than 99%), these three variables can be traded off among each other. For example, CBCs suspend the soil particles in a fluidized bed, resulting in excellent mixing with combustion air. Consequently, CBCs can be operated at a lower temperature range and a shorter residence time than rotary kilns.

Most rotary kilns have a fixed air rate, set by the air blower design and head loss in the exhaust gas cleanup train. If a rotary kiln has a variable slope or a variable speed

Table 9-1 Advantage/disadvantage analysis of each incinerator type.

Incinerator Type	Advantages	Disadvantages
Rotary Kiln	Will incinerate a wide variety of liquid and solid wastes	High capital cost for installation
	Feed capability for drums and bulk containers	Operating care necessary to prevent refractory damage; thermal shock is a particularly damaging event
	Adaptable to wide variety of feed mechanism designs	Spherical or cylindrical items may roll through kiln before complete combustion
	Can be operated at temperatures in excess of 2,500°F making them well suited for the destruction of toxic compounds that incinerate poorly and incompletely at lower temperatures	
Liquid Injection	Capable of incinerating a wide range of liquid wastes	Only wastes which can be atomized through a burner nozzle can be incinerated
	No continuous ash removal system is required other than for air pollution control	Burners susceptible to plugging; therefore particle size is a critical parameter for successful operation
	Low maintenance costs	
Fluidized Bed	Combustion design is fairly simple and its maintenance cost is low	Bed diameters and height are limited by design technology
	Comparatively low gas temperatures and excess air requirements minimize the formation of NOx	Difficult to remove residual materials from the bed
		Operating cost are relatively high
	Has a high combustion efficiency	Formation of eutectics is a serious problem
Multiple Hearth	Residence time is higher for low volatile material than in other incinerator configurations	Maintenance costs are high due to the moving parts
	Because of its multizone configuration, fuel efficiency is high and typically improves with the number of hearths used	Susceptible to thermal shock resulting from frequent feed interruptions and excessive amounts of water in the feed
	Able to utilize many fuels including natural gas, reformer gas, propane, butane, oil, coal dust, waste oils, and solvents	Due to the longer residence time, temperature response throughout the incinerator when burners are adjusted is slow
Fixed Hearth	Will handle a wide variety of wastes with different chemical properties	Supplemental fuel must be added for many of the solid wastes that are typically incinerated in these units
	Low maintenance costs because of no moving parts inside the incinerator chamber	A secondary chamber is generally necessary for the required destruction of hazardous waste
	Small size makes these units favorable for on site treatment of small quantities of	These units are not applicable for incineration of large volumes of hazardous waste
Infrared	Solids are not volatilized allowing for lower energy requirements and less particulate carryover into the gas stream	Free liquids can not be incinerated; they must be dewatered before incineration
	Ceramic-fiber insulation provides the ability to heat up the furnace to operating temperature in under 2 hr	Particle size must be in the range between 5 μm and 5 cm in order to fit on conveyor belt
		Mobility of heavy metals can not be reduced, thus further furnace ash processing is required
	Heat generated by the electric elements does not produce additional flue gas, as does the burning of fossil fuel	Belt operation and maintenance

of rotation, the soil residence time can be varied. However, in many instances the residence time is not varied, and temperature is the only control parameter. This makes temperature the prime independent variable that affects contaminant destruction efficiency during operations.

Some incineration systems are not dedicated to contaminant destruction but are used for waste destruction while carrying out another process. The most common example of such a system applied to hazardous wastes is the use of cement kilns. Rotary kilns used for producing cement can accommodate a certain percentage of their fuel as a sludge, drummed waste, or waste organic liquid. Fuel savings are substantial when high-Btu wastes are included in the fuel mixture of a cement kiln. (A special concern with cement kilns is the heavy metals content of the kiln dust that is generated. Some of the metals come from wastes used as fuel, and an even greater portion come from materials used to make the cement, according to I. Kim (*Chemical Engineering*, pp 41-45, April 1994).

Another example of a process thermal system utilizing waste materials is the cofiring of petroleum-contaminated soil with coal in a steam boiler. Boilers fired with lump coal are set up to feed solids onto a grate that provides good contact with combustion air and long residence times at high temperatures. In such an application, the soil feed rate should represent only a small percentage of the coal feed rate. The treated soil comes out of the boiler mixed with the ash. High soil feed percentages could overload the ash-handling system and would cause excessive slag and clinker formation. Inorganic contaminants in the soil, such as heavy metals, may cause a reclassification of the ash from a nonhazardous waste to a hazardous waste. If the soil contains fines, the flue gas particulate cleanup system may become overloaded. Safety protection and workers' breathing zone monitoring are needed when introducing petroleum hydrocarbons into the coal handling equipment. Even with petroleum contamination, the soil will probably have a lower Btu content than the coal. If all of these concerns can be appropriately addressed, cofiring with coal can be an attractive, low-cost option if slightly reduced steam production or boiler efficiency can be tolerated.

A summary of the temperature range, residence time, and some features of various incinerator types (adapted from ENVEST^{TM}, Talisman Partners, Englewood, Colorado, *see* Table 4.30-1) is as follows:

- Rotary kiln, 1,500° to 2,900° F, 20 to 60 min — Rotation promotes good turbulence.
- Circulating bed, 1,380° to 1,470° F, 2 to 3 min — Heat transfer is through a fluidized inert granular bed.
- Liquid injection, 1,300° to 3,000° F, 0.5 to 2 sec — This is for liquids up to 10,000 Saybolt seconds universal (SSU) viscosity.
- Advanced electric, 4,000° to 5,000° F, 5 sec — This is needed for the pyrolysis process and radiant heat transfer.
- Infrared furnace, 1,500° to 1,830° F, 10 to 180 min — This differs from the rotary kiln in the primary unit.

Table 9-2, which is adapted from ENVEST^{TM} (Talisman Partners, Englewood, Colorado) Table 4.30-2 and based on incinerator vendors' information and US EPA documents, summarizes what types of contaminants and wastes can be treated in each incinerator type.

Table 9-2 Effectiveness of incinerators for various contaminants.

Compound Class	Rotary Kiln	Circulating Bed Combustor	Infrared Furnace
Halogenated Volatiles	DE	DE	DE
Halogenated Semivolatiles	DE	DE	DE
Nonhalogenated Volatiles	DE	DE	DE
Nonhalogenated Semivolatiles	DE	DE	DE
PCBs	DE	DE	DE
Pesticides	DE	PE	DE
Organic Cyanides	DE	PE	PE
Organic Corrosives	DE	PE	PE
Volatile Metals	AI	AI	AI
Inorganic Corrosives	PE	PE	PE
Dioxins and Furans	DE	PE	DE
Solids	APP	APP	APP
Liquids	APP	APP	N/A
Sludges	APP	APP	N/A

DE - Demonstrated Effectiveness
PE - Potential Effectiveness
NE - No Effectiveness
APP - Applicable
N/A - Not Applicable
AI - Adverse Impact

For sludges and organic liquids and slurries, their moisture and Btu contents are significant. For soils, moisture content and Btu content generally are lower, and their effect on fuel consumption is not significant at the high temperatures involved. Thermal desorbers run at much lower temperatures than incinerators. Moisture content usually has the greatest effect on fuel consumption, whereas the Btu content of a waste is of lesser importance than is moisture content for desorbing soils.

9.1.2 Low-Temperature Thermal Desorption Basics

Desorption technology competes with incineration as a process for treating soil contaminated with organics. It does not apply to many high-molecular-weight compounds that can be destroyed by incineration. For organic molecules that will volatilize without breaking carbon/carbon bonds by cracking or pyrolysis, desorbers generally operate at less than half the temperature level of incinerators. Hence the term "low-temperature thermal desorption" applies. The materials of construction are lighter and less costly, and the costs per ton of soil treated are significantly less than with incineration. Desorption generally has a better chance of community acceptance than does incineration, because the public often associates incineration with destruction of garbage and associated emissions to the atmosphere.

The two main categories of thermal desorbers are:

- Direct fired
- Indirect fired

With direct-fired units, the soil is in intimate contact with flame from a burner and/or burner exhaust. Most direct-fired units do not include recovery of the contaminants as liquids. The total mix of vaporized water, contaminants, and burner exhaust is filtered and passed through an afterburner. Most direct-fired units are either asphalt aggregate dryers that have been converted to thermal desorption service or rotary kilns designed specifically for desorption.

Soil temperatures generally range from 300° F to the 500° to 700° F range with conventional designs. Higher temperatures are achieved with special designs of indirect-fired units. Because soils in direct-fired units are in contact with hot burner products (without any heat transfer surface), these units can generally achieve higher soil temperatures than indirect-fired units with comparable heating duties.

With indirect-fired units, the soil is in contact with a heat-exchange surface, and the burner exhaust is discharged separately from vaporized soil moisture and contaminants. Electrically heated units are similar to indirect-fired units — the vaporized water and contaminants are not mixed with burner exhaust, because there is no burner. The burner exhaust from indirect-fired units generally can be discharged to the atmosphere without treatment, resulting in a relatively small volume of only vaporized water and contaminants needing treatment. Because indirect-fired systems have heat-exchange surfaces, except for the vapor treatment portion they are heavier, more complex, more costly to transport, and more costly to build than are direct-fired units.

Recovery of hydrocarbons and solvents as liquids is feasible with indirect-fired and electrically heated units. Without dilution from burner exhaust (except for a small volume if used as sweep gas), the concentration of organic vapors may be high enough to make recovery by direct condensation or by adsorption and regeneration-condensation practical.

Most indirect-fired units are either modified rotary kilns or fixed, horizontal cylinders with internal hollow screw augers containing a heat-exchange fluid that is heated in an external fired unit. The indirect-fired rotary kilns are generally of two types:

- The burner flame and exhaust are in an annular space surrounding an inner rotating shell containing the soil, or exhaust from an external burner is passed through the annular space.
- The burner exhaust is passed through an inner cylinder surrounded by the soils; in some designs, the exhaust exiting from the inner cylinder is routed through hollow flights on the inside walls of the main cylinder containing the soil. (Rotary kilns have flights that tumble the soil and help convey the soil toward the outlet).

Electrically heated and indirect-fired units usually use some sweep gas to convey vaporized soil moisture and contaminants out of the unit. This sweep gas can be air or an inert gas, such as nitrogen, or a portion of the burner exhaust. The volumetric flow rate of this sweep gas is small compared to that of the burner exhaust in a direct-fired unit. If inert gas or burner exhaust gases are used, there is a deficiency of oxygen, and the lower explosive limit (LEL) can be safely exceeded in the vapor phase by operating at contaminant vapor concentrations higher than the upper explosive limit.

With direct-fired units, the LEL of the organic vapors in air cannot be safely exceeded. Operations must be controlled at some fraction of the LEL — usually 25%. The LEL is usually greater than 10,000 ppmv for many organic vapors. Gasoline vapor, for example, has an LEL of approximately 13,000 to 14,000 ppmv. LEL values are listed in the National Fire Protection Association (NFPA) Standard 325. Some LEL values from NFPA (1994) are provided in Table 9-3.

Shelton (1995) tabulates the heating values, autoignition temperatures, and LEL values of vapors and gives the equations for estimating heating values (*see* Section 9.6.3) and LEL values for hydrocarbons and for organic vapor mixtures containing hydrogen, carbon, sulfur, oxygen, chlorine, and nitrogen in their molecular structure.

The maximum weight percent of vaporizable organics that can be safely handled in a direct-fired desorber can be readily calculated if the molecular weights and gas flow rate are known. A numerical example will illustrate how to perform this calculation:

Given: LEL values for the contaminants are 10,000 ppmv or higher; the explosive vapor concentration must stay less than 2,500 ppmv (dry basis), which is 25% of the LEL; soil throughput is a maximum of 20 ton/hr; gas flow rate in the primary chamber is 15,000 dry scfm (standard temperature of 60° F); and the average molecular weight of the contaminants is 128 lb/lb mole.

Assumption: More than 99% of the contaminants are vaporized.

Find: Maximum weight percent of organic contaminants that will be allowed in soil fed to a direct-fired absorber

Step 1: Calculate the organic mass flow rate corresponding to 2,500 ppmv in 15,000 scfm, knowing that 1 mole of any gas has a volume of 379 ft^3 at standard temperature and pressure, as shown in Equation 9-1.

Organics flow rate = $(2,500/10^6)$ (15,000 scfm)/(379 scf/mole) (60 min/hr)
(128 lb/mole) = 760 lb/hr (9-1)

Step 2: Divide the organics flow rate by 20 ton/hr soil flow rate and convert to a percentage, as shown in Equation 9-2.

(760 lb/hr)/[(20 ton/hr (2,000 lb/ton)] 100% = 1.9% (9-2)

If the soil has more than this amount of organics contamination, it must be mixed first with uncontaminated soil, or some treated soil must be recycled through the desorber. If the organics contamination is slightly higher than this calculated amount, an alternative allowed by some fire protection officials is to continuously monitor the organics concentration in the flowing gas and shut off the soil feed automatically at

Table 9-3 LELs (Reprinted with permission from NFPA 325. "Fire Hazard Properties of Flammable Liquids, Gases, and Volatile Solids," National Fire Protection Association, Quincy, Massachusetts, 1994. This reprinted material is not the complete and official position of the NFPA on the referenced subject, which is represented only by the standard in its entirety).

Compound	LEL (ppmv)	Compound	LEL (ppmv)
Acetone	25,000	Isohexane	10,000
Aniline	13,000	Isopentane	14,000
Benzene	12,000	Isopropyl Alcohol	20,000
Chlorobenzene	13,000	Kerosene	7,000
Chloroethane	38,000	Jet Fuel JP-4	8,000
m-Cresol	11,000	Methylcyclohexane	12,000
Cumene	9,000	Methylisobutyl Ketone	12,000
Cyclohexane	13,000	Naphtha V.M.&P.	9,000
Decane	8,000	Nicotine	7,000
1,1-Dichloroethane	54,000	Nonane	8,000
1,2-Dichloroethane	62,000	Octane	10,000
1,1-Dichloroethylene	65,000	Pentane	15,000
Dimethyl Sulfoxide	26,000	Pentene (Amylene)	15,000
Ethyl Alcohol	33,000	Phenol	18,000
Ethylbenzene	8,000	Phthallic Anhydride	17,000
Ethylcyclopentane	11,000	Propylene Glycol	15,000
Ethylene Glycol	32,000	Styrene	9,000
Ethyl Ether	19,000	Toluene	11,000
Gasoline	14,000	Trichloroethane	75,000
Gasoline, Aviation Grade 115-145	12,000	Trichlorethylene	80,000
Heptane	10,500	Trimethylpentane	11,000
Hexane	11,000	o-Xylene	9,000
Isoheptane	10,000	p-Xylene	11,000

40% of the LEL. Otherwise, a direct-fired unit designed for a greater gas flow rate or an indirect-fired unit should be selected.

9.1.3 Heat Recovery

In the design of incineration systems, an important feature that affects fuel consumption is the possible use of heat recovery devices. A common method of heat recovery is the use of a heat exchanger that transfers heat from exhaust gas to incoming combustion air. It is also possible to transfer exhaust heat to incoming soil. A popular method of recovering heat is the use of a steam generator (a boiler) to capture some of the exhaust heat. An example that illustrates the design of such a system is given in Section 9.6.

9.2 Incinerators

9.2.1 Rotary Kilns

Figure 9-1 shows a conceptual schematic for a rotary kiln system.

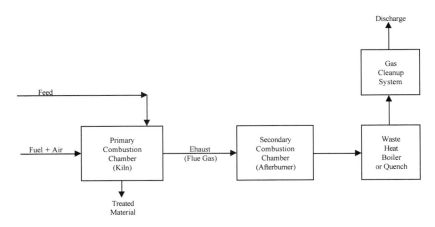

Figure 9-1 Schematic of a rotary kiln system.

Johnson (1994) describes a portable rotary kiln incineration system in detail, including a system equipment schematic and scale drawings that show a complete plan view and an elevation view. Rotary kilns are used for a wide variety of materials — soils, solid wastes, sludges, slurries, containerized wastes, and liquids.

9.2.1.1 Features

A rotary kiln is a refractory rotating cylinder inclined slightly from horizontal. Feed enters the upper end of the incinerator. Rotation aids in mixing and in moving solids through the unit. In some designs, flights attached to the inside shell of the cylinder aid in tumbling the solids with the rotation. This provides the mixing needed for thorough combustion. Combustion heat is provided from burning flammable contaminants and auxiliary fuel. Treated soil exits from the low end of the kiln, and it may be cooled by spraying water on it.

Residence time for the soil may be 20 min or more. Transportable units generally have heating duties in the range of 8 to 100 million Btu/hr. Some fixed-base units exceed this range, with duties up to 250 million Btu/hr.

Exhaust gases pass through an afterburner. The primary combustion process does not completely burn all of the combustible gases that form. Once these substances are in the gas phase, good mixing with combustion air can be readily achieved. Only a fraction of a second is needed for complete combustion (more than 99.99% destruction in some designs) at the high temperatures occurring in the afterburner section. For conservative design, some afterburners operate with 1 to 2 sec of residence time. The afterburner uses auxiliary fuel with additional combustion air. Exhaust gases leaving the afterburner may be treated to remove particulate matter and acid gases.

9.2.1.2 Oxygen Lancing and Oxygen Enrichment

When sludges with high organics content are introduced, there may be momentary oxygen depletion, causing incomplete combustion within the incinerator. Incinerators sometimes have automated waste feed cutoffs that stop the feed flow when excessive organics emissions occur until oxygen concentrations increase.

To minimize such emissions and feed cutoff incidents, the kiln and/or the afterburner can be fitted with lances for injection of oxygen, as described by Fouhy and Ondrey (1994).

Oxygen-fuel burners, as well as lances, may be used to provide oxygen enrichment and a reduction in combustion air rate. A summary of oxygen enrichment in hazardous waste incinerators and cement kilns given in *The Hazardous Waste Consultant* (1995) is based on papers by Prakash Acharya and L. L. Shafer and by E. R. Hansen et al., published in the *1994 International Incineration Conference Proceedings*, University of California, Irvine. Benefits noted from oxygen enrichment as compared to conventional firing include:

- Higher flame temperature will corresponding improved heat transfer, shorter residence time and increased throughput capacity.
- Reduced gas volume (less nitrogen brought in with air) with corresponding less heat input being wasted on heating nitrogen, resulting in reduced capacity needed for emissions abatement equipment, and less particulate carryover from the incinerator. The reduction in particulate carryover results in less solids loading in the abatement equipment and less slagging in the afterburner. Less slagging, in turn, results in reduced downtime and maintenance requirements.
- Less mass emissions of oxides of nitrogen. However, the reduced gas volume results in a higher concentration of oxides of nitrogen (NOx).

An important feature of a case history reported by Acharya (1994) was the use of an oxygen fuel burner designed with a relatively low recirculation rate. This design minimized turbulence, with correspondingly less particulate carryover than what would be experienced with usual recirculation rates. The incineration unit also had an oxygen lance system in the afterburner that maintained stoichiometric combustion air/fuel conditions.

9.2.2 Fluidized CBCs

Because CBCs use a fluidized bed, the soil must be first subjected to double screening and sometimes to grinding before entering the CBC. The particulates are suspended by combustion air. CBCs are most often used for steam boilers, not for soil incineration. For boilers, substances such as waste wood and other forms of biomass, and coal are used for fuel. In such applications, sand is used as a fluidizing medium. With soils incineration, there may be enough sand in the soil so that no sand has to be added.

Residence time for soil particulates in the combustor is a few minutes per pass; and it may exceed an hour for some particles in the circulating loop. The heating duty for a typical transportable unit is approximately 10 million Btu/hr.

As shown in Figure 9-2, CBCs have a somewhat gradual decrease in particle density as solids flow upward through the combustor. This is in contrast to bubbling fluidized beds, which have a distinct dense phase and a dilute phase. Primary combustion air is admitted through a perforated plate or grid; treated soil and ash are removed from the bottom of the bed. Secondary air may be admitted above the plate or grid. The air/flue gas velocity is so high through a CBC that a large fraction of the particulate mass flows up and out of the combustion unit, and it must be returned to the combustor following separation from the combustion air.

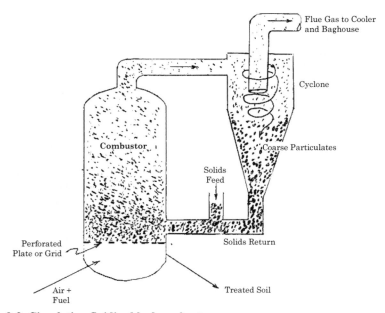

Figure 9-2 Circulating fluidized bed combustor.

Brunner (1991) points out that the air/flue gas velocity in a CBC is 15 to 20 ft/sec, whereas a bubbling fluidized bed is generally designed for 1.5 to 4.5 ft/sec.

The main separation of particulate mass from flue gas takes place in a cyclone separator. The upper part of the cyclone is a vertical cylinder with a tangential inlet.

This arrangement results in a circular, swirling action. As the particles fall downward, centrifugal force keeps almost all but the small particle-size fractions near the wall. The lower part of the wall is conical, which helps maintain the swirling velocity. Flue gas with entrained fine particulates goes out the top through the central exit duct. The fine particulates are subsequently removed from the flue gas with an additional abatement device, such as a Venturi scrubber, an electrostatic precipitator, or a baghouse.

The arrangement in Figure 9-2 shows the cyclone external to the combustor. In some arrangements, the cyclone is integral with the combustor, forming an internal circulating fluidized bed combustor.

Fluidized bed combustors produce lower NOx emissions than do rotary kilns. NOx production is higher at the higher temperatures associated with rotary kilns. CBCs operate at less than 1,800° F, frequently at less than 1,600° F. Rotary kilns generally operate at or greater than 1,800° F and frequently have an afterburner operating at greater than 2,000° F. Fluidized bed combustors are often operated with staged combustion, with some of the air introduced above the bottom air distributor. This arrangement, using secondary air, also reduces NOx emissions.

By mixing limestone with the incoming soil, emissions of acid gases, such as hydrogen chloride and sulfur oxides, and to a very limited extent oxides of nitrogen, can be controlled. Less limestone is needed with a CBC than with a bubbling fluidized bed, because the limestone has a longer residence time in a CBC. The removal efficiency for hydrogen chloride and sulfur oxides can exceed 99% with limestone addition. Solid calcium chloride and calcium sulfate are produced with the treated soil and are not considered hazardous contaminants. Wet scrubbing is not needed when limestone additions are used with fluidized bed combustors.

The good mixing of combustion air with the organic contaminants on the soil particles that is inherent in fluidized bed combustors results in very low carbon monoxide (CO) emissions. In contrast, with rotary kilns, long residence times, high temperatures, and sometimes oxygen lancing are needed to minimize CO emissions.

CBCs can be designed for operating under reducing conditions. An example, for destruction of chlorofluorocarbons, is described in *Chemical Engineering* (April 1995). A test unit combusting wood chips had a reducing atmosphere (because of the presence of methane and hydrogen) that destroyed the chlorofluorocarbons, forming hydrogen chloride and hydrogen fluoride gases. The unit was modified to blow air and chlorofluorocarbons into the bottom of the incinerator. Calcium carbonate was fed to the unit to absorb hydrogen chloride and hydrogen fluoride gases.

The main equipment components included in a fluidized bed combustion system are as follows (adapted from TERA, 1982):

- Soil/limestone handling system
    - Soil-receiving piles or bunkers
    - Limestone-receiving hoppers

- Limestone storage silo
- Conveyors, screens, crushers
- Dust control system — hoods, ducting, cyclone, fan, baghouse
- Bunkers for screened soil and limestone
- Soil/limestone blender
- Soil/limestone feeder
- Fluidized bed combustor
  - Forced-draft fan
  - Hot air generator
  - Fuel system and burners
  - Fluidized bed unit (can incorporate steam boiler tubes)
  - Heat-recovery system or quench cooler, steam drum, economizer, and steam superheater
  - Treated soil removal and cooling system
- Flue gas cleaning
  - Cyclone
  - Baghouse
  - Fine particulate hoppers and conveyors
  - Induced-draft fan and stack

If steam generation is included, additional equipment would be needed, such as chemical water treatment units, feed water heater, deaerator, blowdown system, and pumps.

9.2.3 Infrared Furnace Systems

Infrared heating can be accomplished with either electric resistance elements or radiant metal plates with gas burners behind the plates. Final combustion of gases exiting the primary chamber is accomplished in a gas-fired afterburner — generally at temperatures greater than 2,000° F. The soil can be conveyed through the primary chamber on an alloy belt. With this design, the soil does not contact the walls. Instead of lining the walls with dense refractories, light-weight insulating materials can be used. The savings in weight makes the design of mobile units simpler and less costly.

In some designs, rotary rakes are used to turn the soil as it passes along the belt conveyor through the primary chamber. Systems with belt conveyors are used for solids destruction only, not for watery sludges.

The primary chamber residence time varies widely among various designs and, with a given unit, can be varied easily by changing the conveyor belt speed. Residence times from 10 min to 3 hr have been reported. Transportable units have heating duties in the range of 10 to 30 million Btu/hr.

Hay and McCartney (1991) describe a 10-ton/hr system with electric-resistance heating in a primary chamber with a 10- to 60-min residence time discharging treated soil at 760° C (1,400° F). A number of zone temperatures are controlled as the stainless alloy conveyor belt transports soil. Combustion air is injected through jets along the way. The soil is stirred at six points. The secondary chamber operates at temperatures up to 982° C (1,800° F) with a 2-sec residence time, fired with four burners fueled by propane, natural gas, or fuel oil. Combustion air is fed from a forced-draft blower rated at 32 in. w.c. and excess air is fed from a second forced-draft blower. An induced-draft fan downstream of the exhaust gas treatment system rated at 35 in. w.c. maintains negative pressure in the primary and secondary chambers.

9.3 Thermal Desorbers

The contaminants that each thermal desorber design can handle depend on the temperatures that can be attained and whether direct contact with burner exhaust can be tolerated. Most designs operate in the range of 150° to 370° C (300° to 700° F) soil temperature. With certain designs, temperatures of 538° C (1,000° F) have been attained, and one electrically heated unit can operate at up to 760° C (1,400° F). At these higher temperatures, pyrolysis and combustion of organics may occur, as well as volatilization. If there is an oxygen deficiency, tars and coke can form, impeding heat transfer and conveyance of the soil. Two methods of treating soils with high boiling point compounds are exemplified by these designs:

- One method of achieving vaporization with minimum tar or coke formation is to operate the desorber under vacuum conditions. High boiling point compounds can be volatilized at lower temperatures with vacuum than at atmospheric pressure. The Industrial Recovery Systems International (Charlotte, North Carolina) portable system uses six parallel batch desorption chambers. After a batch of soil is loaded into an open chamber, the lid is sealed, and a vacuum is drawn on the unit while heat is applied. Vapors are treated with a cyclone particulate collector, cooler, and an activated carbon adsorption unit with regeneration-condensation.
- The SoilTech ATP (Porter, Indiana) portable unit operates at atmospheric pressure in an oxygen-deficient retort at 900° to 1,150° F, forming cracked gases and coked soil. The coked soil is subsequently combusted with air at 1,200° to 1,450° F. Vapors and combustion exhaust are treated as separate streams. Vapors from preheated soil are treated with a cyclone particulate collector and condenser. Cracked gases and vapors from the retort section are treated with cyclone particulate collectors, a wet scrubber, fractionator, and condenser. Finally, combustion exhaust is treated with a cyclone particulate collector, fabric filter baghouse, acid gas wet scrubber, and activated carbon adsorption unit.

The most common type of thermal desorber is a rotary kiln. Air flow is either cocurrent with soil movement through the kiln or countercurrent, at the discretion of the designer.

Thermal desorption is most commonly applied to petroleum-contaminated soils and refinery wastes. It is also used for soils contaminated with wood preservatives, paint wastes, and a variety of semivolatile organic compounds, including pesticides, chemicals, polychlorinated biphenyls (PCBs), and dioxins. Higher operating temperatures and longer residence times are needed to desorb PCBs and dioxins than for many other contaminants.

Asphalt aggregate dryers that have been converted to desorber duty outnumber all other desorber types and are essentially rotary kilns that are dedicated primarily to petroleum-contaminated soils. Other types of desorbers — rotary kilns not originally manufactured for asphalt aggregate drying, thermal screws, infrared furnaces — are sometimes used for petroleum-contaminated soils but are also used for hazardous wastes.

Troxler et al. (1992) tabulated 43 case histories of thermal desorption systems used for petroleum-contaminated soil treatment with the soil discharge temperatures given. The percentage contaminant removal is also given for some of the rotary kiln and thermal screw cases. For rotary kilns, the following data are tabulated: three cases for gasoline with more than 99.5% removal at 500° to 580° F; four cases with diesel (No. 2 fuel oil) with more than 99.9% removal at 550° to 625° F; two cases with No. 6 fuel oil with more than 99.4% removal at 725° to 750° F; one case with crude oil with 93.56% removal at 650° F; and another with 98.65% removal at 855° F. For thermal screws, one case is tabulated for gasoline at 400° to 550° F with 96.36% removal; and four cases for diesel (No. 2 fuel oil) with 91.28% to 98.36% removal at 400° to 550° F.

9.4 Handling of Feed and Treated Soils

Moisture content directly impacts operating costs — thermal desorption dries the soil to below 10% moisture and consumes Btus in the process. If methods can be applied at ambient temperature to dewater the feed, they are generally much more economical. Generally, the moisture content must be low enough so that the feed can be handled as a solid. Desorption system vendors may require a minimum solids content, such as 30 wt%.

If the soil has clods, crushing is needed. Finally, screening out of clods and cobbles larger than approximately 2 in. is practiced. Usually, larger cobbles do not need treatment, because contaminants generally are adsorbed on small particles.

The treated soils are quenched with water for cooling and dust control. If a wet scrubber is used as part of off-gas treatment, bleed-off water from the scrubbing system is used for cooling and dust control.

The application of water for dust control must be carefully controlled for fine-grained soils. At greater than a certain moisture content, such soils may become plastic. This condition inhibits screening and conveying in the feed handling system and inhibits mixing and heat transfer in the desorber. Keeping soil in the proper condition is sometimes achieved by recycling some of the desorbed soil and mixing it with untreated soil entering the desorber.

9.5 Air Pollution Control

9.5.1 Use of Afterburners (Thermal Oxidizers)

Products of incomplete combustion (PICs) in incinerator emissions, including oxygenated organics and carbon monoxide, are controlled by combusting with excess air at temperatures of 1,000° to 1,260° C (approximately 1,800° to 2,300° F). The organics destruction efficiency with good mixing at 1,000° C and 2-sec residence time is generally more than 99.99%, whereas at greater than 1,200° C it can exceed 99.9999%. Carbon monoxide can be combusted in such afterburners to carbon dioxide to the extent that carbon monoxide becomes almost nondetectable. Incinerators and direct-fired desorbers emit carbon monoxide. The generation of carbon monoxide in the primary combustion chamber (such as an incineration rotary kiln) is discussed in Section 9.5.5. Troxler et al. (1992) tabulated measured carbon monoxide concentrations from rotary kilns desorbing diesel contamination from soils. One unit without an afterburner emitted 1373 ppmv; two units with afterburners operating at or greater than 1,500° F emitted less than 5 ppmv. An example of afterburner calculations is given in Section 9.6.

When afterburners are used for combusting desorbed vapors, they generally operate at 650° to 1,000° C (approximately 1,200° to 1,800° F) and are designed for at least 0.5-sec residence time. The efficiency of destroying organic vapors at 760° C (1,400° F) is 96% to more than 99%. The destruction efficiency at 1,000° C (1,832° F) with 2-sec residence time can be 99.99%. Significant quantities of fuel, usually natural gas or propane, are consumed in direct thermal afterburners. Fuel consumption can be reduced by heat exchanging the vapors with hot exhaust from the afterburner.

Catalytic oxidizers can be used to destroy desorbed vapors instead of using direct thermal afterburners. Catalytic units typically operate at 315° to 540° C (600° to 1,000° F). Organics destruction efficiency is approximately 95% if the temperature entering the catalyst block is 315° to 370° C (600° to 700° F). Because the operating temperature with catalytic units is relatively low, little or no auxiliary fuel is needed in afterburner service. The usual catalyst is platinum on an alumina substrate. This catalyst can be deactivated or poisoned if chlorides, arsenic, mercury, or phosphorus is present. A special catalyst that is chloride resistant can be used.

Incinerator systems that include afterburners almost universally use the direct thermal type. For treating vapors from thermal desorbers, direct thermal afterburners are used more frequently than catalytic units. Generally, direct thermal afterburners have a more reliable, higher organic destruction efficiency; however, they produce much more oxides of nitrogen than do catalytic units.

9.5.2 Recovery of Organic Fluids from Indirect-Fired Desorbers

Methods other than using afterburning for controlling organics emissions from desorbers include activated carbon adsorption, condensation, and combinations of these methods. Carbon adsorption is often cost-effective with indirect-fired desorbers. The vapors from the thermal desorber are filtered, usually via a baghouse, and scrubbed or partially condensed before entering the carbon adsorption unit. Sometimes refrigerated condensers are used. Condensing can be an energy-intensive process. In order to minimize the duty of a condenser, some designs include a water quench upstream of the condenser. Directly spraying water into the vapors reduces their temperature because the quench water removes heat from the flue gas when the water evaporates. Any type of condensing system that causes liquid water to form will result in 100% relative humidity. In that event, heating the uncondensed moist vapor approximately 13° C (23° F) will provide optimum relative humidity for carbon adsorption. The used carbon can be regenerated on-site for recovery of the organics or shipped off-site for reactivation. One problem with on-site regeneration is that the higher boiling point organic compounds that vaporize with some thermal desorbers do not readily desorb from carbon.

High boiling organics emissions from an indirect-fired desorber are best recovered directly by using condensation.

9.5.3 Abatement of Particulate Emissions and Acid Gases

The exhaust from a thermal treatment step passes through a series of emission abatement devices. A comparison of various abatement devices as published by the U.S. EPA's predecessor agency in Publication AP-51 (January 1969) is given in Table 9-4.

Sequencing of these devices within a gas treatment train is subject to a variety of choices. Quench chambers are sometimes used in series with wet scrubbers. High-energy wet scrubbers, or fabric filter baghouses following dry cyclone collectors, are commonly used for particulate collection. With incinerators and direct-fired desorbers, the position of a baghouse in the treatment train depends on gas temperature. Low-energy acid-gas scrubbers often follow baghouses.

Fine particulates are more difficult to abate than coarse ones. Particulates smaller than 10 μm, especially the fraction smaller than 2.5 μm, are a particular health concern that is the basis of recent air pollution control regulations. Abatement devices that can capture particulates ranging in size from submicron to 10 μm are of prime interest. Brunner (1991) gives the following average collection efficiencies for 5-μm diameter particulates:

- High-efficiency cyclone, 73%
- Low-energy scrubber, 83% to 94%
- Medium-energy Venturi scrubber, 98% to 99%
- High-energy Venturi scrubber, more than 99%
- High-efficiency precipitator, more than 99%
- Fabric filter, more than 99%

The most efficient control of particulate emissions, including fine particulates, can be achieved with a baghouse. Baghouses can be operated with woven glass fabric filter bags at temperatures up to 260° to 290° C (500° to 550° F). Woven nylon can be used up to 230° C (450° F), and other synthetic fibers, up to 150° to 175° C (the low 300° F). Ceramic fibers have been developed in recent years that are fabricated into bag filters or cartridge filters that are rated up to 538° C (1,000° F).

A number of compartments, each containing multiple filter bags, are built into the baghouse. At any one time, one compartment is taken out of service with automatic valves for bag cleaning (filter cake removal). Baghouse designs that use shakers for

Table 9-4 Advantages and disadvantages of collection devices.

Collector	Advantages	Disadvantages
Gravitational	Low pressure loss, simplicity of design and maintenance	Much space required. Low collection efficiency
		Much head room required
Cyclone	Simplicity of design and maintenance	Low collection efficiency of small particles
	Little floor space required	Sensitive to variable dust loadings and flow rates
	Dry continuous disposal of collected dusts	
	Low to moderate pressure loss	
	Handles large particles	
	Handles high dust loadings	
	Temperature independent	
Wet collectors	Simultaneous gas absorption and particle removal	Corrosion, erosion problems
		Added cost of wastewater treatment and reclamation
	Ability to cool and clean high-temperature, moisture-laden gases	Low efficiency, on submicron particles
		Contamination of eluent stream by liquid entrainment
	Corrosive gases and mists can be recovered and neutralized	Freezing problems in cold weather
	Reduced dust explosion risk	Reduction in buoyancy and plume rise
	Efficiency can be varied	Water vapor contributes to visible plume under some atmospheric conditions
Electrostatic Precipitator	99+ percent efficiency obtainable	Relatively high initial cost
	Very small particles can be collected	Precipitators are sensitive to variable dust loadings or flow rates
	Particles may be collected wet or dry	
	Pressure drops and power requirements are small compared to other high efficiency collectors	Resistivity causes some material to be economically uncollectable
	Maintenance is nominal unless corrosive or adhesive materials are handled	Precautions are required to safeguard personnel from high voltage
	Few moving parts	Collection efficiencies can deteriorate gradually and imperceptibly
	Can be operated at high temperatures (550° to 850°F)	
Fabric Filtration	Dry collection possible	Sensitivity to filtering velocity
	Decrease of performance is noticeable	High-temperature gases must be cooled to 200° F to 550° F
	Collection of small particles possible	
	High efficiencies possible	Affected by relative humidity (condensation)
		Susceptibility of fabric to chemical attack
Afterburner, Direct Flame	High removal efficiency of submicron odor-causing particulate matter	High operational cost
		Fire hazard.
	Simultaneous disposal of combustible gaseous and particulate matter	Removes only combustibles
	Direct disposal of non-toxic gases and wastes to the atmosphere after combustion	
	Possible heat recovery	
	Relatively small space requirement	
	Simple construction	
	Low maintenance	
Afterburner, Catalytic	Same as direct flame afterburner	High initial cost
	Compared to direct flame: reduced fuel requirements, reduced temperature, insulation requirements, and fire hazard	Catalysts subject to poisoning
		Catalysts require reactivation

filter cake removal should not be used with fiberglass bags, as shakers would cause excessive wear of bag cloth. With woven fiberglass cloth, bag cleaning is accomplished by reversing the air flow through the bags.

Some baghouses historically have used felt cloth instead of woven fabrics, especially when flue gas temperatures are less than 175° C (350° F). Bag cleaning is accomplished with a short pulse of compressed air periodically admitted to a row of bags. Recent technological advances in the design of such pulse-jet baghouses include use of bags made of a woven composite of fiberglass and fluoropolymer good to 260° C (500° F).

Because of fabric temperature limitations, flue gas temperature reduction is needed when particulate material is removed using a baghouse. Use of a flue gas heat exchanger for preheating combustion air not only saves fuel but helps reduce flue gas temperature. Or, as shown in Figure 9-1, the flue gas temperature can be reduced by generating steam in a waste heat boiler.

To help ensure that the baghouse is never overheated, the baghouse inlet temperature can be automatically controlled with a water cooler or admitting dilution air. However, any dilution air that is admitted decreases the potential for heat recovery and increases the volume of gases that must be handled.

If the exhaust gas in a desorber flows countercurrent to soil, it is cooler than with cocurrent designs. This exhaust may not have to be cooled before entering a baghouse. The baghouse may or may not be preceded by a cyclone collector to remove coarse particulates. With cocurrent desorber designs, the exhaust gas from the desorber usually goes through a cyclone, afterburner, gas cooler, baghouse, and blower, in that order.

Acid gases mainly include hydrogen chloride, hydrogen fluoride, oxides of sulfur, and oxides of nitrogen (NOx). Control of NOx is discussed in the next subsection. If acid gases such as hydrogen chloride or sulfur oxides are in the exhaust, a low-energy wet scrubber is used in series after a baghouse. Sometimes caustic soda or other alkaline agents are added to the scrubber water to enhance capture of acid gases.

For controlling hydrogen chloride emissions, just enough caustic soda is added to keep the pH at 7 or slightly below, as described by Anderson (1993). (Excessive elevation of pH might cause calcium carbonate to precipitate, resulting in plugging caused by scale formation.) When calcium carbonate is not a potential problem, Leite (1996a) suggests that up to 60% excess caustic soda be used for once-through scrubbers and 5% to 30% excess be used for scrubbers with solution recirculation.

The chemical reactions between the acid gases and the caustic soda are:

$$HCl + NaOH \rightarrow NaCl + H_2O \tag{9-3}$$

$$HF + NaOH \rightarrow NaF + H_2O \tag{9-4}$$

Based on molecular weights, each 36.5 lb of HCl or 20 lb of HF present requires 40 lb of caustic soda addition. An example of a packed wet scrubber design for control of hydrochloric acid is given at the end of Section 9.6.

Most of the scrubbing solution is recirculated through the scrubber. A small fraction is removed as blowdown, which is pumped to a wastewater treatment system. De Gesaro, Teringo, and McIlhenny (1993) give details of using magnesium hydroxide and caustic soda for treating incineration scrubber blowdown containing acids and dissolved heavy metal ions.

Lime is frequently used for scrubbing sulfur oxides or mixtures of HCl and sulfur oxides. The sulfur oxides that form when sulfur-containing compounds are incinerated are sulfur dioxide and sulfur trioxide. Lime (calcium oxide) is first slaked with water:

$$CaO + H_2O \rightarrow Ca(OH)_2 \qquad (9\text{-}5)$$

The main reactions in the scrubber are:

$$CO_2 + Ca(OH)_2 \rightarrow CaCO_3 + H_2O \qquad (9\text{-}6)$$
$$2\ HCl + Ca(OH)_2 \rightarrow CaCl_2 + 2\ H_2O \qquad (9\text{-}7)$$
$$SO_3 + Ca(OH)_2 \rightarrow CaSO_4 + H_2O \qquad (9\text{-}8)$$
$$SO_2 + Ca(OH)_2 \rightarrow CaSO_3 + H_2O \qquad (9\text{-}9)$$

Some of the calcium sulfite is converted to calcium sulfate:

$$CaSO_3 + 0.5\ O_2 \rightarrow CaSO_4 \qquad (9\text{-}10)$$

The calcium sulfate that forms precipitates as a hydrate, gypsum. Based on molecular weights, each 36.5 lb of HCl requires 28 lb of dry lime. Each 64 lb of sulfur dioxide requires 56 lb of dry lime. Limestone, $CaCO_3$, slurried with water, is sometimes used instead of lime for scrubbing sulfur oxides and is a cheaper reagent. Sometimes a higher sulfur oxides removal efficiency can be achieved by using a mixture of caustic and limestone slurry.

Rinaldi (1995) compared lime scrubbing systems to limestone systems. Although operating expenses are lower for scrubbing with limestone, overall capital and operating costs are higher, considering the need for costly reagent feed components, fly ash abatement/handling/disposal equipment, and operating costs.

Low- and medium-energy wet scrubbers are also used to control certain organic compounds and volatile metals. Low-energy wet scrubbers have low efficiencies for capturing fine particulates. They are used in conjunction with other devices that remove fine particulates. Efficiencies for various wet scrubbers and certain other abatement devices are reprinted in Table 9-5 from Niessen (1978) by courtesy of Marcel Dekker, Inc.

Scrubbers designed solely for acid gas abatement are usually relatively low-pressure-drop devices, such as plate towers or packed towers. Scrubbers designed for particulate control are high-energy devices using either high-pressure water or high gas pressure drop. The most common high-energy scrubber is the Venturi type. (Although high-energy Venturi scrubbers are efficient for particulate removal, they are not efficient enough to remove 99% of HCl or HF, as typically required by permit conditions (Leite, 1996a)). As the gases pass through the narrow throat of the

Table 9-5 Efficiency of abatement devices. (From Niessen, 1978, courtesy of Marcel Dekker, Inc.).

Abatement Devices	NOx	HCl	PAHs	Volatile Metals
Spray chamber	25%	40%	40%	
Wetted-wall chamber	25%	40%	40%	
Wetted, closely-spaced baffles	30%	50%	85%	
Medium-energy wet scrubber	65%	95%	95%	80%
Electrostatic precipitator			60%	90%
Fabric filter			67%	99%

Venturi at high velocities, water directed into the throat is sheared into fine droplets that provide intimate contact with the particulate matter in the gases. The higher the velocity, the higher is the pressure drop through the Venturi throat, and the smaller is the particle size that can be captured. Anderson (1993) reports that gas velocities in a Venturi throat are 200 to 600 ft/sec, with 2 to 12 gal of water/1,000 scfm of gas, and that high-efficiency Venturi scrubbers have a pressure drop of 10 to 30 in. w.c. Even higher pressure drops would be needed to attain high efficiency in the collection of submicron particles. Ollero (1984) gives equations and describes a calculator program for determining Venturi scrubber efficiencies and pressure drops.

In some Venturi designs, movable vanes form a rectangular throat; other designs have a movable, tapered disc in a circular throat. Thus, the velocity, and corresponding pressure drop through the throat, can be controlled to be constant if there are gas flow rate fluctuations. Vertical downflow Venturi scrubbers must have an elbow in the ducting, which is kept flooded with scrubber water to retard erosion of the elbow.

Gases exiting scrubbers entrain water mist that must be eliminated. Plate towers and packed towers usually include a mist elimination section near the top gas exit, above the water inlet. Gases exiting Venturi scrubbers usually are directed tangentially into a large vertical vessel that eliminates mist entrainment by slowing the gas velocity to less than 10 ft/sec. Scrubbing solution droplets entrained in the flue gas are coalesced and dropped out in the mist separator vessel. Final mist elimination is done with baffles in the upper part of the vessel, where droplets impinge on metal or plastic baffle surfaces. The droplets coalesce into large drops that readily separate from the gas stream by gravity.

If the mist separator vessel is not of sufficient size, drops of scrubbing solution (containing captured particulate matter) may fall out on the area around the vessel. The system would not pass stringent air pollution control limitations on particulate emissions under these conditions and must be modified to eliminate the mist emission problem.

Hay and McCartney (1991) describe an infrared furnace system that uses a horizontal gas flow Venturi scrubber followed by chevron-shaped baffles that serve as mist eliminators. The Venturi removes coarse and fine particulate matter and some of the acid gases such as HCl. A final low-energy alkaline scrubbing step to remove remaining HCl is accomplished with a cross-flow packed scrubber. The gas flows horizontally through packing, and the alkaline solution is sprayed vertically downward through the packing. Another set of baffles performs final mist elimination.

A common dry scrubber system that removes both fine particulates and acid gases includes passing the flue gas through a spray drying vessel and then through a baghouse. A spray of limestone slurry (or other alkaline agents) is injected into the drying vessel. The water in the slurry rapidly evaporates, leaving an alkaline dry powder that is in intimate contact with the flue gas and that neutralizes the acid gases. The baghouse then filters out the powder, as well as fine soil particulate matter carried over with the flue gas.

For hot exhaust gases, a wet scrubber is often preceded by quenching with direct injection of water through spray nozzles. Evaporation of the water spray droplets provides cooling to temperatures less than 212° F. Calculations for designing quench systems are given in Section 9.6.

Dry cyclones are low-energy collectors that capture coarse particulates. Low-energy scrubbers capture coarse particulates and some fine particulates. High-energy scrubbers capture very fine particulates (e.g., down to or below 1 μm in average diameter with a Venturi operating at a pressure drop of 50 in. w.c.), as well as coarser particulates. Baghouses capture very high percentages of all particulates, including submicron sizes. Trends in emissions control regulations in recent years have been toward more stringent limits on emissions of particulates smaller than 10 μm and on emissions of certain metals. (Some regulations limit emissions smaller than 2.5 μm.) Many metal contaminants stay adsorbed on the finer particulates. Consequently, cyclones are not usually used alone; they are best used ahead of a device designed to remove fine particulates. Some metals may vaporize in the desorber, and water-quench chambers and high-energy scrubbers may be needed to condense and control these metals.

A wide variety of gas treatment devices and their arrangements are used. Combinations including cyclones, baghouses, and afterburners are common with direct-fired desorbers. Desorbers with indirect heating sometimes use a quench chamber, wet scrubber, mist eliminator, condenser, and an activated carbon adsorber.

The use of scrubbers and condensers with thermal desorbers leads to production of mixed organic liquids and water. Oil/water separators are sometimes used for handling scrubber water blowdown, but the separation of phases is not always simple. Sludges and froths form, and several stages of treatment are sometimes needed to separate solids, break emulsions, and produce separate liquid phases.

The liquid products from condensers can usually be phase-separated by decanting; the water is subjected to aqueous-phase activated carbon treatment or biological treatment.

Many systems use water withdrawn from a scrubber system for cooling treated soil and for dust control, so that zero water discharge is attained. Typically, water is mixed with treated soil in a continuous pug mill or is sprayed on the soil at the exit conveyor.

Another type of device for collecting fine particulate matter is the electrostatic precipitator. Precipitators contain vertical electrically grounded metal plates with vertical electrically charged wires spaced between the plates. Flue gas is passed between the plates. Fine particulate matter, which develops an electrostatic charge from the charging on the corona wires, migrates to and deposits on the plates. Periodically, the plates are automatically rapped or shaken to release the collected fines, which drop into a hopper. Only a very small percentage of contaminated soils and hazardous waste incinerators use precipitators or ionizing wet scrubbers for particulate removal.

9.5.4 Emissions of NOx

Unless the burners are designed or arranged for staged combustion or unless the flue gas is recycled, the formation of NOx at a given temperature is approximately proportional to the percent of excess air. Formation of NOx is highly dependent on temperature and increases rapidly at incineration temperatures greater than $1,042°$ C ($1,700°$ F). NOx is formed from both nitrogen in the combustion air and nitrogen in the fuel, so fuel nitrogen content is a factor. The amount of nitrous oxide and nitrogen dioxide formed per mole of stoichiometric oxygen or per million Btu of heat release is shown as a function of excess air and incineration temperature in ASME (1974).

Haroutunian (1995) lists a number of firms that offer computerized simulations of burner dynamics and combustion chemistry. Some of these simulations will more accurately predict the amount of NOx that forms than will ASME graphs.

Control of NOx emissions can be achieved to some extent with combustion techniques, wet scrubbing, and ammonia reduction. The combustion techniques used include temperature reduction, excess air reduction, use of oxygen instead of atmospheric combustion air, flue gas recycle, and staged combustion. The staging of combustion achieves a relatively low temperature in the first stage without excess air, thereby minimizing incipient NOx formation. The final stage uses excess air to complete combustion of the fuel. Staged combustion can be achieved within the burners (so-called low-NOx burners) or by using stoichiometric combustion in the primary section of an incineration chamber and excess air in a subsequent section.

Because scrubbing does not achieve high efficiencies for removal of oxides of nitrogen, ammonia reduction is sometimes applied. Brunner (1991) indicates that more than 80% NOx conversion to nitrogen gas can be achieved in the presence of platinum, palladium, vanadium oxide, or titanium oxide catalysts. The process is termed selective catalytic reduction and proceeds with these reactions:

$$4\ NH_3 + 4\ NO + O_2 \rightarrow 4\ N_2 + 6\ H_2O \tag{9-11}$$

$$4\ NH_3 + 2\ NO_2 \rightarrow 3\ N_2 + 6\ H_2O \tag{9-12}$$

Ammonia is injected upstream of the catalyst bed at temperatures between 500° and 800° F. The reactions will proceed in the absence of catalyst at a much higher, narrower temperature range in a process termed thermal deNOx. Not only is the narrow thermal deNOx temperature range difficult to control, but the efficiency is reduced, and ammonia consumption is higher if HCl or sulfur oxides are present.

More information on selective catalytic reduction is given in *Environmental Engineering World* (January-February 1995) and in *Environmental Engineering* (a supplement to *Chemical Engineering*, November 1994).

9.5.5 CO Emissions

CO forms from incomplete combustion of organic contaminants and of auxiliary hydrocarbon fuels. CO formation generally decreases with the amount of excess air used and the degree of turbulence. If contaminated soils are not thoroughly mixed with combustion air, carbon monoxide emissions can be excessive. The amount of carbon monoxide formed per pound of stoichiometric air from combusting hydrocarbon fuels can be derived from data given in ASME (1974) relating excess air, incineration temperature, and fuel hydrogen/carbon ratio.

For example, determine the amount of carbon monoxide generated per mole of fuel for a system operating at 1,000° C with propane fuel and 50% excess air. The procedure consists of interpolating in the ASME table to find the carbon monoxide generated per pound of stoichiometric air and multiplying that amount by the stoichiometric air rate, determined from the combustion reaction.

The ASME (1974) table gives the carbon monoxide generation rates at 1,000° C (1,832° F) with 50% excess air as follows (lb CO/lb stoichiometric air):

5.896×10^{-8} for fuel H/C ratio of 1.33/1 \hfill (9-13)
3.997×10^{-8} for fuel H/C ratio of 4/1 \hfill (9-14)

Propane (C_3H_8) has a hydrogen/carbon ratio of 8/3, or 2.67/1. Interpolation gives a carbon monoxide generation factor equal to 4.95×10^{-8}. Considering that the molar (or volumetric) ratio of nitrogen/oxygen in air (used for combustion) is 79 to 21, the combustion reaction for propane is:

$$C_3H_8 + 5\ O_2 + 5\ (79/21)\ N_2 \rightarrow 3\ CO_2 + 4\ H_2O + 18.8\ N_2 \tag{9-15}$$

Each mole of propane uses $(5 + 18.8)$ moles of air for stoichiometric combustion, or (23.8×28.9) lb of air, considering the molecular weight of air. The amount of carbon monoxide generated per mole of propane is calculated, as shown in Equation 9-16.

$$4.95 \times 10^{-8}\ (23.8)\ (28.9) = 3.40 \times 10^{-5}\ \text{lb} \tag{9-16}$$

The concentration of carbon monoxide in the discharge, on a dry basis, is calculated from the combustion reaction considered with excess air. At 50% excess air (1.5 times stoichiometric air), the moles of oxygen per mole of propane is 1.5 x 5. The combustion reaction is:

$$C_3H_8 + 7.5\ O_2 + 7.5\ (79/21)\ N_2 \rightarrow 3\ CO_2 + 4\ H_2O + 28.2\ N_2 \qquad (9\text{-}17)$$

From this equation, a total of 35.2 moles of flue gas are formed from each mole of propane combusted. Included in the flue gas are 4 moles of water, so on a dry basis 31.2 moles of flue gas are formed per mole of propane. The molecular weight of carbon monoxide is 28, so the moles of carbon monoxide generated per mole of propane is calculated, as shown in Equation 9-18.

$$3.40 \times 10^{-5}\ \text{lb}\ (1\ \text{mole CO}/28\ \text{lb}) = 1.2 \times 10^{-6}\ \text{mole CO} \qquad (9\text{-}18)$$

Because volume is proportional to moles, the parts per million by volume for 31.2 moles of flue gas is as shown in Equation 9-19.

$$10^6\ [(1.2 \times 10^{-6}/(31.2)] = 0.038\ \text{ppmv} \qquad (9\text{-}19)$$

Haroutunian (1995) indicates that computerized burner simulations can predict CO produced by burners. However, when combusting solids CO is produced anywhere in the kiln where good mixing is not taking place. CO emissions are best quantified by measuring them during a trial burn of a sample of the soil or sludge. If the emissions are excessive and mixing cannot be improved, then kiln or afterburner residence time and/or temperature must be increased.

9.6 Main System Design Parameters for Thermal Treatment

Consideration of soil or sludge physical properties that affect feed handling and determination of optimum desorption conditions from treatability studies are discussed in Section 9.7.

At a given heat input, the maximum soil temperature attained will vary approximately inversely with the throughput rate. The residence time, in turn, is inversely proportional to the throughput rate. With screw auger desorption units, the residence time is adjusted by varying the rotational speed of the augers. In rotary kiln units, the residence time is determined both from the degree of inclination (usually fixed by the manufacturer) and the speed of rotation (variable). Typical residence times in direct-fired desorption units are of the order of several minutes, whereas they are a half hour or more in indirect-fired units. Anderson (1993) gives the rotational speed of desorption kilns as typically 0.25 to 10 rpm, with a maximum heat input as 25,000 Btu/hr/ft^3 of internal kiln volume.
Section 9.6.2 discusses vapor pressure effects and the role of vaporized water (steam) in thermal desorbers.

9.6.1 Characterization of the Waste for Thermal Treatment

Characterization is the process of sampling and analyzing the material to determine its physical and chemical properties. Certain analyses are needed to meet regulatory requirements — for example, to determine if the material is a RCRA or Toxic Substances Control Act (TSCA) waste and how to classify it if RCRA applies. Some of the analyses also help determine combustion characteristics and emissions control requirements. Combustion characteristics have a bearing on incinerator fuel consumption. In addition, the presence of certain substances, such as chlorides, fluorides, or sulfur, may determine that scrubbing of the emissions is needed.

The heating value (heat of combustion) of the material, measured in Btu/lb, directly impacts the fuel requirements. Garvin (1998) gives equations and data for determining the heat of combustion for organic compounds, including compounds containing oxygen, nitrogen, sulfur, fluorine, chlorine, bromine, iodine, and mixtures of such compounds. An oily waste with a heating value greater than 5,000 Btu/lb may require only a pilot flame for ignition and no auxiliary fuel. If the presence of a high oil content or carbonaceous material is suspected, a proximate analysis, as well as an ultimate analysis, is useful. A proximate analysis determines the amounts of volatiles, fixed carbon, ash, and moisture that are present in the waste material. An ultimate analysis includes the determination of carbon, halogen, hydrogen, metal, nitrogen, oxygen, phosphorus, and sulfur content in the waste material.

Halogens, especially chloride and fluoride, sulfur, and to some extent nitrogen, form acid gases. Metals that are somewhat volatile, such as mercury, can affect air emissions. Other metals that are toxic will concentrate in the fly ash, or in the solid portion of a sludge fed to the incinerator, often resulting in further treatment by fixation being required for collected fly ash or for treated solids. Large amounts of sodium can cause defluidization of the bed in a fluidized CBC. Knowing the ash content of the waste helps predict whether slag formation will be a major problem.

Other tests such as instability (or reactivity), toxicity, and corrosivity may be in order. Instability can affect how the material can be safely handled, especially if it is pyrophoric, can spontaneously ignite, chemically breaks down, is an oxidizer (e.g., nitrates), is sensitive to shock, or is thermally unstable.

If VOCs are present, special air monitoring for feed handling and protection of workers' breathing may be needed, and precautions should be taken against uncontrolled ignition.

The following is a list of analyses to be performed to characterize wastes for thermal treatment:

- Percentages of carbon, hydrogen, oxygen, nitrogen, and sulfur — These are used to calculate combustion stoichiometry and air requirements and to estimate combustion gas flow and main composition.
- Halogens — Along with sulfur, these are used to estimate the formation of acid gases and the extent of acid gas treatment required.

- Salts (especially alkali earth metals) — The cause of concern is relative to slagging and to defluidization of fluidized bed reactors.
- Moisture — Moisture is used to determine the need for auxiliary fuel and/or the need for drying of waste prior to incineration.
- Trace metals — These are of concern because of the toxicity of metal oxides generated during combustion.
- Ash — Ash is of concern because of excess slag and particulate emissions that could be generated. Ash also reduces the effective energy content and heating value of the waste; this is especially of concern for incineration of combustible sludges. When incinerating soils, ash is the main constituent.
- Heating value — Heating value is used to determine the need for auxiliary fuel and the economics of incineration.
- Special characteristics (hazardous nature of the waste, i.e., toxicity, mutagenicity, carcinogenicity, corrosiveness, pyrophoric properties, thermal instability, shock sensitivity, chemical instability) — These characteristics are used to determine alternatives, if any, for safe disposal of the waste. Wastes with certain special characteristics may have few economic options, other than incineration, for their proper disposal.

9.6.2 Vapor Pressure Considerations for Thermal Desorbers

An initial approach to estimating a suitable operating temperature for thermal desorbers involves a determination of the vapor pressure of the organic contaminant with the highest boiling point present in the soil. The boiling point of a pure compound is the temperature at which the vapor pressure above the liquid is 1 atmosphere. Vapor pressures at various temperatures for many organic compounds are published in chemical handbooks and other technical literature. Troxler et al. (1992) use the Antoine equation for estimating vapor pressures, as shown in Equation 9-20.

$$\ln(\text{vapor pressure}) = A - B/(T+C) \tag{9-20}$$

in which A, B, and C are constants published in the literature for a number of compounds, and T is absolute temperature. If vapor pressures are tabulated at two temperatures other than the boiling point, this equation can be used to evaluate the constants so that the vapor pressures at other temperatures can be predicted. For nonvacuum operations, the ideal vapor pressure is near 1 atm at the exiting soil temperature. In practice, vapor pressures ranging down to 0.5 atm may be acceptable. Components present that have a vapor pressure higher than 1 atmosphere will vaporize faster than heavier components. Compounds with much higher vapor pressures, such as greater than 2 atm, can be readily volatilized by lower-cost methods than thermal desorption, such as venting with air ex situ or soil vapor extraction in situ without heating.

Moisture present in the soil results in more fuel being needed to attain a target soil exit temperature. However, the moisture becomes steam, which has an effect similar to operating under a vacuum, sometimes called a steam stripping effect. The target temperature can be lowered accordingly. For a unit with cocurrent flow of exhaust gases and soil, instead of a target temperature corresponding to a vapor pressure of 0.5 to 2 atm, it should correspond to a vapor pressure ranging from 0.5 to 2 atm multiplied by the factor given in Equation 9-21.

Factor = 1 - volume fraction of steam (9-21)

in which the volume fraction of steam is that anticipated to be in the exhaust.

An example showing how this relationship can be applied to a cocurrent desorber is as follows:

Given: Soil moisture content is 25 wt% and organic contaminants are 1.9%; organic contaminant with lowest vapor pressure is naphthalene; and burner fuel is propane, consumed at the rate of 1,250,000 Btu/ton of soil treated, with 50% excess air. Note that this heat input can be approximated as the sum of the following heating duties, using average soil heat capacity and heats of vaporization from handbooks, or in this example given by Troxler et al. (1992):

- Soil heating — assume a heat capacity of 0.25 Btu/(lb-°F)
- Organics vaporization — 380 Btu/lb
- Water vaporization — 1,057 Btu/lb
- Gases sensible heat increase — See Section 9.6.3 and the following equations for propane combustion, considering that the molal (or volumetric) ratio of nitrogen/oxygen in air (used for combustion) is 79/21:

Reaction A, stoichiometric:

$$C_3H_8 + 5\ O_2 + 5\ (79/21)\ N_2 \rightarrow 3\ CO_2 + 4\ H_2O + 18.8\ N_2 \quad (9\text{-}22)$$

Reaction B, 50% excess air (note that 2.5 moles of oxygen are excess and are not consumed):

$$C_3H_8 + 7.5\ O_2 + 7.5\ (79/21)\ N_2 \rightarrow 3\ CO_2 + 4\ H_2O + 2.5\ O_2 + 28.2\ N_2 \quad (9\text{-}23)$$

- Heat loss caused by radiation from the outside shell of the desorber — assume 1% of the total heating duty, or use the heat loss factors given by Brunner (1991)

Assumptions: Desirable soil exit temperature corresponds to a vapor pressure near 1 atm — say in this example, 1.2 atm; virtually all moisture evaporates to form steam before the soil exits (this implies good mixing and adequate residence time). For a conservative approach, assume that the ambient air used for combustion is dry. The actual air moisture content usually does not exceed 2 wt% of the air and depends on relative humidity and ambient temperature.

Find: Reduced temperature corresponding to corrected vapor pressure, 1.2 atm times the factor calculated from Equation 9-23

Step 1: Determine volume fraction of steam in flowing gas. Sources of water vapor (steam) are:

- Water vapor naturally in combustion air — Water vapor in the air is zero, as given in the assumptions.
- Moisture vaporized from soil — Steam derived from each ton (2,000 lb) of soil with 25% moisture is as given in Equation 9-24.

$$0.25 \ (2,000 \ lb) \ (379 \ scf/mole)/(18 \ lb/mole) = 10,530 \ scf \quad (9-24)$$

- Steam formed as a product of combustion of fuel — The water produced by combustion is 4 moles per mole of propane burned, as can be seen from Reaction B. Propane has a heating value of approximately 21,000 Btu/lb and a molecular weight of 44. At the given propane combustion rate of 1,250,000 Btu/ton of soil, the propane combusted is calculated as shown in Equations 9-25 and 9-26.

$$1,250,000/(21,000 \ Btu/lb) \ (1 \ mole/44 \ lb) = 1.35 \ moles \quad (9-25)$$
$$\text{Steam from combustion} = 4 \ (1.35 \ moles) \ (379 \ scf/mole) = 2,050 \ scf \quad (9-26)$$

From Reaction B, the total moles of gas produced from combustion is (3 + 4 + 2.5 + 28.2), or 37.7 moles/mole of propane. The total volume of gas in the desorber, from Reaction B and Equations 9-25 and 9-26, is as given in Equation 9-27.

$$(37.7) \ (1.35 \ mole) \ (379 \ scf/mole) + 2,050 \ scf = 21,340 \ scf \quad (9-27)$$

The volumetric fraction of steam is the steam from the soil (*see* Equation 9-23) plus the steam from combustion (*see* Equation 9-26), all divided by the total gas volume (*see* Equation 9-27), as given in Equation 9-12.

$$(10,530 + 2050)/21,340 = 0.59 \quad (9-28)$$

Step 2: Evaluate naphthalene vapor pressure at various temperatures directly from the literature or from Antoine constants (either published or derived from minimal vapor pressure data), and apply Equation 9-20 or use Figure 7 from Troxler et al. (1992). For this example, the vapor pressure data given by Troxler et al. will be used. Applying the volume fraction of steam given by Equation 9-28 to Equation 9-23, the vapor pressure correction factor is as given in Equation 9-29.

$$1 - 0.59 = 0.41 \quad (9-29)$$

Multiplying the assumed originally desired vapor pressure of 1.2 atm by this factor results in a corrected vapor pressure of 0.49 atm. The corresponding temperature is 370° F for the soil exiting from a desorber with hot gas flowing in the same direction as the movement of soil (cocurrent flow).

Some conclusions can be drawn from the example above. Without the steam correction being applied to the cocurrent desorber, a vapor pressure of 1.2 atm for naphthlalene would correspond to a temperature of 460° F for the soil exiting the desorber. A unit with hot gas flowing countercurrent to the soil would be operated accordingly. Also, with countercurrent units, the gas exits the primary chamber in contact with cold incoming soil, so the gas is colder entering the afterburner chamber. Therefore, the primary burner duty is lower and the afterburner combustion duty is higher than with a cocurrent design. An advantage with countercurrent designs is that the gas from the primary chamber can pass directly into a fabric filter baghouse without first being cooled.

Items A, B, and C of the given data are the important heating duties to consider for the primary burner. (Item D, for sensible heat, affects the primary duty, but the higher that duty D is, the less burner duty is needed for the afterburner. This holds for systems that do not have a fabric filter upstream of the afterburner.) For soil moisture content greater than about 15 wt% (most soils), water vaporization (item C) is much larger than soil heating duty (item A) and organics vaporization duty (item B).

9.6.3 Examples of Design Calculations

An important factor in the design of incineration is the heating value of the organic compounds in the soil or sludge being treated. If the heating value is not given, it can be estimated from chemical formulae. Shelton (1995) and Leite (1996b) give Equation 9-30 for the high heating value (HHV) with units of Btu/lb, based on weight percentages of hydrogen, chloride, oxygen and nitrogen.

$$HHV = 15{,}410 + (323.5)(\%H) - (162)(\%Cl) - 200.1(\%O) - (120.5)(\%N) \tag{9-30}$$

Leite (1996b) gives an alternative Dulong correlation (*see* Equation 9-31) based on weight fractions of carbon, hydrogen, oxygen, sulfur, and chloride.

$$HHV = (14{,}544)(C) + (62{,}028)(H-O/8) + (4{,}050)(S) - (760)(Cl) \tag{9-31}$$

The volumetric air/fuel ratio required for stoichiometric combustion is given by Leite (1996b), as shown in Equation 9-32.

$$V(air)/V(fuel) = (2.38)(CO + H_2) + (9.53)(CH_4) + (11.91)(C_2H_2) + (14.29)(C_2H_4) + (16.68)(C_2H_6) + (23.9)(C_2H_8) + (31.1)(C_4H_{10}) - (4.76)(O_2) \tag{9-32}$$

For hydrocarbon fuels, a simplified correlation for determining the stoichiometric volumetric flow rate (standard ft³/hr, or scfh) of air is given by Leite (1996b) in Equation 9-33 and is related to the Btu/hr of total heat released, based on the fuel HHV:

$$V(air) = (Btu/hr)/100 \tag{9-33}$$

Reynolds, Dupont, and Theodore (1991) give an example showing the amount of methane fuel required for stoichiometric incineration of a waste sludge at 2,000° F. That example has been adapted as presented here. The sludge in the example is fed to the incinerator at 60° F and has a heating value of 1,000 Btu/lb. The specific heats involved are available in appendices in Reynolds, Dupont, and Theodore (1991), as shown in Table 9-6.

Table 9-6 Example specific heats of exhaust gases (From Reynolds, J.P., Dupont, R.R., Theodore, L. 1991, "Hazardous Waste Incineration Calculations — Problems and Software," John Wiley & Sons, Inc. Reprinted by permission).

S_i	Substance	Btu/(mol-°F) at 60°F	Btu/(mol-°F) at 2000°F	Btu/(mol-°F) Average C_p
S_1	Water vapor	8.00	9.33	8.67
S_2	Carbon dioxide	8.60	11.94	10.27
S_3	Hydrogen chloride	6.93	7.38	7.155
S_4	Nitrogen	6.94	7.53	7.235
S_5	Inerts			1.50

The number of moles of each component based on their weight fraction in the sludge and on their atomic weights is given next. (Note that each mole of chlorine gas derived from the sludge forms 2 moles of HCl, resulting in less hydrogen being available for combustion to form water vapor.)

Water: A_1 = fraction H_2O/18 + fraction H/2 - fraction Cl_2/[2 (35.5)]
Carbon dioxide: A_2 = fraction C/12
Chlorine: A_3 = fraction as Cl/[2 (35.5)])
Nitrogen: A_4 = fraction as N/[2 (14)]
Oxygen: A_5 = fraction as O/[2 (16)]

The combustion reactions are as follows, using methane (CH_4) for auxiliary fuel and considering that the molar (or volumetric) ratio of nitrogen/oxygen in air (used for combustion) is 79 to 21.

$$C + O_2 \rightarrow CO_2 \tag{9-34}$$

$$2H + 0.5\, O_2 \rightarrow H_2O \tag{9-35}$$

$$2\,Cl + H_2O \rightarrow 2\,HCl + 0.5 O_2 \tag{9-36}$$

$$CH_4 + 2\,O_2 + 2\,(79/21)\,N_2 \rightarrow CO_2 + 2\,H_2O + 2\,(79/21)\,N_2 \tag{9-37}$$

The moles of each flue gas component per pound of sludge, derived from these combustion equations, are shown below where Y is the number of moles of methane fuel required per pound of sludge.

Water: $B_1 = 2Y$
Carbon dioxide: $B_2 = Y$
Chlorine: $B_3 = 0$ (methane combustion does not affect chlorine)
Nitrogen in air: $B_4 = 2\,(3.76Y) + (79/21)\,B_5$
Oxygen in air: B_5 = fraction H/2 + A_2 - $0.5 A_3$ - A_5

Note that for stoichiometric conditions, oxygen associated with Y is accounted for as carbon dioxide and water in B_1 and B_2. The value of B_5 is needed only for deriving B_4. It does not affect the heat balance directly, because the oxygen is all reacted.

For substances S_1, S_2, S_3, and S_4, the heat required to raise their temperature from 60° to 2,000° F is, for a unit mass of sludge, as shown in Equation 9-38.

$$Q_1 = \sum (A_i + B_i)(C_p)_i (2{,}000 - 60) \text{ Btu/lb of sludge} \tag{9-38}$$

For substance S_5, the heat required is as shown in Equation 9-39.

$$Q_2 = \text{weight fraction of inerts } (1.5)(2{,}000 - 60) \text{ Btu/lb of sludge} \tag{9-39}$$

The heat produced, Q, from the Btu value of sludge and from the methane fuel is, for each pound of sludge, as shown in Equation 9-40.

$$Q = 1000 \text{ Btu} + 379 \times 950Y \text{ Btu} \tag{9-40}$$

in which 379 is the standard volume of any gas (in this example methane) in units of ft^3 per mole, and 950 is the methane net heating value in Btu/scf Standard conditions here are 60° F and 1 atm.

With adiabatic conditions (i.e., heat losses from the insulated incinerator walls are assumed to be not significant), Y can be derived from Equation 9-41.

$$Q = Q_1 + Q_2 \tag{9-41}$$

The heat release at the methane burner is as shown in Equation 9-42.

$$\text{Burner duty, Btu/hr} = (\text{sludge feed ton/hr})(2{,}000 \text{ lb sludge/ton})(379 \text{ scf/mole})(950 \text{ Btu/scf})(Y \text{ mole/lb sludge}) \tag{9-42}$$

Reynolds, Dupont, and Theodore (1991) point out that the incinerator must be rated at least for the sum of this burner duty, plus Q_2 times the sludge feed in ton/hr times 2,000 lb sludge/ton.

The usual incinerator arrangement would include a primary chamber operating as just described plus an afterburner for cleanup of PICs in the exhaust flue gas. The afterburner would consume additional methane fuel. If the primary chamber is fired with excess air, more methane fuel would be needed than the amount computed in this example with stoichiometric firing.

For the process design of an incinerator for burning sludge containing chlorinated hydrocarbons, Reynolds, Dupont, and Theodore (1991) give an example that covers the following items:

- Stoichiometric amount of air required for complete combustion
- Flow rate (in ft^3/min) of required methane auxiliary fuel

- Heat released from methane fuel
- Heat released from Btu value of the sludge
- Heat required to increase the temperature of the products of combustion, nitrogen in the sludge, and nitrogen in the air to the incineration temperature

The combustion reaction equations are:

$$C + O_2 \rightarrow CO_2 \qquad (9\text{-}43)$$
$$2H + 0.5 O_2 \rightarrow H_2O \qquad (9\text{-}44)$$
$$2Cl + H_2O \rightarrow 2HCl + 0.5 O_2 \qquad (9\text{-}45)$$
$$0.5 CH_4 + O_2 + (79/21) N_2 \rightarrow 0.5 CO_2 + H_2O + (79/21) N_2 \qquad (9\text{-}46)$$

in which the term (79/21) is the molar ratio of nitrogen to oxygen in the air used for combustion.

Another example from Reynolds, Dupont, and Theodore (1991) derives the incineration temperature or, more precisely, the adiabatic flame temperature required for stoichiometric combustion of a chlorinated hydrocarbon mixed with propane.

Brunner (1991) gives an example of an incinerator design with 125% excess air and with fuel oil as an alternative to methane fuel. The example includes a correction factor that accounts for the humidity of the combustion air. The heat of vaporization of humidity moisture is added to the heat capacity calculated for the flue gas, at the rate of 970 Btu/lb of air moisture. The example also accounts for heat losses, including the heat content of the solids discharged and radiation losses. Radiation losses for refractory-lined incinerators operating near 2,000° F are given as ranging from 3% of the incinerator Btu/hr rate at or less than 10 million Btu/hr down to 1.5% greater than 35 million Btu/hr.

The example includes a determination of the quench water injection rate for cooling the flue gas, considering these factors:

- Quenching is an adiabatic process. The sensible heat reduction in the flue gas equals the change in enthalpy of the injected water that evaporates.
- Flue gas will become saturated if a wet scrubber follows the quench.
- The adiabatic temperature can be determined from a table or derived from a psychrometric chart.
- The water injection rate is equal to the saturated moisture content of the flue gas minus the prequench moisture content of the flue gas.

The example determines the scrubber water discharge temperature and the amount of cooling needed for a scrubber water circulation system. The scrubber water discharge temperature cannot be predicted accurately and is taken as equal to the adiabatic quench temperature, °F, divided by 1.2. Brunner (1991) points out that, in practice, water in excess of the amount calculated for adiabatic quench must be injected and recirculated. This is because quench chambers have inadequate residence times and heat transfer rates for all the quench water to be evaporated in one pass through quench sprays and chambers. In Brunner's example, the calculated quench water flow rate is multiplied by a factor of 8. Leite (1996a) suggests that quench water be recirculated at a rate that is eight to 10 times the calculated adiabatic quenching water flow rate and that the quench tower diameter correspond to a gas superficial velocity of 5 to 7 ft/sec.

An example of afterburner calculations is given here:

Given and assumptions: The afterburner for the above sludge incineration example raises the temperature of the exhaust flue gas from 2,000° to 2,300° F. Excess combustion air is added at the afterburner chamber such that the oxygen content of the final discharge to atmosphere is 3 vol%. Let Z represent the moles of methane fuel needed for the afterburner per pound of sludge fed to the primary chamber. Assume that the design of the primary chamber is such that the degree of turbulence and the amount of residence time for the sludge results in only a small amount of PICs. The effect of final combustion of these products in the afterburner will be neglected in the equations, because combustion of additional methane fuel will be much more significant. Assume that the chamber is well insulated.

Find: The afterburner heating duty.

Step 1: The equation for the combustion reaction is:

$$CH_4 + 2\ O_2 + \text{excess}\ O_2 + [2\ (79/21) + \text{excess}\ O_2\ (79/21)]\ N_2 \rightarrow$$
$$CO_2 + 2\ H_2O + \text{excess}\ O_2 + [2\ (79/21) + \text{excess}\ O_2\ (79/21)]\ N_2 \quad (9\text{-}47)$$

Step 2: Additional exhaust gases will result from afterburner operation, as shown in Equations 9-48, 9-49, 9-50, 9-51, and 9-52 (moles/lb of sludge fed to the primary chamber).

Water: $C_1 = 2\ Z$ (9-48)
CO_2: $C_2 = Z$ (9-49)
Chlorine: C_3 = zero (9-50)
Nitrogen: $C_4 = 79/21\ (2 + \text{excess}O_2)\ (C_5)$ (9-51)
Oxygen: $C_2 = 2\ Z + \text{excess}O_2$ (9-52)

Step 3: Calculate the sensible heat duty, Q_3, and the heat of combustion, Q_4. Note that Z is moles of methane per pound of sludge and that the net heating value of methane is 950 Btu/lb. The heat required to increase the temperature of these additional gases and the temperature of the gases exhausting from the primary combustion chamber from 2,000° to 2,300° F is, for each pound of sludge, as shown in Equation 9-53.

$$Q_3 = \sum (A_i + B_i + C_i)\ (Cp)_i\ (2{,}300 - 2{,}000)\ \text{Btu/lb sludge} \quad (9\text{-}53)$$

in which the C_p values are determined, similar to Table 9-6, for the 2,000 to 2,300° F range, and i ranges from 1 to 4. The heat released at the methane burner for afterburner operations is as shown in Equation 9-54.

$Q_4 = 379 \, (950Z)$ Btu/lb sludge (9-54)

We now have two unknowns, Z and the moles of excess oxygen. These unknowns can be evaluated by simultaneously solving the heat balance equation and the excess oxygen equation, as given in Equations 9-55 and 9-56, given that the oxygen content of the final exhaust is 3 vol%.

$Q_3 = Q_4$ (heat balance for adiabatic conditions, in a well-insulated unit) (9-55)
ExcessO$_2$ = 0.03 $\sum(A_i + B_i + C_i)$ (9-56)

On evaluating Z, Q_4 is readily determined from Equation 9-59, and the afterburner heating duty is as shown in Equation 9-57.

Duty, Btu/hr = sludge feed ton/hr (2,000 lb sludge/ton) (Q_4) (9-57)

If the percentage of excess air is given, then the corresponding oxygen can be determined directly. For example, if an afterburner operates with methane fuel at 10% excess air, the molar oxygen/methane ratio is 2.2 (equal to 2 x 1.1) instead of 2. The methane combustion reaction is:

$CH_4 + 2.2 \, O_2 + 2.2 \, (79/21) \, N_2 \rightarrow CO_2 + 2 \, H_2O + 0.2 \, O_2 + 2.2 \, (79/21) \, N_2$ (9-58)

Brunner (1991) gives an example of an incinerator with an afterburner using waste hydrocarbon liquid as the fuel. The combustion equations are based on the hydrogen/carbon ratio of the waste liquid. The example includes these features:

- A rotary kiln is used for primary combustion of wastes considered hazardous and consisting of plant trash and waste hydrocarbon liquid.
- The volume of the primary chamber is based on a heat release of 25,000 Btu/ft^3. The length and diameter are calculated based on a 4/1 aspect ratio.
- The thickness of an insulating castable against the steel shell, protected from the incineration process by 6-in. thick firebrick, is calculated, based on a graph giving the thermal resistance required of the refractory layers needed to achieve a steel shell temperature not exceeding 200° F. For this example with the primary combustion at 1,600° F, a 7-in. thick insulating castable is chosen. These calculations also determine the outside diameter of the kiln.
- A thermal resistance graph gives the heat loss factor corresponding to a steel shell temperature of 200° F in terms of Btu loss per square foot of cylindrical surface. (A simplification sometimes used in this step is to assume that the diameter is large enough to treat the steel shell as a vertical surface.) The heat loss is determined by multiplying the outside surface area, using the outside diameter and length of the cylinder, by the heat loss factor, 263 Btu/(hr-ft^2).
- The heating value of the waste hydrocarbon is calculated by two alternative methods: (a) a rigorous method based on the heat of formation of the hydrocarbon; (b) an approximation for hydrocarbons, 184,000 Btu/lb of oxygen needed for stoichiometric combustion. The oxygen needed can be either interpolated from a table of hydrocarbon combustion relationships or determined from a stoichiometric reaction equation for the fuel hydrogen/carbon ratio.

- The afterburner chamber volume is taken as the larger of two alternative values: (a) based on a heat release of 60,000 Btu/(hr-ft^3); or (b) based on residence time. For this example, a residence time of 2 sec is assumed. The length/diameter ratio is arbitrarily chosen in the 1.6 to 3 range.
- The afterburner in this example is designed for operating at 2,200° F, with an outside steel shell not exceeding 250° F. The refractory design is simpler than the two-layer design applied to the primary chamber, because only hot gases are involved. Using a thermal resistance graph, a 9-in. thick insulating castable is chosen, with a corresponding heat loss factor of 418 Btu/(hr-ft^2).
- The final oxygen content is then calculated. The oxygen content of flue gas leaving the primary chamber depends on how much excess air is used for primary combustion. For the afterburner, 10% excess air was assumed. The final discharge is checked for oxygen content, with a minimum of 2% being required.
- An alternative set of afterburner combustion calculations is completed using fuel oil as an alternative to waste liquid hydrocarbon fuel.

Another example from Brunner (1991) shows steps required for the calculation of the amount of steam that can be generated with a waste heat boiler, considering the following factors:

- In practice, the approach temperature (final flue gas temperature minus boiling temperature) is approximately 100° F for efficient designs or 150° F for standard, economical construction. (Note that the final flue gas temperature should be maintained above the acid dew point if large amounts of halogens or sulfur are present. Leite (1996b) gives equations for determining acid dew points for hydrogen chloride, hydrogen bromide, sulfur trioxide, and sulfur dioxide as functions of their partial pressure and water vapor partial pressure. The partial pressure of each substance in the vapor phase in mm of Hg is the mole fraction multiplied by 760).
- The enthalpy change of the flue gas can be calculated assuming that the flue gas properties can approximated by air properties.
- Saturated steam is generated at various pressures; for each pressure, the heat of vaporization and enthalpy of water are obtained from steam tables.
- The heat recovered is proportional to the flue gas temperature drop.
- Steam production depends on heat in the steam, blowdown, and feedwater and on whether condensate is returned. A deaerator heat balance is needed.

Brunner's incinerator, wet scrubber, and waste heat boiler calculation methods are available as computer programs from Incinerator Consultants Incorporated (Reston, Virginia). Included are computations for design of rotary kiln incinerators that can be applied for hazardous wastes.

9.6.4 Contaminant Destruction Efficiency and Emission Limitations

The minimum percentage of contaminant destruction efficiency is often set by regulations. These requirements are usually met by providing adequate temperature and residence time in the afterburner. If RCRA applies to the material, 99.99% destruction of Principal Organic Hazardous Constituents (POHCs) is required. If

PCBs are present at 50 mg/kg or greater, TSCA requires 99.9999% destruction, at least 3% flue gas oxygen, and 2 sec of gas residence time at 1,200° C (2,192° F).

Hydrochloric acid emissions must be at least 99% removed when they exceed 1 lb/hr. There are also limits on emissions PICs, carbon monoxide, hydrocarbons, and particulate matter. For carcinogenic metals (arsenic, cadmium, chromium, and beryllium), the dispersion in the atmosphere must be modeled mathematically and a risk assessment completed.

9.6.5 Limitations on Particulate Emissions and Plume Opacity Correlations

Particulates cause the stack emission plume to be visible, reduce the clarity of the downwind atmosphere, cause dust fallout, and may contain toxic metals. Sometimes the particulates in incinerator emissions have a higher concentration of certain metals than the soil or sludge has. This effect is more pronounced at higher incineration temperatures. Tillman et al. (1991) describe a case history in which most metal concentrations in the particulate from an 1,800° F incinerator were highly enriched except in chromium. Arsenic and cadmium were especially enriched, as was lead somewhat, at incineration temperatures as low as 1,460° F.

For high-temperature incineration, Brunner (1991) suggests that the worst case for potential emissions toxicity is to assume that all of the lead, cadmium, nickel, and mercury are emitted from the incineration process as oxides. With this assumption, none of these metals remains with the treated soil or sludge.

There may be a federal or local regulatory limit on total particulate emissions or on the portion up to 10 μm in size (PM-10) or up to 2.5 μm (PM-2.5). Regulations may impose limits on particulate emissions in various ways, such as:

- Concentration of the total particulate or of PM-10 expressed in milligrams or micrograms per cubic meter, grains per dry standard cubic foot, or similar units and corrected to what the concentration would be at a set percentage of excess air, oxygen, or carbon dioxide.
- Mass rate of the total particulate, of PM-10 or PM-2.5, in g/sec or lb/day or ton/yr.
- Opacity of visible plumes (usually excluding the effect of water condensing in the plume, forming fog).

Reynolds, Dupont, and Theodore (1991) give an example that derives the particulate concentration (grains per dry standard ft^3) emitted in the flue gas, considering the following items:

- Stoichiometric equations for combustion of multiple chlorinated hydrocarbons
- Amount of water in the flue gas derived from humidity in the combustion air
- Amount of water in the flue gas derived from combustion reactions
- Amount of hydrogen chloride produced
- Flue gas volume, accounting for actual excess air used for combustion
- Flue gas volume corrected to 50% excess air

9.6.5.1 Plume Opacity

The amount of total particulate, PM-10 or PM-2.5, and the corresponding amount of toxic metals in the discharge to the atmosphere can usually be estimated by the abatement equipment vendors. However, a prediction of plume opacity is very difficult, especially for plumes in which water condenses and forms fog.

If fog is in the plume, the opacity is measured either by observing the residual plume remaining after the fog dissipates or with an instrument operating with a light wavelength that penetrates fog. Opacity is measured either in percentage obscuration of light passing through the plume or in Ringelmann numbers as follows:

- 20% opacity is Ringelmann No. 1.
- 40% opacity is Ringelmann No. 2.
- 60% opacity is Ringelmann No. 3.
- 80% opacity is Ringelmann No. 4.
- 100% opacity is Ringelmann No. 5.

Thus, a Ringelmann No. of 0.5 corresponds to 10% opacity. A number of state and local air pollution control districts have limited opacity to 20% or 40%. A stringent limitation would be 10%.

It is difficult for the designer to predict the percent opacity from the particulate concentration (e.g., grain loading). This is especially true for aerosol emissions — the very small particulate sizes that are the most difficult to abate. Particulate matter that is near 1 µm in diameter or smaller is near the wavelength of visible light. This causes light scattering, with the effect of magnifying light obscuration over that which might be calculated from the area blanked out by the particles.

The scattering ratio, K, accounts for this magnifying effect. For most substances, K can be determined from graphs as a function of particle radius, index of refraction, and wavelength of light (van de Hulst, 1957). For evaluating K, a wavelength of 0.55 µm (5,500 Angstrom units) is used. This wavelength is the average for the visible spectrum, and the human eye is highly sensitive to light at this wavelength. The index of refraction can be determined from experiments or from handbooks such as the *CRC Handbook of Chemistry and Physics* (CRC Press, Boca Raton, Florida) or *Lange's Handbook of Chemistry* (McGraw-Hill, New York). Examples of refraction index values include:

- Silica
- Alumina
- Calcium oxide
- Ferric oxide

The main problem in evaluating K is the determination of particle radius, because particulate emissions have a size distribution. Ensor and Pilat (1971) concluded that K is primarily a function of particle size for diameters greater than about 1 µm and is a function of refractive index for smaller particles. Ensor and Pilat correlated K with an assumed log-normal particle size distribution and presented graphs of K values for water, white aerosol emissions, black aerosol (carbon) emissions, and iron oxide emissions.

Rather than attempting to measure the particle size distribution near 1-µm diameter or assuming a log-normal distribution, a conservative approach is given by Hyman (1970). Assume that the particle size that causes maximum light scattering is totally present. This is a practical approach, because at 0.55-µm light wavelength, for each substance (or for each refractive index), only one particle size applies — generally near 0.5 µm radius. For design of abatement equipment, this approach recognizes that the particles that escape to the atmosphere will be small ones. Also, because all particulate is taken to be of the radius that causes maximum light scatter, this approach is independent of changing conditions in the soil or sludge properties, in incinerator operations, or in abatement system conditions.

Based on equations given by Hyman (1970) and by Elmer Robinson in Stern (1962) and based on communications with D.S. Ensor (a co-author of Ensor and Pilat, 1971), the correlation between opacity and particulate mass concentration at maximum light scattering is as given in Equations 9-59 and 9-60.

$W = (0.021/L)$ [specific gravity/(m-1)] (1/m) @ 20% opacity　　　　(9-59)
$W = (0.049/L)$ [specific gravity/(m-1)] (1/m) @ 40% opacity　　　　(9-60)

in which W is the grains/acf; L is the length of the light path through the emission plume at the stack tip, ft; specific gravity is the individual particle density divided by the density of water; and m is the refractive index. For nonrectangular stacks, L is the square root of stack area. Thus, for a round stack, L is the diameter multiplied by the square root of (π divided by 4). Weir et al. (1976) emphasize that other factors besides mass concentration and plume width affect perceived opacity. These factors include the angle of the sun, time of day, and geographic location, but they are not covered in the design approach presented here. Whenever possible, the constants 0.021 and 0.049 in Equations 9-59 and 9-60, respectively, should be adjusted by calibrating them against actual measured grain loadings and observed opacities.)

Cooper (1976) describes how light transmission through plumes with mixed particulates should be averaged.

A simplified approach for mixed substances in the plume is given by Equation 9-61.

$W = \sum(F_i)(W_i)$　　　　(9-61)

in which F_i is the weight fraction of each substance present as emitted particulate, and W_i is the the value of W calculated for each substance. If not all of weight fractions are known, then before applying the correlations, each known weight fraction should be divided by a normalizing factor equal to the sum of the known weight fractions. If not all of refractive indices are known, use the largest known index as the value for the unknown indices.

Equations 9-59, 9-60, and 9-61 were applied in the early 1970s for work on industrial stack emissions conducted in conjunction with W.C. Frederiksen (Frederiksen Engineering Co., Oakland, California). Application of the correlations is exemplified as follows:

Given and assumptions: A stack with a 4.0-ft diameter stack tip has emissions derived from sandy soil. The allowable opacity of the particulate emissions plume is 20% (Ringelmann No. 1). The particulates will contain silica, alumina, and calcium oxide weight fractions estimated from analyses of similar emissions, as shown in Table 9-7, and relatively small amounts of other substances.

Table 9-7 Example of correlating plume opacity to grain loading.

Substance	Weight Fraction	Normalized Weight Fraction	Specific Gravity	Refractive Index, m	Grain/acf 20% Opacity
Silica	0.6	0.706	2.3	1.5	0.0127
Alumina	0.2	0.235	3.7	1.8	0.0036
Calcium Oxide	0.05	0.059	3.4	1.8	0.0008
Σ	0.85	1.0			0.017

Find: The grain loading at which 20% opacity will never be exceeded, no matter what the particle size distribution or operating conditions might be.

Step 1: Determine the significant dimension of the plume width. The significant dimension L for the path length of light transmitted through the plume is, for this round stack, as given in Equation 9-62.

$$L = 4.0 \text{ ft } (\pi/4)^{1/2} = 3.54 \text{ ft} \tag{9-62}$$

Step 2: The grain loading value is given in the last column in Table 9-7, as derived from Equation 9-59. The table shows that a grain loading at or less than 0.017 grains/acf in this example will result in always meeting the opacity limitation.

Regulatory limits on particulate mass concentration are usually higher than the grain loading determined in this example. In this situation, the plume opacity limitation will control the particulate abatement system design. Narrowing the stack tip diameter would result in reduced opacity at any given particulate mass concentration. Tapered stacks are more stable structurally than straight stacks, and their installation provides both reduced opacity and stability. The particulate mass concentration corresponding to a particular opacity is inversely proportional to stack tip width. For a given opacity limitation, the particulate mass concentration need be calculated only once in order to determine the effect of changing the stack tip width. Hyman (1970) has shown how this relationship can be graphed with straight lines, as indicated in Figure 9-3. In this figure, L is the path length of light transmitted through the plume,

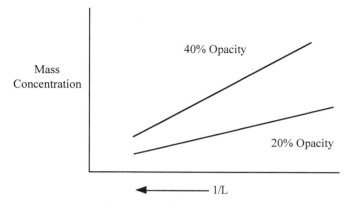

Figure 9-3 Relationship among plume opacity, stack width, and plume mass concentration.

which is the plume width at the tip of a rectangular stack or the diameter of a round stack tip multiplied by $(\pi/4)^{0.5}$.

9.6.5.2 Emissions Source Testing

An emissions source test determines the mass of particulate and water vapor in a measured volume of exhaust gas sample, along with the exhaust gas velocity. The oxygen content must be determined, and the measured particulate concentration must be corrected to 50% excess air. (Other parameters may be measured, such as particle size distribution and the content of carbon dioxide, carbon monoxide, acid gases, and organic compounds.) From the particulate and water mass figures and sample volume, the concentration on a wet basis and on a dry basis can be calculated. From the gas velocity and cross-sectional area of the stack, the volumetric flow rate can be calculated. The mass emission rate is the product of concentration and volumetric flow rate.

Particulate emissions-abatement equipment is usually designed for a desired reduction in concentration or for a final discharge concentration limit. The abatement efficiency is determined from the reduction in concentration or from the corresponding reduction in mass emission rate.

9.6.6 Baghouse Design Parameters

The most important parameter in selecting the filter fabric area for a baghouse is the air/cloth ratio, expressed as acfm of air flow per square foot of fabric area. The units reduce to ft/min, and this is sometimes called the filtration velocity. Typical conservative designs use a filtration velocity less than 6 ft/min. Filtration at velocities of 15 ft/min have been reported for pulse-jet units, especially for secondary collection downstream of a primary particulate abatement unit. Pressure drop across filter fabrics is typically a maximum of several inches w.c. Croom (1996) discusses other baghouse design parameters that are summarized as follows:

- Typical duct velocities of 4,000 to 5,000 ft/min (adequate to keep particulate matter in suspension in the flowing gas stream) should be reduced to 2,000 ft/min or less at the inlet. This lowers fan energy requirements and allows immediate drop out of larger particulates into the dust collection hopper. (Filter bags are usually oriented vertically above the hopper, and initial gas flow is upward. At the cloth, the flow direction takes on a horizontal component).
- Inlet velocity reduction can be accomplished with a duct transition piece with an included angle of 20° to 65°. Croom (1996) illustrates five different transition pieces and gives their friction loss coefficients.
- The can velocity is the velocity of the gas as it passes around the bags. It is determined as the (gas volumetric flow rate)/(baghouse enclosure area - total cross-sectional area of the bags). The can velocity for upward dirty gas flow is typically 200 ft/min, and it may be less than 100 ft/min for filtering fine particulates. It is desirable to have as uniform a can velocity as possible for all the bags. Croom (1996) illustrates nine different collector inlet designs that use a variety of baffles to help achieve a uniform can velocity.
- A low value for the can velocity is important with fine particulates because the collected dust must flow downward when the bags are cleaned and may be re-entrained with upward-flowing dirty gas.
- Discharge velocity from the clean-gas plenum can be 4,000 to 5,000 ft/min, and a 45° transition piece is recommended.

9.6.7 Wet Scrubber Power Requirements

If a wet scrubber is used for control of particulate emissions, the blower power requirements are significant. Predicting the required blower horsepower depends largely on the flue gas pressure drop through the scrubber. The required pressure drop depends on how small the particle sizes are and the limit on emissions mass concentration. Particulate sizes smaller than 10 μm are of most concern and require more pressure drop for successful scrubbing than do larger particles. The particulate size range of concern is similar to blast furnace dust and fumes.

A preliminary estimate of blower horsepower requirements for wet scrubber can be made using Figure 9-4, adapted from Figure 9 in Semrau (1960), which was developed for scrubbing blast furnace dust and fumes. The horsepower derived from this curve should be divided by blower efficiency to obtain the minimum blower motor rating.

If the scrubber pressure drop is known from experience or vendor information, and if the flue gas pressure drop through other devices and through the ducting is accounted for, the blower horsepower requirement is calculated as shown in Equation 9-63.

$$bhp = scfm\ (h)\ (0.000157/eff) \qquad (9\text{-}63)$$

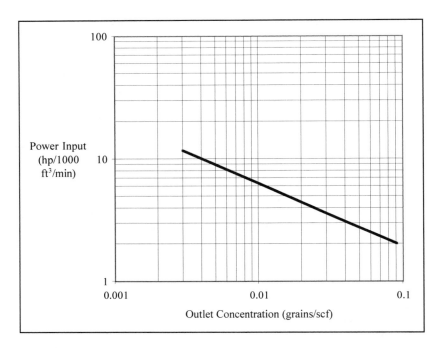

Figure 9-4 Scrubber blower horsepower.

in which bhp is brake horsepower; h is the pressure rise through the blower (in. w.c.), equal to the total scrubber and system pressure drop; and eff is the fractional blower efficiency. The efficiency of centrifugal blowers is usually in the 0.6 to 0.8 range.

An example of applying this equation to a system that includes a high-energy Venturi scrubber with a 35 in. w.c. pressure drop, plus a total system pressure drop of 45 in. for a 68% efficient blower operating at 5,000 actual ft^3/min at 174° F (634° R), is shown in Equations 9-64, 9-65, 9-66, and 9-67.

scfm = acfm (standard temperature)/(actual temperature) (9-64)
standard temperature = 60° F + 460° R = 520° R (9-65)
scfm = 5,000 acfm (520)/(634) scfm/acfm = 4,100 scfm (9-66)
bhp = 4,100 (45) (0.000157/0.68) = 43 bhp (9-67)

The motor selected for this service should be rated at least at 50 hp, the closest larger standard rating. An even larger motor might be selected, depending on the blower manufacturer's horsepower requirements at the upper end of the blower's ft^3/min range, with the largest wheel ever expected to be installed and considering the highest rotational speed anticipated.

Brunner (1991) indicates that low-energy scrubbers operate at less than 7 to 12 in. w.c.; medium-energy is at 7 to 20 in.; high-energy is at greater than 20 in. An illustration is given by Brunner (1991) for a high-energy Venturi scrubber in which water is circulated to the throat at a rate of 5 to 7 gal/1,000 ft^3 of flue gas. This example given above shows how flue gas flow rate and gas pressure drop through a scrubber and other equipment lead to deriving the blower horsepower. The water circulation rate, along with scrubber spray nozzle pressure drop, elevation rise, and system water pressure drop, lead to estimates of the pump horsepower. The applicable relationship is Equation 9-68:

bhp = gal/min (SG_L) (0.433) (TDH)/[(1,713) (eff)] (9-68)

in which bhp is the brake horsepower; gal/min is the rate at the flowing temperature; SG_L is the specific gravity of the water solution at flowing temperature; TDH is the total dynamic head for the pump (ft of water column); and eff is the fractional pump efficiency. Note that the term (SG_L) (0.433) (TDH) is the pressure rise through the pump, psi. The equation is approximate in that the effect of suspended solids content is neglected. The efficiency for a centrifugal pump is usually in the 0.55 to 0.75 range. (Pumps are often rated at higher efficiencies, but they are not usually operated at their maximum efficiency point.)

Consider an example of applying this equation to circulating water at 135° F, with a 72% efficient pump at a pressure rise of 50 psi, through a Venturi scrubber operating at 5,000 ft^3/min with 6 gal/min/1,000 ft^3 flue gas. The specific gravity of pure water at the flowing temperature can be readily found in a number of handbooks, but dissolved minerals raise the specific gravity an unknown amount. The water flow rate is 5,000 ft^3/min (6 gal/min)/1,000 ft^3/min, or 30 gal/min. The specific gravity can be approximated by finding the density of a salt solution at a standard temperature in a handbook and multiplying that value by the specific gravity of pure water at flowing temperature. Assuming for this example that SG_L is approximately 0.99, then Equation 9-69 applies.

bhp = 30 (0.99) (50)/[(1,713) (0.72) bhp] = 1.2 bhp (9-69)

A 1.5-hp motor might be selected for this service if a pump were dedicated to serve the Venturi sprays only. If the pump is also used to serve quench sprays, the water flow rate and horsepower would be much higher.

9.6.8 Design of Vertical Packed Acid Gas Scrubbers

Leite (1996a) suggests the following design parameters for packed scrubbers:

- Gas residence time in the scrubber, 0.4 to 0.6 sec
- Superficial gas velocity of 7 to 10 ft/sec, equivalent to approximately 2,000 lb/hr/ft^2 of tower cross-sectional area
- Liquid-to-gas ratio of 20 to 50 gal/acf

Theodore (1996) gives the Equations 9-70 and 9-71 for determining the packing height, Z:

$N_{OG} = \log_e$ of (C_{in}/C_{eff}) (9-70)
$Z = H_{OG}N_{OG}(SF)$ (9-71)

in which N_{OG} is the number of gas transfer units; C_{in} is the contaminant inlet concentration; C_{eff} is the effluent concentration; H_{OG} is the height of a gas transfer unit; and SF is a safety factor, ranging from 1.25 to 1.5. If H_{OG} is in meters or feet, then Z is in meters or feet, respectively. For water scrubbers absorbing an acid gas such as HCl, Theodore (1996) gives a table of H_{OG} values (in feet) for plastic packing and for ceramic packing. For plastic packing up through 2 in. nominal diameter:

H_{OG}, ft = (packing diameter, in.)$^{0.6}$ (9-72)

For 3-in. packing, H_{OG} is 2.25; for 3.5-in. packing, H_{OG} is 2.75. Leite (1966) states that the height of a transfer unit is typically 1 to 1.7 ft.) The H_{OG} values for ceramic packings are double these values for plastic packings.

Theodore (1996) gives these additional steps for approximating the other required process design parameters:

- Cross-sectional area — Use a superficial gas velocity of 5 to 6 ft/sec for plastic packings and 3 to 4 ft/sec for ceramic packings. These velocities hold for HCl. Use lower velocities for gases that are hard to absorb.
- Liquid flow rate — Use 1,500 to 2,000 lb/hr (3 to 4 gal/min of water)/ft^2 of cross-sectional area for plastic packings and 500 to 1,000 lb/hr (1 to 2 gal/min of water)/ft^2 for ceramic packings. These rates hold for HCl; use higher rates for gases that are hard to absorb.
- Gas pressure drop varies from 0.145 to 0.40 in. w.c./ft of packing height. The average for plastic packings is 0.2 in.; for ceramic packings, 0.25 in.
- Packing diameter — Use 1-in. diameter packing for 3-ft diameter towers. Use smaller than 1 in. for smaller towers. Use larger diameter packings for larger towers.

These guidelines from Theodore (1996) are illustrated in the following example.

Given: Ninety-nine percent of the HCl must be removed from 7,585 scfm of gas flowing at a temperature of 160° F; ceramic packing is used in a round, vertical tower; tower pressure is very close to 1 atm.

Find: The tower diameter and nominal packing diameter, water flow rate, packing height, and pressure drop.

Step 1: Convert the given scfm (at a standard temperature of 60° F, or 520° R) to acfm at a temperature of 160° F, or 620° R, and use superficial velocity applicable to ceramic packing (say, 3 ft/sec) to determine the cross-sectional area and tower diameter. Then select a packing diameter suitable for that tower diameter.

(7,585 scfm) (620° R/520° R) = 9,044 acfm (9-73)

Cross-sectional area = (9,044 acf/min) (1 min/60 sec)/(3 ft/sec) = 50.24 ft² (9-74)

For a round tower, the corresponding diameter is 8 ft. Select a packing diameter suitable for tower diameters greater than 3 ft (packing diameter larger than 1 in,), say, 2 in.

Step 2: Use up to 2 gal/min/ft² of cross-sectional tower area for liquid flow rate with ceramic packing. Equation 9-75 gives the liquid flow rate.

(2 gal/min/ft²) (50.24 ft²) = 100 gal/min (9-75)

Step 3: Determine H_{OG} corresponding to 2 in. of ceramic packing and N_{OG} from Equation 9-75; determine packing height Z from Equation 9-71 using a safety factor between 1.25 and 1.5. Note that for the given contaminant removal efficiency of 99% and for 100 units of contaminant inlet concentration, only one unit of contaminant effluent concentration is allowed to be released.

H_{OG} (2 in. plastic packing) = $2^{0.6}$ ft = 1.5 ft (9-76)
H_{OG} (ceramic packing) = 2 times value for plastic = 3 ft (9-77)
N_{OG} = $\log_e$ (C_{in}/C_{eff}) = $\log_e$(100/1) = 4.6 (9-78)
Z = $H_{OG}N_{OG}$(SF) = (3 ft) (4.6) (1.38) = 19 ft (9-79)

Step 4: Determine the pressure drop through the packing, using 0.25 in w.c./ft of packing height (the average value for ceramic packings), as shown in Equation 9-80.

Pressure drop = (0.25 in. w.c.) (19 ft) = 4.8 in. w.c. (9-80)

9.6.9 Venturi Scrubber Design Parameters

Based on Theodore and Reynolds (1987) and Leite (1996a), the following correlations are used in the design of Venturi scrubbers. Gas velocities through the throat are typically 200 to 400 ft/sec, and the liquid flow rate is 7.5 to 15 gal/acf of gas flow. With the throat velocity V_T (ft/sec) and liquid/gas ratio R (gal/acf) chosen, the pressure drop at high velocities is as shown in Equation 9-81.

Pressure drop, in. w.c. = 5 x 10^{-5} (V_T^2R) SG_G (9-81)

in which SG_G is the specific gravity of the gas relative to air. The pressure drop over a wide range of velocities is given by the Hesketh equation for Venturi scrubbers that have liquid injected before the throat, as shown in Equation 9-82.

Pressure drop, in. w.c. = (V_T^2 $R^{0.78}$ $A^{0.133}$ W_G)/1,270 (9-82)

in which A is the throat area, ft²; and W_G is the gas density, lb/acf.
The Johnstone equation predicts the particulate collection efficiency at a given liquid/gas ratio R, as shown in Equation 9-83.

Fractional efficiency = 1 - exp(-K R $I^{0.5}$) (9-83)

in which K is a correlation factor that depends on the Venturi geometry and operating conditions (typically 100 to 200 acf/gal); and I is a dimensionless inertial impaction parameter, as given in Equation 9-84.

$$I = (C)(d_P)^{2}(W_P V_T)/(9\mu_G d_l) \tag{9-84}$$

in which C is the Cunningham factor (about 1.4 for 0.5-μm diameter particles); d_P is the particle diameter, ft, W_P is particle density, lb/ft^3, μ_G is the gas viscosity. lb/ft-sec; and d_l is the mean droplet size, ft, of the scrubbing water in the Venturi. Each of these parameters is evaluated as follows, starting with Equation 9-85.

$$C = 1 + 0.22/d_P \text{ for air at atmospheric pressure, } 212° F \tag{9-85}$$

Factor C is significant only for particles smaller than 1 μm in diameter; for such sized particles, C for air at 70° F is equal to the value given by Equation 9-85 multiplied by $(d_P)^{0.12}$. If the particle size distribution is unknown, then an initial approximation of the particle diameter can be made by assuming that the particles are 0.25 μm in diameter (8.2 x 10^{-7} ft).

If the particle density is not available from a handbook, use the specific gravity values given in Table 9-7 multiplied by 62.4.

Theodore and Reynolds (1987) give an example with gas viscosity equal to 1.5 x 10^{-5} lb/ft-sec.

The mean droplet size is given by the Nukiyama and Tanasawa relationship, as shown in Equation 9-86.

$$d_l = 16,400/V_T + 1.45 R^{1.5} \tag{9-86}$$

Schifftner and Hesketh (1983) indicate that for injecting water tangentially into annular Venturi throats, the water velocity is 6 to 8 ft/sec. The pressure in the water header is 5 to 10 psig.

9.7 Treatability Studies and Trial Burns

The temperature and residence time needed to achieve a soil cleanup goal depend on soil type, contaminant vapor pressures, contaminant concentration, and soil moisture content. Soils with a high clay content are difficult to feed and convey, and they form clods that trap contaminants. High volatility (measured as vapor pressure, a function of temperature for each contaminant compound) and low inlet concentration result in reduced temperature and residence time requirements. These requirements increase with soil moisture content. Pilot tests that can aid in planning thermal treatment include soil screening and crushing, while measuring organic vapor emissions, and dewatering of sludges. Pilot tests used to determine optimum thermal desorption conditions and trial burn requirements are described in the following section.

9.7.1 Testing Thermal Desorption from Soils

Pilot testing is often needed for thermal treatment systems because of variability in contaminant concentrations, soil particle size, and adsorptive tendencies of different contaminants. Also, soil moisture content may cause effects that cannot always be predicted without testing. Although moisture generally affects the required heat input directly, it aids desorption from fine-grained soils by preferentially occupying adsorption sites. Pilot testing can also help determine if a practical desorption temperature can be used to remove a significant portion of the residual contaminant that is most strongly held on soil surfaces.

The optimum temperature and residence time are found from pilot runs, sometimes accomplished in a laboratory with drum-sized samples of the soil. A unit similar to an electrically heated laboratory oven is fitted with a device that conveys soil samples horizontally. A sample of soil is analyzed for organic contaminant concentration and moisture content. Soil temperature and contaminant concentration are measured partway through and also at the exit of the pilot unit. The sample weight loss is also noted. The pilot runs are repeated at a number of different temperatures. The number of test runs (and a corresponding number of costly laboratory analyses) can be reduced if a straight-line correlation is established between temperature and concentration in the soil. A straight-line relationship can readily be interpolated to the target cleanup level to determine the required temperature.

The higher the temperature, the higher is the vapor pressure of the contaminant, with corresponding improved desorption and reduced soil concentrations. The relationship between vapor pressure and temperature is given by the Clausius-Clapeyron equation, given here as Equation 9-87.

$$\ln(p) = -(\Delta H)/RT + k \tag{9-87}$$

in which p is vapor pressure; ΔH is the heat of adsorption for the contaminant; R is the universal gas constant; T is the absolute temperature; and k is a constant.

If the contaminant concentration in the soil is inversely proportional to the vapor pressure after sufficient residence time in the desorber, then the relationship between concentration, C, and temperature, T, can be estimated using Equation 9-88.

$$\ln(C) = (\Delta H)/RT - K \tag{9-88}$$

in which K is a constant. A plot of the logarithm C versus 1/T is a straight line, as illustrated in Figure 9-5.

Two curves of concentration-versus-temperature can be plotted on the same graph: one based on the soil temperature partway through the pilot desorption unit; the other based on the soil exit temperature. Temperature points are selected where each of the two curves intersect the concentration that represents the cleanup goal. Thus, two temperatures are selected, between which is the design range for the full-scale operating unit.

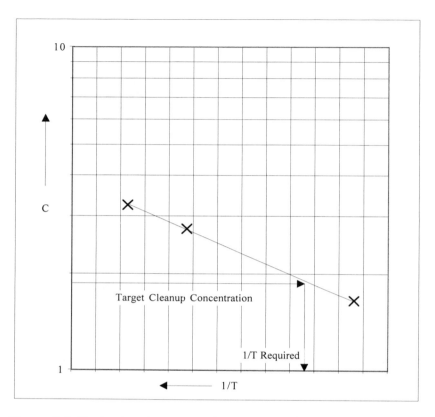

Figure 9-5 Contaminant concentration versus a function of desorption temperature.

A less sophisticated method of conducting laboratory tests is described by the US EPA (1992b), wherein an unmodified laboratory oven is used to warm a batch of a few grams of soil on a fixed tray. This type of testing does not yield data adequate for design of a soil cleanup process, but it is useful for testing the potential for success of using thermal desorption. If the maximum possible temperature for a given full-scale desorber is used for the initial oven test, and the target cleanup contaminants concentration is not achieved, thermal desorption is screened out from consideration of remediation alternatives.

9.7.2 Trial Burns

Incineration of RCRA and TSCA wastes may be done after a trial burn plan is approved, trial burns are conducted and reported, and results are approved. The main concerns addressed during a trial burn are destruction of organics (especially POHCs) and control of hydrogen chloride and particulates emissions. Other concerns and procedures are given in Section 3 of the US EPA (1986a) and in Gorman (1989).

Depending on individual air quality agency permitting requirements, a typical trial burn is run for a number of hours at one temperature and a constant combustion air rate, and it is repeated so that a total of three runs are done. Another three runs may be done at a different set of conditions.

9.8 Cost Estimation for Thermal Soil Treatment

9.8.1 Incineration Costs

9.8.1.1 Incineration Costs per Ton

Excluding excavation and hauling costs, incineration costs for fixed-base units are generally $700 to $1,100 per ton. Oily sludges and liquids cost 30% to 65% less because their high Btu content reduces auxiliary fuel requirements. If heavy metals are also present and the system is not designed purposely to vitrify the soil, fixation agents such as cement may be needed for post-treatment at an additional cost.

Costs for transportable units brought onto a site vary widely because of highly variable setup, transport, and demobilization costs. These costs are in addition to the labor, utilities, equipment use fees, analyses costs, and backfill costs, all of which are comparable to fixed-base charges (which often include landfilling or disposal of the treated soil). The setup, transport, and demobilization costs must be obtained from vendors of the mobile units and compared to costs of hauling the soil to a competing fixed-base unit. DuTeaux (1996) describes a case history with a cost of $173/ton for incinerating 75,000 tons of soil with a transportable system.

As a general rule, it is more economic to haul the contaminated soil to a fixed-base unit if the quantity is less than 2,000 tons. For transportable units, setup, transport, and demobilization costs generally range from $2 million to $4.5 million. Setup costs include site preparation, design and permitting, trial burns, erecting the equipment, and connecting utilities. Permitting and trial burn costs amount to $1 million to $1.5 million if RCRA hazardous wastes or PCBs are involved. Without permitting and trial burn costs, setup, transport, and demobilization costs are therefore in the $1 million to $3 million range.

Johnson (1994) gives the treatment costs for a portable rotary kiln unit as $150 to $250 per ton in addition to the fixed costs described above, depending on the following variables:

- Contaminant levels
- Chemical characteristics
- Whether soil, sludge, or liquid is involved
- Safety considerations
- Moisture content

Hay and McCartney (1991) report overall costs with a portable 10-ton/hr infrared furnace system at $250 to $350 per ton, including mobilization and demobilization, for soil quantities exceeding 10,000 tons.

9.8.1.2 Computerized Cost Estimating for Incineration

Costs for using commercial disposal incineration and for thermal desorption for remediating soils can be estimated with COMPOSER GOLD software available from Building Systems Design (Atlanta, Georgia).

The RACER/ENVEST computer software marketed by Talisman Partners Ltd. (Englewood, Colorado) includes cost estimates for these types of mobile units brought to a site:

- Rotary kiln
- Fluidized CBC
- Liquid injection incinerator
- Infrared furnace

The cost-estimating model can also be used for predicting the cost of using a fixed-base incinerator by the program user specifying zero for the distance the unit must be transported. Another ENVEST model, for loading and hauling soil, can then be used to find total fixed-base costs, excluding final disposal of the treated soil.

The ENVEST model was developed for combustible solids (not necessarily soils) with approximately 20% ash content. When soil is incinerated, it sometimes is almost all ash. However, the ENVEST model is useful for estimating soil incineration costs in instances in which a mobile unit is brought to the site. The model includes the options of estimating the costs for shredding and screening the soil, for dewatering solids with very high moisture contents, and for destruction of drummed wastes, and it takes into account the distance that the mobile units have to be transported. Trial burn costs are not included in the ENVEST incineration model. Except for installing a concrete slab, site preparation and utilities costs are excluded, but they can be estimated using other ENVEST models.

The program user designates the type of incinerator and distance from the vendor's yard to the application site and evaluates the following characteristics:

- Whether the material is nonhazardous.
- Physical form — liquid, solid, or sludge.
- Volume — cubic yards, at 1.5 tons/yd^3 (the density of liquid wastes is assumed to be that of water — 0.00417 tons/gal).
- Number of drums — If one or more drums are indicated, the program includes in the estimate the cost of using a drum shredder.

A reasonable estimate can be made by specifying only incinerator type, transport distance, hazardous designation, physical form, and volume. The program user has the option of specifying the values of certain parameters if it is also desired to include the cost of pretreatment. For solids incineration, the user may specify whether the particle size exceeds 1 in. average diameter and whether moisture exceeds 20%. If some particles are larger than 1 in., operation of an electric shredder is included. If moisture exceeds 20%, operation of a belt filter press is included for dewatering to 20%. This reduces the volume of solids being incinerated.

The user also has the option of having the cost of other pretreatment equipment added to the estimate, including:

- Hopper, for transferring solids to a conveyor.
- Screens, for separating particles larger than 1 in.
- Conveyors — The user may specify the length of each conveyor, or the model will automatically select certain conveyor lengths for incoming and outgoing materials and for conveyors connecting screens and shredders.
- Electric shredder — Purchase, versus rental, may be economical if more than 11,000 tons will be incinerated.
- Concrete slab — If the user does not specify the thickness but wants a slab included, 8 in. is assumed.

Other ENVEST models are available for estimating potential associated costs for sitework and utilities:

- Access roads
- Parking lots
- Fencing
- Clearing
- Loading/hauling
- Overhead electrical distribution
- Gas distribution
- Water distribution
- Storm sewer

9.8.2 Desorption Costs

9.8.2.1 Desorption Costs per Ton or per Cubic Yard

Costs are usually expressed in terms of dollars per ton or per cubic yard of feed soil, and they depend on whether a fixed-based desorber or a mobile (portable) unit is used. If the tonnage is high enough, say greater than 2,000 tons, it may be least costly to bring a mobile unit to the site. This avoids paying hauling charges for taking contaminated soil to a remote fixed-base desorber. The operator of a fixed-base unit cannot quote precise costs per ton without first examining samples of the contaminated soil and having the contaminant cleanup concentrations defined. The cost for using a mobile unit are even more difficult to arrive at, because there is a setup charge and a demobilization charge that must be estimated. In addition to the costs of desorption being volume dependent, the type of soil and its moisture content have a significant influence on costs.

Also, the soil bulk density must be taken into account because the mass of soil treated is the product of density times its volume. Because of bulking when excavating and handling soils, the volume fed to the screening equipment is significantly higher than the in-place volume. The volume fed to the desorber is reduced by the amount of large rocks and debris first removed during screening. The parameters that affect fuel cost — heat capacity of the soil, heat of vaporization, moisture content, and Btu value — are mass dependent.

The cost of treatment by thermal desorption is best expressed in dollars per ton at a given moisture content. In estimating costs, a mobile-unit vendor might determine costs per dry ton at various total tonnages for equipment transport, setup, and demobilization; power; water, chemicals, or supplies; fuel; replacement parts; equipment amortization; labor for operations, maintenance, and supervision; and overhead and fees. Then a factor is applied that is proportional to soil moisture content in excess of 10%. The factor depends on the fuel price. At 20% moisture, the factor is typically 1.15 to 1.2; at 30% moisture, it is approximately 1.35. Troxler et al. (1992) present detailed information on the cost per ton of petroleum-contaminated soil and the effect of moisture on cost, as summarized in Table 9-8.

Table 9-8 Thermal desorption treatment costs.

	Costs ($/ton)	
Mobile Units	**5,000 Tons**	**10,000 Tons**
10% Moisture	52	40 to 48
20% Moisture	60	48 to 56
30% Moisture	72	56 to 68
Stationary Units	**Soil Transport Distance 100 Miles**	**Soil Transport Distance 200 Miles**
10% Moisture	47	57
20% Moisture	54	64
30% Moisture	62	72

Based on data in Troxler et al. (1992) and in Cudahy and Troxler (1992), Anderson (1993) tabulated contractor prices (excluding remedial investigation and excavation costs). Soils contaminated both with petroleum and with hazardous substances were considered. For treatment with mobile units, the tabulated prices are summarized in Figure 9-6. From this figure, at quantities greater than 3,000 tons, the cost for petroleum-contaminated soils is less than $100/ton. At 3,000 tons, the cost for hazardous wastes is $250 to $440 per ton.

Baker and Moore (2000) give a comparison of costs for thermal desorption, soil venting, land farming, and bioventing for projects ranging from 500 to 20,000 yd^3 of soil remediated.

DuTeaux (1996) tabulates case histories with costs for low-temperature thermal desorption, summarized as shown in Table 9-9.

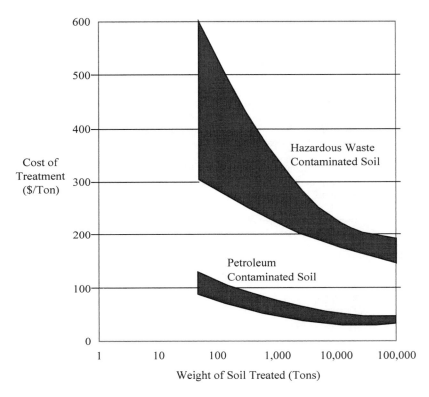

Figure 9-6 Thermal desorption costs.

Table 9-9 Thermal desorption costs from case histories.

Cost	Quantity	Contaminants
$250/yd^3	11,500 yd^3	Halogenated VOC
$197/ton including $59 for mobilization, monitoring, treatability study	4,300 tons	Chlorinated Pesticides
$194/ton including $71 for mobilization, site prep., monitoring, testing	12,755 tons	PCBs
$535/yd^3	10,000 yd^3	(not given)
$352/yd^3	100,000 yd^3	(not given)
$373 to $725/ton	3,000 yd^3	VOC, SVOC
$276/ton including $22 for mobilization, preparation, monitoring including $81 after treatment	(not given)	PCBs
$80/yd^3	3,000 yd^3	TPH, BTEX, PAHs

9.8.2.2 Computerized Equipment Sizing and Cost Estimating for Desorption

Software for estimating desorber applications is included in COMPOSER GOLD marketed by Building Systems Design (Atlanta, Georgia). The RACER/ENVEST™ computer software developed by Talisman Partners Ltd. (Englewood, Colorado) includes cost estimates for direct-fired and indirect-fired units and is described in detail here. The computer model is designed for mobile units, but it can be applied to stationary units by the user, specifying zero for the distance the unit must be transported. Another ENVEST model, for loading and hauling soil, can then be used to find total fixed-base costs (excluding final disposal of the treated soil).

Trial burn costs are not included in the ENVEST model. Except for installing a concrete slab, site preparation and utilities costs are excluded, but they can be estimated using other ENVEST models.

The program user designates the type of firing (direct or indirect) and distance from the vendor's yard and evaluates the following characteristics:

- Whether the material is nonhazardous
- Whether particle size exceeds 2 in., in which case the program includes use of a rotary shear electric shredder
- Volume, in cubic yards, at 1.5 ton/yd^3
- Number of drums — If one or more drums is indicated, the program includes in the estimate the cost of using a drum shredder.
- Moisture content — If this value exceeds 60%, the program assumes that pretreatment includes dewatering with a belt filter press.
- Safety level, ranging from no hazard to Level A (which requires complete worker personal protective equipment, including supplied air for breathing)

A reasonable estimate of system costs can be made by specifying only the following items: whether firing is direct or indirect; transport distance; hazardous designation; whether particles are larger than 5 cm (2 in.); moisture content; safety level; and soil volume. The program user has the option of specifying the values of certain parameters if it is also desired to include the cost of pretreatment.

The user has the option of having the cost of each of these types of pretreatment equipment included in the estimate:

- Hopper, for transferring solids to a conveyor
- Screens, for separating particles larger than 5 cm (2 in.)
- Conveyors— The user may specify the length of each conveyor, or the model will automatically select certain conveyor lengths for incoming and outgoing materials and for conveyors connecting screens and shredders.
- Electric shredder — Purchase, versus rental, may be economical if more than 11,000 tons will be desorbed.

The user does not have the option of specifying whether or not a concrete slab should be added, as in the ENVEST model for incineration. The desorption model automatically includes the cost of a slab.

Other ENVEST models are available for estimating potential associated costs for sitework and utilities:

- Access roads
- Parking lots
- Fencing
- Clearing
- Loading/hauling
- Overhead electrical distribution
- Gas distribution
- Water distribution
- Storm sewer

9.8.3 Total Project Costs for Ex Situ Soil Remediation

In addition to costs entailed in processing soil with materials handling equipment (crushers, screens, conveyors, water mixers, loaders) and with the desorber, an entire soils remediation project includes these other elements:
Fixed or semi-fixed charges (not proportional to tons processed):

- Characterizing the site to determine the nature and extent of contamination
- Characterizing soil samples to determine physical properties, as well as chemical contamination
- Regulatory analysis and permitting
- Project management, planning, scheduling, and reporting
- Pilot testing and analyses
- Trial burn, if RCRA or TSCA wastes are involved, with special plans, sampling/analysis, and reporting costs
- Safety analysis and planning
- Monitoring/analysis of ambient air at the site perimeter
- Transporting and setting up a mobile office and lab, if treatment is on-site
- Geophysical testing to locate potential underground obstructions that could affect excavation

Variable costs (proportional to tons of soil processed):

- Excavating
- Sampling, packaging of samples, and analyses of:
    - Soil at banks or perimeter of excavation
    - Excavated soil
    - Treated soil
    - Process water
    - Stack emissions
    - Residues
- Loading, hauling, and stockpiling excavated soil
- Disposing of residues other than treated soil
- Loading, hauling, and stockpiling treated soil
- Backfilling of or disposing of treated soil

- Post-treating soil, if applicable, such as using fixation of metals

The fixed costs must be estimated for each project, based on regulatory requirements and quotations from consultants, laboratories, and vendors. Most of the variable costs per ton of soil can be predicted from contractors' experience and cost estimating unit prices. Large-scale excavation costs can be as low as $1/ton if no special worker personal protection equipment is needed. Anderson (1993) reports much higher costs. Without obtaining specific bids from contractors, the computerized estimating methods for excavation marketed by Talisman Partners Ltd. (Englewood, Colorado) or developed by the Army Corps of Engineers and marketed by Building Systems Design (Atlanta, Georgia) are recommended. Backfilling costs approximately $3/ton, including spreading in lifts, moistening, and compacting.

Based on data from the US EPA (1992) Anderson (1993) reports hauling costs to be $0.08 to $0.15 per ton-mile for petroleum-contaminated soil and $2 to $4 per ton-mile for hazardous wastes.

Sampling of treated soil is usually done in batches. In cases in which RCRA wastes are involved, a batch might typically be 24 hr of production. A composite sample made up of samples taken every 6 hr is analyzed. If the same laborers who operate the treatment equipment also perform the sampling, packaging, documentation, and sample transport, then the main monitoring cost is for the laboratory analysis costs (and for verification and validation, which can add as much as 40% to analysis costs). Thus, the sampling/analysis cost per ton might be equal to the laboratory analysis fee per sample times 1.4 divided by the daily tons of desorber throughput.

The various contractors with mobile units estimate their costs in different ways. The operating, supervision, and routine maintenance labor and overhead costs may be estimated on a weekly basis. The amount of weekly tons of throughput is a fraction of the treatment capacity, allowing for downtime and repeated treatment of off-specification batches of soil. Also, there might be several weeks of reduced throughput near the startup, awaiting official review of trial burn analysis results. A contractor may be able to estimate average mobilization costs from experience but must consider for each site the following items: installation of an electric power supply or a concrete pad for a portable generator and fuel storage; water piping; natural gas piping, if applicable; site grubbing, clearing, roadways, fencing, and signs; erection of bunkers for batches of treated soil awaiting analysis results; grading and paving with secondary containment and storm runoff control; equipment transport; and equipment erection and later disassembly.

9.9 Summary of Important Points for Thermal Treatment

- Thermal treatment is often followed by another treatment technology, such as fixation if nonvolatile metals are among the soil contaminants of concern at a site.
- The most common soil incinerator type is the rotary kiln. Rotary kilns are used for soils, solid wastes, sludges, slurries, containerized wastes, and liquids.
- The tradeoff among temperature, residence time, and mixing with adequate combustion air determines the organic destruction efficiency of incinerators.

- Temperature is the prime independent variable that affects incinerator fuel consumption and contaminant destruction or removal during field operations.
- Other important parameters that affect fuel consumption are the soil or sludge moisture content and its Btu content.
- Thermal desorption is the vaporization of VOCs and semivolatile organic soil contaminants at temperatures below the ignition point of the contaminants.
- Some incineration systems are not designated specifically for contaminant destruction, but they are used to perform this function while carrying out another process (e.g., cement kilns).
- Rotary kiln incinerators, CBCs, and infrared furnaces have demonstrated effectiveness for destroying a wide variety of organics.
- Infrared furnaces are not used for remediating liquids and sludges.
- Thermal desorption competes with incineration for cleanup of soils contaminated with VOCs and semivolatile organic compounds. (Incineration can achieve higher destruction efficiencies and is needed for heavier, or less-volatile, organic compounds).
- Desorbers generally operate at less than half the temperature of incinerators, use lighter and less-costly equipment, and cost less to operate per ton of soil than conventional incinerators.
- Some desorber systems recover and liquify the organic compounds, but most desorber systems include an afterburner that converts organic vapors to carbon dioxide and water vapor.
- The two main types of desorbers are direct fired and indirect fired, or electrically heated. With direct firing, burner exhaust is in intimate contact with the soil or sludge, producing a larger volume of emissions requiring treatment.
- Indirect-fired desorber units have heat exchange surfaces and therefore are heavier, more complex, and more costly to build and transport than direct-fired units.
- Indirect-fired desorber units can readily be designed to operate with oxygen levels too low for ignition of contaminant vapors.
- Rotary kiln incinerators have residence times of 20 min or more, whereas rotary kiln units for desorbing petroleum contaminants have residence times of approximately 5 min.
- A conservative design for residence time in afterburners is 1 to 2 sec.
- Some incineration systems have automated waste feed cutoffs that stop the feed flow during periods of momentary oxygen depletion. Oxygen lancing minimizes feed cutoff incidents.
- Oxygen-fuel burners result in higher flame temperature (with corresponding improved heat transfer, shorter residence time requirements, and increased throughput capacity) and in reduced gas volume (with less heat input wasted on heating nitrogen, smaller abatement equipment, and less particulate carryover).
- In CBCs, the main separation of particulate mass from flue gas is in a cyclone separator.
- In CBCs, residence time is a few minutes per pass and may exceed 1 hr for some soil particles in the circulating loop.
- Fluidized bed combustors operate at temperatures significantly below those of rotary kilns and many afterburners, thereby producing less NOx.
- Staged introduction of air in fluidized bed combustors results in less NOx.

- Mixing limestone with soil fed to a fluidized bed combustor controls potential acid gases emissions, sometimes negating the need for a wet scrubber.
- When heat recovery is accomplished by generating steam, additional equipment is needed, including chemical water treatment units, a feedwater heater, a deaerator, blowdown systems, and pumps.
- Infrared furnace systems use either electric resistance elements or radiant metal plates heated by gas burners.
- Indirect-fired desorbers are favored over direct-fired units for organic compounds that form unwanted intermediate compounds when in contact with burner products and for soils with organics concentrations high enough to form explosive vapor mixtures with air.
- Thermal desorption is generally applied at soil temperatures of 300° to 700° F for vaporizing organic compounds that have a vapor pressure between 0.5 and 2 atm.
- Rotary kilns are either cocurrent or countercurrent designs. Cocurrent operation can take better advantage of steam stripping effects. With countercurrent operation, a rotary kiln desorber can discharge gases directly into a baghouse without cooling the gases.
- Of the heating duties for the primary burner — warming the soil, vaporizing organics, and vaporizing water — the largest duty with most soils is for vaporizing water.
- Vapor pressure data are used for initial estimates of the final temperature to which soil must be heated; pilot experiments are used to determine temperatures and residence times for reducing contaminant concentrations to the target level.
- With full-scale units, throughput/residence time and temperature are the main parameters that can be controlled.
- Concentrations of soil contaminants, soil particle size, adsorptive tendencies of contaminants, and moisture content affect desorption.
- Feed preparation may include crushing and/or removal of particles larger than 5 cm (2 in.), dewatering, and dust control.
- Treated soils are quenched with water for cooling and dust control.
- Recycling some treated soil through the desorber may be needed if the soil tends to become plastic, such as with moist clays.
- Many vapor treatment schemes are used for recovery or combustion of organic compounds and abatement of acid gases and particulate emissions.
- The most efficient control of particulate emissions, including fine particulates, can be achieved with a baghouse containing fabric filter bags. Because of fabric temperature limitations, flue gas temperature reduction is needed for most incineration systems.
- Baghouse inlet temperatures can be controlled with a water cooler or by admitting dilution air.
- Low-energy wet scrubbers, such as packed towers, control acid gas emissions by contacting the flue gas with water or an alkaline solution.
- Low-energy wet scrubbers have low efficiencies for capturing fine particulates.
- The most common wet scrubber used to control fine particulates is the high-energy Venturi scrubber.
- With an increase in gas pressure drop in a wet scrubber, more flue gas blower energy is consumed, and more fine particulate capture can be attained.

- If a mist elimination vessel is not large enough, drops of scrubbing solution containing captured particulate matter will rain down on the area surrounding the vessel, and particulate emissions limits are likely to be exceeded.
- Other devices used for controlling particulate emissions are the electrostatic precipitator and the wet ionizing scrubber.
- If wet scrubbing of emissions is used, water treatment is needed.
- A quench chamber is commonly used to lower flue gas temperature for subsequent flue gas treatment and consists of a water spray in a chamber where the spray water evaporates. Water in excess of the calculated adiabatic amount must be injected.
- Caustic soda solution, slaked lime, and limestone slurry are common alkaline scrubbing agents for controlling acid gases.
- Formation of nitrogen oxides increases rapidly at temperatures greater than 1,042° C (1,700° F), originating from both nitrogen in the combustion air and in the fuel.
- Scrubbing techniques do not achieve high efficiencies for removal of NOx, but ammonia reduction can achieve more than 80% conversion to nitrogen gas.
- Carbon monoxide forms from incomplete combustion of organic contaminants and auxiliary fuels, and it decreases with an increase in the amount of excess air and the degree of turbulence used in an incinerator.
- Soil analyses aid in determining whether it is a RCRA or TSCA waste and in determining combustion characteristics and emissions control requirements.
- Incinerator calculation methods, including computerized programs for incineration of hazardous wastes, are available from consultants.
- Plume opacity can be calculated as a function of particulate concentration, plume width, light wavelength, and particulate index of refraction, and it depends on particle size distribution.
- Graphical methods of correlating opacity and particulate concentration are sensitive to particle size for particles larger than 1 μm and to refractive index for particles smaller than 1 μm. A conservative correlation avoids evaluating particle size by assuming that the particle size that causes maximum light scattering is totally present in the atmospheric emission.
- If wet scrubber flue gas pressure drop is known and other pressure drops are accounted for, the blower power requirement can be calculated directly from the flue gas volumetric flow rate, total pressure differential, and blower efficiency. If the scrubber pressure drop is unknown, a graphical correlation is available for preliminarily estimating wet scrubber blower horsepower requirements.
- The main concerns addressed during a trial burn are destruction POHCs and control of hydrogen chloride and particulates emissions.
- Desorption treatability studies conducted with an electrically heated laboratory oven can be used to correlate temperature and contaminant concentrations in soil.
- A less sophisticated desorption laboratory oven treatability study includes the determination of soil weight loss, temperature versus time, and contaminant concentration change for a sample on a static tray that is warmed until a target temperature is reached.
- Cost considerations determine whether excavated soil should be hauled to an off-site, fixed-base (stationary) desorber or treated in a portable (mobile) system that

is transported to and set up at the site. Larger volumes or tonnages of soil (e.g., greater than 2,000 tons) favor mobile units.
- Costs depend on volume or mass of soil treated; distance from equipment vendor/contractor; nature of contaminants; vapor pressures and required operating temperature and residence time; type of soil and its particle size; use of indirect versus direct firing; and soil moisture content.
- Computerized cost-estimating programs are available for incineration and desorption that can be applied to both fixed-base and mobile units. For incineration, a reasonable estimate can be made by specifying only the incinerator type, transport distance, hazardous designation, physical form of the soil or sludge, and volume. For desorption, the parameters needed are whether desorption is direct or indirect, the transport distance, hazardous designation, whether particles are larger than 5 cm (2 in.), the safety level, and the volume of soil to be treated.

Chapter 10

Soil Washing

Soil flushing and soil washing remove contaminants from the outer surfaces and pores of soil particles. Soil flushing is an in situ process by which water in large quantities is infiltrated into the soil and recovered by extraction from trenches or wells for subsequent processing. Soil washing is an ex situ process, using water (usually with added agents) or a solvent to remove contaminants from excavated soil. Additives can be used to enhance removal of contaminants from the soil with either in situ soil flushing or soil washing. Water is used in two processes:

- Separation of coarse soil particles from fines
- Removal of contaminants from soil particle surfaces

Water may also be used to slurry the soil so that it can be conveyed by pumping. The technology can be applied for the aqueous removal of either metals or organic contaminants. Alternatively, solvents are used to extract organics from contaminated media. This chapter covers in situ soil flushing and ex situ soil washing and solvent extraction.

The US EPA (1989a) reports removal efficiencies from four different soil washing systems for coarse-grained soil fractions, as given in Table 10-1.

Table 10-1 Contaminant removal efficiencies with soil washing.

Contaminant	Removal Efficiency (%)	Contaminant	Removal Efficiency (%)
Mineral oil	98	Aromatics	81 to 99.8
Cyanides	94	PAHs	95
Zinc	67 to 83	Crude oil	97
Cadmium	92	Hydrocarbons	96
Nickel	66 to 89	Chlorinated hydrocarbons	100
Lead	75	Phenol	100

Note that for the organic compounds, biodegradation may account for some of the reported removal efficiency. For soluble organics, such as phenol, removal to nondetect levels is expected. Part II of Table 3 in the US EPA (1990c) gives the percentage of contaminant removal for a number of United States and European soil washing schemes that have been marketed or pilot tested for sandy soils. Some of the reported removal efficiencies and residual contaminant concentrations are as given in Table 10-2.

Table 10-2 Contaminant removal efficiencies and residual concentrations.

Contaminant	Efficiency (%)	Residual Concentration (mg/kg)
Oil & Grease	50 to 83	250 to 600
PCP	90 to 95	<115
Arsenic trioxide	50 to 80	0.5 to 1.3
VOCs	98 to greater than 99	<50
Semivolatile organics	98 to greater than 99	<250
Most fuel products	98 to greater than 99	<2,200
Aromatics	>81	45
PAHs	95	15
Crude oil	97	2,300
PCBs	84 to 88	0.5 to 1.3
Cyanides	95	5 to 15
Heavy metal cations	Approximately 70	<200
Oil	95 to 99	20

Exner (1995) indicates that with soil washing, coarse soil fractions generally have cleanup efficiencies of 90% to 99% for volatiles, 80% to 95% for semivolatiles, and 50% to 90% for metals. With solvent extraction, organics removal efficiencies generally range from 90% to 99%.

10.1 Basic Principles of Soil Washing

Fine soil particles have large pore-surface areas relative to volume, and contaminants tend to stay adsorbed, so fine-grained soils are difficult to wash. This tendency is stronger when removing organic contaminants with water than with a nonaqueous solvent.

According to Anderson (1993), soil particles smaller than 63 μm tend to be attached loosely to the coarser particles. When water is used as the extractant, the physical attachment forces (adhesion and compaction) are effectively broken by using attrition mills (see Section 10.4.2). The contaminants mainly stay with the fine fraction when separated from sand and gravel fractions. Separating out the readily washed coarse fractions reduces the volume of soil that must be further treated or disposed.

A wide variety of contaminated soils exist, containing 30% to 60% or more of coarse-fraction sands and gravels that adsorb relatively small amounts of contaminants and are readily washed. The volume of soil that remains contaminated after washing is much less than the original volume. With the particle size separation that aqueous soil washing can accomplish, sandy soils may undergo a volume reduction of more than 80% for final cleanup or disposal. However, successful metals removal from the coarse fraction may require acid or chelant additions to the wash water. And thorough organics removal from sand fractions containing iron oxide may not be possible (Dove, Bhanduri, and Novak, 1992,1993).

Soil washing is not often used for soils with different types of contaminants, such as mixtures of metals, volatile organics, and nonvolatile organics, unless sequential washing steps are used with different additives.

The fundamental process involved in the removal of contaminants from soil particle surfaces, once particle separation has been accomplished, is the use of an extracting fluid with a higher affinity for the contaminants of interest than the soil organic matter. This drives the contaminants off the soil surface and into the fluid. The relative concentration of a contaminant at equilibrium between two phases in contact with one another is described quantitatively by an equilibrium distribution coefficient, K, by the following definition:

$$K = \frac{\text{Concentration in Phase 1, mass/volume or mass}}{\text{Concentration in Phase 2, mass/volume or mass}} \tag{10-1}$$

The equilibrium concentration between a soil and water system is described by the distribution coefficient, K_d:

$$K_d = \frac{\text{Concentration in Soil, mass/volume}}{\text{Concentration in Water, mass/volume}} \tag{10-2}$$

The equilibrium concentration between water and a representative organic solvent, octanol, is described by the the octanol/water partition coefficient, K_{ow}:

$$K_{ow} = \frac{\text{Concentration in Octanol, mass/volume}}{\text{Concentration in Water, mass/volume}} \tag{10-3}$$

The extent of sorption of many contaminants to soil surfaces is highly correlated to the amount of organic carbon in the soil, leading to an organic carbon normalized soil-water coefficient, K_{oc}, described mathematically as:

$$K_{oc} = K_d/\text{decimal fraction of soil organic carbon} \tag{10-4}$$

Standard methods exist for the determination of these distribution coefficients from a variety of sources. These standard methods provide details of experimental methods and procedures that should be used to determine these coefficients reliably, and they include recommended experimental mixing vessel sizes and configurations, methods for determining equilibration times, recommended volumes of each phase, etc. Readers are referred to American Society of Testing and Materials (ASTM) methods (e.g., ASTM Standard E 1147-87 for K_{ow} determinations) for reference procedures for measuring these coefficients.

The efficiency of solvent extraction should be directly related to the nature of the organic contaminant in terms of its K_{ow} and K_{oc} values (i.e., relative affinity for an organic solvent and soil surface, respectively). For highly hydrophobic compounds with large K_{ow} values (i.e., greater than 10,000), the use of an organic washing solution takes advantage of the contaminant's organic-phase affinity and, by mass action, can effectively remove contaminant from the soil organic matter. For hydrophilic organic compounds (phenols, aromatics, and low molecular weight halogenated hydrocarbons) with low K_{ow} values (i.e., from 1 to 1,000), aqueous-based washing solutions should be selected to optimize recovery of these compounds from contaminated soil because of these compounds' high aqueous-phase affinity.

The organic carbon fraction of soils tends to concentrate contaminants and makes soils difficult to wash. For contaminants with a large K_{ow} in a soil high in organic carbon, the contaminant equilibrium distribution shifts to the soil phase, and aqueous-based washing solutions become ineffective. To increase the efficiency of soil washing in these soils, a more aggressive washing solution that is organic solvent-based must be used to drive the contaminant into this organic extraction fluid. Generally, soils with a high organic content are not good candidates for soil washing.

10.2 In Situ Soil Flushing

Flushing is accomplished by flooding the land surface within a berm, by using injection wells, by heavily spraying water on the land, or by applying water through an infiltration gallery. The water infiltrates the soil and desorbs contaminants from soil particles. Horizontal flushing is possible in a shallow soil layer just above an aquitard (Anderson, 1993). In situ flushing is not generally applicable if more than one type of contaminant is encountered, nor is it applicable in fine-grained or highly heterogeneous soils or at sites where hydraulic control of injected water cannot be assured.

The wash water is recovered by pumping from wells screened below the water table or from trenches and subsurface drains above the water table. The recovered water is treated, and usually most of it is recycled by reapplying it to the soil.

Anderson (1993) indicates that flushing is applicable for removing hydrocarbons, chlorinated hydrocarbons, metals, salts, pesticides, herbicides, and radioisotopes. The technology is less effective if pockets of soil with low hydraulic conductivity exist or if the contaminants are relatively insoluble or tightly bound to the soil.

Contaminants that have a low K_{ow} (i.e., less than 10) are good candidates for removal by flushing (Hyman and Bagaasen, 1997). Additives that might be used include sulfuric, hydrochloric, nitric, phosphoric, or carbonic acid; alkaline agents; and surfactants (detergents) (US EPA, 1990c). Table 10-3 summarizes the properties of various surfactants that can be used for hydrophobic organic contaminants.

Nutrients can be added to recycled water for biodegradation of organic contaminants that can take place in the vadose zone soils and/or in an aquifer. However, permitting problems may be formidable even for adding just water, as concerns often arise over the potential of the flush water to transport contaminants into an aquifer or to uncontaminated soil. Using acids, bases, surfactants, or nutrients raises even more concerns.

Table 10-3 Surfactant characteristics.

Surfactant Type		Selected properties and uses	Solubility	Reactivity
Anionic	1) Carboxylic acid sails	Good detergency	Generally water-soluble	Electrolyte tolerant
	2) Sulfuric acid ester sails	Good wetting agents		Electrolyte-sensitive
	3) Phosphoric and polyphosphoric acid esters	Strong surface tension reducers	Soluble in polar organics	Resistant to biodegradation
	4) Perfluorinated anionics			High chemical stability
	5) Sulfonic acid salts	Good oil-in-water emulsifiers		Resistant to acid and alkaline hydrolysis
Cationic	1) Long chain amines	Emulsifying agents	Low or varying water solubility	Acid stable
	2) Diamines and polyanilines	Corrosion inhibitor		
	3) Quaternary ammonium salts		Water-soluble	Surface adsorption to silicaceous materials
	4) Polyoxyethylenated long-chain amines			
Nonionic	1) Polyoxyethylenated alkylphenols alkylphenol ethoxylates	Emulsifying agents	Generally water-soluble	Good chemical stability
	2) Polyoxyethylenated straight chain alcohols and alcohol ethoxylates	Detergents	Water insoluble formulations	Resistant to biodegradation
	3) Polyoxyethylenated polyoxypropylene glycols	Wetting agents		Relatively nontoxic
	4) Polyoxyethylenated mercaptans	Dispersents		
	5) Long-chain carboxylic acid esters	Foam control		Subject to acid and alkaline hydrolysis
	6) Alkylamine "condensates," alkanolamides			
	7) Tertiary acetylenic glycols			
Amphoterics	1) pH-sensitive	Solubilizing agents	Varied (pH-dependent)	Nontoxic
	2) pH-insensitive	Wetting agents		Electrolyte tolerant
				Adsorption to negatively charged surfaces

The US EPA (1990b) lists these information needs for considering potential application of soil flushing:

- Characterization and concentration of contaminants
- Depth and vertical and aerial distribution of contaminants
- Partitioning of contaminants between solvents and soil
- Effects of washing agent on physical, chemical, and biological properties of soil
- Suitability of site for flooding and for installation of wells or subsurface drains
- Site-specific groundwater flow rate and direction
- Trafficability of soil and site

Anderson (1993) tabulates these factors for predicting success of soil flushing:

- Soil hydraulic conductivity at least 10^{-5} cm/sec, preferably greater than 10^{-3} cm/sec
- Soil carbon content less than 10%, preferably less than 1%
- Contaminant water solubility at least 100 mg/L, preferably greater than 1,000 mg/L
- Soil sorption constant, Kd, less than 10,000 L/kg, preferably less than 100 L/kg
- Contaminant vapor pressure less than 100 mm Hg, preferably less than 10 mm Hg
- Contaminant liquid viscosity less than 20 cp, preferably less than 2 cp
- Contaminant liquid specific gravity greater than 1, preferably greater than 2

10.3 Soil Washing and Solvent Extraction

Traditional ore refining techniques in the mining and metals industries include separating fines, which tend to be relatively enriched in metals. Equipment used in those industries has been adapted to soil washing with water. The coarse materials are separated by the washing process, producing clean coarse sands and gravel, thereby reducing the volume of soil that needs further treatment.

Besides the size of the soil particles, other soil characteristics that affect washability include humic content and cation exchange capacity. Contaminants partition strongly to humic material, making washing of high humic content soils difficult. A potential candidate soil for washing is at least 60% greater than a 63-μm particle size and has less than 20 wt% organic matter content. The US EPA (1991b) indicates that soils with at least 50 wt% sand/gravel (particle size greater than 200 μm) wash the best. Soils with high clay and silt content are not good candidates for soil washing.

Equilibrium distribution studies with shaking of a soil sample in water, or with other extractant fluid, determine contaminant partitioning. Another important laboratory test is the determination of soil cation exchange capacity. The lower the cation exchange capacity, the better soil washing can succeed, especially for removal of cationic metals.

Equilibrium distribution data help assess whether soil washing is feasible and what extractant might work best. Table 10-4, adapted from the US EPA (1990c), summarizes physical and chemical characteristics that affect soil washing.

Table 10-5, from the US EPA (1989a), indicates that for sand, silt, and clay, respectively, for three types of contaminants, the following equipment is needed: some candidate wash media (extractants); applicable extraction equipment and solid-liquid separation equipment; and potential methods of treating wash media for recycle (regeneration"). Note that the suggested equipment for extraction is not a proven technology in cases in which silts and clay soils are involved. Some examples will illustrate how Table 10-5 is used. The extractor-type inclined screw is applicable for washing sand contaminated with hydrophobic and hydrophillic nonvolatile organics but not with heavy metals/inorganics. Looking at the bottom three rows of the table, which apply to clay, an inclined screw is not applicable, whereas a stirred tank extractor is; water with pH control is an applicable extractant for removing

Table 10-4 Physical and chemical characteristics that affect soil washing.

Key Physical Parameter	
Particle size distribution	
>2mm	Oversize pretreatment requirements
0.25 – 2 mm	Simple soil washing
0.063 – 0.25 mm	Complex soil washing
<0.063 mm	Clay/silt fraction – unsuited for soil washing
Other Physical Parameters	
Type, specific gravity, physical form, handling properties	Affects pretreatment and transfer requirements
Moisture content	Affects pretreatment and transfer requirements
Key Chemical Parameters	
Organics concentration, volatility, partition coefficient	Determine contaminants and assess separation and washing efficiency, hydrophobic interaction, washing fluid compatibility, changes in washing fluid with changes in contaminants. May require preblending for consistent feed. Use the jar test protocol to determine contaminant partitioning.
Metals	Concentration and species of constituents (specific jar test) will determine washing fluid compatibility, mobility of metals, post-treatment.
Humic Acid	Organic content will affect adsorption characteristics of contaminants on soil. Important in marine/wetland sites.

hydrophilic nonvolatile organics but not for removing hydrophobic nonvolatile organics.

The best washing medium to use at a specific site depends on the type of contaminant being removed. A summary of washing media (extractants) is as follows:

- Cationic metals — acid or chelating agent
- Amines, ethers — acid
- Anionic metals — hydrogen peroxide
- Insoluble organic compounds — water with surfactant or organic solvent
- Polar organic compounds — water or liquid carbon dioxide

In cases in which acids are used for removing cationic metals, the pH of the wash water solution is usually just below 3. Chelants are ligands that are applied only to remove cationic metals by forming, at the molecular level, an organic ring structure around and then bonding to positively charged metal ions.

Metals in anionic states, such as arsenite or arsenate, or in an anionic complex, such as a metal cyanide ion, are difficult to remove by soil washing. Acid will not work unless the metal is first treated with a strong oxidizing agent, such as chlorine or peroxide, to change its oxidation state or to destroy the anion complex.

Table 10-5 Soil-contaminant technology matrix.

Category	Subcategory	15.3	2.3	12.5	3.4	0.6	5.1	2.3	0.6	1.7
SOIL TYPE	• SAND	X	X	X						
	• SILT				X	X	X			
	• CLAY							X	X	X
CONTAMINANT TYPE	• NONVOLATILE ORGANIC HYDROPHOBIC	X			X		X			
	• NONVOLATILE ORGANIC HYDROPHILLIC		X			X			X	
	• HEAVY METALS/INORGANICS			X		X				X
	• FREQUENCY OF OCCURRENCE (%)	15.3	2.3	12.5	3.4	0.6	5.1	2.3	0.6	1.7
EXTRACTANT	• WATER		X		X			X		
	• WATER/SURFACTANT	X			X					
	• WATER/CHELATE			X			X			X
	• WATER/ACID/BASE	X		X	X		X	X		X
	• WATER/pH CONTROL		X			X			X	
	• SOLVENT WASH	X			X					
EXTRACTOR TYPE	• STIRRED TANK	X	X	X	X	X	X	X	X	X
	• INCLINED SCREW	X	X							
	• ENDLESS BELT		X			X				
SOLID-LIQUID SEPARATION	• CORRUGATED PLATE GRAVITY SEPARATOR				X	X	X	X	X	X
	• HYDROCYCLONE				X	X	X			
	• BASKET/PUSHER CENTRIFUGE	X	X	X						
	• VACUUM FILTER	X	X	X						
EXTRACTANT REGENERATION	• CHEMICAL/BIOLOGICAL OXIDATION	X	X		X	X		X	X	
	• HYDROLYSIS	X			X			X		
	• CARBON FILTRATION	X	X		X	X		X	X	
	• ION EXCHANGE			X			X			
	• PRECIPITATION/ELECTROLYSIS			X			X			
	• CONDENSATION/INCINERATION									

For water washing organic compounds such as hydrocarbon oils, surfactants and alkaline agents such as sodium hydroxide (caustic soda) are used. Just as oil can be obtained from tar sands by extracting it with sodium hydroxide (caustic soda) solution, hydrocarbons can also be extracted from sand or gravel with alkaline agents. Also, alkaline agents improve extraction efficiency when washing soils with high organic content, for which aqueous surfactant washing is not applicable. Table 10-6 from the US EPA (1989a) gives more details on what reagents are used in extractant fluids.

When washing soil with aqueous solutions, heating the water usually will aid in mobilizing contaminants and will improve removal efficiency.

After selecting potential extractants and performing equilibrium distribution tests, pilot tests or treatability studies (discussed in detail in Section 10.5) can be conducted using the selected reagents to determine:

- Contaminant removal efficiency
- Volume fraction of soil fines that are not cleaned
- Ratio of wash fluid to soil and contact time needed during wash steps
- Treatment requirements for used wash fluid
- Percent of wash fluid that can be recycled

For extraction of contaminants with nonaqueous solvents, tests should be conducted to check for residual solvent in the washed soil.

Multi-stage wash steps and size classification steps can be applied. For example, soil with oil and metals contaminants might be subjected to:

- Coarse screening to remove large rocks and debris
- Magnetic separation of ferrous materials
- Screening of 2-in. and larger cobbles, which are rinsed with water applied at the screen
- Fine screening or classifying to separate out particles larger than 5 mm (3/16 in.)
- Attrition scrubbing of these larger particles with surfactant to remove oil
- Classifying to separate out particles ranging from 0.25 to 5 mm (0.01 to 3/16 in.)
- Washing these medium-coarse particles with acid or chelants to remove metals
- Filtering and water rinsing the filter cake
- Centrifuging (dewatering) and vacuum drying the fine soil fractions
- Solvent extracting the remaining organic contaminants from the fine soil particles

Many of the surfactants used for soil washing can be treated by bacterial degradation. Acid wash water and rinse water may need treatment with neutralization, precipitation, or evaporation before disposal.

10.3.1 Aqueous Soil Washing for Particle Size Separation

Figure 10-1 shows a soil washing scheme with size separation equipment. The bulk of the soil fractions produced is clean gravel and sand. The fine-grained materials

Table 10-6 Summary of process parameters and reagents for various contaminant groups.

Contaminant Group	Contaminant Examples	Process Parameters	Potential Reagents
Hydrophilic Organic Compounds	Alcohols (e.g., methyl, isopropyl, butyl)	Wash fluid pH	Alkaline pH
	Phenols (e.g., picric acid, pentachlorophenol, creosote)	Humic content in soil	Caustic or Na_2CO_3 for dispersing humus
		Degree of agitation	Wetting agent
	Dioxane	Time, soil loading, and staging	
	Urethane	Wetting agent	
	Rocket fuel		
Slightly Hydrophilic Organic Compounds	Aromatics (e.g., benzene, toluene, xylene)	Wash fluid pH	Alkaline pH
		Humic content in soil	Caustic or Na_2CO_3 for dispersing humus
	Halogenated hydrocarbons (e.g., trichloroethylene, ethylene dichloride vinyl chloride, methylene chloride)	Degree of agitation	Wetting agent
		Time, soil loading, and staging	
		Wetting agent	
	Chloroform		
	Trichloroethane		
Hydrophobic Organic Compounds	Oil and grease	Use of surfactants	Surfactants (e.g., Adse 799, Hyonic NP-90, P & G Institutional Formula Tide®)
	Chlorinated hydrocarbons (e.g., endrin, lindane, DDT, dieldrin)	Caustic agent	
		Extraction stages	
	Polynuclear aromatics (PAHs)	Degree of agitation	Caustic or Na_2CO_3 for dispersing humus
		Temperature	
		Reactor configuration	Water-soluble solvents (e.g., acetone, ethanol)
		Soil:solution ratio	
Volatile Organics	Hexane	Extraction stages	Water (hot)
	Ethylbenzene	Degree of agitation	Coal
		Temperature	
		Reactor configuration	
Heavy Metals	Mercury	Effects of other metal cations	Chelants (e.g., EDTA, DTPA)
	Nickel	Effect of other anions	Acids (e.g., hydroxyl-amine hydrochloride, citric, nitric, aqua regia, acetic, fluosilicic)
	Zinc	Soil classification	
	Lead	Temperature	
	Chromium	Chelant or acid concentration	
	Arsenic	Chelation duration	
	Cadmium	Soil loading	
	Copper	Wash fluid pH	
Polychlorinated Biphenyls	PCBs	Use of surfactants	Surfactants (e.g., Adse 799, Hyonic NP-90, P & G Institutional Formula Tide®)
		Extraction stages	
		Degree of agitation	
		Temperature	
		Reactor configuration	
		Soil:solution ratio	
Radioactive Materials	Uranium mining and purification wastes	Wash fluid pH	Water
		Soil:solution ratio	Inorganic salts (e.g., NaCl, KCl)
	Radium	Time	Acids (e.g., HCl, HNO_3)
	Tritium	Temperature	Chelants (e.g., EDTA, DTPA)
Other Inorganics	Cyanides	Depends on contaminant(s) present	Depends on contaminant(s) present
	Acids (e.g., sulfuric)		
	Alkalis (e.g., lime, ammonia)		
	Sulfides		
	Beryllium		

and sludge remaining would typically need a different type of treatment, such as fixation with cement.

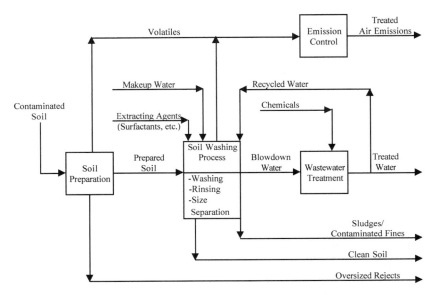

Figure 10-1 Aqueous soil washing process (From the US EPA 1990b).

Depending on how high the sand content is, soil washing can reduce the volume needing more expensive treatment (or permitted disposal) by a very large percentage. A demonstration unit operated at 5 to 10 ton/hr by Bechtel Hanford, Inc. (Richland, Washington) produced an 85% clean fraction. Alternative Remedial Technologies (Tampa, Florida) reports a metals removal case history with 83% of the volume cleaned with the following treatment scheme:

- Screening out oversized material
- High-pressure spraying to break clods and form a slurry
- Separating coarse- from fine-grained materials using a hydrocyclone
- Using air flotation for the coarse fraction from the bottom of the hydrocyclone
- Using Lamella (slant-plate) clarifiers dosed with polymers to settle fine-grained materials
- Filtering, which produced 55% dry solids cake

Hydrocyclones use a combination of centrifugal and gravitational forces to separate coarse fractions. Other particle separation techniques are described in Table 10-7.

Some of the major soil dewatering techniques are described in Table 10-8.

10.3.2 Solvent Extraction for Removing Organic Contaminants

Solvent extraction is more effective than aqueous extraction for removing insoluble organics. When solvent extraction is used instead of water washing, contaminant

Table 10-7 Particle separation techniques (From Eagle, M.C., et al. "Soil Washing for Volume Reduction of Radioactively Contaminated Soils," *Remediation*, Summer 1993. Copyright 1993 John Wiley & Sons, Inc. Reprinted by permission of John Wiley & Sons, Inc.).

Technique	Sizing	Settling Velocity	Specific Gravity	Magnetic Properties	Flotation
Common Name	screening	classification	gravity separation	magnetic	flotation
Basic Principle	various diameter openings and effective particle size	faster vs. slower settling, particle density, size, shape of particles	differences in density, size, shape, and weight of particles	magnetic susceptibility	suspend fines by air agitation, add promoter/collector agents, skim oil froth
Major Advantage	inexpensive	continuous processing, long history, reliable, inexpensive	economical, simple to implement, long history	simple to implement	very effective for some particle sizes
Major Disadvantage	screens can plug, fine screens are fragile, dry screens produce dust	difficulty with clayey, sandy, and humus soils	ineffective for fines	high operating costs	contaminant must be small fraction of total volume
General Equipment	screens, sieves	mechanical, non-mechanical hydrodynamic classifiers	jigs, shaking tables, troughs, sluices	magnetic separators	flotation machines
Lab Test Equipment	vacuum sieve/ screen, trommel screen	elutriation columns	jig, shaking table	lab magnets	agitair laboratory unit

Table 10-8 Dewatering techniques (From Eagle, M.C., et al. "Soil Washing for Volume Reduction of Radioactively Contaminated Soils," *Remediation*, Summer 1993. Copyright 1993 John Wiley & Sons, Inc. Reprinted by permission of John Wiley & Sons, Inc.).

Technique	Filtration	Centrifugation	Sedimentation	Expression
Basic Principle	passage of particles through porous medium, particle size	artificial gravity settling: particle size, shape, density, and fluid density	gravity settling: particle size, shape, density, and fluid density; flocculent aided	compression with liquid escape through porous filter
Major Advantage	simple operation more selective separation	fast, large capacity	simple, less expensive equipment, large capacity	handles slurries difficult to pump, drier product
Major Disadvantage	batch nature of operation, washing may be poor	expensive, more complicated equipment	slow	high pressure required, high resistance to flow in cases
General Equipment	drum, disk, horizontal (belt) filters	solid bowl sedimentation and centrifugal perforated basket	cylindrical continuous clarifiers, rakes, overflow, lamella, deep cone thickeners	batch and continuous pressure
Lab Test Equipment	vacuum filters, filter press	bench or floor centrifuge	cylindrical tubes, beaker, flocculents	filter press, pressure equipment

removal is not limited to coarse fractions such as sand and gravel only. Solvent extraction can be applied to a much wider range of soil particle sizes. However, separation of solvent from clay soils may be a problem, and centrifugation or thermal desorption may be needed (US EPA, 1989a). A disadvantage of solvent extraction systems is there is a risk of leaving significant amounts of solvent in the treated soil. Also, some solvents are mildly toxic and/or flammable.

In choosing a solvent, besides toxicity and safety/handling considerations, the following factors are important: ability of the solvent to dissolve the contaminants, especially in the presence of soil moisture; volatility and the potential for air emissions if the soil wash system is not completely enclosed and vapors are not controlled; separability of the solvent/contaminant mix from the soil; and separability of the contaminants from the solvent so that the solvent can be recycled.

The common organic solvents that are used to extract hydrocarbon oils and other nonpolar organic contaminants are hydrocarbons with a boiling temperature range selected so that the range is lower than the boiling temperature range of the contaminants. Then the solvent can be recovered for reuse by distilling it or by stripping it from the contaminants with steam or hot nitrogen. The US EPA (1992a) reports that solvent extraction is effective for removing petroleum wastes, halogenated solvents, and PCBs from sediments, sludges, and soils. The preferred hydrocarbon solvents are alkanes. Alcohols are effective solvents that do not suffer from moisture interference as much as hydrocarbons do, but they are expensive.

If the contaminated soil is wet, solvent extraction with hydrocarbons is improved by first evaporating the water, preferably under vacuum. Under vacuum conditions, the water will evaporate rapidly without raising the temperature to 212° F. An example of such a solvent extraction system is the Carver-Greenfield system marketed by Dehydro-Tech Corporation (East Hanover, New Jersey). Note that in the water vacuum evaporation step, some oil may be vaporized with the water. The water vapor and any evaporated oil are condensed, and the oil is decanted from the water in an oil/water separator.

Solvent extraction may have to be repeated for a given batch of soil in order to achieve effective contaminant removal. This can be accomplished by using an extraction system with multiple stages, by recycling treated soil, or by repeated mixing of soil with solvent and successive solvent withdrawals.

An emerging technology is the use of liquified petroleum gas (LPG) or carbon dioxide as the extracting solvent. These two compounds liquify by being placed under pressure, by chilling, or by a combination of pressure and cooling. The LPG or carbon dioxide solvent can readily be recovered in the vapor phase by depressuring, or warming the solvent/contaminant mixture drained from the soil. Such solvent extraction systems are marketed by C. F. Systems (Woburn, Massachusetts). Liquified propanes and butanes extract nonpolar or hydrophobic organics. Residual solvent, which contains contaminants, is displaced from the soil with warm water. Liquified carbon dioxide extracts polar organics. (Fluid extraction using supercritical carbon dioxide is not limited to polar organics.) The high volatility of propane or carbon dioxide solvent minimizes residual solvent concentrations in the treated soil. The process is effective for removing PCBs, PAHs, dioxins, and TPHs (Anon., 1993).

An innovative method of recovering solvent has been developed by Resources Conservation Co. (Bellevue, Washington), with a system that uses triethylamine (TEA). This compound's water miscibility changes markedly over a relatively small temperature range. At temperatures less than 60° F, TEA is used to dissolve organic compounds that are not soluble in water. The mix is chilled to a somewhat lower temperature at which TEA is miscible with water. A mixture of TEA, organic contaminants, and water is removed from the washing process and warmed to at least 70° to 160° F. The process is described in the EPA Office of Solid Waste and Emergency Response publication "Tech Trends," No. 11, January 1993 (EPA/542/N-93/001) as follows:

> "Contaminated material is screened to less than 1/2-inch diameter (1/8-inch for this demonstration) and added to a refrigerated premix tank with a predetermined volume of 50% sodium hydroxide. After the tank is sealed and purged with nitrogen, chilled triethylamine solvent is added. The chilled mixture is agitated and then allowed to settle, creating the non-homogenous mixture of moisture-free solids and the solution of solvated oil, water and solvent. The solution is decanted from the solids and centrifuged. The solvent and water are removed from the solvent/water/oil mixture by evaporation and condensation of the solvent and water. Solids with high moisture content may require more than one cold extraction. For example, for this demonstration, a sediment containing 41% moisture required two cold extractions.
> Once a sufficient volume of moisture-free solids is accumulated, it is transferred to a steam jacketed extractor/dryer where warm triethylamine is added to the solids. The mixture is heated, agitated, settled and decanted to separate any of the organics not removed during the initial cold extraction. The solids remaining in the extractor/dryer contain triethylamine following decanting. A small amount of steam is injected to volatilize this remaining triethylamine. The hot extraction process can be repeated, when necessary, to further remove contaminants.
>
> The products from the process are: (1) solids, (2) water and (3) concentrated oil containing the organic contaminants. The recovered oil fraction can be dechlorinated or incinerated to destroy the organics. The triethylamine is recovered and reused in further extractions."

10.4 Main System Design Parameters for Soil Washing

10.4.1 Conceptual Designs

The first important steps in applying soil washing are developing a flow scheme and balancing the mass flow rates of wash solution and soil. Because most soil washing schemes involve separating fractions of soil by particle size range and forming slurries, the mass balance should be done for each size fraction, and the percentage of solids content (or water content) should be determined for each step of the process. Then the equipment can be sized.

Equilibrium distribution studies and treatability test data are needed to estimate the amount of contaminant reduction and the particle size separations that might be attained, the water/soil ratio needed, additives requirements, and water treatment conditions. The US EPA (1991b) describes applicability of jar tests and gives details for planning pilot tests.

Some of the equipment is selected based on producing washed product size fractions that do not have excessive water content. A rough approximation of the particle size distribution that may result from a given washing and separation step can be predicted from the type of equipment selected. For example, Besendorfer (1996) correlates separation percent for various size particulates when solids in water are treated with hydrocyclones.

Minimizing the water/soil ratio is important, as this results in smaller equipment size and pumping horsepower requirements. Water treatment steps are chosen that allow maximum recycle of wash water. It is also desirable to minimize the volume of water discharge that will need metals removal, organics destruction, or other final treatment.

10.4.2 Mass Balances

An example of a flow scheme and mass balance developed for a soil washing system is shown in Figure 10-2. Refer to Table 10-9 for equipment identification.

The following figures show major components (as listed in Table 10-9) of this scheme:

- Figure 10-3 shows the mass balance for the trommel screen Y-1 and coarse flat-deck screens SC-2. A trommel screen is a water-washed cylindrical screen rotating within a drum. This step breaks down agglomerates, somewhat classifies the soil by particle size, and washes the coarse fraction.
- Figure 10-4 shows the mass balance for the spiral classifier Y-2 and a flat-deck screen. A spiral classifier is an inclined device that is effective in separating out particles larger than approximately 2 mm average diameter.
- Figure 10-5 shows the mass balance for a series of attrition mills Y-3 and hydrocyclones Y-4 and Y-5. Attrition mills wash the medium-sized particles with water using two opposed-pitch, relatively high-speed mixers in each mill. Cyclones separate larger-sized particles by centrifugal and gravitational action, with no moving parts — just a high-velocity, horizontal, tangential inlet for the water/soil slurry.
- Figure 10-6 shows the mass balance for flotation cells Y-7 and a belt filter Y-8. Surfactant is first mixed in a conditioner, and froth is formed in the cells. The finer particles are trapped by the froth bubbles and are skimmed off the tops of the cells.

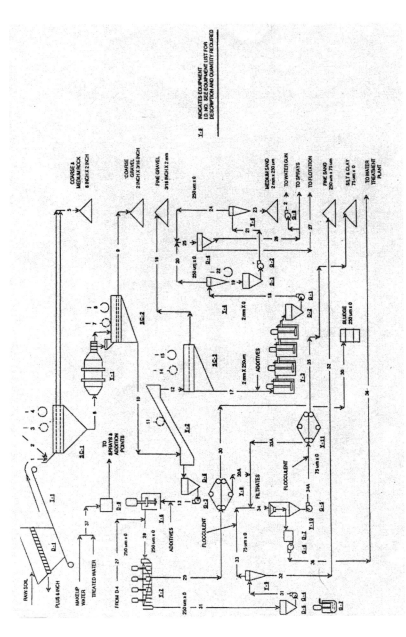

Figure 10-2 Example of a soil washing scheme.

Table 10-9 Generic soil washing major equipment list.

SI.NO.	Title	I.D. No.	Description	Qty
1	Dump Hopper	D-1	Volume: two times the dumping bucket Material: CS with liners for sloping sides Accessories- Bar grid with 6-in. opening and slide gate at outlet	1
2	Feeder Conveyor	T-1	Capacity: 25 tph Width and Length: to suit	1
3	Rock Screen	SC-1	Type: Horizontal Vibrating No. of Decks:- Two Feed Capacity: 25 tph Dock Openings: 4 inch and 2 inch	1
4	Rotary Scrubber	Y-1	Feed Capacity: 20 tph Size (dia x length): 4 ft x 12 ft (inside liners Liners: Mn steel Internals: Lifters Feed and Discharge Sleeves: Lined and with spiral flights Discharge trommel: Required Retention time: 3 minutes	1
5	Coarse Gravel Screen	SC-2	Type: Horizontal Vibrating No. of Decks: Two Feed Capacity: 25 tph Deck Openings: 1 inch and 3/16 inch	1
6	Spiral Classifier	Y-2	Flow rate (Slurry): 165 gpm Solids: 22 tph Solid Sp. Gr.: 2.6 Water: 135 gpm Sands: 10 tph (dry basis) Classification: 250 microns Accessories: Water Sprays at the beach Manual hydraulic lift for screw	1
7	Fine Gravel Screen	SC-3	Type: Horizontal Vibrating No. of Decks: One Feed Capacity: 10 tph (dry basis) Deck Openings: 2 mm- wedge wire	1
8	Attrition Scrubber	Y-3 (A through D)	Flow rate (Slurry): 100 gpm Solids percentage: 25 to 30 % Number Required: 4 Arrangement: 4 in Series Retention Time: 20 mins (total) Approximate Size: (Dia x hi): 4 h x 5 h Impellers: Turbine Liners: Tanks and Shafts lined for abrasion resistance- Include 'Sand Relief' Maximum Particle Size: 2 to 3 mm silica	4
9	Pri. Sand Cyclone Feed Sump	D-2	Volume: 600 gals Type: Conical with cylindrical top Material: CS Accessory: Mixer for keeping contents in suspension	1

Table 10-9 Generic soil washing major equipment list (continued).

Sl.NO.	Title	I.D. No.	Description	Qty
10	Pd. Sand Cyclone Feed Pump	G-1	Type: Centrifugal, Horizontal, Slurry Flow rate (Slurry): 100 gpm Sp. Gr. of Slurry: 1.2 TDH: to suit Arrangement Speed Control: Mechanical Variable -Remote	1
11	Primary Sand Cyclone	Y-4	Flow rate (Slurry): 100 gpm Solids: 7.2 tph Solid Sp. Gr.: 2.6 Sp. Gr. of Slurry: 1.2 Classification: 250 microns	1
12	Sec.Sand Cyclone Feed Sump	D-3	Volume: 850 gals Type: Conical with cylindrical top Material: CS Accessory: Mixer for keeping contents in suspension	1
13	Sec. Sand Cyclone Feed Pump	G-2	Type: Centrifugal, Horizontal, Slurry Flow rate (Slurry): 160 gpm Sp. Gr. of Slurry: 1.1 TDH: to suit Arrangement Speed Control: Mechanical Variable -Remote	1
14	Sec. Sand Cyclone	Y-5	Flow rate (Slurry): 160 gpm Solids: 6 tph Solid Sp. Gr.: 2.6 Sp. Gr. of Slurry: 1.1 Classification: 250 microns	1
15	Recycle Water O'Head Tank	D-4	Volume: 550 gals Type: Conical with cylindrical top Material: CS	1
16	Not Used			0
17	Not Used			0
19	Spiral Overflow Sump	D-5	Volume: 850 gals Type: Conical with cylindrical top Material: CS Accessory: Mixer for keeping contents in suspension	1
20	Conditioner Feed Pump	G-3	Type: Centrifugal, Horizontal, Slurry Flow rate (Slurry): 180 gpm Sp. Gr. of Slurry: 1.2 TDH: to suit Arrangement Speed Control: Mechanical Variable -Remote	1
21	Conditioner	Y-6	Type: Double Impeller high speed Retention time- 1 min Volume: 180 gals Power Draw: 40 HP	1
22	Flotation Cell	Y-7 (A through D)	Flow rate (Slurry): 180 gpm Retention Time: 6 mins Approximate Size- 40 cft each- hog trough	4
23	Sludge Filter	Y-8	Type: Double Bell Press filter Capacity: 1 tph (Based on Dry Solids)	1

Table 10-9 Generic soil washing major equipment list (continued).

SI.NO.	Title	I.D. No.	Description	Qty
24	Fine Sand Cyclone Feed Sump	D-6	Volume: 850 gals Type: Conical with cylindrical top Material: CS Accessory: Mixer for keeping contents in suspension	1
25	Fine Sand Cyclone Feed Pump	G-4	Type: Centrifugal, Horizontal, Slurry Flow rate (Slurry): 180 gpm Sp. Gr. Of Slurry: 1.2 TDH: to suit Arrangement Speed Control: Mechanical Variable -Remote	1
26	Fine Sand Cyclone	Y-9	Flow rate (Slurry): 180 gpm Solids: 12 tph Solid Sp. Gr.: 2.6 Sp. Gr. of Slurry: 1.2 Classification: 75 microns	1
27	Thickener	Y-10	Type: Delta Stack or Equal Flow rate (Slurry): 210 gpm	1
28	Sill & Clay Filter Feed Pump	G-5	Type: Centrifugal, Horizontal, Slurry Flow rate (Slurry): 75 gpm Sp. Gr. Of Slurry: 1.3 TDH: to suit Arrangement Speed Control: Mechanical Variable -Remote	1
29	Silt and Clay Filter	Y-11	Type: Double Belt Press filter Capacity: 8 tph (Based on Dry Solids)	1
30	Thickener Overflow Sump	D-7	Volume: 1000 gals Material: CS	1
31	Thickener Overflow Pump	G-6	Type: Centrifugal, Horizontal, Water Flow rate: 150 gpm TDH: to suit Arrangement	1
32	Spillage pump	G-7	Type: Vertical Sump Pump Flow rate (Slurry): 75 gpm TDH: to suit Arrangement	1
33	Treated Water Tank	D-8	Volume: 1500 gals Material: CS	1
34	Booster Pump	G-8	Type: Centrifugal, Horizontal, Water Flow rate: 150 gpm Discharge Pressure: 45 psig	1
35	Additive Metering Pumps	Y-12 (A through F)	Flow Rate: 0-10 gpm	6

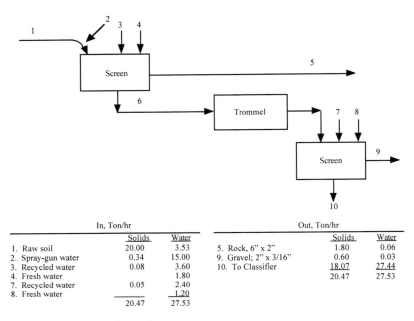

	In, Ton/hr			Out, Ton/hr	
	Solids	Water		Solids	Water
1. Raw soil	20.00	3.53	5. Rock, 6" x 2"	1.80	0.06
2. Spray-gun water	0.34	15.00	9. Gravel; 2" x 3/16"	0.60	0.03
3. Recycled water	0.08	3.60	10. To Classifier	18.07	27.44
4. Fresh water		1.80		20.47	27.53
7. Recycled water	0.05	2.40			
8. Fresh water		1.20			
	20.47	27.53			

Figure 10-3 Mass balance for screens.

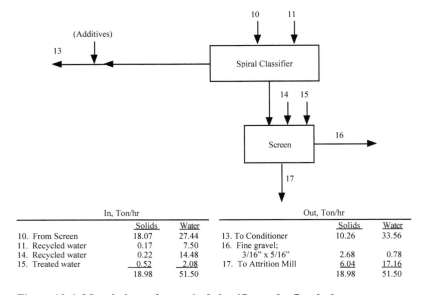

	In, Ton/hr			Out, Ton/hr	
	Solids	Water		Solids	Water
10. From Screen	18.07	27.44	13. To Conditioner	10.26	33.56
11. Recycled water	0.17	7.50	16. Fine gravel;		
14. Recycled water	0.22	14.48	3/16" x 5/16"	2.68	0.78
15. Treated water	0.52	2.08	17. To Attrition Mill	6.04	17.16
	18.98	51.50		18.98	51.50

Figure 10-4 Mass balance for a spiral classifier and a flat-deck screen.

446

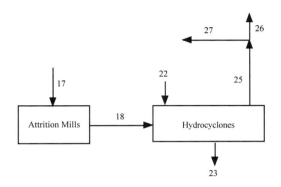

	In, Ton/hr			Out, Ton/hr	
	Solids	Water		Solids	Water
17. From Screen	6.04	17.16	23. Medium sand	5.11	3.41
22. Fresh water		27.27	26. Water to Sprays	0.87	38.50
	6.04	44.43	27. Water to Flotation	0.06	2.52
				6.04	44.43

Figure 10-5 Mass balance for attrition mills and hydrocyclones.

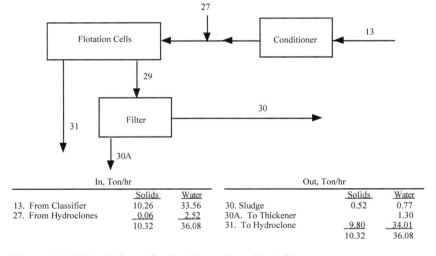

	In, Ton/hr			Out, Ton/hr	
	Solids	Water		Solids	Water
13. From Classifier	10.26	33.56	30. Sludge	0.52	0.77
27. From Hydroclones	0.06	2.52	30A. To Thickener		1.30
	10.32	36.08	31. To Hydroclone	9.80	34.01
				10.32	36.08

Figure 10-6 Mass balance for flotation cells and belt filter.

- Figure 10-7 shows the mass balance for a hydrocyclone Y-9 and a thickener filter system Y-10 and Y-11 for the relatively coarse material that does not float in the cells.

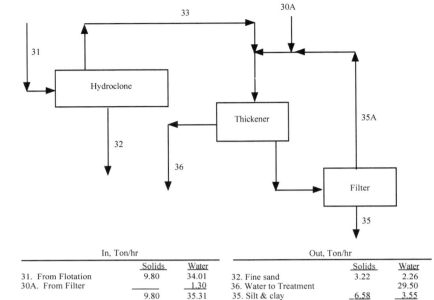

Figure 10-7 Mass balance for hydrocyclone and thickener filter system.

	In, Ton/hr			Out, Ton/hr	
	Solids	Water		Solids	Water
31. From Flotation	9.80	34.01	32. Fine sand	3.22	2.26
30A. From Filter		1.30	36. Water to Treatment		29.50
	9.80	35.31	35. Silt & clay	6.58	3.55
				9.80	35.31

The mass balance for each major stream is summarized in Table 10-10, along with the particle size distribution estimated for each numbered stream. In Table 10-10, TPH is short tons per hour, and GPM is US gallons per minute.

When chelants are used for removal of metal cations, the US EPA (1989a) indicates that the following process parameters should be considered:

- Naturally occurring, noncontaminant metals compete for the chelant. Proper pH control and chelant selection can minimize excessive chelant consumption that occurs with such competition. For example, raising the pH to 7 to 9 favors chelation of divalent lead over trivalent iron.
- Separation of soil particle size fractions before applying chelant washing may be needed for removing some metals.
- Extended contact time in the chelant washing step is needed to chelate the maximum amount of metal.

10.4.3 Treatment of Wash Water

- Anderson (1993) lists the following materials and contaminants that may be in the wash water: coarse sand, silt and clay (with adsorbed contaminants), dissolved

Table 10-10 Generic soil washing material balance.

	SIZE FRACTION	COARSE ROCK	MED ROCK	COARSE GRAVEL	FINE GRAVEL	MED SAND	FINE SAND	SILT AND CLAY	TOTAL	∑TOTAL
					SIZE DISTRIBUTION OF SOIL (%)					
A	6 in. to 3 in.	4							4	4
B	3 in. to 2 in.		5						5	9
C	2 in. to 3/16 in.			3					3	12
D	3/16 in. to 2 mm				10				10	22
E	2 mm to 250 µm					26			26	48
F	250 µm to 75 µm						16		16	64
G	<75 µm							36	36	100
	TOTAL	4	5	3	10	26	16	36	100	

STREAM NO.		1	2	3	4	5	6	7	8	9
DESCRIPTION		RAW SOIL	WATER GUN RECYCLED	SPRAY WATER RECYCLE	SPRAY WATER TREATED	COARSE & MED ROCK 6 in. to 3 in.	ROTARY SCRUBBER FEED	SPRAY WATER RECYCLE	SPRAY WATER TREATED	COARSE GRAVEL 2 in. to 3/16 in.
GRAIN SIZE NOMINAL		6 in. X 0								
A 6 in. to 3 in.	TPH	0.8				0.8				
B 3 in. to 2 in.	TPH	1				1				
C 2 in. to 3/16 in.	TPH	0.6					0.6			0.6
D 3/16 in. to 2 mm	TPH	2					2			
E 2 mm to 250 µm	TPH	5.2					5.2			
F 250 µm to 75 µm	TPH	3.2	0.17	0.04			3.41	0.025		
G <75 µm	TPH	7.2	0.17	0.04			7.41	0.025		
TOTAL DRY SOLIDS	TPH	20	0.34	0.08	0	1.8	18.62	0.05	0	0.6
WATER	TPH	3.53	15.0	3.6	1.8	0.06	23.87	2.4	1.2	0.03
WATER	GPM	14.12	60.0	14.4	7.2	0.22	95.49	9.6	4.8	0.13
DRY SOLIDS	%	85	2.2	2.2		97	43.8	2.2		95
TOTAL STREAM	TPH	23.53	15.34	3.68	1.8	1.86	42.49	2.45	1.2	0.63
TOTAL SLURRY	GPM	45	60.1	15	7	3	124	10	5	1

Table 10-10 Generic soil washing material balance (continued).

STREAM NO.		10	11	12	13	14	15	16
DESCRIPTION		SCREEN-2 UNDERFLOW	SPRAY WATER RECYCLE	SPIRAL SANDS	SPIRAL OVERFLOW	SPRAY WATER RECYCLE	SPRAY WATER TREATED	FINE GRAVEL
GRAIN SIZE NOMINAL		3/16 in. to 0		3/16 in. to 250 µm	250 µm to 0			3/16 in. to 2 mm
A	6 in. to 3 in.	TPH							
B	3 in. to 2 in.	TPH							
C	2 in. to 3/16 in.	TPH							
D	3/16 in. to 2 mm	TPH	2		2.0				2.0
E	2 mm to 250 µm	TPH	5.2		5.1	0.10	0.11		
F	250 µm to 75 µm	TPH	3.44	0.085	0.34	3.09	0.11		
G	<75 µm	TPH	7.44	0.085	0.37	7.07	0.22		2.00
TOTAL DRY SOLIDS		TPH	18.07	0.17	7.81	10.26	0.22		2.00
WATER		TPH	27.44	7.5	1.38	33.56	10	6.0	0.22
WATER		GPM	109.77	30	5.51	134.25	40	24	1
DRY SOLIDS		%	39.7	2.2	85	23.4	2.2		90
TOTAL STREAM		TPH	45.51	7.67	9.19	43.82	10.22	6.0	2.22
TOTAL SLURRY		GPM	138	30	18	150	40	24	4

STREAM NO.		17	18	19	20	21	22	23
DESCRIPTION		SCREEN-3 UNDERFLOW	PRIM CYCL FEED	PRIM CYCL UNDERFLOW	PRIM CYCL	SEC CYCL FEED	WATER ADDITION	SEC CYCL UNDERFLOW
GRAIN SIZE NOMINAL		2 mm to 250 µm	2 mm X 0	2 mm X 0		2 mm X 0		2 mm to 250 µm
A	6 in. to 3 in.	TPH							
B	3 in. to 2 in.	TPH							
C	2 in. to 3/16 in.	TPH							
D	3/16 in. to 2 mm	TPH							
E	2 mm to 250 µm	TPH	5.1	5.1	5.1		5.1		5.1
F	250 µm to 75 µm	TPH	0.46	0.46	0.05	0.41	0.05		0
G	<75 µm	TPH	0.48	0.48	0.07	0.41	0.07		0.01
TOTAL DRY SOLIDS		TPH	0.00	6.04	5.21	0.82	5.21		5.11
WATER		TPH	17.16	17.16	3.48	13.68	30.75	27.27	3.41
WATER		GPM	68.63	68.63	14	54.72	123	109	14
DRY SOLIDS		%	26	26	60	5.7	14.5		60
TOTAL STREAM		TPH	23.19	23.19	8.69	14.5	35.96	27.27	8.52
TOTAL SLURRY		GPM	7.8	7.8	22	56	131	109	21

Table 10-10 Generic soil washing material balance (continued).

STREAM NO.		24	25	26	27	28	29	30
DESCRIPTION		SEC CYCL OVERFLOW 21-23	COMBINED OVERFLOW 20-24	TOTAL RECYCLE 2+3+7+11+14	FLOTN ADDITN RECYCLE, 25-26	FLOTATION FEED, 13+27	FLOATS	FILTERED
GRAIN SIZE NOMINAL		250 μm X 0	250 μm X 0	250 μm X 0	250 μm X 0	250 μm X 0		250 μm X 0
A	6 in. to 3 in.	TPH						
B	3 in. to 2 in.	TPH						
C	2 in. to 3/16 in.	TPH						
D	3/16 in. to 2 mm	TPH						
E	2 mm to 250 μm	TPH				0.1		
F	250 μm to 75 μm	TPH	0.05	0.46	0.435	0.02	3.11	0.16	0.16
G	<75 μm	TPH	0.06	0.47	0.435	0.04	7.11	0.36	0.36
TOTAL DRY SOLIDS		TPH	0.10	0.93	0.87	0.06	10.32	0.52	0.52
WATER		TPH	27.34	41.02	38.5	2.52	36.08	2.08	0.78
WATER		GPM	109	164.08	154	10.08	144.33	8.31	3.12
DRY SOLIDS		%	0.4	2.207	2.3	2.3	22.2	20	40
TOTAL STREAM		TPH	27.44	41.95	39.37	2.58	46.4	2.6	1.3
TOTAL SLURRY		GPM	110	166	155	10	160	9	4

STREAM NO.		30A	31	32	33	34	34A	35
DESCRIPTION		SLUDGE FILTRATE	FINE SAND CYCL FEED 28-30	FINE SAND CYCL UFLOW	FINE SAND CYCL OFLOW 31-32	TAIL THICK FEED 33+30A+35A	TAIL THICK UNDERFLOW	SILT AND CLAY
GRAIN SIZE NOMINAL			250 μm X 0	250 μm to 75 μm	75 μm x 0	75 μm x 0		75 μm x 0
A	6 in. to 3 in.	TPH						
B	3 in. to 2 in.	TPH						
C	2 in. to 3/16 in.	TPH						
D	3/16 in. to 2 mm	TPH						
E	2 mm to 250 μm	TPH		0.1	0.1				
F	250 μm to 75 μm	TPH		3.04	2.74	0.3	0.3	0.3	0.3
G	<75 μm	TPH		6.83	0.55	6.28	6.28	6.28	6.28
TOTAL DRY SOLIDS		TPH	0.00	9.97	3.39	6.58	6.58	6.58	6.58
WATER		TPH	1.3	34.01	2.26	31.75	41.73	12.23	3.55
WATER		GPM	5.19	136.02	9.03	126.99	166.91	48.91	14.18
DRY SOLIDS		%		22.4	60	17.2	13.6	35	65
TOTAL STREAM		TPH	1.3	43.81	5.48	38.33	48.31	18.81	10.13
TOTAL SLURRY		GPM	5	151	14	137	177	59	24

Table 10-10 Generic soil washing material balance (continued).

STREAM NO.		35A	36	37	TOTAL IN	TOTAL OUT
DESCRIPTION		SILT FILTER FILTRATE 34A-35	WATER TO TREATMENT 34-34A	TREATED WATER ADDITIONS 4+8+15+22	1+37	5+9+16+23+32+ 35+30+36
GRAIN SIZE NOMINAL						
A 6 in. to 3 in.	TPH				0.8	0.8
B 3 in. to 2 in.	TPH				1	1
C 2 in. to 3/16 in.	TPH				0.6	0.6
D 3/16 in. to 2 mm	TPH				2	2
E 2 mm to 250 μm	TPH				5.2	5.2
F 250 μm to 75 μm	TPH	0	0		3.2	3.2
G <75 μm	TPH	0	0		7.2	7.2
TOTAL DRY SOLIDS	TPH	0.00	0.00		20.00	20.00
WATER	TPH	8.68	29.5	36.27	39.8	39.8
WATER	GPM	34.73	118	145.08	159.2	159.2
DRY SOLIDS	%					
TOTAL STREAM	TPH	8.68	29.5	36.27	59.8	59.8
TOTAL SLURRY	GPM	35	118	145	190	190

salts, naturally occurring organic matter, undesirable pH conditions, solubilized heavy metals, and other contaminants such as hydrocarbons.

A frequently used scheme for washing petroleum-contaminated soil uses biodegradable surfactants. The spent wash water is treated in a bioreactor. Final cleanup of the soil can be done biologically in a slurry bioreactor or by spreading the soil in a land treatment system and adding nutrients.

Some soil washing systems use activated carbon for removal of dissolved organics from the wash water.

Fine materials that stay suspended in wash water can be removed using two alternative methods: by coagulation, flocculation, and sedimentation; or by ultrafiltration. A batch ultrafiltration scheme is shown in Figure 3-4.

When acid is used to leach metals from soil, the wash water can be neutralized with sodium bicarbonate or caustic soda, and the pH can be raised to precipitate metal hydroxide sludge. With polymer additions, the sludge can be separated from the water by using a clarifier or an ultrafiltration unit. Acid addition is then used for final wash water pH correction prior to discharge.

Ion exchange can be used for wash water cleanup. The conventional method uses fixed-bed ion exchange resins or zeolites following water filtration. Some innovative soil washing techniques use ion exchange resin beads suspended in the wash water.

10.5 Treatability Studies for Soil Washing

The US EPA (1991b) recommends treatability studies to determine the following:

- The recoverable clean soil fraction
- The volume and characteristics of fine soil and sludge fractions requiring treatment or disposal
- The efficacy of additives
- The degree to which additives can be recovered and recycled
- The ratio of additives to soil
- The ratio of soil to wash water

The first three factors provide information regarding the necessity and costs of downstream treatment options. The last three factors help estimate costs of supplies and utilities and the sizes of major equipment components. Treatability studies can also help determine contaminant removal efficiency, the processing conditions needed for recycling or disposal of wash water or solvent, and the percentage of each stream that can be recycled. For solvent extraction, treatability studies are used to determine how much solvent remains with the treated soil.

The US EPA (1991b) recommends that the tests in Table 10-11 be done on soil samples before considering washing as a candidate process.

Table 10-11 Laboratory tests for soil washing.

Parameter	Test	Analysis Method
Particle size distribution	Sieve with #10 and #60 screens or equivalent	ASTM D422
Cation exchange capacity	Ammonium acetate	EPA 9080
	Sodium acetate	EPA 9081

For the particle size distribution test, the US standard sieve series No. 10 and No. 60 provide grain-size separation into three fractions approximately as follows: more than 2 mm, 0.25 to 2 mm, and less than 0.25 mm.

The recommended equilibrium distribution tests include the following: additions of surfactants, acid, base, or chelants; use of heated water and cold water; wash step; and rinse step. Treated soil should be grain-size separated using a No. 10 sieve. In general, at least 50% contaminant reduction should be experienced in the coarse fraction if washing is to be considered further. If further consideration and testing are indicated, in addition to the basic test requirements listed above, remedy selection tests should be done that include separating grain sizes into finer fractions, varying wash times, varying rinse water-to-wash water ratios, and analyzing the wash water.

A good starting ratio for soil to water is 1/3, using 10 to 30 min of shaking. The US EPA (1989a) suggests that a 1/1 soil/water ratio is preferred, but ratios up to 1/3 are more practical.

Laboratory equilibrium distribution tests can be carried out to determine for each contaminant the partitioning coefficient — the ratio of these two parameters: (1) the mass concentration that remains in the soil sample, usually expressed as mg/kg; (2) the liquid concentration, e.g., mg/L. If the concentration in the soil is below the cleanup target level, a washing process has the potential for success without another treatment process being needed in series with washing. The concentrations of contaminants in the liquid are determined so that liquid treatment requirements can be designed.

The equilibrium distribution tests are carried out at various soil/water ratios. There may be a certain ratio above which the cleanup target level can be met. The ratio finally selected for full-scale washing operations is critical. The lower the soil/water ratio, the larger the treatment equipment must be for a given soil batch size.

If removal of cationic heavy metals by water washing is to be attempted, equilibrium distribution tests at lowered pH levels and/or with chelating agents should be carried out. A weak acid, such as acetic acid, should be tried first. Dilute sulfuric acid or hydrochloric acid would be a second choice. A starting point for testing chelants is to use 20% EDTA (ethyl diamine tetraacetic acid) at a pH of 2 to 3. For metal-forming anions (e.g., anion complexes of arsenic or selenium), hydrogen peroxide solution or a chlorine-based oxidizing agent may prove effective.

The water from the equilibrium distribution tests can be used for other experiments on treatability of the water.

The octanol/water partition coefficient should be determined from laboratory tests for any contaminants for which the coefficient is not known from literature sources. A high coefficient favors solvent extraction, whereas a low coefficient favors washing or flushing with water solutions.

For testing in situ soil flushing schemes, a column of soil can be flooded in a laboratory setup as shown by Anderson (1993). Or a column of soil can be used in an apparatus similar to a dynamic ion exchange bench-scale system. The reduction in soil contaminant concentrations can be measured with water as the extracting fluid and with various agents (e.g., acids, alkaline agents, surfactants) added to the water.

Pilot testing of size separation processing can be carried out by utilizing small-scale equipment developed in the mining and metallurgical industries for the classifying and benefaction of ores. Steps such as screening, spiral classification, flotation, hydrocyclonic separation, and hydraulic classification or elutriation may be carried out in series in a soil washing scheme. Pilot testing can be used to determine the particle size distribution that can be attained with each step and the level of contamination in each size fraction. The required water/soil ratio can be confirmed, and water recycling can be evaluated to reduce water consumption and the volume of water needing final treatment. For chelation processes, the contact time for washing that is needed to attain maximum removal of contaminant metal cations can be determined.

Anderson (1993) gives the following examples of equipment components that can be used in conducting treatability studies:

- Grizzly bar screen for removal of debris larger than 50 mm (2 in.)
- Tramp iron and steel separator
- Density media separator for removal of leaves, twigs, roots, plants, shells, etc., based on specific gravities of these materials
- Rotary trommel screen for initial soil breakup and classification, with screened coarse product greater than 9.52 mm (0.375 in.)
- Attrition scrubber that contacts soil particulate smaller than 9.52 mm with chemical wash solution to mobilize fines smaller than 74 μm (200 mesh)
- Hydrocyclone for separating less than 74-μm particulate from sand and gravel in the underflow, which is 70% to 75% solids
- Reverse-slope dewatering module, which is a screening system with high-frequency shaking for final rinsing, dewatering, and desliming
- Wash water clarifier system for flocculation, sedimentation, and densification of soil particulate less than 74 μm
- Oil and grease separator for wastewater
- Dissolved air flotation unit for removal of undissolved hydrocarbons
- Continuous belt filter for dewatering materials less than 63 μm (230 mesh), producing filter cake with 40% to 60% solids.

Some soil washing systems incorporate adsorption and/or ion exchange media suspended in the wash water. Bench-scale or pilot-scale tests can be used to evaluate various media for their effectiveness and capacity for the contaminants of concern.

10.6 Cost Estimating for Soil Washing

The US EPA (1991b) data indicate that after simple equilibrium distribution tests are conducted, bench-scale treatability studies for remedy selection cost $20,000 to $100,000, and field pilot tests, $100,000 to $500,000.

In situ soil flushing costs range from $80/yd^3 of soil using surface flooding to $165/yd^3 using subsurface injection (Anderson, 1993). These costs include treatment of used wash water but exclude excavation, debris removal, and treatment/disposal of residuals.

Studies by Bechtel Environmental Inc. in 1994 indicated that ex situ soil washing costs are typically $100 ± $30/ton of soil for physical separation. Additional leaching or further processing adds $25/ton to $150/ton. These figures exclude costs for transport, setup, and demobilization of equipment, analyses, excavation, hauling, and backfill. Higher unit costs are encountered if the quantity of contaminated soil is less than a few thousand tons or if removal of relatively high concentrations of heavy metals is involved. Exner (1995) indicates that soil washing costs are in the range of $80/ton to $200/ton, whereas solvent extraction is in the range of $150/ton to $400/ton. Anderson (1993) indicates that soil washing costs are in the range of $150/ton to $250/ton.

Remediation cost estimating/process design software is available for various types of soil washing from a number of companies (e.g., COMPOSER GOLD from Building Systems Design, Atlanta, Georgia, and RACER/ENVEST^{TM} from Talisman Partners Ltd., Englewood, Colorado). COMPOSER GOLD includes estimating software for in situ soil flushing and ex situ soil washing and solvent extraction.

For in situ soil flushing, Talisman Partners Ltd. use a computerized ENVEST model that estimates the cost of these items:

- Installing and removing an infiltration gallery with 2-in. PVC piping arms in 12-in. deep by 12-in. gravel-filled trenches lined with filter fabric, at 5-ft spacing, fed by a PVC header in a 24-in. deep by 24-in. trench — The program output defines the cost of dismantling; the user can delete this cost if it is not applicable at a particular site.
- An earth berm around the site 12 in. high and 24 in. wide at the top
- Optional use of additives to the flushing water, such as alkylbenzene sulfonate detergent at 0.05 wt%, sulfuric acid from 220-lb drums at 4.1 lb/1,000 gal, or sodium hydroxide (caustic soda) solution from 100-lb drums at 1.5 lb/1,000 gal — The program does not include the cost of makeup water, wells, or extraction well pumps, and it assumes recirculation of water pumped from a downgradient well. The program does not include the cost of treating recirculated water. Costs

for a mixing tank and a pump for feeding the infiltration gallery are estimated by the program.
- The quantity of water — The user can choose the number of soil pore volumes, ranging from one to 20. If this parameter value is unknown, the program defaults to 10 pore volumes. The program assumes that the soil porosity is 30%. The program user can adjust the number of pore volumes in proportion to porosity if the actual soil porosity is known to be different than 30%.
- Operation costs, including electric power, acid or caustic addition, and surfactant — Another model is available for estimating sampling and analysis costs. Because water treatment costs are excluded, no operating labor, other than what is accounted for in the sampling/analysis model, is estimated.
- Area of contaminated soil — The value of this parameter must be given by the user.
- Depth to groundwater — The model assumes that the flushing solution will infiltrate through and past the soil contaminant plume and down to an aquifer (or to a hydrological confining formation), from which it can be extracted for treatment and recycling.
- The level of protective clothing/equipment needed by site operators
- Soil vertical permeability — The water application rate is proportional to permeability. If the user does not input the vertical permeability, one can choose from three categories of soil, from which the model infers vertical permeability and water application rate:
    - Silty to fine sand (0.001 cm/sec permeability; apply water at 3 in./day)
    - Fine to coarse sand (0.01 cm/sec permeability; apply water at 6 in./day)
    - Coarse sand to gravel (0.1 cm/sec permeability; apply water at 12 in./day)

The model calculates the theoretical duration of applying flushing based on the number of pore volume flushes, porosity, area of contaminated soil, depth to groundwater, and permeability. For the first flush, an additional 33% volume of solution is assumed, to account for soil retentivity. The model multiplies the theoretical duration by a safety factor of two and then derives the operating costs.

The ENVEST computer model for estimating soil washing costs is based on the BioTrol (Chaska, Minnesota) system. The 1993 costs using this model, with a leased, mobile 20-ton/hr unit, were $72/ton for treating 20,000 tons of soil or $66/ton for 60,000 tons. The scheme includes screening, flotation, attrition scrubbing, classifying, and dewatering of separated sand, and it includes the use of a slurry bioreactor for the silt/clay fraction, followed by dewatering. The model assumes that a mobile, 20-ton/hour system will be used. The program user must state the tons of soil to be processed and the level of protective clothing/equipment needed by the system operators. Most soil washing operations require only coveralls, safety footware, gloves, and eye protection. Breathing protection is required for handling dusty soils feeding the process (Anderson, 1993). The costs estimated with this ENVEST model exclude treatment or disposal of residuals.

The ENVEST model for solvent extraction for removal of organic contaminants is for systems handling up to 18,000 yd^3/mo. The required input parameters are:

- Designation of whether the medium being remediated is soil, sludge, or liquid
- Throughput in cubic yards per month
- Quantity of soil to be treated (not recommended for less than 5,000 yd^3)

For better estimating accuracy, values may be included in the input for these secondary parameters:

- Whether crushing and screening are required pretreatment steps
- Downtime allowed (10% is the default value)
- Treatment difficulty, in which 0% is not difficult and 100% requires extensive treatment

The model assumes that wastewater and solvent are stored on-site in tanks or in mobile tankers and that treated soil is stockpiled. No treatment of liquid wastes and no backfilling of treated soil are included in the estimated costs.

Other ENVEST models that can be applied to arrive at total costs include:

- Water distribution
- Overhead electrical distribution
- Fencing and signage
- Excavation, hauling, backfill, etc.

10.7 Summary of Important Points for Soil Washing

- Contaminants tend to stay adsorbed on fine-grained soil fractions, which are difficult to extract with water.
- Coarse fractions tend to be relatively clean or are readily washed. Separating out coarse fractions with water washing reduces the volume of soil that must be disposed of or treated.
- Solvents can remove organic contaminants from fine-grained soil fractions, as well as from coarse-grained soil fractions.
- With in situ flushing, infiltrating water dissolves contaminants. The water with the contaminants is recovered, treated, and recycled through the soil.
- Various additives (such as acids, chelants, alkaline agents, or surfactants) may be used to improve the extraction efficiency of the soil washing system.
- Permitting for soil flushing may be difficult, because additives and contaminants may be transported to an aquifer or to uncontaminated soil.
- Ore refining techniques and equipment can be adapted for separating fines and water washing in an environmental soil washing application.
- Laboratory equilibrium distribution tests help quantify how much contaminant partitions into the washing fluid versus the amount retained with the soil, as well as what extractant might work best.
- The lower the soil cation exchange capacity, the more efficient soil washing will be, especially for removing cationic metals.

- Alkaline agents are used as additives in aqueous soil washing systems to improve the removal of hydrocarbon oils from soil.
- Extraction solvents can be recovered by distillation or stripping.
- Solvent extraction may have to repeated for a batch of soil, because a single extraction may be insufficient for achieving the remediation goal.
- Liquified propane or carbon dioxide may be the preferred solvent for low-boiling organic contaminants, as well as semivolatile organics.
- A flow scheme and a mass balance for the washing solution and for the soil should be developed for each soil washing project.
- If the treatment scheme includes separation by soil particle size fractions, a mass balance should be conducted for each size fraction.
- The percent solids content should be determined for each process step in a mass balance.
- Aqueous streams from petroleum contamination soil washing systems can be treated using biodegradation or activated carbon adsorption.
- Biotreatment can be used for final cleanup of petroleum-contaminated soil.
- Water treatment may include clarification and/or filtration.
- Treatability tests determine contaminant removal efficiency; volume fraction of fines not cleaned; ratio of soil to wash fluid; additive requirements, efficacy, and recycling ability; washing contact time needed and treatment requirements for used wash fluid; and percent of wash fluid that can be recycled.
- Computerized cost-estimating models are available for in situ flushing, ex situ water washing, and ex situ solvent extraction.

Chapter 11

Stabilization and Solidification

Fixation of soil, sludge, sediment, and oily wastes is achieved by mixing with various agents with the goal of making contaminants immobile, so that they do not leach and travel with infiltrating water or with groundwater. The main categories of fixation discussed in this chapter are:

- Stabilization — addition of a chemical to cause a reaction
- Solidification — mechanical entrapment and, sometimes, sorption
- Combinations of these processes

The US EPA (1986b) states that combined processes are termed "waste fixation" or "encapsulation." This chapter describes stabilization, in situ techniques, ex situ solidification by microencapsulation of soil or sludge containing metals contaminants, sorption of liquids, and thermoplastic microencapsulation of soil or sludge containing organics. Some of the techniques involved apply to fixation of both metal and organic contaminants.

11.1 Basic Principles for Stabilization and Solidification

Stabilization is exemplified by the addition of a base, such as caustic or lime, to produce highly insoluble metal hydroxides or oxides. Mechanical entrapment is exemplified by microencapsulation with sulfur or with siliceous compounds, such as portland cement, to produce a form of concrete. Mechanical entrapment also includes absorption and adsorption, as with silicates (e.g., clay minerals) (Sell, 1988). Other sorbents used when liquids are present include calcium sulfate, bottom ash, and cement kiln dust (Arniella and Blythe, 1990).

As noted by the US EPA (1986b), adsorbents that react with specific contaminants perform chemical scavenging, going beyond adsorption of free liquid. Chemical scavenging agents include certain clays, ion exchange resins, natural and synthetic zeolites, silica gel, and finely divided ferric hydroxide or aluminum hydroxide. These agents can potentially reduce treatment requirements for water discharged from dewatering operations. Cheremisinoff (1992) notes that certain dried clays will absorb large quantities of water, resulting in a solid product that is resistant to leaching metals and some organics. Cheremisinoff advises against using certain adsorbents other than clay unless chemical solidification is also applied.

Stabilization and solidification sometimes occur together. The alkaline nature of portland cement causes hydroxides to form when mixed with soils containing metals as microencapsulation proceeds.

Either a monolith or a granular product is produced with encapsulation. Granular product has the advantage of easier handling, and it is usually the objective in the ex situ fixation of soils. Sometimes more than one agent is used in the formulation for fixation, depending mainly on how fast of a setting time is desired and what contaminants are present. When using multiple agents, one needs to determine if the mix is proprietary and if any licensing fees are required.

11.2 In Situ Applications and Area Mixing

Most of the applications described in this chapter are ex situ. However, either stabilization or solidification can be accomplished in situ. The best in situ method consists of injecting slurried agents through jets in a large auger that mixes the soil. The auger is maneuvered up and down while being rotated or twisted back and forth. The auger penetrations are repeated side-by-side with overlapping circles. Capacities of 600 to 900 yd^3 of soil fixation per day are possible using this method. Some equipment can penetrate to depths of up to 120 ft, with diameters up to 18 ft. Similar technology developed in Japan and now being applied in the United States can penetrate more than 200 ft and can auger through boulders that US systems may not be able to penetrate.

The same rigs used for in situ fixation can be used for in situ bioremediation by injecting air or peroxide solution and bacteriological nutrients, and they can be used for soil venting of semivolatile compounds by injecting steam. When soil venting is carried out, a hood is placed over the area being augered or over a caisson surrounding the auger. The drill stem penetrates the center-top of the hood. A vacuum blower draws air from under the hood, conveying it to an exhaust cleanup system to prevent uncontrolled emissions during soil treatment.

The US EPA (1989c) describes a system for fixation of waste lagoons up to 10- to 18-ft deep. A fixed, 10-ft thick lift can be excavated and then successively deeper 10-ft lifts can be fixed at rates from 500 to 1,500 yd^3/day. The agents, typically cement, lime, or kiln dust, are fed through an injector on a backhoe. A volume increase of 10% is typically observed. Other systems are designed for depths up to 30 ft.

Other in situ methods used for shallow contamination use excavating equipment For example, a backhoe can be used to mix chemical agents and water with soil, but no soil is removed.

Another method applicable to shallow soils (depths to 5 m, or 16 ft) is in situ vitrification. The soil is glassified by electric heating to above its melting point. A high voltage is applied at four electrodes in the soil, and electrical resistance of the soil results in heating. A block of soil 30 ft by 30 ft in area can be vitrified using this method. Further treatment takes place by moving the electrodes to vitrify adjacent blocks. Organics are desorbed and pyrolyzed, and metals are entrapped in the glass that is formed. Plasma torching (*Chemical Engineering*, April 1995) can also accomplish vitrification. Plasma torches being developed at the Georgia Institute of Technology Construction Research Center do not have the depth limitations that presently constrain the use of resistance heating. (Developments in resistance in situ vitrification are underway to achieve greater depths.)

A plasma torch could be lowered to any depth in a predrilled borehole. The soil would be vitrified as a blob several feet wide near the tip of the torch. As the torch is withdrawn upward, a vitrified column forms. A row of columns could be formed with a series of predrilled boreholes. Successive parallel column rows would result in a wide area of vitrified soil. Blundy and Zionkowski (1997) describe a field application of the technology.

Area mixing is done by spreading fixation agents on the surface, adding water, and mixing a thin layer of soil in place. If contaminated soil has been excavated, it can be spread on top of fixed soil in successive layers and fixed. The remediated soil layers can remain in place, be used for on-site backfill, or be removed for off-site disposal.

11.3 Microencapsulation

11.3.1 Cement/Pozzolanic (Siliceous) Solidifiers

Portland cement is the most commonly used solidification agent. Lime plus pozzolans, such as fly ash and blast furnace slag, will microencapsulate soil particles and solidify sludges similarly, except setting time is longer than with cement. Slow setting can be beneficial if transport is required before ultimate disposition. The use of such agents versus cement often depends on availability. Sometimes pozzolans are added with portland cement, resulting in improved compressive strength of the final material. Any of the siliceous solidifiers will immobilize cationic metals.

Frequently, these agents are used with a relatively high agent/soil ratio, with a very much smaller amount of a proprietary agent or a silicate also added to improve the mix. Nehring and Brauning (1992) describe adding sodium or potassium silicate, which are soluble in water, to "flash set" mixes made with portland cement. However, too fast a set can result in incomplete mixing and solids buildup in the mixing and product handling equipment that is difficult to wash off.

Portland cement will not work well with many organic contaminants, but sometimes it will for oily materials. Lime plus pozzolans will work better than portland cement for solidification when organics are present, except for oily materials (Arniella and Blythe, 1990). Sometimes detergents are added to emulsify organics.

With cationic metals contaminants, both stabilization and solidification occur with either portland cement or lime. Lime is especially effective in forming insoluble metal hydroxides. Fly ash is most desirable for solidification because it naturally has fine particle sizes. Such fine-grained solidification agents are more reactive and potentially mix more uniformly with soil than coarse particles. Depending on the source of the fly ash, however, introduction of other contaminants (e.g., polychlorinated dibenzodioxins) could be a problem.

Cement/soil ratios usually are in the range of 1/4 to 1/2.5, although smaller and larger ratios have been used. Lime/soil ratios range from 1/20 to 1/3 (Arniella and Blythe, 1990).

Good mixing is essential for stabilization or solidification. With excavated soil, the process is usually carried out batch-wise in a pug mill and, sometimes, in a ribbon blender. Soil, water, and fixation agents are mixed. The treated batch is stockpiled and sampled for leachate analysis and slump tests (or compressive strength tests). If the treated material fails the leachate test or is too fluid, the batch is recycled (possibly with grinding first) through the process. Both mobile and fixed-based systems are used for ex situ applications. The US EPA (1986b) describes a system in which soil is slurried with water and mixed with fixation agents in a concrete transit mixer while on route to the disposal site.

As described by the US EPA (1986b), special equipment decontamination practices apply in cases in which hazardous wastes are involved. Equipment moving around the site should be decontaminated daily. (If decontamination water cannot be incorporated in the mix, then treatment of the water is needed.) Stationary equipment should be cleaned as needed and decontaminated after project completion.

Compressive strength is important if the product is returned to the earth and structures are built over it or if the material is to be landfilled with significant overburden. Some months are required for the treated material to reach full strength. Arniella and Blythe (1990) report that minimum unconfined compressive strength specifications sometimes are 25 or 50 psi; cement forms a product with a strength ranging from 20 to 1000 psi; pozzolans, from 30 to 200 psi. The strength of the treated material depends on the properties and proportions of the agents and the water/solids ratio used in the mix.

The US EPA (1986b) reports that the Nuclear Regulatory Commission guidelines call for solidified materials to reach a compressive strength of 150 psi. Bituminous materials (asphalt mixes) must show less than 20% deformation at 150 psi. The EPA lists the following procedures for testing compressive strength of cemented wastes:

- ASTM D2166-66 Unconfined
- ASTM D2850-70 Triaxial shear
- ASTM D1194-72 Plate load

Soil or semisolid sludge can also be solidified ex situ as a layer using the area mixing technique. Because area mixing is usually done on-site, with the product sometimes left in place or backfilled into the excavation, it is sometimes considered to be an in situ technique. Additionally, shallow mixing can be done in situ over a broad area, and the fixed material can be hauled away for disposal. The US EPA (1989c) describes a system that uses a rotary auger mounted on the front end of a bulldozer that fixes the soil in 8- to 10-in. lifts and another system that can reach remotely into a lagoon to solidify bottom deposits to depths up to 10 ft. In the latter system, chemical agents are pneumatically transferred to injectors and then transferred with screw conveyors to augers to be mixed with pond or lagoon sediments.

Arniella and Blythe (1990) describe the steps for ex situ area mixing as follows:

- The spread soil is covered with a layer of agents.
- The soil and agents are lifted and turned repeatedly, or they are mixed with a high-speed rotary (or other) mixer.
- The mixed material is compacted.
- The steps are repeated with additional layered batches until the entire volume of contaminated soil (or sludge) is processed.

The area mixing method may fail if weather conditions cause either rapid drying (because water of hydration is needed for siliceous agents) or too much water because of rain or snow. Cold weather inhibits proper cement setting. Successfully treated material can be capped with planting soil and seeded or transported to final disposal.

11.3.2 Thermoplastic Agents

Both thermoplastic and thermosetting agents have been used for fixation of organics contaminants. Thermoplastics are substances that melt when warmed; thermosetting plastics are permanently hardened by warming. Thermoplastic agents are used much more frequently than thermosetting plastics for fixation.

Asphalt is the prime example of a thermoplastic agent that can be used to fix soils containing organic contaminants. Road paving blacktop is asphaltic concrete, prepared by mixing dried aggregate and bitumen ("asphalt"). The same technology will solidify soil with organic contaminants, with soil used in place of aggregate. Petroleum-contaminated soil fixed with bitumen can sometimes be used for paving. If the bitumen is mixed when hot, liquids and volatile compounds evaporate. A cold-mix system is also available. Other thermoplastic agents that will solidify sludges and some soils include:

- Urea-formaldehyde resin
- Polyethylene and other polyolefins (e.g., polybutadiene)
- Epoxies and polyesters, including polyacrylates and polyacrylamides

The prepolymer is mixed with the sludge, and then polymerizing catalyst or an accelerator is added. Although these agents cost more than siliceous agents, such as portland cement, they encapsulate soil particles contaminated with organics such as PCBs and dioxins. However, their potential biodegradability can be a problem (Sell, 1988). On the plus side, many of the thermoplastic agents are more resistant to alkaline degradation than silicate sorbents (which are used for solidifying metals contaminants, as well as organic wastes).

The thermoplastic resins will solidify soluble contaminants that siliceous agents cannot; the resins resist deterioration by most aqueous solutions; leachability is a very small fraction of the leachability with cement; and because thermoplastics are often applied hot, water may evaporate resulting in a volume decrease (Arniella and Blythe, 1990). Heating may cause dangerous emissions of volatile and semivolatile organic compounds, and it may be necessary to enclose the fixation operations and capture organic vapors. Soils with volatile or semivolatile organics can be pretreated with soil venting (soil vapor extraction) or thermal desorption, thereby lessening the danger of emissions that would otherwise be generated by adding hot thermoplastics to these materials.

The US EPA (1986b) reports that temperatures ranging from 130° to 230° C (266° to 446° F) are used during mixing with thermoplastics. Screw extruders used in the plastics industry are adapted for hazardous waste microencapsulation by adding fume control and equipment safety interlocks and by minimizing worker exposure. The soil or sludge must be predried, and the matrix material must be melted before fixation using these thermoplastic systems. Energy costs are high compared to siliceous systems.

11.4 Silicate Sorbents

Another way to fix contaminants and solidify organic liquids is to use agents that absorb and/or adsorb. Sorbents that are acceptable for landfilling are silicate compounds found in volcanic ash, clay minerals, feldspars (e.g., $KAlSi_2O_4$) and zeolites (e.g., $NaAlSi_2O_4 \cdot H_2O$). Sell (1988) describes these sorbents and methods of combining sorbent and microencapsulation techniques in successive steps for solidifying organic wastes.

With sorption combined with microencapsulation, the wastes are first mixed at pH 7 to 8 with a silicate sorbent that is organophillic (and repels water). Then water and microencapsulating agents are added — silica, calcium silicate, calcium aluminate, and aluminum oxide — at a pH of 11 to 12. Metals are mostly precipitated, and the high-strength, impermeable monolith that forms is not biodegradable.

11.5 Main System Design Parameters

An important factor with some encapsulation agents is the volume increase that they produce. Soil solidified with some agents has a much larger volume (and weight) than the originally excavated soil. Also, excavated soil has a lower density and thus a larger volume than it has in place. Of the solidification agents described in this chapter, cement causes the largest volume and weight increase (Arniella and Blythe, 1990).

The addition of some chemical agents (i.e., lime) with water to soil may volatilize organic compounds, because of heat of hydration that is generated. If there is a potential explosivity problem, the mixing equipment should be enclosed and vented to an emissions abatement system or diluted with fresh air to less than 25% of the LEL of the generated vapor.

The US EPA (1989c) summarizes the potential for contaminants to interfere with setting, as shown in Table 11-1.

Arniella (1990) discusses how remediation effectiveness can be gauged by running tests for unconfined compressive strength and for permeability.

11.6   Treatability Studies for Stabilization and Solidification

Bench-scale and pilot tests are worthwhile for accomplishing a number of objectives when evaluating the use of this technology, including:

- Measuring the volume increase following chemical treatment
- Determining whether using a selected fixation agent will result in a product that passes a leachate test
- Determining the minimum agent/soil ratio, mainly based on nonleachability, durability as measured with wet-dry and freeze-thaw tests, and compressive strength
- Measuring viscosity changes during mixing and checking uniformity of the mix
- Determining mix time
- Determining setting time and curing time, minimum temperature, and moisture addition requirements (or dewatering that may be required before processing)
- Determining temperature rise and potential for volatilization of flammables and checking for odors and other potential safety concerns
- Estimating costs for full-scale implementation

References for standard methods for carrying out some of these tests are given in the bench- and pilot-scale screening section of the "Handbook for Stabilization/Solidification of Hazardous Wastes" (US EPA, 1986b). Test methods are cited for measuring or evaluating temperature increases during mixing, concrete mixer performance, pumpability, and water retentivity and separation.

When lime and water are added to soil during the mixing step, the heat of hydration of the lime can be sufficiently intense that sufficient VOCs evolve to form an explosive mixture with air. The 1986 EPA handbook states that a typical bench-scale procedure for determining temperature increases during mixing is ASTM-C-186, Test for Heat of Hydration of Hydraulic Cements. An organic vapor analyzer, a gas chromatograph, or a combustible gas monitor can be used to check the explosivity of the gases evolved.

Treatability tests can also help determine the efficacy of additives. For example, soluble silicates added in relatively small amounts to portland cement may reduce interferences to binding caused by certain metals.

Pilot tests can also help determine whether contaminants will directly interfere with setting.

The US EPA (1989c) mentions the following additional items that are important and can be measured during testing:

Table 11-1 Effect of waste components on stabilization/solidification.

Waste component	Cement-based	Pozzolan-based	Thermoplastic	Organic polymer
Nonpolar as: oil and grease, aromatic hydrocarbons, halogenated	May impede setting. Decreases durability over a long time period. Volatiles may escape upon mixing. Demonstrated effectiveness under certain conditions.[a]	May impede setting. Decreases durability over a long time period. Volatiles may escape upon mixing. Demonstrated effectiveness under certain conditions.[b]	Organics may vaporize upon heating. Demonstrated effectiveness under certain conditions.[c]	May impede sailing. Demonstrated effectiveness under certain conditions.[d]
Polar as: alcohols, phenols, organic acids, glycols	Phenol will significantly retard setting and will decrease durability in the short run. Decreases durability over a long period.[a]	Phenol will significantly retard setting and will decrease durability. In the short run. Alcohols may retard setting. Decreases durability over a long time period.	Organics may vaporize upon heating.	No significant effect on setting.
Acids as: hydrochloric acid, hydrofluoric acid	No significant effect on setting. Cement will neutralize acids. Types II and IV portland cement demonstrate better durability characteristics than Type I. Demonstrated effectiveness.[f,g]	No significant effect on setting. Cement will neutralize acids. Demonstrated effectiveness.[f,g]	Can be neutralized before incorporation.	Can be neutralized before incorporation. Ureaformaldehyde demonstrated to be effective.[f]
Oxidizers as: sodium hypochlorite, potassium permanganate, nitric acid, potassium dichromate	Compatible	Compatible	May cause matrix breakdown, fire.	May cause matrix breakdown, fire.
Salts as: sulfates, halides, nitrates, cyanides	Increase setting times. Decrease durability. Sulfates may retard setting and cause spalling unless special cement is used. Sulfates accelerate other reactions.	Halides are easily leached and retard setting. Halides may retard setting, most are easily leached. Sulfates can retard or accelerate reactions.	Sulfates and halides may dehydrate and rehydrate, causing splitting.	Compatible[h]
Heavy metals as: Pb, Cr, Cd, As, Hg	Compatible. Can increase set time. Demonstrated effectiveness under certain conditions.[i]	Compatible. Demonstrated effectiveness on certain species (Pb, Cd, Cr).[d,j]	Compatible. Demonstrated effectiveness on certain species (Cu, As, Cr).[d]	Compatible. Demonstrated effectiveness with As.[d]
Radioactive materials	Compatible	Compatible	Compatible	Compatible

Table 11-1 Effect of waste components on stabilization/solidification (continued).

References identified in Table 11-1:

1. Federal Register, August 17, 1988, "First Third," 40 CFR268-10.
2. JACA Corporation.1985. Unpublished report prepared for the US EPA Hazardous Waste Engineering Research Laboratory, Cincinnati, Ohio.
3. Kolvites, B., Bishop, P.1987. Column Leach Testing of Phenol and TCE Stabilized/Solidified with Portland Cement. Presented at the ASTM 4th International Hazardous Waste Symposium, Atlanta, Georgia, p. 27.
4. Kyles, J.H., Malinowski, K.C., Stanczyk, T.F. 1987. Proceedings of the 19th Mid-Atlantic Industrial Waste Conference, June 21-23, Jeffrey C. Evans, ed., pp 554-568.
5. Musser, D., ENRECO, Smith, R., Raba-Kistner Consultants. 1984. Proceedings of the 39th Industrial Waste Conference, Purdue University, West Lafayette, Indiana, May.
6. Tittlebaum, M.E., et al. 1985. CRC Critical Reviews in Environmental Control, 15(2):191-193.
7. US EPA. 1984. Trammel Crow Site and College Point Site, EPA 540/2-84-002, Washington, D.C.
8. Van Keuren, E., et al. 1987. Proceedings of the 19th Mid-Atlantic Industrial Waste Conference, June 21-23, Jeffrey C. Evans, ed., pp 330-341.

- Permeability of the product, which should be two orders of magnitude below that of the materials that will surround the product
- Evolution of gases during predrying, processing, or curing

Testing methods that have been developed to optimize concrete mix ratios, mix times, set times, and curing conditions can be applied to soil fixation pilot tests.

Before launching full-scale remediation of excavated soil by stabilization or solidification, at the least a bucket-mix treatability test should be conducted to check the leachability and the increase in volume of the soil for various ratios of selected agents and the soil.

11.7 Cost Estimating for Stabilization and Solidification

Arniella and Blythe (1990) suggest that bringing mobile equipment such as augers and portable pug mills to the site should be considered for quantities of contaminated soils larger than 5,000 yd^3. For such mobile applications, they give costs (excluding mobilization and demobilization, engineering, safety gear, decontamination of equipment, and delivering reagents) for a process using 30% portland cement and 2% sodium silicate, as shown in Table 11-2. Data from the US EPA (1986b) indicate that labor, overhead, and profit are approximately $30/yd^3 higher than the amounts given at the top of Table 11-2, making the total approximately $70/yd^3 of soil.

Table 11-2 Costs of fixation with cement, using mobile equipment (From *Chemical Engineering*, February 1990).

Description	Fixation Method Cost ($/yd^3)		
	In Situ	Pug Mill	Area Mixing
Labor, overhead, profit	1.40	1.10	3.00
Equipment & metering	1.60	0.70	3.00
Conveyance	-	1.40	-
Pre-screen & size reduce	-	0.50	-
Monitoring & testing	4.00	3.10	5.10
Reagents & mixing materials	31.10	31.10	31.10
Offsite disposal	-	3.10	-
Supplies	0.60	0.80	1.30
TOTAL	38.70	41.80	43.50
Production rates assumed (yd^3/d)	500	650	400

Estimates by the author for pug mill type of applications indicate costs (in 1995 dollars) as follows:

- 10,000 yd^3 — $73/ yd^3
- 50,000 yd^3 — $58/ yd^3
- 100,000 yd^3 — $54/ yd^3

DuTeaux (1996) tabulates cost data from three 1994 case histories involving inorganics or metals contamination as follows:

- In situ stabilization using an auger — $111/ton to $194/ton
- Ex situ solidification with alumina, calcium, and silica — $73/ton
- Ex situ solidification with cement and soluble silicates — $85/ton

If a step-by-step detailed estimate is attempted, the US EPA (1989c) lists the following items that should be accounted for in addition to processing:

- Decontamination of transport, handling, and processing equipment and, in some cases, treatment and disposal of the decontamination water
- Grading, with a 2% to 3% slope; berming; and lining of a cleared area with a subdrainage system for storing untreated material
- Covering of untreated material
- Installing a paved loading ramp
- Bins, hoppers, or pallet storage areas that are weatherproofed for storing reagents
- Backfilling of treated soil in lifts of 8 to 10 in., adding moisture, and compacting — Moisture is needed to provide lubrication for the soil particles to slide past each other during compaction, and the optimum moisture content is the amount that results in maximum bulk density. Too much moisture lowers the soil's bulk density.

The US EPA (1986b) gives the basis for estimating costs using examples for two mixing methods:

- Using a mobile mixing plant with a capacity of 180 yd^3/8-hr day, using a ribbon blender for near-surface soil or unpumpable sludge
- Performing area mixing in 12-in. lifts using a high-speed rotary mixer at a rate of 250 yd^3/8-hr day.

Both examples assume the use of portland cement with a soluble silicate additive for fixation of 2,850 tons (2,500 yd^3) with a bulk density of 85 lb/ft^3 and disposal of the product on-site.

The California Department of Health Services (1988) gives a cost estimate for a large stabilization/solidification unit (2,000 yd^3/day design) for fixation of metals in incinerated soil with equal parts cement and sodium silicate. Table 11-3 gives the estimated capital investment costs (excluding contingency), and Table 11-4 gives the estimated operating and maintenance costs.

The RACER/ENVEST computer software marketed by Talisman Partners Ltd. (Englewood, Colorado) gives updated costs for ex situ on-site pug mill solidification of:

- Solids
- Sludges
- Incineration ash

Table 11-3 Estimated capital costs for stabilization/solidification.

Description	Quantity	Unit	Unit Price ($)	Total Cost (Rounded $)	Assumption	Reference
Cement Storage Bins	2	each	600,000	1,200,000	100,000 ft³ total capacity with inlet and outlet conveyors, foundations and other appurtenances. Two are needed, one for cement storage, the other for sodium silicate	Best Professional Judgement
Water Supply	1	lump sum	500,000	500,000	Facilities, pipelines, storage, etc. Assume extracted groundwater is used and available from downgradient remedial action	Best Professional Judgement
Mixers	6	each	250,000	1,500,000	Install mixers for soil, chemicals and water. Assume that a total of six mixers are required, each sized to handle 125 yd³/hour	Mfg. estimate
Building for mixers	1	each	1,100,000	1,100,000	Building for mixing and mixing activity; 14,000 ft²	Building cost estimate guide
Conveyors, piping, pumps, and other appurtenances	1	lump sum	500,000	500,000	- - -	Best Professional Judgement
Raw material and treated material stockpiles	1	lump sum	400,000	400,000	- - -	Best Professional Judgement
Total				$5,200,000		

Table 11-4 Estimated stabilization/solidification operation and maintenance costs (for 5 yr).

Description	Quantity	Unit	Unit Price ($)	Total Cost (Rounded $)	Assumption	Reference
Labor	50,000	hr	110	5,500,000	5 people, 10 hours per day, $50/hr base pay. Level C Protection (2.2 cost impact multiplier)	REM IV Cost Guidance
Protective equipment	1,000	days	275	300,000	Level C cost $55/day/person (5 x 55 = $275/day)	REM IV Cost Guidance
Administration/supervision	12,000	hr	50	600,000	Includes senior project manager (Chem Eng.) at 8 hrs/day and 4 hr/day support	Best Professional Judgement
Chemical analysis	1,740,000	ton	2	3,480,000	Assumes analysis at $2/ton of soil. Total weight of soil 3.48 billion lb (1.74 million tons)	Treatability Study
Power	1,740,000	ton	0.60	1,044,000	Assume costs at $0.6/ton	Best Professional Judgement
Parts/supplies	1	lump sum	1,300,000	1,300,000	Assume parts/supplies are 25% of capital costs (.25 x 5,200,000)	Best Professional Judgement
Cement	350,000	ton	12	4,200,000	$60/ton, which equals $12/ton at 20% dose	Treatability Study
Sodium silicate	350,000	ton	30	10,500,000	$150/ton, which equals $30/ton at 20% dose	Treatability Study
	Total			$26,924,000		

Note: Assumes plant operates 1000 days (200 days/year), and 1,090,000 cu yd (in-situ) of soil is treated. Dry soil weighs 1.6 tons/cu yd, so weight of soil treated = 1.74 million tons

The user must designate the total waste volume and the batch capacity of the mobile pug mill. The options provided for pug mill capacity are 2 yd³, 5 yd³, 10 yd³, and 15 yd³.

The capacity can be chosen based on how many working days or months are to be allowed for the cleanup project. If the program user does not specify a batch cycle time, the model assumes 20 min. For example, the duration of a 20,000-yd³ cleanup project using a 10-yd³ pug mill would be calculated as follows:

Batches/day = (8 hr) (60 min/hr)/20 min = 24   (11-1)
yd³/day = 24 batches (10 yd³/batch) = 240   (11-2)
Project duration = (20,000 yd³/240 yd³/day)(21.67 day/mo.) = 3.85 mo.   (11-3)

The 20-min cycle time allows 5 min for loading a batch, 14 min of mixing, and 1 min for unloading.

If choosing the pug mill capacity is not based on how many working days or months are to be allowed, the user could try specifying one capacity at a time and finding the minimum cost option.

The user may specify the waste bulk density. If not, the model assumes 100 lb/ft³ for solids, 80 lb/ft³ for sludges, and 52 lb/ft³ for ash. The model assumes that waste moisture content does not exceed 30% for solids and is between 30% and 70% for sludges. If the user does not specify the actual moisture content, the model assumes these values:

- 15% for solids
- 60 % for sludges
- 20% for ash

If the user does not specify how much water is to be added to the solids or ash, the model assumes a water/cement ratio of 0.40. (Sludges do not need water added). If the user does not specify how much proprietary agent is to be added, the model assumes an additive/waste ratio of 0.01. If the user does not specify how much cement is to be added, the model assumes a cement/waste ratio as follows:

- 0.15 for solids
- 0.40 for sludges
- 0.10 for ash

An important feature of the model is that it will compute the volume of the product. For estimating the product volume, if the user does not specify densities of agents added, the model assumes the following bulk density values:

- Portland cement   94 lb/ft³
- Proprietary binders   100 lb/ft³
- Fly ash   65 lb/ft³
- Cement kiln dust   90 lb/ft³

- Hydrated lime   94 lb/ft^3
- Bitumen   81 lb/ft^3
- Activated carbon   97 lb/ft^3

Product volume is needed for estimating costs of final disposal. Other ENVEST models can be used for estimating hauling, backfill, or landfill disposal costs once the product volume is known. Other ENVEST models can also be used for estimating related costs for:

- Excavation
- Permitting
- Pilot testing
- Sampling and analysis

11.8 Summary of Important Points for Stabilization and Solidification

- Stabilization and solidification do not remove toxic or hazardous substances, but they do immobilize them with mixing processes that apply to soils and a variety of hazardous wastes.
- Stabilization involves adding a chemical reagent, usually to form a precipitate. Solidification processes include microencapsulation and sorption.
- For ex situ remediation of contaminated soils, the usual objective is not to form a monolith but to produce a granular product that can be handled easily.
- Fixation can be done in situ with hollow augers with ports for injection or with conventional mixing devices and backhoes at shallow depths.
- Most fixation is done ex situ with plant mixing setups that use equipment such as pug mills. Another technique (area mixing) processes soil that is spread in layers in the open.
- Fixation usually results in a significant volume increase, especially when siliceous solidifiers, such as cement, are used as chemical additives.
- Area mixing is accomplished with successive multiple lifts.
- Portland cement or lime plus pozzolans is widely used, especially for fixation of metals and sometimes for organics.
- Soluble silicates or proprietary chemical agents can be added in relatively small amounts to improve the mix.
- Critical requirements for the treated soil include resistance to leaching and adequate compressive strength.
- Compressive strength of the product is especially important if it will be placed where structures will be built or in a landfill subject to significant overburden.
- Decontamination of equipment affects scheduling and costs. If decontamination water cannot be utilized in the mix, water treatment may be necessary.
- Thermoplastic agents, including asphalt, resins, polyolefins, and polyesters, are used for organics and potentially can biodegrade.
- Treatability tests are used to select fixation agents and are needed to determine what ratio of agents/soil will result in a durable product that passes a leachate test.
- Treatability tests also provide information on mixing conditions, moisture requirements, curing conditions, cooling requirements, and safety concerns.

- For cost estimating, the main processing costs usually include reagents, labor, equipment rental, mobilization, and contractor's profit and overhead.
- Computer software is available for estimating volume increases and costs for processing, excavating, hauling, backfill or disposal, permitting, pilot testing, sampling, and analysis.

Chapter 12

Cost Estimating and Life Cycle Analysis

This chapter covers methods of preliminarily estimating investment cost; definitive estimating of investment cost; estimating of annual expenses; computer software available for estimating costs of remediation; and life cycle analysis for determining the present value of total investment cost plus annual expenses, closure costs, and post-closure costs.

12.1 Basic Principles

For a number of reasons, including budget allocations and tax effects, costs for treatment are categorized as being either capital investments or expenses. The capital investment category typically includes design engineering, equipment, fabrication, erection, commissioning/debugging/startup, and, in some accounting systems, dismantling. Usually, most investment occurs before startup. However, some equipment may be installed years after the initial startup, and closure costs occur at the end of active remediation.

Expenses are usually accounted for annually and include all resources used to keep treatment systems operating, plus overhead, property taxes, insurance, monitoring, and reporting. Post-closure costs, such as ongoing monitoring, occur after active remediation and, if applicable, are an expense.

The total of investment costs (including closure costs), dismantling costs, and expenses (including post-closure costs) is the life cycle cost for the project. The US Department of Energy's Los Alamos National Laboratory published in 1996 "A Compendium of Cost Data for Environmental Remediation Technologies" that gives life cycle analyses for remediation processes. Access to this report (No. LA-UR-96-2205) is available on the Web at http://lib-www.lanl.gov/la-pubs/00326055.pdf. Other Los Alamos National Laboratory reports are catalogued at lib-www.lanl.gov.

By dividing the total life cycle cost of a treatment system by the volume or mass treated a unit cost for treatment is derived. Unit costs are sometimes expressed in dollars (or cents) per 1,000 L or per 1,000 gal of groundwater treated or in dollars per ton or per cubic yard of soil treated. If estimates are made of life cycle costs for competing remediation alternatives in order to help choose an alternative, the present worth (present value) of each alternative should be determined. Present worth determinations take into account the time value of money when some costs, such as postponed investments and multiyear operating expense, do not occur during the initial year of the project. A method for calculating the present worth of remediation projects and examples are provided at the end of this chapter.

Calculating life cycle costs on a present-worth basis is straightforward if the investment costs and expenses can be estimated and the duration of active remediation can be predicted. Fortunately, computer programs are available that aid in estimating investment costs and some expenses for proven remediation technologies, as well as for wells, water distribution/storage, roads, site preparation, utilities, excavating/loading/hauling, fencing, and demolition. Some of the available software systems perform such estimating tasks with minimal input from the program user. For projects that are barely into the conceptual design phase, these programs have default values for design parameters that accomplish much of the preliminary process design for the user.

12.2 Investment Costs

12.2.1 Preliminary Estimates for Investment Cost

Frequently, budget estimating is needed before a project has been well defined, and estimating is needed based on only a conceptual design or on the beginnings of a conceptual design. Estimating methods traditionally applied to process plants can also be applied to remediation treatment plants.

Two methods are described:

- Ratioing the known cost of a similar, previous project with different equipment sizes
- Factoring costs of principal equipment

12.2.1.1 Ratioing Costs With Exponents

If o designates an old (or previous) processing project with a known cost and n designates a new, similar project, the following relationship holds for preliminarily estimating treatment plant investment costs:

$$\text{Cost}_n/\text{Cost}_o = (\text{Size}_n/\text{Size}_o)^{\text{exp}} \tag{12-1}$$

The size is usually expressed as throughput capacity. For example, with a groundwater pump-and-treat remediation project, the size might be expressed in terms of gallons per minute of throughput. The exponent (exp) would be different for the aboveground processing equipment than for wells and for extraction pumps.

For the installed processing equipment, the exponent has traditionally been chosen to be 0.6 (Chilton, 1950; Williams, 1947; Thuesen and Fabrycky, 1984; and Peters and Timmerhaus, 1991) for ratioing upwards in size. Some estimators use 0.8 for ratioing downwards. Remer and Chai (1990) and Remer, Low, and Heaps-Nelson (1994) suggest 0.7 for ratioing up or down. If the previous project is old enough that inflation has significantly affected costs, then the calculated new cost should be multiplied by the ratio of the projected cost index to the old cost index. Remer, Low, and Heaps-Nelson (1994) suggest that cost index data be obtained from the Marshall and Swift Equipment Cost Index, the *Chemical Engineering* Plant Cost Index, or the *Engineering News Record* Index.

As an example, consider a new air stripping system with a capacity of 100 gal/min for removing gasoline hydrocarbons from groundwater. An old system sized for 50 gal/min had an installed cost (excluding wells) of $110,000. The cost index ratio is determined to be 1.12 if the new equipment were to be installed now, but the cost index ratio is projected to be 1.16 when the new equipment is scheduled to be procured next year. The cost of the new system is estimated as follows, by rearranging Equation 12-1 and multiplying by the projected cost index ratio:

$$\text{Cost}_n = \$110,000 \ (100 \text{ gal/min}/50 \text{ gal/min})^{0.7} \ (1.16) = \$207,000 \qquad (12\text{-}2)$$

Note that if inflation were not taken into account, the cost of doubling the capacity is 1.6 times the old cost.

A more accurate cost estimate can be made if a different exponent is used for each type of equipment. Desai (1981) gives the following values:

- Air fins (coolers)   0.80
- Agitators   0.50
- Blowers   0.65
- Compressors   0.75
- Drums   0.65
- Dryers   0.50
- Heaters   0.80
- Heat exchangers   0.60
- Pumps   0.60
- Refrigeration units   0.75
- Tanks   0.70
- Towers   0.70

Remer, Low, and Heaps-Nelson (1994) give exponents for air pollution control equipment, including cyclone dust separators, ductwork, fans and fan motors, hoods, pumps, rotary air locks, screw conveyors, spray chambers, quenchers, stacks, electrostatic precipitators, fabric filters (baghouses), wet scrubbers, dry scrubbers, carbon adsorbers, oxidizers, and refrigerated condensers. Garnett and Patience (1993) suggest that the exponent depends on $1/n$, in which n is the number of dimensions applicable to the equipment being scaled. For example, towers or stacks being scaled by height have one dimension; storage tanks being scaled by volume have three dimensions.

When applying ratios with exponents it is important to distinguish between installed equipment within the treatment plot (within the plant boundary limits) and facilities outside the plant boundary limits (e.g., utilities interconnections). The method of using ratios with exponents does not work well if fixed costs are a high proportion of the total system cost, if there are significant associated costs outside the plant boundary limits, or if auxiliary equipment is required for the new plant that was not in the old plant. Methods of handling these situations using ratios with exponents are described by Haseltine (1996). For example, if the fixed costs (investment requirements that are the same for all sizes of plants) can be estimated separately, they can be added to the ratioed costs of the non-fixed portion of the total system costs.

12.2.1.2 Factoring Costs of Principal Equipment

Factored estimates are developed by determining the cost of major equipment and then applying a factor or factors to account for installation of equipment and labor and for minor components. Minor components include common instruments, piping, insulation, electrical services, foundations, supports, structures, and paint. Major equipment costs are obtained from vendors or from equipment cost curves.

Cost curves have been published for towers, vessels, heat exchangers, furnaces, pumps, blowers, and compressors. Certain items such as buildings and large transformers are best estimated separately, based on unit costs such as cost per square foot, per kilovolt amps, etc. Equipment cost curves are plots of cost versus size, capacity, or weight. Certain factors are sometimes applied to account for various materials of construction. See Figure 12-1 for an example. When using cost curves, the costs must be indexed to current prices.

Deriving cost estimates by factoring the cost of major equipment does not work if the equipment is purchased pre-assembled with supports, piping, and controls. In that event, the costs for minor components and installation must be estimated individually using definitive methods (*see* Section 12.2.2) and/or published unit prices (e.g., cost per foot for piping to and from the packaged assembly).

Desai (1981) summarizes two methods for deriving factored cost estimates that are rapid and easy to apply to nonpackaged major equipment total cost: the Lang (1948) method, and the Hand (1958) method.

With the Lang method, the major equipment costs are totaled, and a factor ranging from 3.1 to 4.74 is applied. The smaller Lang factor is used for solids processing units such as soils handling systems. An example using the Lang factor 3.63, which applies to units processing solids and fluids follows for a hypothetical soil washing system. Costs of principal equipment have been obtained from vendors for this hypothetical soil washing system as follows:

Conveyors	$21,000
Classifiers	$32,000
Slurry pumps	$11,000
Water pump	$4,000
Chemicals injection systems	$14,000
Thickener	$7,000

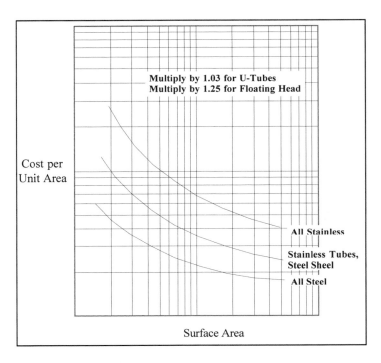

Figure 12-1 Unit costs of fixed-tubesheets heat exchangers (From *Chemical Engineering*, July 27, 1981).

Tank and hopper	$21,000
Filter	$19,000
TOTAL PRINCIPAL EQUIPMENT	$170,000
	x 3.63
TOTAL INSTALLED COST =	$617,000

The Hand method can be more accurate, because individual factors are applied for each type of major equipment. Desai (1981) gives these examples from Hand (1958):

Equipment	Hand factor
Towers, such as strippers	4
Pressure vessels, such as carbon adsorbers	4
Heat exchangers	3.5
Fired heaters	2
Pumps	4
Compressors	2.5
Miscellaneous equipment	2.5

Table 12-1 provides a hypothetical example using the Hand factors for a unit for removing metals, nitrates, and semivolatile and volatile organics from groundwater. Note that the example in Table 12-1 excludes costs for wells and for a steam boiler.

Table 12-1 Example of a cost estimate using Hand factors.

	Equipment Cost	Hand Category and Factor	Total Installed Cost
Cation exchange vessel	$7,000	Press. vessels, 4	$28,000
Anion exchange vessel	$7,000	Press. vessels, 4	$28,000
Continuous backwash sand filter	$49,000	Miscellaneous, 2.5	$22,500
Heated stripper column	$57,000	Towers, 4	$228,000
Air blower	$3,000	Miscellaneous, 2.5	$7,500
Heat exchangers	$39,000	Heat exch., 3.5	$36,500
Pumps	$14,000	Pumps, 4	$56,000
Air compressor	$4,000	Compressors, 2.5	$10,000
		TOTAL INSTALLED COST	$616,500

There are a variety of factored cost-estimating methods that are more complex and more accurate than these two simplified methods. One method is to multiply the total principal equipment cost by individual factors that depend on the type of installation work required. The example above using a Lang factor is used to illustrate this detailed factored method in Table 12-2. The factors are derived from information given in Peters and Timmerhaus (1991), from *Chemical Engineering* (1963), and from recent experience. Note that for item F_2, the factor applies to $E+F_1$ if the contractor purchases principal equipment, or the factor applies to $C+D+F_1$ if the owner does.

Note that the factors used may be adjusted to fit particular projects. A smaller factor for site preparation or site improvement can be chosen for a flat site that already has road access and fencing and no structures in the way. The factor for piping would be higher if stainless steel alloys are specified. This example is for a plant processing both solids (soil) and fluids (for washing). The factor for piping would be half that given in item C-7 in Table 12-2 for an all-solids plant and double for an all-fluids plant, as given in Peters and Timmerhaus (1991). The factor for insulation would be high if process heating were involved (as in the example in Table 12-1) or if winterization is required.

It is of interest to compare the overall factor for the total contract cost, G/A, to the Lang factor of 3.73. In this example, G/A is $787,000/$204,000, or 3.86.

Note that the factors for engineering, supervision, and contingency are significantly higher than the other items included in the cost estimate. These items are estimated as a percentage of item A or of item G, as shown in Table 12-2. However, traditional percentages for engineering design of process plants and of civil engineering work are too low for remedial design. Also, permit applications require significant

Table 12-2 Example of a detailed factored cost estimate.

		FACTOR	AMOUNT
A.	**MAIN COMPONENTS**		
1.	Total principal equipment (from Table 12-1)	1.0	$170,000
2.	Freight and offloading	.10	17,000
3.	Contingency on principal equipment	0	17,000
		Subtotal A =	$204,000
B.	**SALES TAX ON MAIN COMPONENTS**	.07(A1)	B = $11,900
C.	**INSTALLATION COSTS (incl. sales tax on materials)**		
1.	Site preparation	.05 x A	
2.	Site improvements	.05 x A	
3.	Civil/structural work	.15 x A	
5.	Buildings	.15 x A	
6.	Sewers, drains	.03 x A	
7.	Treatment system piping	.31 x A	
8.	Electrical work	.10 x A	
9.	Instrumentation	.13 x A	
10.	Insulation	(not applicable for this example)	
11.	Painting	.03 x A	
12.	Other	.03 x A	
		Subtotal C =	$210,100
D.	**CONSTRUCTION EQUIPMENT**	0.10 x C	D = $210,100
E.	**TOTAL DIRECT COSTS**	A+B+C+D	E = $447,000
F.	**INDIRECT COSTS**		
1.	Overhead and general & administrative costs	.40 x C	$84,000
2.	Fees and profit*	.05(E+F1)*	$26,600
		Subtotal F =	$110,600
G.	**TOTAL CONTRACT COST**	E + F	G = $557,600
H.	**OTHER COSTS**		
1.	Engineering and supervision	.33 x E	$147,500
2.	Contingency	.25(C+D+F1) x 1.05	$82,700
		Subtotal H =	$230,200
I.	**TOTAL CAPITAL INVESTMENT**	G + H	I = $787,800

engineering effort and sometimes require additional hydrogeologic expertise. For remediation projects, the percentage of project costs allowed for engineering should be approximately double the values used on traditional projects.

Contingency is an allowance to cover unknown or unanticipated costs. Because remediation projects usually deal with nonvisible underground problems, the likelihood of encountering unanticipated costs is high. The contingency percentage should be comparable to the accuracy of the estimate for the known items. This should be 10% or more for definitive estimates and approximately 30% for estimates based only on conceptual designs.

12.2.2 Definitive Estimating of Investment Cost

The further a remedial design has progressed, the more accurate an investment cost estimate can be. If the design is 90% or more complete, it is possible to estimate the total installed cost of the treatment equipment with an accuracy of ± 10%. The method used for definitive estimates is the same method used by many installation contractors for buildings and commercial plants, so experience with the method can result in an accuracy approaching that of the contractors' accuracy. Two alternative reference sources are widely used by contractors for definitive estimates:

- R.S. Means (Kingston, Massachusetts)
- Richardson Engineering Services (Mesa, Arizona)

The Army Corps of Engineers also has a system for making definitive cost estimates. The Corps of Engineers data can be accessed by using the software MCACES GOLD from Building Systems Design (Atlanta, Georgia). Means and Richardson also have computerized aids available for definitive cost estimating.

These methods are most accurate for estimating costs for site work, buildings, utilities, and mechanical and plumbing systems within buildings, but they have been adapted to estimating costs for treatment or processing equipment as well. With Means or Richardson, the user utilizes a set of manuals that give material costs, labor hours, and construction equipment costs.

It is best to use vendors' quoted prices for major treatment equipment and to use a reference such as Means or Richardson for site work and soils handling with heavy construction equipment and for minor materials and installation costs. These references are very good for estimating costs of grading, excavating, driveways, foundations, structures, sewers, fencing, insulation, piping, electrical power service, control buildings, etc.

For equipment rentals, rates can be obtained from the "Rental Rate Blue Book" published by K-III Directory Corp (San Jose, California).

12.2.2.1 Associated Costs

The estimating systems offered by Means and by Richardson give data on labor rates for various geographical areas, overhead and indirect costs, and contractor's profit. Usually there are local and/or sales taxes on materials. If vendors have prefabricated and/or assembled equipment, the sales taxes apply to the entire package, including labor. However, for field fabrication and erection work, sales taxes in some states apply to materials only (excluding freight) and not to labor. The freight costs must be added to the basic equipment costs.

If all of these costs are accounted for, the total investment cost can be arrived at by adding costs for design and permitting, acceptance testing, and contingency. Contingency is included to account for unknowns. It should be at least 20%; higher percentages apply if the detailed design is not complete or if significant amounts of work involve subsurface items.

12.2.2.2 Engineering Design Costs

Frequently, costs for the design of construction projects have been estimated as a percent of the constructed cost or have been derived by estimating the number of engineering drawings involved and multiplying by the average historical cost per drawing. If the design has just been started, either of these two methods can be applied. However, the percent or cost per drawing for remediation projects is usually much higher than for other construction projects. This may be caused by the detailed attention that must be paid when hazardous substances are involved or by the relatively small size of the equipment involved for many remediation projects. With small equipment, P&ID development, equipment sizing and rating, engineering needed for permitting, designing utilities interconnects, specifying instruments, producing drawings, and developing erection specifications require the same effort as with larger (more costly) equipment. Thus, the percent of total project costs for engineering is relatively high with small equipment. Use of vendor packages and applicable parts of designs from previous projects helps decrease engineering design costs.

The COMPOSER GOLD^{TM} and RACER/ENVEST^{TM} software programs described in the section on Computer Applications to Cost Estimating aid in estimating remedial design costs.

12.3 Estimating Annual Expenses

This discussion covers the following items, which all play a role in the annual expense of running a treatment system:

- Utilities consumption
- Operating labor and overhead
- Maintenance
- Chemicals, adsorbents, and supplies
- Property tax and insurance
- Monitoring and reporting
- Other direct costs

If remediation or emission abatement equipment has been rented, rental costs should be treated as an additional expense. Even if the equipment is rented for use only during part of the first year of operation, the rent should not be lumped with investment. For budgeting, property tax, and income tax considerations, it is important to distinguish precisely between capital investment and expense.

12.3.1 Utilities Consumption

Electric power and fuel are the main utilities used for operating remediation facilities. Sewage disposal fees are sometimes significant and must be determined for each locality if applicable. Fresh water supply costs must be determined in a similar fashion. Estimating methods are given here for electric power based on known charges from the electric utility in dollars per kilowatt hour and for fuel gas based on known charges per therm (10^5 Btu).

The main electrical costs are for motors driving pumps, blowers, mixers, and soils handling equipment (conveyors, classifiers, etc.) and for special electrical equipment, such as water treatment systems using UV light and electrically energized thermal treatment systems for soils. The kilowatt consumption of the special electrical equipment must be obtained from the vendors. For induction motors (the type usually used), the relationship between brake horsepower (bhp) and kilowatts is approximately one to one when power factor and motor efficiencies are taken into account. Thus, the yearly expense for operating motors is estimated by totalling the bhp values and multiplying by predicted hours per year of actual use and by the cost of electric power.

Annual power expenses for motors (\$/yr) = bhp x hr/yr x \$/kWh   (12-3)

The actual running horsepower for each motor should be used for totaling the horsepower requirements and not motor nameplate ratings. The brake horsepower for a liquids pump is:

bhp = gal/min(SG_L) (0.433) (TDH)/[1,713 (eff)]   (12-4)

in which the gal/min are determined at flowing temperature; SG_L is the specific gravity of the liquid at flowing temperature; TDH is the total dynamic head for the pump (ft w.c.); and eff is the fractional pump efficiency. Note that the term (SG_L) (0.433) (TDH) is the pressure rise through the pump, with units of psi. The efficiency for a centrifugal pump is usually in the 0.55 to 0.75 range. (Although centrifugal pumps are often rated at higher efficiencies, they are usually not operated at their optimal efficiency point.)

The brake horsepower for blowers depends on the absolute gas pressure and temperature, flow rate, and pressure rise. It is convenient to express gas flow rate in standard cubic feet per minute. Standard conditions are 1 atm (760 mm Hg or 29.92 in. Hg absolute pressure) and a stated temperature, such as 0° C or 70° or 60° F (520° R absolute). Using absolute terms, the conversion from actual cubic feet per minute is:

scfm = acfm [(actual pressure)/(standard pressure)] x
[(standard temperature)/(actual temperature)]   (12-5)

The blower horsepower is:

bhp = scfm (SG_G) (h) (0.000157)/eff   (12-6)

in which the flow in scfm is at 1 atm and in at the 60° to 70°F standard temperature; SG_G is the specific gravity of the gas mixture relative to air; h is the pressure rise through the blower (in. w.c.); and eff is the fractional blower efficiency. The specific gravity is equal to the average molecular weight of the gas mixture divided by the molecular weight of air, 28.9. The efficiency of centrifugal blowers is usually in the 0.6 to 0.8 range.

An example will illustrate use of these equations. A flow rate of 261 acfm of air with 5.0 vol% toluene at 7 in. Hg vacuum and 50° F (510° R) with a 68% efficient blower at 108 in. w.c. pressure rise will have a horsepower requirement calculated as follows:

Step 1. Calculate the molecular weight and specific gravity of the air, based on the molecular weight (MW) of toluene of 92:

$$MW = 0.05 (92) + 0.95 (28.9) = 32.06 \tag{12-7}$$
$$SG_G = (32.06)/(28.9) = 1.109 \tag{12-8}$$

Step 2. Calculate flow rate at standard conditions of 1 atm and 520° R, based on 7 in. Hg vacuum being equal to 22.92 in. Hg absolute:

$$scfm = (261 \text{ acfm})(22.92 \text{ in. Hg}/29.92 \text{ in. Hg}) (520° \text{ R}/510° \text{ R}) = 204 \text{ scfm} \tag{12-9}$$

Step 3. Calculate brake horsepower using Equation 12-6:

$$bhp = (204 \text{ scfm}) (1.109) (108 \text{ in. w.c.}) (0.000157)/0.68 = 5.65 \text{ hp} \tag{12-10}$$

The total electric power consumption is calculated by summing the individual pump horsepower and individual blower horsepower requirements and equating the total bhp to kilowatts, adding the kilowatt consumption of any special equipment that is electrically energized, and adding an allowance for lighting and controls. An appendix to this chapter gives a detailed example of this total power consumption calculation.

For fuel gas (normally natural gas or propane), the usual duty in remediation systems is associated with auxiliary fuel for the combustion of flammable vapors in an air stream. An acceptable approximation of auxiliary fuel requirements is to assume that exhaust gases have properties similar to those of air, with an average specific heat of 0.24 Btu/(lb-°F) and a density of 0.076 lb/scf at 60° F standard temperature. The fuel consumption is calculated in three steps:

1. Determine what the Btu (or equivalent therms) requirement would be if the gases were all air raised from ambient temperature to the desired operating temperature.
2. Correct the heating duty determined in Step 1 by any savings anticipated from heat exchange if the combustion equipment includes heat recovery.
3. Subtract the heat release from burning the flammable vapors.

An example illustrates this procedure. It is given that 204 scfm of air contains 100 ppmv of petroleum vapors exiting from a blower at 150° F. This gas stream then enters a thermal oxidizer operating at 1,400° F with 50% heat recovery. The petroleum components have an average molecular weight of 105 and a heating value of 19,000 Btu/lb. The 1,400° F exhaust is heat exchanged with the 150° F air, thereby warming the air to a temperature approximately halfway between 150° and 1,400° F. Auxiliary fuel gas combustion heats the air the rest of way to 1400° F. Find the auxiliary fuel gas heating duty in Btu/day or therms/day (1 therm is equivalent to 10^5 Btu).

Step 1. Calculate the heating duty from 150° to 1,400° F using the average specific heat and the density of air at standard temperature and pressure.

(204 scf/min) (0.076 lb/scf) (1440 min/day) (0.24 Btu/(lb-°F)) (1,400 – 150)° F
= 6,697,728 Btu/day (66.98 therm/day) (12-11)

Step 2. Account for heat recovery.

(0.50) (6,697,728 Btu/day) = 3,348,864 Btu/day (12-12)

Step 3. Determine the heating value of the petroleum vapors, given that the vapor concentration is 100 ppmv ($100/10^6$), the vapor molecular weight is 105, and the molecular weight of air is 28.9.

Specific gravity of petroleum vapors = 105/28.9 = 3.63 (12-13)
Mass rate of flow = (204 scf/min) ($100/10^6$) (3.63) = 0.074 lb/min (12-14)
Heating value = (0.074 lb/min) (1440 min/day) (19,000 Btu/lb)
= 2,024,640 Btu/day (12-15)

Step 4. Subtract the heating value of flammable petroleum components.

Net heating duty = 3,348,864 Btu/day - 2,024,640 Btu/day
= 1,324,224 Btu/day (13.24 therm/day) (12-16)

If the system operates 24 hr/day with a 90% on-stream factor (equivalent to operating 328 days/yr) and if auxiliary fuel costs $0.80/therm, the yearly fuel gas cost in this example is:

(13.24 therm/day) (328 days/yr) ($0.80/therm) = $3,474/yr (12-17)

12.3.2 Operating Labor and Overhead

The hourly rate for operators or technicians must be known or assumed and adjusted to account for benefits to obtain the burdened hourly rate. The burdened rate for each class of operator or technician should be multiplied by the hours per year each position is filled. If the plant is attended 24 hr/day for 5 days/wk, then 3 times the number of operator positions must be accounted for. If the plant is attended 24 hr/day for 7 days/wk, then 4.2 times the number of operator positions must be accounted for.

The following is an example of operating labor costs for a plant with two operators per shift for 24 hr/day, 7 days/wk, and a technician in attendance 40 hr/wk. The plant operates 90% of the year, but these 3 positions are paid for 50 wk/yr for 8 hr each per shift, so each person is paid for 2,000 hr/yr. The burdened hourly rates are given as follows:

- Lead operator $55/hr
- Operator helper $40/hr
- Technician $48/hr

The annual operating labor cost is:

(2,000 hr/yr) (4.2) ($55 + $40)/hr + (2000 hr/yr) ($48/hr) = $894,000/yr   (12-18)

If the overhead costs, which might include administrative labor, supervision, and engineering help, are known, they should be added. However, overhead costs usually apply to more functions than plant operations, and an average percent of wages is applied to arrive at overhead costs. If the percent is unknown, an allowance of approximately 40% should added to the calculated burdened labor cost. If significant premium wages are anticipated to be paid to the operators because they may work overtime to ensure full coverage during all shifts, the premium pay should be estimated based on unburdened wages (with no benefits). The overhead percentage applies only to the labor cost calculated, as in the example, and not to the premium pay.

12.3.3 Maintenance Expense

If special major maintenance costs are predictable, such as overhaul of internal combustion engines used for soil vapor extraction, or the yearly inspection/repair of a steam boiler, these expenses should be accounted for individually. If experience makes maintenance labor and parts predictable, the yearly maintenance expense can be estimated accordingly. In order to account for overhead and parts re-ordering costs, Wilke and Pecar (1995) suggest that an average hourly rate in 1995 dollars of $50/hr be used for maintenance labor. Without detailed knowledge of predictable maintenance labor and parts requirements, maintenance of treatment units can be estimated for each year at 4% of total investment costs.

12.3.4 Chemicals, Adsorbents, and Supplies

Processes requiring chemical usage periodically for years, such as for precipitation or UV oxidation of groundwater, must be examined individually. The best way to determine how much precipitation or neutralization chemicals will be used is to conduct laboratory flask tests or field pilot tests. Calculating chemical dosage from stoichiometry alone is not accurate because groundwater usually has buffering capacity. Unknown groundwater constituents will result in less than 100% efficient reactions, often requiring excess chemical doses to provide required treatment efficiency.

For determining the peroxide consumption in conjunction with UV light treatment, a stoichiometric calculation will provide a figure for the minimum amount of chemical dosage. However, for estimating yearly expenses, a consumption figure based on the UV system vendor's experience is needed, so that the excess dosage needed in practical applications is accounted for.

For ion exchange regeneration chemicals, the frequency of regeneration and dosage are best determined from dynamic flow-through testing of the adsorption/elution cycle and adsorbent vendor's experience.

For ion exchange media and adsorbents, the life of a bed for a nonregenerative system should be determined from flow-through dynamic testing. For GAC, such testing is needed if a mixture of organic constituents is in the groundwater and there is no experience with that mixture or if naturally occurring organics may be present. For a single organic component to be adsorbed in a dynamic system with an empty-bed contact time of 15 min, the amount that carbon can adsorb can be approximated as one-half the amount determined from published equilibrium isotherm data.

In any event, once the consumption rate of a chemical, ion exchange medium or adsorbent is determined, the type of packaging and delivery, and corresponding prices, must be established. The price varies depending on the amount of a chemical in each unit delivered, the percent of chemical content, whether it is packaged in individual containers or shipped in bulk form, and the travel distance from which it is shipped.

The cost of supplies, which may include lubricants, housekeeping materials, disposable safety gear, paper goods, and a variety of other materials, is difficult to predict. In the absence of experience with tracking such costs, an arbitrary allowance should be included in the cost estimate.

12.3.5 Property Taxes and Insurance

Unless treatment equipment, wells, buildings, and other components associated with installation are exempt from state and local property taxes, a yearly tax is assessed against the value of those items. (Such taxes may also apply to land, but in most instances the land tax is paid whether or not treatment equipment is installed.) If the tax rate is not known precisely, assume that it is 1% of the total installed cost and is paid at that rate each year.

A similar amount can be assumed for insurance expense. Thus, in the absence of known tax and insurance expense, assume that the total paid each year for these two items is 2% of the total capital investment.

12.3.6 Monitoring and Reporting

The costs for monitoring include much more than laboratory analysis fees. Monitoring costs must be estimated for each project, and they usually depend on what sampling and analyses are required by an environmental quality agency.

The frequency of sampling and the parameters of analysis must be known in order to estimate these expenses. Also, frequency and parameters may change in future years. For example, a control agency may require that a certain group of wells be sampled quarterly and that the samples be analyzed for an extensive list of characteristics and species concentrations. It is often possible to negotiate deletion of some parameters or to reduce the frequency and/or number of wells sampled after a number of quarterly results are reported with the concentrations within limits or nondetectable.

A complete estimate of expenses for groundwater and soil monitoring and reporting should include these elements:

- Mobilization for sampling
- Travel to and from the site for the sampling crew
- Field sampling, packaging, and documentation labor
- Shipping costs for samples
- Laboratory fees
- Costs for verification and validation of analysis results
- Checking for completeness of validated results
- Preparing summaries of results, drafting reports, checking report drafts, and issuing reports.

In this context, verification means checking that the values of the parameters in a laboratory report are the same as the values noted by the analyst. This can be accomplished manually or automatically with computer programs plus manual checking of a fraction of the results. Validation includes checking whether analyses were done within prescribed maximum sample holding times; checking whether instrument calibrations were within acceptable tolerances; and flagging results that exceeded such limits or did not meet other quality control parameters. Completeness in this context is a measure of whether enough validated results for a group of samples are unflagged for the results to be considered adequate. These quality controls may be required for monitoring CERCLA and RCRA sites and for other sites for which legally defensible analysis data are needed. Such quality control requirements can impact costs by adding approximately 40% of the laboratory fees.

Cressman (1991) gives unit costs for analyzing water, soil, and sludges, and he estimates sampling labor to be 50 hr/well for groundwater and 20 hr/sample for soil borings. The cost for each boring is $105/ft plus 2 hr technical labor/ft. Updated costs for laboratory fees are usually readily available from local laboratories. Cressman gives costs as of mid-1991 for air sampling as follows:

- One-time labor cost per sampler for preparation, transport, setup, and teardown — $185/sampler
- Particulate matter less than 10 μm (PM-10), per sample — $90 field cost, $20 lab fee
- Metals, pesticides, and organics per sample — $70 to $90 field cost, $360 to $450 lab fee
- Meteorological data — $180

(Field costs include labor and equipment rental charges).

Most of these costs have come down since 1991, especially for well sampling and laboratory analyses, because of competition and use of low-volume sampling and field analyses techniques.

12.3.7 Other Direct Costs

If certain costs are not included in the above categories, it may be necessary to add such costs based on known amounts from experience or by applying an arbitrary allowance. Examples include mileage expense for technicians or operators visiting an automated treatment system, copying charges applied to reports, and communications or shipping charges.

Other miscellaneous expenses may be direct costs or in a category of their own. Examples include licensing or royalty expenses, and periodic (5-yr) site reviews.

12.4 Computer Applications to Cost Estimating

Computerized cost-estimating programs can be obtained for detailed estimates from Means and from Richardson (*see* Section 12.2.2). Means' computerized spreadsheet applications include databases for costs of buildings, electrical work, and civil work. Richardson's RACE system provides a spreadsheet that can be easily edited. Versions of RACE are available for either single computers or for network systems. Richardson databases include process equipment, as well as general construction costs. Capital investments can be estimated from the Richardson database with the Engineer's Aide program marketed by Epcon International (Houston, Texas). Pricing information for a variety of components and construction steps and the Corps of Engineers database for maintenance and repair of certain facilities is contained in MCACES GOLD, available from Building Systems Design (Atlanta, Georgia).

Preliminary cost-estimating information on some remediation processes can be derived from the EPA computerized database CLU-IN. Although this program is not designed for cost estimating, it does provide information on processes being developed and being marketed commercially, including some cost information. The program is free of charge; ordering information is obtainable by telephoning 301-589-8366 or obtaining publication EPA/542/N-93/004 from the EPA in Cincinnati (fax 513-891-6685). Internet access is available at www.clu-in.org/techfocus. The data are directly from vendors.

For cost estimates based on conceptual designs, CORA, COMPOSER GOLD^{TM}, and RACER/ENVEST^{TM} have been developed specifically for remediation systems and have proven to be very valuable.

CORA (Cost of Remedial Action) was developed by CH2M-Hill (Reston, Virginia) for the EPA and was available until 1994. It can still be used effectively, and a number of environmental consulting firms have the software.

COMPOSER GOLD, accompanied by a software system called ECHOS, is marketed by Building Systems Design (Atlanta, Georgia) and includes the Hazardous Waste Database from the Corps of Engineers and the EPA. Personalized training is offered, with continuing telephone help via an 800 number.

RACER/ENVEST was developed by Delta Research Group for the Air Force. The ENVEST models are available from Talisman Partners Ltd. (Englewood, Colorado). This software was developed for remediation projects, and the cost includes personalized training and continuing telephone help via an 800 hotline. The program can also be used to develop a conceptual design when only minimal information is available.

Primary parameter values or qualitative descriptions (e.g., the rate of groundwater flow in gal/min, or whether the aquifer is shallow or deep) must be furnished by the user. Secondary parameter values should be furnished if known, resulting in improved accuracy of the cost estimate. If the values are not furnished, the program assumes default values, thereby advancing the conceptual design. For some parameters, RACER/ENVEST will suggest a range of values and a default value. For example, a sand-bed filter can be sized for suggested values of 2 to 8 gal/min/ft^2 of cross-sectional area. If the user does not choose such a value, the default value is applied. With either a chosen suggested value or the default value, the cross-sectional area (and consequently the diameter) of the filter vessel becomes known.

RACER/ENVEST can be used to estimate the costs of extraction and injection wells, pumping systems, electrical service, piping, excavation, soil loading/hauling/disposal in landfills, backfilling, installed equipment for remedial process options, remedial design, treatability studies, startup, operations/maintenance, and monitoring/analysis.

The developers of RACER/ENVEST have also produced a program called TANK RACER for the EPA and the Air Force for estimating costs of leaking underground storage tank cleanups.

For specially designed treatment processes that are not already in COMPOSER GOLD, RACER/ENVEST, or HAZRISK programs, process simulation software that does cost estimating is available from Icarus Corp. (Rockville, Maryland).

12.5 Life Cycle Analysis

The total cost for constructing a treatment system and running it for more than 2 yr is best expressed in terms of present worth (present value) of the expenditures. This is because expenditures that are to occur in future years of running the plant have less value than initial expenditures because of the interest value that money has. This present value approach to costing the life cycle of a project is especially valuable when comparing alternative methods of remediation. One method may have higher initial capital investment requirements and lower annual expense than an alternative method. In this instance the other method essentially postpones the need for expenditures. The party paying for the project could invest the postponed expenditures until the time comes for making payment and could collect interest or some other form of return on investment. That party's net outflow of money is the total expenditure minus the return.

The same type of life cycle analysis can be applied when choosing among alternative components that could apply to a selected method of treatment. For example, a direct thermal oxidizer will use more fuel than a catalytic oxidizer, but it may have a lower life cycle cost when considering that the higher fuel costs occur in future years.

Wilke and Pecar (1995) give a detailed analysis of the factors that affect the present value of alternatives when selecting from among alternative equipment choices.

12.5.1 Investment, Expense, Closure, and Post-Closure Costs

Items that impact present value include:

- **Initial investment** — This includes the total cost for engineering, permitting, procurement, erection, and startup.
- **Postponed investment** — This includes costs similar to the initial investment but for major components that are planned to be installed at some time after startup. For example, a group of extraction wells and a groundwater stripper may be included in the initial investment, and more wells plus a parallel stripper are planned to be installed after 2 yr of hydrologic data are obtained from operating the initial group of wells. The data will be used to help determine the optimum location of the future wells.
- **Annual expense**.
- **Dismantling costs** — Dismantling costs, salvage value, and closure costs are those that apply at the end of running the plant.
- **Post-closure costs** — These costs usually involving monitoring and maintenance of items such as fencing and caps.

Methods of estimating investment costs and annual expenses have been covered in this chapter and are illustrated in examples in Section 12.2.1.2. The net amount of dismantling costs plus closure costs minus salvage value should be treated in the same way as postponed investment. The salvage value of remediation equipment is very significant in some instances. Soil venting equipment, for example, is often trailer or skid mounted and can be moved to another site on completion of remediation.

Closure costs in mid-1989 dollars are shown in Table 12-3.

If control agencies require that an independent professional engineer certify closure, Cressman estimates that the labor involved includes the following:

- Initial review-closure plan   8 hr
- Inspection during closure process   192 hr
- Preparation of final documentation   4 hr

For estimating the cost of post-closure activities, Cressman gives the following information: The post closure phase at the DOE sites lasts 30 yr or longer depending on potential health impacts and location. Quarterly activities include groundwater air monitoring, inspection of landfills and caps (with costs approximately equal to monitoring costs), and leachate removal if needed. The mid-1989 costs for annual activities, including mowing, seeding, fertilizing, landfill and/or cap maintenance, road maintenance, and fence repair, averaged $3,500/acre or $0.16/yd^3. Typical unit costs for these annual activities are shown in Table 12-3.

Table 12-3. Example of typical closure and post closure costs (From Cressman, K.R."Cost Estimating of the Closure/Post-Closure Phase," *Remediation*, Summer 1991. Copyright John Wiley & Sons, Inc., reprinted by permission of John Wiley & Sons, Inc.).

Activity	Unit Cost
Removal of a 4- in or 6 –in. asphalt-paved roadway	$5/yd^2
Removal of an unpaved roadway and finish grading	$1/yd^2
Spreading and compaction of topsoil	$4/yd^3
Mechanical seeding at a rate of 215 lb/acre	$1,300/acre
Fine grading and mechanized application of lime, fertilizer, and seed	$2/yd^2
Drilling and installation of monitoring wells	$6,000 each*
Erection of fencing with barbed wire	$11/ft
Erection of fencing with no barbed wire	$10/ft
Erection of a gate	$85 each
Erection of warning signs	$22 each
Mowing	$30/acre
Seeding — routine erosion repair	$1,300/acre
Reseeding of final cover	$1,200/acre
Fertilizing	$300/acre
Mulching	$3,100/acre
Cover repair by hand with crushed stone	$36/yd^3

*Wells can also be estimated at $150/ft for subcontracted drilling and installation; plus 2 labor hr/ft for a geologist, technician, health and safety monitoring, quality assurance, and project management; plus 10 hr of technical labor per well for aquifer testing.

Although not included by Cressman, maintenance of monitoring wells is another post-closure activity that should be included in the cost estimate.

12.5.2 Present Value Factors

For investment costs that are postponed or for future annual expenses in the second year or beyond, the present value is the cost multiplied by $1/(1 + \text{rate})^n$, in which rate is the discount rate or interest value of money and n is the year of the expenditure. The discount rate depends on how the party paying for the remediation values money. For example, that party may be borrowing money at an interest rate of 10% per year, so the discount rate is considered to be 0.10. Another party may be able to invest in chemical plant construction with a 4-yr payback and therefore values money at 25% per year, so the rate is 0.25. Another party may only invest in money market certificates and receive interest at 4% per year, so the rate is 0.04. Typically, discount rates used in life cycle analyses are in the 0.05 to 0.10 range.

As an example, suppose wells and a second stripper will be added to a pump-and-treat system in the second year at a cost of $65,000, and annual expenses projected for the second year are estimated to be $35,000. If the discount rate were 0.08, the present value of the total second-year expenditures of $100,000 is:

$$\$100,000/(1 + 0.08)^2 = \$85,700 \tag{12-19}$$

For this party, expenditures that can be postponed 2 yr are valued at 0.857 times the amount of the expenditures.

Because the annual expense often varies from year to year, investments are sometimes postponed, and dismantlement/closure costs occur in a future year, the present value should be computed year by year. The total present value (present worth) is the total of the initial investment plus the successive present values for the life of the system plus the present values of the post-closure costs. Some present value factors for year- by- year application are shown in Table 12-4.

Table 12-4 Selected present worth factors.

Years	Rate		
n	0.05	0.08	0.1
2	0.907	0.8573	0.8262
3	0.8638	0.7938	0.7513
4	0.8227	0.7350	0.6830
5	0.7835	0.6806	0.6209
6	0.7462	0.6302	0.5645
7	0.7107	0.5835	0.5132
8	0.6768	0.5403	0.4665
9	0.6446	0.5002	0.4241
10	0.6139	0.4632	0.3855

For other discount rates or for additional years, the present value factor is $1/(1 + \text{rate})^n$.

An example of life cycle analysis for comparing two alternative soil treatment schemes, plus the no-action alternative, will illustrate the use of present value factors. In this hypothetical example, the no-action alternative requires 5 yr of monitoring a site with periodic soil borings, sampling, and analysis. The control agency has agreed that if natural degradation of the organic contaminants in the soil (as proven from the analyses) is such that the cleanup goal has been reached within 3 yr, less sampling and analysis are needed in the fourth and fifth years. The site has already been fenced and signed. The treatment alternatives are ex situ bioremediation of the soil and ex situ soil washing. The bioremediation alternative will extend into the second year. The soil washing scheme can all be done in the first year. The treated soil will then be backfilled on-site. The life- cycle analysis with a discount rate of 8% with the present value (PV) factors listed and costs expressed in thousands of dollars is shown in Table 12-5. (Note: even though bioremediation has a slightly higher total cost than does soil washing, its present value is less, because some of the expenditure is postponed until the second year.)

Table 12-5 Example of comparing present values for three alternatives.

	No Action			Bioremediation			Soil Washing
Yr	$k	P.V. Factor	P.V. $k	$k	P.V. Factor	P.V. $k	P.V. $k
1	40		40	80		80	108
2	40	0.8573	34	30	0.8573	25.7	
3	40	0.7938	32				
4	20	0.7350	15				
5	20	0.6806	14				
Total			$135			$105.7	$108

Another example, given in the appendix to this chapter, illustrates the estimation of annual expenses, closure costs and post-closure costs, as well as application of present value factors, using the installed equipment cost estimate in Table 12-1. The investment cost in Table 12-1 is $616,500 plus wells; for this hypothetical example, it is assumed that an existing steam boiler is on-site devoting a small fraction of its steam production to this heated stripper. The total installed costs for the principal equipment listed in Table 12-1 includes minor components and piping within the plot. However, additional piping is needed for long runs from two extraction wells, and a long run of insulated piping from the steam boiler is needed.

The total installed cost for this system is estimated to be $733,000, including the two extraction wells, a monitor well needed in addition to existing monitor wells, and the additional piping. The example also accounts for yearly expenses and postponed investment costs including an additional monitoring well to be installed at closure and other closure activities. This example was developed in 1995 for a project planned for mid-1996 construction. For this example, the effect of inflation on the initial investment cost is accounted for, because changes in cost indices affect the basic equipment cost estimates. However, this example does not account for the effect of inflation on future costs beyond the initial investment. Another simplification is that no amortization, depreciation, nor income tax effects are accounted for. The investment costs, year- by- year expense, and present values are summarized in Table 12-6 for running the plant 8 yr and continuing monitoring and certain maintenance activities during a 10-yr post-closure phase.

12.6 Summary of Important Points for Cost Estimating

- Life cycle costs account for initial investments, postponed investments, dismantling or closure costs, annual expenses, and post-closure costs (e.g., ongoing monitoring, maintenance of fences and landscaping).
- Costs per 1,000 L of groundwater treated or per ton of soil treated are derived by dividing the life cycle cost by the volume or mass of contaminated media treated.
- When comparing life cycle costs of different remediation alternatives, a present value analysis should be applied in order to account for the value of money when expenditures will occur in future years or are postponed.

Table 12-6 Example of a present value summary.

Year	Capital Investent	Expense	Total Expenditure	Present Value Factor	Present Value
1	$733,000	$745,000	$1,478,000		$1,478,000
2		745,000	745,000	0.8572	639,000
3		728,000	713,000	0.7938	578,000
4		728,000	713,000	0.7350	535,000
5		728,000	713,000	0.6806	495,000
6		728,000	713,000	0.6302	459,000
7		728,000	713,000	0.5835	425,000
8	126,000	728,000	854,000	0.5403	461,000
9		33,000	33,000	0.5002	17,000
10		33,000	33,000	0.4632	15,000
11		33,000	33,000	0.4289	14,000
12		33,000	33,000	0.3971	13,000
13		33,000	33,000	0.3677	12,000
14		33,000	33,000	0.3405	11,000
15		33,000	33,000	0.3152	10,000
16		33,000	33,000	0.2919	10,000
17		33,000	33,000	0.2703	9,000
18		33,000	33,000	0.2502	8,000
				Total:	$5,189,000

- Preliminary estimating methods include using ratios of treatment plant capacities or multiplying the cost of total principal equipment by factors.
- The cost of a new plant that is derived from ratioing the cost of an old plant should be multiplied by the ratio of the cost indices if the previous project is old enough that inflation has significantly affected costs.
- Cost estimates based on old pricing information should be brought up to current values using price indices.
- Simple ratioing does not work well if fixed costs are a high proportion of the total costs, if there are significant costs involved outside the plant battery limits, or if additional auxiliary equipment is required for the new plant.
- Principal equipment costs can be obtained from vendors, published cost curves, and databases.
- A variety of factored cost-estimating methods are available. The more complex methods generally are more accurate.
- In accounting for investment costs, indirect costs such as contractors' overhead, general and administrative costs, engineering (including remedial design), and supervision should be included.
- A contingency should be added to a cost estimate to account for unanticipated costs, which are highly likely when dealing with subsurface cleanup projects.
- Definitive cost-estimating data are available from Means and from Richardson that are best applied when detailed design is near completion.
- Annual expenses include utilities, operating labor and overhead, maintenance, chemicals, property tax, insurance, monitoring, reporting, and other direct costs.
- A variety of vendors and the Corps of Engineers have computerized systems that aid in cost estimating. Computerized systems that apply specifically to

remediation projects include COMPOSER GOLD™ from Building Systems Design, RACER/ENVEST™ from Talisman Partners, and HAZRISK from Independent Project Analysis, Inc.
- The net cost for dismantling and closures should account for the salvage value of treatment equipment.
- To account for the value of money in analyzing life cycle costs, a discount rate or interest rate must be applied, typically in the 0.05 to 0.10 range. An example at a discount rate of 8% shows that $100,000 expended 2 yr hence has a present value of $85,700.
- Present value life cycle analysis results in a credible comparison of alternatives that have different durations and different requirements for capital and ongoing operating expenses. The no-action alternative with long-term monitoring may be more costly than short-term cleanup. A high capital investment spread through future years may have less present value than a lower capital investment all expended in the first year.

Appendix to Chapter 12
Investment Costs and Yearly Expenses Example

Basis: Construction in mid-1996 includes an air stripper at 75 gal/min.

- Plant life — 8 yr
- Post-closure phase — 10 yr

A. Total installed cost for equipment, from Table 12-2 — $616,500

- Equipment cost index from *Chemical Engineering* (1995) — 428.1
- Projected cost index for mid-1996 — 448

Adjusted cost = (448/428.1) ($616,500) = $645,200   (A-1)

B. Add for two extraction wells and an additional monitoring well.

- Monitoring well (small bore) 70 ft deep
- From Cressman (1991) — 70 ft x ($150/ft + $50/hr x 2 hr/ft) = $17,500
- Extraction wells (8-in. finished diameter) 60 ft deep, drilling cost 50% more than $150/ft for monitoring well:

60 ft [$225/ft + $50/hr (2 hr/ft)] (2 wells) = $19,500   (A-2)

Subtotal for wells, mid-1989 basis per Cressman = $37,000

- 1989 Plant Cost Index from *Chemical Engineering* (1995) — 355.4
- Projected Plant cost Cost index Index for mid-1996 — 392

Adjusted cost = (392/355.4) ($37,000) = $40,800   (A-3)

C, Piping outside of treatment plot

- Buried runs from two extraction wells and in-well piping, estimated from Means or Richardson — $20,000
- Insulated steam line on supports, estimated from
- Means or Richardson — $27,000
- Condensate return line — Assume a small-diameter, short run is negligible in cost.

D. Total initial investment — $733,000

E. Yearly expense, based on a 24-hr/day, 7,500-hr/yr operating time

<u>Utilities</u> — electric power at assumed cost of $0.08/kWh
    Two well pumps, 200 ft head, 75 gal/min total, 0.60 efficient

- From Equation 12-4:

bhp = 75 (1.0) (0.433) (200)/[1,713 (0.60)] = 6.3 bhp (A-4)

- Two process pumps, 50 psi pressure rise each, 75 gal/min each, 0.55 efficient:

(2) (75) (50)/[1,713 (0.55)] = 8.0 bhp (A-5)

- Stripper air blower at 50:/1 air/water A/W ratio, 25 in. w.c., 0.65 efficiency (eff)

- From Equation 12-6:

bhp = 500 (1.0) (25) (0.000157)/0.65 = 3.0 bhp (A-6)
Total bhp = 6.3 + 8.0 + 3.0 = 17.3 (A-7)

- The number of kilowatts is approximately equal to the brake horsepower:

17.3 hp (0.746 kW/hp)/[0.90 motor eff (0.90 power factor)] = 15.9 kW (A-8)

- Add allowance for lighting, controls, room air conditioning, and an air compressor - say total electrical load is at 20 kW

Total power consumption = 20 kW ($0.08/(kWh) (7500 hr/yr) = $12,000/yr (A-9)

<u>Utilities</u> — steam, at incremental cost of $12/1000 lb at existing boiler
Total temperature rise of groundwater fed to stripper, 100° F assumed, with 70° F rise in heat recovery exchangers

- Duty of steam heater:

75 gal/min (8.34 lb/gal) (60 min/hr) (100° F - 70° F) (1 Btu/(lb-°F)
= 1,126,000 Btu/hr (A-10)

- Steam consumption, at approximately 1 lb of steam for 1,000 Btu:

(1,126,000 Btu/hr)/(1,000 Btu/lb) = 1,126 lb/hr (A-11)
Steam cost = 1,126 lb/hr ($12/1,000 lb) (7,500 hr/yr) = $101,000/yr (A-12)

<u>Operating labor</u> — operator plus technician, 8 hr/day, 7 day/wk
Assume burdened labor rates--- $55/hr – for the operator; $48/hr – for the technician; each works 40 hr/wk for 50 wk/yr and rotates with main plant personnel for weekend coverage. The treatment plant is fully automated and fail-safe for operating at night unattended.

($55 + $48)/hr (8 hr/day) (7 day/wk) (50 wk/yr) = $288,400/yr (A-13)
Total operating labor cost including overhead at 40% of labor
= $288,400/yr (1.40) = $404,000/yr (A-14)

<u>Maintenance</u> — Estimated at 4% of investment
= 0.04/yr ($733,000) = $29,000/yr (A-15)

Note: This figure is in addition to the technician services included in operating labor.

Chemicals, adsorbents, and supplies

The main cost item in this category for this example is ion exchange resin. The main plant has larger ion exchange units and regeneration capacity, and it charges the treatment plant $2,000 for each batch of resins. A tradeoff study has indicated that this method costs less than having regeneration in place or sending spent resin off-site for regeneration or disposal. The resins are rinsed each week with treated water, slurried out of their vessels, and traded for regenerated resin. Costs are:

$2000/batch (50 batches/yr) = $100,000/yr (A-16)

- In this example, minor amounts of chemicals are used for pH adjustments. Allowing $20,000/yr for supplies and this use of chemicals:

Total chemicals, adsorbents, and supplies
= $100,000/yr + $20,000/yr = $120,000/yr (A-17)

Property tax and insurance — The plant is located in a state with a uniform tax rate (e.g., California) of 1.1%/yr of value. The main plant pays tax on the land whether the treatment plant is there or not, so only the installed equipment value of $733,000 is used for calculating the property tax. Allowing 1%/yr of value for insurance coverage:

Total property tax and insurance = $(0.011 + 0.01)$/yr ($733,000) = $15,000/yr (A-18)

Monitoring and reporting — assume four total monitor wells

Analyses are for metals, nitrates, semivolatile organic compounds and VOCs. Plant influent and effluent are sampled twice per month for the first 2 yr and once per month for 6 yr by on-site technician for outside lab analyses (and more frequently for field screening analyses). No quality control charges or reporting labor costs apply to this process monitoring (but are accounted for below for outside services applied to groundwater monitoring of wells). Shipping and lab fees for each of two samples sets are $700; the total is:

$700/set (2 sets) (2/mo.) (12 mo./yr) = $33,600/yr for first 2 yr (A-19)
$700/set (2 sets) (1/mo.) (12 mo./yr) = $16,800/yr for 6 yr (A-20)

- Outside services include a sampling crew once every 3 mo. for the five monitoring wells, analyses, and reporting, with costs as follows for one set of samples per well, with three sample containers per set:

- Mobilization and travel time each 3 mo., 8 labor hr
- Technician sampling/packaging labor:

0.7 hr/sample (3 samples/well) (5 wells) = 10.5 labor hr (A-21)
Preparation of data summaries, draft report, final report, 8 hr
Total labor per quarter = 26.4 hr ($50/hr) = $1,330 (A-22)

- Charges for vehicle use and safety equipment, $150 each quarter
- Lab fees = $629 each quarter per set of samples:

$629/set (1 set/well) (4 wells/quarter) = $2,520 each quarter (A-23)

- Verification, validation, and completeness check, estimated at:

40% of basic lab fees: 0.40 ($2,520) = $1,010 (A-24)

Total outside services for monitoring of wells = $5,010/quarter x 4 quarter/yr
= $20,000/yr (A-25)
Total monitoring expense for first 2 yr:
= $33,600/yr for process monitoring + $20,000/yr for groundwater monitoring
= $54,000/yr for 2 yr (A-26)
Total, next 6 yr = $16,800/yr + $20,000/yr = $37,000/yr for 6 yr (A-27)

Other direct costs — allowance $10,000/yr

F. Total yearly expense

Items	First 2 Years Expense/yr	Next 6 Years Expense/yr
Utilities--- - electric power	$12,000	$12,000
Utilities--- - steam	$101,000	$101,000
Labor and overhead	$404,000	$404,000
Maintenance	$29,000	$29,000
Chemicals, adsorbents, supplies	$120,000	$120,000
Property tax and insurance	$15,000	$15,000
Monitoring and reporting	$54,000	$37,000
Other direct charges	$10,000	$10,000
Total expense each year	**$745,000**	**$728,000**

G. Closure costs (at 8th year)

- Dismantling cost minus salvage value: allowance — $100,000
- Add monitor well: same as in B ---17,500
- Other investment, based on Cressman (1991) — 5,000 yd^2 site
- Fine grading, add lime and fertilizer and seed:

5000 yd^2 ($1.67/yd^2) = $8,400 (A-28)

H. Total dismantlement and closure costs (at 8th year) — $126,000

I. Post-closure annual expense (for 10 yr)

- Maintenance of five monitoring wells:

Allowance, $500/well (5 wells) = $2,500/yr (A-29)

- Monitoring and reporting, based on Cressman (1991):
  Outside services include a sampling crew once every 3 mo. for the five monitoring wells, analyses, and reporting, with costs as follows for three samples per well:

- Mobilization and travel time each 3 mo., 8 labor hr
- Technician sampling/packaging labor:

0.7 hr/sample (3 samples/well) (5 wells) = 10.5 labor hr (A-30)
Preparation of data summaries, draft report, final report, 8 hr:
Total labor per quarter = 26.5 hr ($50/hr) = $1,330 (A-31)

- Charges for vehicle use and safety equipment — $150/quarter

- Lab fees are $650/quarter per set of samples:

$650/set (1 set/well) (5 wells/qtrquarter) = $3,250/quarter (A-32)

- Verification, validation, and completeness check:

0.40 ($3,250) = $1,300 (A-33)

Total outside services for monitoring
= $6030/quarter x 4 quarters/yr = $24,100/yr (A-34)

- Landscaping maintenance per acre, per Cressman (1991):

  - Mowing — $30/mowing (2 mowings/yr) = $60/yr
  - Erosion repair — $1,306/yr
  - Reseeding — $1,230/yr
  - Fertilizing — $310/yr
  - Mulching — $3,090/yr

Total landscaping maintenance for a 1-acre plot — $6,000/yr (A-35)

J. Total post-closure annual expense

Activity	Annual Expense
Maintenance of monitor wells	$2,500
Monitoring	$24,100
Landscaping maintenance	$6,000
Total per year	**$33,000**

References

Adams, J.N. 1997. Quickly estimate pipe sizing with Jack's cube. *Chemical Engineering Progress* **93**(12):55-58.

Aglitz, J., R.K. Bhatnagar, D.E. Bolt, and T.L. Butzbach. 1995. Installing liquid-ring vacuum pumps. *Chemical Engineering* **102**(11):132-138.

Alther, G. 1997. Working hand in hand with carbon for better results. *Soil & Groundwater Cleanup* (5):17-19.

Amdurer, et al. 1986. *Systems to accelerate in situ stabilization of waste deposits*, US EPA Office of EPA/540/2-86/002.

American Society for Testing and Materials. 1995. *Standard guide for risk-based corrective action applied at petroleum release sites*. ASTM E-1739-95. American Society for Testing and Materials, Philadelphia, PA.

American Academy of Environmental Engineering. 1995. *Iinnovative site remediation technology. vol. 1. bioremediation*. W.C. Anderson (Ed.), WASTECH®, Annapolis, Maryland. 228 pp.

Anderson, W.C., (Ed.) 1993. *Thermal desorption*, WASTECH/American Academy of Environmental Engineers, Annapolis, Maryland.

Anderson, W.C. 1993. *Innovative site remediation technology - soil washing/soil flushing*, Amer. Academy of Environmental Engineers, Annapolis, Maryland.

Anon. 1993. Liquified Gases remove organic contaminants from soil. *Chemical Engineering* **100**(6):29.

Arniella, E.F. and L. Blythe, 1990. Hazardous wastes. *Chemical Engineering* **97**(2):93-102.

ASME. 1974. *Combustion fundamentals for waste incineration*. The American Society of Mechanical Engineers, New York.

Atlas, R.M. 1995. Bioremediation. *Chemical and Engineering News* (4):32-42.

Bader, C.D. 1997. In situ soil remediation. *Remediation Management* (Second Quarter):22-31.

Bar Ilan, A. et al. 1994. Extend the life of pollution-control catalysts. *Chemical Engineering* **101**(9):EE-22 to EE-24.

Battelle Memorial Institute. 1995. *ReOpt Version 3.1*, Battelle, PNL-7840, Rev. 3, December.

Battelle Memorial Institute. 1996. *RAAS Version 1.1*, Battelle, PNL-8751, Rev. 3, October.

Beeman, R. E., J. E. Howell, S. H. Shoemaker, E. A. Salazar, and J. R. Buttram. 1993. A field evaluation of in situ microbial reductive dehalogenation by the biotransformation of chlorinated ethenes. In R.E. Hinchee, A. Leeson, L. Semprini, and S.K. Ong (Eds.). *Bioremediation of Chlorinated and Polycyclic Aromatic Compounds*. Lewis Publishers, Ann Arbor, Michigan.

Bennedsen, M.B. 1987. Vacuum VOC's from soil. *Pollution Engineering* **19**(2):66-68.

Besendorfer, C. 1996. Exert the force of hydrocyclones. *Chemical Engineering* **103**(9):108-114.

Bhattacharya, S.K. 1992. *Remediation* (Spring):199-210.

Bianchi-Mosquera, G.C., R.M. Allen-King, and D.D. Mackay. 1994. Enhanced degradation of dissolved benzene and toluene using a solid oxygen-releasing compound. *Ground Water Monitor. Rev.* **14**:120-128.

Bigda, R.J. 1996. Fenton's chemistry. *Environmental Technology* (5/6):34-39.

Blundy, R.F. and P.G. Zionkowski. 1997. Passing the torch from lab to site, *Soil & Groundwater Cleanup* (7):22-24.

Bohn, H.L. 1992. Vapor extraction rates for decontaminating soils. *Pollution Engineering* **24**(3):52-56.

Bohn, H.L. 1992. Consider biofiltration for decontaminating gases. *Chemical Engineering Progress* **88**(4):34-40.

Bolton, J.R., S.R. Cater, S.R., and A. Safarzadeh-Amiri. 1992. *The Use of reduction reactions in the photodegradation of organic pollutants in waste streams*, presented April 6 at the American Chemical Society Conference, San Francisco, California.

Bonilla, J.A. 1993. Don't neglect liquid distributors. *Chemical Engineering Progress* **89**(3):47-61.

Bonner, T., et al. 1981. *Hazardous waste incineration engineering*. Noyes Data Corporation, Park Ridge, NJ.

Bouwer, E. J. 1994. Bioremediation of chlorinated solvents using alternate electron acceptors. In Norris, R.D., R.E. Hinchee, R. Brown, P.L. McCarty, L. Semprini, J.T. Wilson, D.H. Kampbell, M. Reinhard, E.J. Bouwer, R.C. Borden, T.M. Vogel, J.M. Thomas, and C.H. Ward (Eds.), *The Handbook of Bioremediation*. Boca Raton, FL: Lewis Publishers.

Bravo, J.L. 1994. Design steam strippers for water treatment. *Chemical Engineering Progress*, **90**(12):56-63.

Brown, R. A. and R. D. Norris. 1988. *Nutrients for stimulating aerobic bacteria*. U.S. Patent 4,727,031.

Brown, R.A., R.D. Norris, and R.L. Raymond. 1984. Oxygen transport in contaminated aquifers. In *Proceedings of the Petroleum Hydrocarbon and Organic Chemicals in Groundwater: Prevention, Detection, and Restoration*, National Water Well Association, Houston, Texas.

Brubaker, G. R., M. S. Westray, D. V. Nakles, J. W. Lynch, M. Schlauch, and C. P. L. Barkan. 1994. *A guide for railroad industry use of in-situ bioremediation*. Report No. R-857. American Association of Railroads, Washington, D.C.

Brunner, C.R. 1991. *Handbook of incineration systems*, McGraw-Hill, New York.

Bryant, J.D., and J.T. Wilson. 1998. Field Demonstration of in situ Fenton's reagent destruction of DNAPLs. *Environmental Technology* (5/6):55-59.

Buck, F.A.M. and C.W. Hauck. 1992. Vapor extraction and catalytic oxidation of chlorinated VOC, *Proceedings of 11th Annual Incineration Conference* (sponsored by University of California), Albuquerque, New Mexico, May.

Buscheck, T.E., K.T. O'Reilly, and S.N. Nelson. 1993. Evaluation of intrinsic bioremediation at field sites. *Proceedings of the Conference on Petroleum Hydrocarbons and Organic Chemicals in Ground Water*. Houston, TX: National Ground Water Association. November. p. 15.

California Department of Health Services. 1988. *Stringfellow Hazardous Waste Site Feasibility Study, Appendix A*. Toxic Substances Control Division, California Department of Health Services, Sacramento, California. June 30.

Chemical Engineering. 1963. New ratios for estimating plant costs. *Chemical Engineering* **70**(20):120-122.

Chemical Engineering. 1995. Chemical engineering plant cost index. *Chemical Engineering* **102**(10):176.

Cheremisinoff, P.N. 1992. Stabilization and fixation of hazardous wastes. *The National Environmental Journal* (5/6):36-40.

Chidgopkar, V. R. 1996. Applying Henry's law to groundwater treatment. *Pollution Engineering* **28**(3):48-49.

Chilton, C.H. 1950. Six-tenths factor applies to complete plant costs. *Chemical Engineering* **57**(4):112.

Chow, V.T., D.R. Maidment, L.W. and Mays. 1988. *Applied hydrology*. New York, NY: McGraw-Hill.

Chowdhury, J. and K. Fouhy. 1994. Soil cleanup: The best of all possible worlds? *Chemical Engineering* **101**(2):33-37.

Chu, W. and H. Windawi. 1996. Control VOCs via catalytic oxidation. *Chemical Engineering Progress* **92**(3):37-43.

Cichon, E., P. Mantovani, and T. McKeon. 1996. Rising bubbles lower costs. *Soil and Groundwater Cleanup* (11):40-43.

Clark, D.G. and R.W. Sylvester. 1996. Ensure process vent collection system safety. *Chemical Engineering Progress* **92**(1):65-77.

Clarke, A.N., D.J. Wilson, and R.D. Norris. 1996. Using models for improving in-situ cleanup of groundwater. *Environmental Technology* (8/9):34 and 36.

Clement, T.P., Y. Sun, B.S. Hooker, and J.N. Petersen. 1998. Modeling multispecies reactive transport in ground water. *Ground Water Monitoring & Remediation* **18**(2):79-92.

Connolly, M.B. Gibbs, and B. Keet. 1995. Bioslurping applied to a gasoline and diesel spill in fractured rock. In R.E. Hinchee, J. A. Kittle, and H. J. Reisinger (Eds.). *Applied Bioremediation of Petroleum Hydrocarbons*. Columbus, OH: Battelle Press. p. 371- 377.

Cookson, J.T. 1995. *Bioremediation engineering: design and application*. New York, NY: McGraw-Hill. p. 247-262.

Cooper, D.W. 1976. Significant relationships concerning exponential transmission or penetration. *J. Air Pollution Control Association* **26**(4):366-367.

Copps, R.W. 1995. Select the optimum pipe size. *Chemical Engineering* **98**(7):128-132.

Cressman, K.R. 1991. Cost estimating of the closure/post-closure phase. *Remediation* Summer:331-339.

Crittenden, J.C. et al. 1991. Predicting GAC performance with rapid small-scale column tests. *J. Amer. Water Works Assoc.* **83**(1):77-87.

Croom, M.L. 1996. New developments in filter dust collection. *Chemical Engineering* **103**(2):80-84.

Cudahy, J.J., and W. L. Troxler. 1992. 1991 Thermal remediation industry survey. *J. Air & Waste Management Assoc.* **42**:844.

Davis J. and S. Madsen. 1991. The biodegradation of methylene chloride in soils. *Env. Tox. Chem.* **10**:463-474.

Davis, K.J. and D.J. Russell. 1993. Soil treatment technologies combined. *Pollution Engineering* (7):54-58.

DeGesaro, R., J. Teringo, and B. McIlhenny. 1993. Treatment of hazardous waste incineration wastewater. *Remediation* Summer:301-308.

Desai, M.B. 1981. Preliminary cost estimating of process plants. *Chemical Engineering* **88**(7):65-70.

Dilzell, J. 1996. Control fouling in groundwater pump-and-treat systems. *Environmental Engineering World* (9/10):10-12.

Dolenc, J.W. 1996. Choose the right flow meter. *Chemical Engineering Progress* **92**(1):22-32.

Domenico, P.A. 1987. An analytical model for multidimensional transport of decaying contaminant species. *J. Hydrology* **91**:49-58.

Dove, D., A. Bhandari, and J. Novak. 1992/1993. Soil washing: practical considerations and pitfalls. *Remediation* Winter:55-67.

Downey, D.C. and J.F. Hall. 1994. *Addendum one to test plan and technical protocol for field treatability test for bioventing - using soil gas surveys to determine bioventing feasibility and natural attenuation potential*. Report to U.S. Air Force Center for Environmental Excellence, Brooks AFB, TX.

Dupont, R.R. 1986. Evaluation of air emission release rate model predictions of hazardous organics from land treatment facilities. *Environmental Progress* **5**(3):197-206.

Dupont, R.R. 1991. *Application of treatability studies in management of fuels/petroleum waste impacted soils*. 84th Annual Meeting of the Air and Waste Management Association, Vancouver, BC. June 16-21. Preprint # 91-17.5.

Dupont, R.R., W.J. Doucette, and R.E. Hinchee. 1991. Assessment of in situ bioremediation potential and the application of bioventing at a fuels-contaminated site, In R.E. Hinchee and R.F. Olfenbuttel (Eds.), *In Situ Bioreclamation: Application and Investigation for Hydrocarbons and Contaminated Site Remediation*, Butterworth-Heinemann, Boston, MA. pp 262-282.

Dupont, R.R., D.L. Sorensen, M.W. Kemblowski, K. Gorder, and G. Ashby. 1996. An intrinsic remediation assessment methodology applied at two contaminated groundwater sites at Eielson AFB, Alaska. *Proceedings of the First International Conference on Intrinsic Bioremediation*, IBC Technical Services, Ltd. London, England, March.

Dupont, R. R., C. J. Bruell, D. Downey, S. G. Huling, M. Marley, R. D. Norris, and B. Pivetz. 1998a. *Innovative site remediation technology, bioremediation: design and applications*. American Academy of Environmental Engineers, Annapolis, MD.

Dupont, R.R., D.L. Sorensen, M.W. Kemblowski, M. Ma, D. McGinnis, and M. Bertleson. 1998b. *Monitoring and assessment of in situ biocontainment of petroleum contaminated ground water plumes*. EPA/600/R-98/020. U.S. EPA National Exposure Research Laboratory, Las Vegas, NV.

Durand, A.A. 1996. A shortcut for designing evaporators. *Chemical Engineering* **89**(1):123-126.

DuTeaux, S.B. 1996. *A compendium of cost data for environmental remediation technologies*. Report LA-UR-96-2205, Los Alamos National Laboratory, August.

Eagle, M.C., W.S. Richardson, S.S. Hay, and C. Cox. 1993. Soil washing for volume reduction of radioactively contaminated soils. *Remediation* Summer:327-344.

Eckert, J.S. 1975. How tower packings behave. *Chemical Engineering* **82**(4):70.

Ely, D.L. and D.A. Heffner. 1988. *Process for in situ biodegradation of hydrocarbon contaminated soil*, Patent No. 4,765,902, U.S. Patent Office.

Ensor, D.S. and M.J. Pilat. 1971. Calculation of smoke plume opacity from particulate air pollutant properties. *J. Air Pollution Control Association* **21**(8):496-501.

Enyedy, G. 1997. How accurate is our cost estimate. *Chemical Engineering* **104**(7):106-111.

Exner, J.H. 1995. Alternatives to incineration in remediation of soil and sediments assessed. *Remediation* (Summer):15-16.

Flathman, P.E., K.A. Khan, D.M. Barnes, J.H. Caron, S. J. Whitehead, and J.S. Evans. 1991. Laboratory evaluation of hydrogen peroxide for enhanced biological treatment of petroleum hydrocarbon contaminated soil. In R.E. Hinchee and R.F. Olfenbuttel (Eds.), *In Situ Bioreclamation: Application and Investigation for Hydrocarbons and Contaminated Site Remediation*. Butterworth-Heinemann, pp 125-142.

Fletcher, D.B. 1991. Successful UV/oxidation of VOC-contaminated groundwater. *Remediation* (Summer):353-357.

Fouhy, K. and G. Ondrey. 1994. *Chemical Engineering* **101**(5):39-43.

Freeman, H.M., (Ed.) 1989. *Standard handbook of hazardous waste treatment and disposal*. New York, NY: McGraw-Hill.

Garnett, D.I. and G.S. Patience. 1993. Why do scale-up power laws work. *Chemical Engineering Progress* **89**(8):76-78; letter and authors' reply, **89**(12):10.

Garvin, J. 1998. Calculate heats of combustion for organics. *Chemical Engineering Progress* **94**(5):43-45.

Geselbracht, L.D., T.A. Donovan, and R.J. Greenwood. 1986. Realistic cost estimates for alternative remedial actions of contaminated, unsaturated soils and underlying aquifers, In *Proceedings NWWA/API Conference on Petroleum Hydrocarbons and Organic Chemicals in Ground Water: Prevention, Detection and Restoration*. Worthington, Ohio, National Water Well Association.

Gomez-Lahoz, C., J.M. Rodriguez-Maroto, and D.J. Wilson. 1991. Soil cleanup by in situ aeration, VI, effects of variable permeabilities. *Separation Science and Technology* **26**(2):133-163.

Gorder,K.A., R.R. Dupont, D.L. Sorensen, M.W. Kemblowski, and J.E. McLean. 1996. Application of a simple groundwater model to assess the potential for intrinsic remediation of contaminated groundwater. *Proceedings of the First International Conference on Intrinsic Bioremediation*, IBC Technical Services, Ltd. London, England, March.

Gorman, P.G. 1989. Guide for incinerator trial burns. *In Standard Handbook of Hazardous Waste Treatment and Disposal*. New York, NY:McGraw-Hill.

Graham, J.R. 1992. *Carbon applications for groundwater cleanup*. Arizona Water and Air Pollution Control Association 65th Annual Conference, Mesa, AZ, May 6-8.

Graves, D. and M. Leavitt. 1991. Petroleum biodegradation in soil: the effect of direct application of surfactants. *Remediation* (Spring):147-166.

Haarhoff, J. and J.L. Cleasby. 1990. Evaluation of air stripping for the removal of organic drinking-water contaminants. *Water South Africa* **16**(1):13-22.

Hairston, D.W. 1996. Acid neutralizers calm the wastewaters. *Chemical Engineering* **103**(12):57-60.

Hand, W.E. 1958. From flow sheet to cost estimate. *Petroleum Refiner* (9):331-334.

Hansen, M.A., M.M. Gates, and S.P. Sittler. 1998. Using high-vacuum technology. *Environmental Technology* (March/April):16-21.

Haroutunian, V. 1995. For better combustion, try computer simulation. *Environmental Engineering World* (July-August):26-31.

Haseltine, D.M. 1996. Improve your capital cost estimating. *Chemical Engineering Progress* **92**(6):26-32.

Hay, G.H., and G.J. McCartney. 1991. Mobile infrared incineration of PCB-contaminated soils. *Remediation* (Spring):211-226.

The Hazardous Waste Consultant. 1995. Elsevier Science Inc. (January/February):1.12-1.19.

The Hazardous Waste Consultant. 1992. Elsevier Science Inc. (January/February):1.16-1.20.

Helling, R.K. and M.A. DesJardin. 1994. Get the best performance from structured packing. *Chemical Engineering Progress* **90**(10):62ff.

Hinchee, R.E., D.C. Downey, R.R. Dupont, P. Aggarwal, and R.N. Miller. 1991. Enhancing biodegradation of petroleum hydrocarbons through soil venting. *J. Haz. Mat.* **27**:315-325.

Hinchee, R. E., S.K. Ong, R.N. Miller, D.C. Downey and R. Frandt. 1992. *Test plan and technical protocol for a field treatability test for bioventing.* Prepared for the U.S. Air Force Center of Excellence, Brooks Air Force Base, Texas. 80 pp.

Hutchins, S.R. G.W. Sewell, D.A. Kovaks, and G.A. Smith. 1991. Biodegradation of aromatic hydrocarbons by aquifer microorganisms under denitrifying conditions. *Environ. Sci. Technol.* **25**(1).

Hutchins, S.R. and J.T. Wilson. 1991. Laboratory and field studies on BTEX biodegradation in a fuel-contaminated aquifer under denitrifying conditions. In R.E. Hinchee and R.F. Olfenbuttel (Eds.). *In Situ and On-Site Bioreclamation.* Stoneham, Massachusetts: Butterworth-Heinemann.

Hyman, M.H. and L. Bagaasen. 1997. Select a site cleanup technology. *Chemical Engineering Progress* **93**(8):22-43.

Hyman, M.H. 1970. Computer applications in air pollution control have advantages. *Oil & Gas Journal* (7).

Johnson, N. 1994. On-site remediation of PCB contamination using transportable incineration systems. *Remediation* (Spring):223-234.

Johnson, P.C., C.C. Stanley, M.W. Kemblowski, D.L. Byers, and J.D. Colthart. 1990. A practical approach to the design, operation, and monitoring of in situ soil-venting systems. *Ground Water Monitoring Review* (Spring). (Reprinted in EPA 500-C-B-001, HyperVentilate Users Manual. March 1992 and in US EPA 1991a as referenced below.)

Kao, C.M., and R.C. Borden. 1994. Enhanced aerobic bioremediation of a gasoline contaminated aquifer by oxygen-releasing barriers. In R.E. Hinchee, B.C. Alleman, R.E. Hoeppel, and R.N. Miller (Eds.), *Hydrocarbon Bioremediation*. Boca Raton, FL: Lewis Publishers.

Katz, M. 1997. Investiga. *Environmental Technology* (May/June):67-69.

Kavanaugh, MC. and R.R. Trusell. 1980. Design of aeration towers to strip volatile contaminants from drinking water. *J. Amer. Water Works Assoc.* **72**(12):684-692.

Keller, R.A. and J.A. Dyer. 1998. Abating halogenated VOCs. *Chemical Engineering* **105**:100-105.

Kim, I. 1996. Harnessing the green clean. *Chemical Engineering* **103**(2):39-41.

Klemas, L. and J. Bonilla. 1995. Accurately assess packed-column efficiency. *Chemical Engineering Progress* **91**(7):27-44.

Klobucar, J.M. 1995. Choose the best heat-recovery method for thermal oxidizers. *Chemical Engineering Progress* **91**(4):57-63.

Koltuniak, D.L. 1986. In situ air stripping cleans contaminated soil. *Chemical Engineering* **93**(8):30-31.

Koolik, S.I. 1992. Chromium removal from groundwater using simple physical/chemical treatment. *Remediation* (Winter):39 ff.

Koros, W.J. 1995. Membranes: learning a lesson from nature. *Chemical Engineering Progress* **91**(10):68-72.

Kucera, J. 1997. Properly apply reverse osmosis. *Chemical Engineering Progress* **93**(2):54-61.

Lamarre, M.A., F. Foster, and D.H. Lucas. 1997. A tale of three service stations. *Soil & Groundwater Cleanup* (5):12-14.

Lang, H.J. 1948. Simplified approach to preliminary cost estimates. Chemical Engineering **55**(6):112-113.

Lantz, R.M. 1991. Methods to determine the efficacy of in situ bioremediation at hazardous waste sites. *HMC--Northeast '91 Conference Proceedings*, Hazardous Materials Control Research Institute, Greenbelt, Maryland.

Lawes, B.C. 1991. Soil-induced decomposition of hydrogen peroxide. In R.E. Hinchee and R.F. Olfenbuttel, (Eds.), *In Situ Bioreclamation: Application and Investigation for Hydrocarbons and Contaminated Site Remediation.* Stoneham, Massachusetts: Butterworth-Heinemann. pp 143-156.

Leeson, A., R.E. Hinchee, D.C. Downey, C.M. Vogel, G.D. Sayles, and R.N. Miller. 1995. Statistical analyses of the Air Force bioventing initiative results. In R.E. Hinchee, R.M. Miller, and P.C. Johnson (Ed.), *In Situ Aeration: Air Sparging, Bioventing, and Related Remediation Processes.* Columbus, OH: Battelle Press. pp 223-236.

Leeson, A. and R.E. Hinchee. 1995. *Principles and practice of bioventing.* Final report to USAF Environics Directorate, Armstrong Laboratory, Tyndall AFB, FL; Headquarters, U.S. EPA, Washington, D.C. and USAF Center for Environmental Excellence, Brooks AFB, TX.

Leeson, A. and R. E. Hinchee. 1996. *Soil bioventing: principles and practice.* Boca Raton, FL: CRC Press. 272 pp.

Leeson, A., J.A. Kittle, R.E. Hinchee, R.N. Miller, P.E. Haas, and R. Hoeppel. 1995. Test plan and technical protocol for bioslurping. In R.E. Hinchee, J.A. Kittle, and H.J. Reisinger (Ed.), *Applied Bioremediation of Petroleum Hydrocarbons.* Columbus, OH: Battelle Press. pp 335-347.

Leite, O.C. 1996a. Cleaning up incineration exhaust. *Environmental Engineering World* (July-August):6-11.

Leite, O.C. 1996b. Equipment for incineration of liquid hazardous wastes. *Environmental Technology* (May/June):26-33.

Long, G.M. 1993. Clean up hydrocarbon contamination effectively. *Chemical Engineering Progress* **89**(5):58-67.

Liu, B.Y. S.P Pradhan, V.J. Srivastava, J.R. Pope, T.D. Hayes, D.G. Linz, C. Proulx, D.E. Jerger, and P.M Woodhull. 1994. An evaluation of slurry-phase bioremediation of MPG soils, In *The Seventh International IGT Symposium on Gas, Oil and Environmental Biotechnology*, Institute of Gas Technology, Colorado Springs, CO.

Lukchis, G.M. 1973. Part 1: design by mass-transfer-zone concept. *Chemical Engineering* **80**(6):111-116.

Marley, M.C. and Droste, E.X. 1995. Successfully applying sparging technologies. *Remediation* (Summer).

Marley, M.C., E.X. Droste, H.H. Hopkins, and C.J. Bruell. 1996. Use air sparging and vapor extraction to remediate subsurface organics. *Environmental Engineering World* (March-April).

McCarthy, A. and B.R. Smith. 1995. Reboiler system design: the tricks of the trade. *Chemical Engineering Progress* **91**(5):34-48.

McGinnis, D., R.R. Dupont, K. Everhart, and G. St. Laurent. 1994. Evaluation and management of field soil pile bioventing systems for the remediation of PCP contaminated surface soils. *Environmental Technology (Letters)* **15**:729-739.

McHarg, W.H. 1993 and 1994. A steady state model for aerobic biological treatment. *Chemical Engineering* **100**(12):133-134 and **101**(3):153-154.

McLaughlin, H.S. 1995. Regenerate activated carbon using organic solvents. *Chemical Engineering Progress* **91**(7):45-53.

McNulty, J.T. 1997. The many faces of ion-exchange resins. *Chemical Engineering Progress* **93**(6):94-100.

Meekins, K. 1997. Cleaning soil with steam injection. *Environmental Technology* (September/October):64-65.

Metcalf and Eddy, Inc. 1990. *Wastewater engineering, 3rd Ed.* New York, NY: McGraw-Hill.

Miller, R.N., R.E. Hinchee, and C.C. Vogel. 1991. A field-scale investigation of petroleum biodegradation in the vadose zone enhanced by soil venting at Tyndall AFB, Florida. In R.E. Hinchee and R.F. Olfenbuttel, Ed., *In Situ Bioreclamation: Applications and Investigations for Hydrocarbon and Contaminated Site Remediation.* Boston, MA: Butterworth-Heinemann. pp 283-302.

Mobil Oil Corporation. 1995. *A practical approach to the evaluation of intrinsic bioremediation of petroleum hydrocarbons in groundwater.* Prepared by the Groundwater Technology Group, Environmental Health and Safety Department and the Environmental Health Risk Assessment Group, Stonybrook Laboratories.

Monod, J. 1950. *Annals Institut Pasteur* **79**:390.

Namkung, E. and B.E. Rittman. 1987. Estimating volatile organic compound emissions from Publicly Owned Treatment Works. *J. Water Pollution Control Federation* **59**(7):670-678.

National Research Council. 1993. *In situ bioremediation. When does it work?* Washington, DC: National Academy Press.

Nehring, K.W. and S.E. Brauning. 1992. S/S process converts hazwaste for reuse. *Environmental Protection* (4).

Nelson, C.H. and R.A. Brown. 1994. Adapting ozonation for soil and groundwater cleanup. *Chemical Engineering* **101**(11):EE-18 to EE-22.

Newell, C.J. and R.K. McLeod. 1996. *Bioscreen natural attenuation decision support system, user's manual, version 1.3*. Final Report to the Technology Transfer Division, Air Force Center for Environmental Excellence, Brooks AFB, San Antonio Texas.

NFPA. 1994. *National Fire Codes NFPA 325, Guide to fire hazard properties of flammable liquids, gases, and volatile solids*. National Fire Protection Association, Quincy, MA.

Niessen, W. 1978. *Combustion and incineration processes, 1st Ed.* New York, NY: Marcel Dekker.

Noonan, D.C., W.K. Glynn, and M.M. Miller. 1993. Enhance performance of soil vapor extraction. *Chemical Engineering Progress* **89**(6):55-61.

Norris, R.D., R.E. Hinchee, R. Brown, P.L. McCarty, L. Semprini, J.T. Wilson, D.H. Kampbell, M. Reinhard, E.J. Bouwer, R.C. Borden, T.M. Vogel, J.M. Thomas, and C.H. Ward (Eds.). 1994. *The handbook of bioremediation*. Boca Raton, FL: Lewis Publishers.

Nyer, E.K. 1992. *Groundwater treatment technology*. New York, NY: Van Nostrand Rheinhold,.

Ollero, P. 1984. Program calculates venturi-scrubber efficiency. *Chemical Engineering* **91**(5):103-105.

Onda, K., H. Takeuchi, and Y. Okumoto. 1968. Mass transfer coefficients between gas and liquid phases in packed columns. *Journal of Chemical Engineering of Japan* **1**(1):56-62.

Ondrey, G. 1995. A regenerative catalytic oxidation process handles brominated hydrocarbons, too. Chemical Engineering **102**(9):17.

Ong, S.K., A. Leeson, R.E. Hinchee, J. Kittle, C.M. Vogel, G.D. Sayles, and R.E. Miller. 1994. Cold climate applications of bioventing. In R.E. Hinchee, B.C. Alleman, R.E. Hoeppel and R.E. Miller, (Eds.), *Hydrocarbon Bioremediation*, Ann Arbor, MI Lewis Publishers. pp 444-453.

Park, K.S., R.C. Sims, R.R. Dupont, W.J. Doucette, and J.E. Matthews. 1990. Fate of PAH compounds in two soil types: influence of volatilization, abiotic loss and biological activity. *Environmental Toxicology and Chemistry* **9**:187-195.

Parker, S. 1995.On the level. *Chemical Engineering* **102**(5):76-84.

Parker, J.C., R.J. Lenhard, and T. Kuppusamy. 1987. A parametric model for constitutive properties governing multiphase flow in porous media. *Water Resources Res.* **23**:618-624.

Parkinson, G. and G. Ondrey. 1966. Batteries not included. *Chemical Engineering* **73**(5):37-41.

Pennington, R.L. 1996. Options for controlling hazardous air pollutants. *Environmental Technology* (11/12):18-23.

Perry, R.H. and D.W. Green. 1984. *Perry's chemical engineers' handbook. 6th Ed.*, New York, NY:McGraw-Hill.

Peters, M.S. and K.D. Timmerhaus. 1991. *Plant design and economics for chemical engineers.* New York, NY: McGraw-Hill.

Plant, L. and M. Jeff. 1994. Hydrogen peroxide: a potent force to destroy organics in wastewater. *Chemical Engineering* **101**(9):EE-16-EE-20.

Prosen, B. J., W. M. Korreck, and J. M. Armstrong. 1992. Design and preliminary results of a full-scale bioremediation utilizing an on-site oxygen generator system. In R.E. Hinchee and R.F. Olfenbuttle, (Eds.), *In Situ Bioreclamation: Application and Investigation for Hydrocarbons and Contaminated Site Remediation.* Woburn, MA: Butterworth-Heinemann. pp 523-526.

Rast, R.R. 1998. Design tips for precipitating metals. *Chemical Engineering* **105**(4):127-134.

Rast, R.R. 1997. *Environmental remediation estimating methods.* Kingston, MA: R. S. Means Co., Inc..

Ravipaty, A. 1996. *Determination of respiration rate/soil concentration correlations of fuel contaminants in three field soils.* Thesis in partial fulfillment of an MS Degree, Civil & Environmental Engineering Department, Utah State University, Logan, Utah.

Raymond, R.L., R.A. Brown, R.D. Norris, and E.T. O'Neill. 1986. *Stimulation of biooxidation processes in subterranean formations.* U.S. Patent 4,588,506.

Raymond, R.L., V.W. Jamison, J.O. Hudson, R.E. Mitchell, and V.E. Farmer. 1978. *Field application of subsurface biodegradation of hydrocarbon in sand formation.* Project No. 307-77. American Petroleum Institute, Washington, DC. 137 pp.

Raymond, R.L. 1974. *Reclamation of hydrocarbon contaminated groundwaters.* U.S. Patent 3,846,290.

Raymond, R.L., V.W. Jamison, and J.O. Hudson. 1976. *AIChE. Symposium Series* **73**:390-404.

Reinhard, M. 1994. In situ bioremediation technologies for petroleum derived hydrocarbons based on alternative electron acceptors (other than molecular oxygen). In Norris, R.D., R.E. Hinchee, R. Brown, P.L. McCarty, L. Semprini, J.T. Wilson, D.H. Kampbell, M. Reinhard, E.J. Bouwer, R.C. Borden, T.M. Vogel, J.M. Thomas, and C.H. Ward (Eds.), *The Handbook of Bioremediation*. Boca Raton, FL: Lewis Publishers.

Remer, D.S. and L.H. Chai. 1990. Estimate costs of scaled-up process plants. *Chemical Engineering* **97**(4):138-175.

Remer, D.S., B.L. Low, and G.T. Heaps-Nelson. 1994. Air-Pollution control: estimate the cost of scaleup. *Chemical Engineering* **101**(11):EE-10-EE-16.

Renko, R. 1994. Thermal treatment puts a cap on VOCs. *Pollution Engineering* **26**(4):62-63.

Reynolds, J.P., R.R. Dupont, and L. Theodore. 1991. *Hazardous waste incineration calculations - problems and software*. New York, NY: Wiley and Sons.

Rinaldi, N.U. 1995. Best chemical reagent. *Environmental Engineering World* (March-April):19-24.

Riser, E. 1988. *Technology review - in situ/on-site biodegradation of refined oils and fuels*, N68305-6317-7115, Naval Civil Engineering Laboratory.

Robbins, L.A. 1991. Improve pressure-drop prediction with a new correlation. *Chemical Engineering Progress* **87**(5):87-91.

Roote, D.S., et al. 1997. Groundwater clean-up options. *Chemical Engineering* **104**(5):104-112.

Rozich, A.F. 1994. Improved methodology for designing and operating biological treatment systems. *Remediation* (Spring):245-250.

Sawyer, C.N., P.L. McCarty, and G.F. Parkin. 1994. *Chemistry for environmental engineering, 4th ed.* San Francisco, CA: McGraw-Hill. pp 93, 98.

Schifftner, K.C. and H.E. Hesketh. 1983. *Wet scrubbers*. Ann Arbor, MI: Ann Arbor Science. p. 41.

Schrauf, T.W., P.J. Sheehan, and L.H. Pennington. 1993/1994. Alternative method of groundwater sparging for petroleum hydrocarbon remediation. *Remediation* (Winter):93-114.

Sell, N.J. 1988. Solidifiers for hazardous waste disposal. *Pollution Engineering* **20**(8):44-49.

Semrau, K.T. 1960. Correlation of dust scrubber efficiency. *J. Air Pollution Control Association* **10**(6):200-207.

Shelley, S.A. 1997. The battle cry for coagulants and flocculants: charge! *Chemical Engineering* **104**(6):63-66.

Shelton, H.L. 1995. Estimating the lower explosive limits of waste vapors. *Environmental Engineering World* (May-June):22-25.

Shin, B. 1996. Industrial wastewater treatment: designing for seasonal flows. *Pollution Engineering* **28**(4):38-40.

Sims, R. C., J.L. Sims, R.R. Dupont, W.J. Grenney, J.E. McLean and D.L. Sorensen. 1986. *Permit guidance manual for land treatment of hazardous wastes*. US EPA Office of Solid Waste, Washington, D.C. EPA-530/SW86-032.

Sirabian, R., T. Sanford, and R. Barbour. 1994. UV peroxidation with air stripping for optimized removal of VOCs from groundwater. *Remediation* (Spring):189-205.

Sloley, A.W. and G.R. Martin. 1995. Subdue solids in towers. *Chemical Engineering Progress* **91**(1):64-73.

Smyth, D., J. Cherry, and R. Jowett. 1995. *Soil & Groundwater Cleanup* (12):36-43.

Stapps, J.J.M. 1989. *International evaluation of in situ biorestoration of contaminated soil and groundwater*, 738708006, National Institute of Public Health and Environmental Protection (RIVM).

Stenzel, M.H. 1993. Remove organics by activated carbon adsorption. *Chemical Engineering Progress* **89**(4):36-43.

Stenzel, M.H. and W.J. Merz. 1989. Use of carbon adsorption processes in groundwater treatment. *Environmental Progress* **8**(4):257-264.

Stenzel, M.H. and C.R. Perryman. 1997. Treating contaminants with SVE can simplify the remediation process. *Soil & Groundwater Cleanup* (7):25-28.

Stern, A.C. (Ed.) 1962. *Air pollution Vol. I, Chapter 7*. New York, NY: Academic Press.

Stover, E.L. and J.A. Thomas. 1992. Carbon adsorption: a primer. *The National Environmental Journal* (11/12):28-32.

Straitz, J.F. 1995. Use incineration to destroy toxic gases safely. *Environmental Engineering World* (July-August):18-23.

Stumm and Morgan. 1996. *Aquatic chemistry, 3rd Ed.* New York, NY: Wiley-Interscience.

TERA Corporation. 1982. *Atmospheric fluidized bed combustion. Vol. II*, In report to the California Energy Commission on Petroleum Coke-Fueled Cogeneration.

Theodore, L. and J. Reynolds. 1987. *Introduction to hazardous waste incineration.* New York, NY: John Wiley & Sons.

Theodore, L. 1996. Packed-tower absorbers: sizing made easy. *Environmental Engineering World* (July-August):18-19.

Thuesen, G.J. and W.J. Fabrycky. 1984. *Engineering economy.* New York, NY: Prentice-Hall International Series in Industrial and Systems Engineering.

Tillman, D.A., W.R. Seeker, D.W. Pershing, and K. DiAntonio. 1991. Developing incineration process designs and remediation projects from treatability studies. *Remediation* (Summer):251-273.

Troxler, W.L., J.J. Cudahy, R.P. Zink, J.J. Yezzi, and S.I. Rosenthal. 1992. *Treatment of nonhazardous petroleum contaminated soils by thermal desorption technologies.* Presented at 85th annual meeting of Air and Waste Management Association, Kansas City, June 21-26.

US Air Force. 1995. *Test plan and technical protocol for a field treatability test for pol free product recovery - evaluating the feasibility of traditional and bioslurping technologies.* Air Force Center for Environmental Excellence, Technology Transfer Division (AFCEE/ERT), Brooks Air Force Base, Texas. 84 pp.

US Department of Commerce. 1991. *Soil air permeability method evaluation.* prepared for EPA by Camp, Dresser & McKee, Inc., PB92-124239.

US EPA. 1969. *Control techniques for particulate air pollutants*, EPA Publication AP-51, originally published by the National Air Pollution Control Administration, of the US Department of Health, Education, and Welfare.

US EPA. 1985. *Remedial action at waste disposal sites.* Office of Solid Waste, Washington, D.C. EPA/625/6-85/006.

US EPA. 1986a. *Handbook: permit writer's guide to test burn data — hazardous waste incineration.* Office of Research and Development, Cincinnati, OH. EPA/625/6-86/012.

US EPA. 1986b. *Handbook for stabilization/solidification of hazardous waste.* Office of Solid Waste, Washington, D.C. EPA/540/2-86/001.

US EPA. 1988. *Guidance for conducting remedial investigations and feasibility studies under CERCLA. Interim final.* Office of Solid Waste and Emergency Response, Washington, D.C. EPA/540 G-89 004.

US EPA. 1989a. *Cleaning excavated soil using extraction agents: a state of the art review.* Office of Research and Development, Cincinnati, OH. EPA/600/2-89/034.

US EPA. 1989b. *Seminar publication. Corrective action: technologies and applications.* Office of Research and Development, Center for Environmental Research Information, Cincinnati, OH. EPA/625/4-89/020.

US EPA. 1989c. *Stabilization/solidification of CERCLA and RCRA wastes - physical tests, chemical testing procedures, technology and screening, and field activities.* Office of Research and Development, Cincinnati, OH. EPA/625/6-89/002.

US EPA. 1990a. *Engineering bulletin. Slurry biodegradation.* Office of Emergency and Remedial Response, Washington, D.C., and Office of Research and Development, Cincinnati, OH. EPA/540/2-90/016.

US EPA. 1990b. *Handbook on in situ treatment of hazardous waste-contaminated soils.* Office of Solid Waste and Emergency Response, Washington, D.C. EPA/540/2-90/002.

US EPA. 1990c. *Soil washing treatment.* Office of Solid Waste and Emergency Response, Washington, D.C. EPA/540/2-90/017.

US EPA. 1991a. *Air stripping of aqueous solutions.* Office of Solid Waste and Emergency Response, Washington, D.C. EPA/540/2-91/022.

US EPA. 1991b. *Guide to conducting treatablility studies under CERCLA: soil washing.* Office of Solid Waste and Emergency Response, Washington, D.C. EPA/540/2-91/020A.

US EPA. 1991c. *Guide for conducting treatability studies under CERCLA: soil vapor extraction.* Office of Solid Waste and Emergency Response, Washington, D.C. EPA 540/2-91/019A.

US EPA. 1991d. *Soil vapor extraction technology. Reference Handbook.* Office of Solid Waste and Emergency Response, Washington, D.C. EPA/540/2-91/003.

US EPA. 1992a. *Air emissions from the treatment of soils contaminated with petroleum fuels and other substances.* Office of Research and Development, Research Triangle Park, NC. EPA/600/R-92/124.

US EPA. 1992b. *Guide for conducting treatability studies under CERCLA.* Office of Solid Waste and Emergency Response, Washington, D.C. EPA/540/R-92/071a.

US EPA. 1992c. *HyperVentilate user's manual. A software guidance system created for vapor extraction applications*. Office of Solid Waste and Emergency Response (OS-420) WF, Washington, D.C. EPA 500-C-B-92-001. (Original Macintosh HyperCard stack for SVE system design and evaluation).

US EPA. 1992d. *A technology assessment of soil vapor extraction and air sparging*. Office of Research and Development, Cincinnati, OH. EPA/600/R-92/173.

US EPA. 1994. *Remediation technologies screening matrix and reference guide. 2nd Ed.* Office of Solid Waste and Emergency Response, Washington, D.C. EPA/542/B-94/013.

US EPA. 1995a. *How to evaluate alternative cleanup technologies for underground storage tank sites. A guide for Corrective Action Plan reviewers*. Office of Solid Waste and Emergency Response, Washington, D.C. EPA 510-B-95-007.

US EPA. 1995b. *In situ remediation technology status report*. Office of Solid Waste and Emergency Response, Washington, D.C. EPA542-K-94-004, April.

Van Eyk, J. 1994. Venting and bioventing for the in situ removal of petroleum from soil, In R.E. Hinchee, B.C. Alleman, R.E. Hoeppel and R.E. Miller, (Ed.), *Hydrocarbon Bioremediation*. Ann Arbor, MI: Lewis Publishers. pp 243-251.

Waisvisz, H.E. 1987. Quick sizing of restrictive orifices. *Chemical Engineering* 94(8):166.

Weir, A., Jr. 1976. Factors influencing plume opacity. *Environmental Science and Technology* 10(6):539 ff.

Wiedemeier, T.H., J.T. Wilson, D.H. Kampbell, R.N. Miller, and J.E. Hansen. 1996. *Technical protocol for implementing intrinsic remediation with long term monitoring for natural attenuation of fuel contamination dissolved in groundwater*. Air Force Center for Environmental Excellence, Brooks Air Force Base, Texas.

Wikoff, P.M. and D.S. Prescott. 1998. Chromium reduction, heavy metal precipitation with the neutral process. *Environmental Technology* (9/10):22-27.

Wilke, G. and B. Pecar. 1995. Equipment costs: don't ignore tomorrow. *Chemical Engineering* 102(8):74-75.

Williams, D. 1996. Letting bugs be bugs. *Pollution Engineering* (7):44-45.

Williams, R., Jr. 1947. Six-tenths factor aids in approximating costs. *Chemical Engineering* 54(6):102.

Wilson, D.J. 1995. *Modeling of in situ techniques for treatment of contaminated soils, soil vapor extration, sparging, and bioventing*. Lancaster, PA: Technomic Publishing Co.

Wilson, J.T., F.M. Pfeffer, J.W. Weaver, D.H. Kampbell, T.H. Wiedemeier, J.E. Hansen, and R.N. Miller. 1994. Intrinsic bioremediation of JP-4 jet fuel. In *Proceedings of the EPA Symposium on Intrinsic Bioremediation of Ground Water*. Office of Research and Development, Ada, Oklahoma. EPA/540/R-94/515. p. 189.

Windmueller, C.R. and A. Sykes. 1995. Alternative test method guides technology choice. *Pollution Engineering* (11 Supplement):14-15.

Wright, D.K., Jr. 1945. *A new friction chart for round ducts*. Amer. Soc. of Heating and Ventilation Engineers (ASHVE) Research Report No. 1280, ASHVE Transactions 51, p. 303, reprinted in *Amer. Conf. of Governmental Industrial Hygienists, Industrial Ventilation 18th Ed.*, Fig. 6-15A, p. 6-31, Lansing, Michigan, 1984.

Yaws, C.L., L. Bu, and S. Nijhawan. 1995. Determining VOC adsorption capacity. *Pollution Engineering* (2):34-37.

Yeh, G.C. 1996. Advancing the science of dissolved air flotation. *Environmental Technology* (November/December):26-30

A/W flow ratios. See air/water (A/W) flow ratios
absorbents, costs 489–490
acid gases 382–388; scrubbers 409–411
activated carbon 109–111, 343–344. See also powdered activated carbon treatment
activated sludge 217–219
adsorbent systems 97–99, 461
adsorption 72, 82–83, 343–344; three stage 119–120
adsorption breakthrough curves 89
adsorption isotherms 111–112
adsorptive capacity 113
aeration chambers 151
aerobic degradation 278
afterburners. See thermal oxidizers
air blowers 278–280
air diffusers 150–151
air flow 290–291, 299–300, 350, 351; volumetric 352–354
air pollution control 380–390
air pressure 154
air pressure drop 140–141
air sparging 152–154, 183, 335
air stripper, flow scheme 139
air stripping computations 174–175
air/water (A/W) flow ratios 140, 144–147
Air Force Center for environmental Excellence (AFCEE) 287–288
algasorb 83
aliphatic hydrocarbons 257
alkaline agents 73
alkaline precipitation 71, 73–75
alternative injection systems 230–231
analysis instrumentation 57
analyzer indicators, automatic 47
anion-exchange system 86
aqueous oxidation 198–199; costs 199–201; important points 201–202
aqueous phase adsorption 121, 124, 128–129

aqueous phase treatment 217, 311–312, 315–230
aqueous soil washing 435, 437
aquifer plugging 231–232
aquifer volume 234–235
aquifers, homogeneous 227
aromatic hydrocarbons 257
arsenic 79
asphalt 465
asphalt aggregate dryers 380
attrition mills 441, 447
automatic restart 61

backwashing 120
bacteria, absence of 110
baghouse design 406–407
batch equilibrium test 312
batch mode 284
bed-life, duration 114–117
belt filter 441, 447
bench-scale tests 40, 41, 97, 112, 198, 310–311
benzene 208, 238, 249, 352–353
bioacclimation 215–217
bioaugmentation 215–217
biochemical reactions 261
biodegradable chemical classes 205, 206
biodegradation 154
biodenitrification 220–221, 232
biofilters 160
biological processes 204–205
biomounds 270–280, 313–314, 322
bioreactors 311
bioremediation 310–311, 315; controlled variables 328; important points 327–331. See also natural attenuation
bioslurping 304, 314–315, 327
biotrickling filtration 161
bioventing 274, 285–304, 314, 324–327
blower design 175–177
blower ratings 184
blowers 154–156, 161

525

breakthrough curve 98, 113–114
BTEX concentrations 257–258
bubble caps 150
bulking agents 274, 277

calcium hydroxide. See hydrated lime
calcium removal 75–76, 77
capital costs 477, 478–485, 494–494, 500–504; aqueous oxidation 199–201; powdered activated carbon treatment 318; reverse osmosis treatment 101; rotating biological contactor systems 317; soil venting 355–356; stabilization/solidification 472
carbon adsorption 161–163; costs 361–362; important points 134–135
carbon beds, sizing 114–117
carbon monoxide emissions 389–390
catalytic electrochemical oxidation (CEO) 190
catalytic oxidizers 165–166, 345, 381
caustic soda 73
CBC. See fluidized circulating bed combustors
cement/pozzolanic solidifiers 463–465
cement/soil ratios 464
CEO. See catalytic electrochemical oxidation
CERCLA sites 38–39
cesium 82
chelated metals 83
chemical adsorption 110
chemical precipitation 70–71, 73–79, 84–85, 94–96; unit costs 104–105
chemical pretreatment 281
chemicals, costs 489–490
chlorinated solvents 165–166
chlorofluorocarbons 377
chromate 76, 78, 81; removal, unit costs 104–105
chromium 76, 78
clarifiers 79

closure 494–495
coagulation 70
collection devices, particulates 383
cometabolism 203–204
composting 274–276
computer/fax 63
computer functions 60–62, 149. See also software
computer-based control systems 62–63
computerized remediation models 103, 105
concentrated wastes 81
concept design 37, 65, 91–93, 122–125, 169–174; report 66–67
constructed wetlands 220–221, 224
construction material, wells 183
contaminant concentrations 21, 248
contaminant degradation 239–241, 249, 262
contaminant destruction efficiency 401–402
contaminant fate 241–242
contaminant loading 262–264, 267
contaminant mass 20, 239, 240; removal rate 352–354
contaminant mobility 262, 271
contaminant source, lifetime 249–250
contingency plans 41
continuous-flow testing 312
control functions, computers 60
controllers, signals from 49
controls 93, 127, 177–178; computerized 59–60
cooling towers 151–152
corrective action plans. See work plans
Corrective Measures Study 39
cost analysis 1–2
cost estimating 32–33, 39; aqueous phase treatment 315–320; biomound treatment 322; bioslurping 327; bioventing 324–327; carbon adsorption 131–134; desorption 417–421; groundwater

stripping 179–184; important points 497–499; metals removal 99–106; slurry-phase reactors 322–324; soil venting 355–360; soil washing 456–458; solid phase treatment 320–327; stabilization and solidification 470–475; thermal soil treatment 415–417
cuprous ion 95

DAF. See dissolved-air flotation (DAF)
daily report 63
data manipulation 60
data quality objectives 39
denitrification reactors 224–226
design basis 63–64
design costs 484–485
desorption 370–373; costs 417–421
destruction efficiency 165, 367
dewatering 438
dispersivity 243
dissolved ions 71, 79–81
dissolved mass 239
dissolved-air flotation (DAF) 217–218
drilling, wells 183, 326
dry cyclones 387
dual pump/drawdown 306
dynamic test data 97

economic analysis 122. See also cost analysis; cost estimating
EDR. See electrodialysis polarity reversal
electric-driven process pumps 94
electrochemical oxidation 190–191
electrodialysis polarity reversal (EDR) 71, 80–81
electron acceptors 205, 207, 227; alternative 229–230
electrostatic precipitator 388
emission abatement 165, 341, 343, 381–382, 401–402; acid gases 382–388; costs 180–181; devices 386; particulate 382–388; steam strippers 160–161
emmissions source testing 406
empty-bed contact time 114
encapsulation 461, 462
energy balance 43–44
enzymes 203
equilibrium distribution tests 454–455
equipment 37, 455; costs 179–180, 200–201, 478, 480–482; sizing and rating 44, 84–91, 125–127, 175–177, 283
evaporation 73, 94, 99
ex situ treatment 212, 215, 217–226, 261–285
ex situ treatment/in situ treatment compared 329
exchange system, dual–bed ion 93
expenses. See operating costs; maintenance costs
exposure pathways 21
extracting fluids 429
extraction wells 347–352, 353

factored estimates 2
fail-safe designs 36–37
feasibility analysis 1
feasibility studies 37–38
Fenton's Reagent 193, 198
ferrous ion 193
field demonstrations 266
fixation 461, 470
fixed film systems 222–226
flameless thermal oxidation 161
flat-deck screen, 441, 446
flocculation 70–71
flotation cells 441, 447
flow conditions 42
flow control valve 55
flow controls 47
flow instrumentation 55–57
flow rates 115; volumetric 170–171
flow streams 42
flow-through rates 41–42
fluidized bed reactors 224

fluidized circulating bed combustors (CBC) 375–378
forced evaporation 83, 99
Freundlich equation 111–112
fuel consumption 166–169, 361, 485–488

G/L ratio 141
GAC. See granular activated carbon
gasoline tanks 253
general response options. See process options
granular activated carbon 109, 126, 138, 161
granular media filters 103
graphic displays 61
grease 117, 269
groundwater stripping, important points 184–188
groundwater treatment, costs 315–316; options 21, 22

halogenated VOCs 164
hazardous waste 217
heat exchange 166–169
heat exchangers 159; unit costs 481
heat recovery 167–168, 373
heated air 336
heating value, organic compounds 395
heavy metals, effect on catalysts 166
Henry's constant 141–143
hexavalent chromium ion. See chromate
high-molecular weight organic compounds 117
high-pressure switch 93
humidity 162
hydrated lime 73
hydraulic loading rate 114–115
hydrocarbon contamination 21, 276
hydrocarbon degradation rate 296, 299
hydrochloric acid emmissions 402
hydrocyclones 441, 447, 448

hydrogen chloride 384
hydrogen peroxide 193, 228–229, 281–282
hydrogeologic investigation 138
hydroxide precipitation 71

in situ air stripping 152–154
in situ biodenitrification 232
in situ treatment 212–214, 462
in situ treatment/ex situ treatment compared 329
in-well stripping 152–154
incineration 367–370; costs 415–417; temperature 398–399
incinerator types 368, 369, 370, 373–379
indirect-fired desorbers 381–382
infiltration galleries 226
influent concentrations 112
influent volumes 117, 119
informal studies 38
infrared furnace systems 378–379
injection wells 226, 231–232
insoluble compounds 70
installation costs 100
installation specifications 37
instrumentation 36–37; remote 50; symbols 49; temperature 53–55
instrumentation/electrical design 37
insurance 485, 490
internal combustion engines 340–341
investment costs. See capital costs
ion exchange 71–72, 81–82, 85–91, 90, 92, 97–99; unit costs 104–105
ion exchange media, regeneration 72
ion exchange regeneration waste 83
ion exchange spent regenerants, disposal 81
iron coprecipitation 75
iron removal 75–76, 77

jar tests. See laboratory tests

Kavanough's correlation 143

labor costs 484, 488–489
laboratory tests 95, 221, 265–266, 312, 454
lagoons 220
land treatment systems 261–270, 312–313; configuration 264; costs 321–322; operation 267
Langelier saturation index (LSI) 117
leachate collection 271
LEL detectors. See lower explosive limit (LEL)
level controls 46, 51
life-cycle analysis 65, 493–495
life-cycle costs 477–478
liquid distribution 156–157
liquid level instrumentation 51–52
liquid ring compressors 338, 339
liquified petroleum gas (LPG) 439–440
logic diagrams 57–59
long-term performance, modeling 266
low-concentration toxic metals 83
lower explosive limit (LEL) 120–121, 169, 337, 355, 372–373; monitoring 120–121

macronutrients 210–211
macroporous coal carbon 110
magnesium 73
magnesium aluminum oxide 83
magnesium hydroxide 73
maintenance costs 103–106, 477, 485, 489; powdered activated carbon treatment 318; reverse osmosis 100, 102; rotating biological contactors 317; stabilization/solidification 473
manganese removal 75–76
mass balance 42, 169–170, 441, 446–448, 449–452
mass flow rate tabulation 43
mathematical models 154
mechanical design 37
mediated electrochemical oxidation (MEO) 190, 191

membrane separation 161
membrane separation 71, 79–81
MEO. See mediated electrochemical oxidation
mercury 78
metal hydroxides 74
metal precipitation 74
metal sulfides 74
metals, oxidation 154
metals removal 94–99, 99–106; important points 106–107
microbial activity 272
microbial degradation process 205
microbial metabolism 205, 207
microbial populations, indigenous 207
microencapsulation 463–466
microorganism 203
microporous coal carbon 110
mist elimination 175
mist separation 154–156
mixed oxidants (MIOX) 190–191
mixing 462–463, 464–465
model calibration 246–249
models 249
modulating controllers, conventional 59
moisture content, soils 380, 392–393
moisture control 267
monitoring , 234–236, 238, 265, 267, 292, 301, 490–491; bioslurping 309–310; bioventing 301; long-term 259–260; remote 62–63
monitoring wells 183, 277
MTZ 129, 130
multiple-effect evaporation 83

naphthalene 393–394
natural attenuation 3–4, 232; assessment approach 234; decision making 256–259. See also bioremediation
negative pressure air stripper 154–156
nitrate systems 229–220
nitrates 81

nitrogen 210–211
nitrogen oxide (Nox) 384, 388–389
nonradioactive strontium 82
Nox. See nitrogen oxide
nozzles 157
nutrients 209–211, 231, 267

observational approach 39–40
off-gas treatment 282
oil 117, 269
oil-water separator 182
on/off liquid level control systems 52, 53
operating conditions 300
operating costs 477, 485, 488–489; ex situ soil remediation 421–422; groundwater stripping 180; metals removal 103–106; powdered activated carbon treatment 318; reverse osmosis 100, 102; rotating biological contactors 317; stabilization/solidification 473; UV/peroxide systems 200
operations, sequence of 46
organic adsorption systems 118
organic chemicals 205, 206
organic contaminants 437, 439–440
organic fluids, recovery 381–382
organic vapor incinerators 164–166
overhead costs 488–489
overpressure, preventing 118–119
oxidants 193–195; ranking 189
oxidizers 344–346
oxygen 207–209; absence of 110; alternative forms 228; concentration 279
oxygen enrichment 375
oxygen lancing 375
oxygen transfer 285–286; rates 207, 273, 291
oxygenation 275
ozone generators, costs 201

P&ID. See Piping and Instrumentation Diagram

packed strippers 144–149; computer applications 149; cross-sectional area 148–149; pressure drop 148–149
packed towers 222, 223; alternatives 150–154
packing depth 144–147
PACT. See powdered activated carbon treatment
particle grain load 405
particle separation 432, 435–437, 438, 455
particle size 403–404, 428, 435–437
particulate emissions 382–388, 402–406
passive remediation 3–4, 341. See also natural attenuation
payback 64–65
PCP. See pentachlorophenol
pentachlorophenol (PCP) 276, 277
performance evaluation 277, 289, 301–304
permits 63
petroleum contamination 21
petroleum-based hydrocarbons 227
pH 73, 74, 209
pH contol 93
phone dialers 62
phosphorus 211
physical adsorption 110
PICs. See products of incomplete combustion
PID control 59
pilot tests 36, 40, 41–42, 122, 128–129, 304–305, 310, 346–347, 413; soil venting 354–355; soil washing 455
pipes, sizing 47–48
Piping and Instrumentation Diagram (P&ID) 46–50, 92,–93, 123, 173, 174
plume containment 153
plume management, natural attenuation 232
plume opacity 403–406

plumes 3–4, 233–239; atmospheric 402–406; dissolved 249–250, 255–256
pneumatic testing 346–346
polishing carbon 138–140
ponds 220
pore water velocity 243
Portland cement 463
powdered activated carbon treatment (PACT) 219–220, 318
pozzolans 463
precipitates. See insoluble compounds
prefiltering 118–119
preliminary plot plan 45–46
preliminary specifications 37, 63–64, 65–66
presoaking 120
pressure controls 46, 50
pressure drop 93
pressure instrumentation 50–51
pressure- and temperature-swing adsorption 161
prestripping 118, 119
pretreatment 137–138, 262–263
pretreatment, chemical 281
process design 32–33; biomounds 276–280; bioslurper systems 304–305; carbon adsorption 122–125; groundwater stripping 169–174; land treatment systems 265–270; metals removal 91–93; natural attenuation 233; slurry reactors 282–283
process flow diagram 42–44, 123
process limitations 231–232
process monitoring 59–60
process operation 301–304, 308–310
process options 4, 5; comparing 6–9, 10
products of incomplete combustion (PICs) 380
pump motor selection 94
pump-and-treat 69, 153
pumping horsepower 94

quench water, incineration 398–399

radioactive isotopes 82
radionuclides 82
ratioing costs 478–480
Raymond process 226–232
RBC. See rotating biological contactors
RCRA facilities 39, 268
reactivated carbon 109–110, 163
reactor configurations 219–221, 223–224
reagents 435, 436
reboilers 159–160
recovery wells 226
regenerative oxidizers 168–169, 361
regression analysis 293, 294, 297, 298
regulations 63
relative costs 2, 10, 104–105, 357, 358, 359, 360
remedial action plans. See work plans
remediation methods, selecting 4–10, 28–32; flexibility 36
remediation technologies 11–20, 23–26; effectiveness 27, 28
remediation time 299
remote monitoring 62–63
residence time 97, 114, 369, 374, 378
residual phase contaminant mass 252–255
respiration rates 290, 291–295, 297, 298, 302–303
retardation factor 243
return on investment 64
reverse osmosis 71, 81, 86, 96–97, 101, 102
reverse osmosis retentate 83; disposal 81
Ringelmann numbers 403
RO. See reverse osmosis
rotary drum vacuum filters 103, 105
rotary kilns 367–369, 373–375, 379
rotating biological contactors (RBC) 223, 317
rotor concentration 161

SAFER 39–40
sampling 183, 285, 304
saturated zones 130, 231
scrubbers 384–387, 409–412; power requirements 407–409
sequence of operations 46, 57, 123
sequencing batch reactors 219
shallow soil venting system 152–153
sieve trays 150
silicate sorbents 466
simulation 305–306
site assessment 3
site characteristics 21
site models 20
site plan 44–45
slurry reactors 280–285, 314, 322–324
sodium hydroxide. See caustic soda
software 1–2, 4, 28, 85, 316, 319–320, 321
software, cost estimating 100, 103, 362–363, 416–417, 420–421, 456–458, 471, 474–475, 492–493; equipment sizing 420–421; process design 103, 105–106, 182–184, 362–363
soil air permeability 347–352
soil cores 251, 252
soil flushing 430–432, 456–457
soil piles. See biomounds
soil respiration rates 277–278, 292
soil screening 280
soil treatment, options 21
soil vapor extraction (SVE) 274, 285–286, 287; basic principles 334–337; important points 363–365; passive 341
soil vapor probes 301
soil venting wells 340
soil venting. See soil vapor extraction
soil wash water 199, 430; treatment 448, 453
soil washing 280–281; basic principles 428–430; chemical characteristics 433; contaminant removal efficiencies 427–428; equipment 443–445; important points 458–459; physical characteristics 433
soil/waste mixture 269, 313
soil water 209
solid phase biological treatment 261, 312–313
solid-liquid separation step 79
solidification, basic principles 461–462; important points 475–476
solubility constant 95
solvent extraction 432–435, 437, 439–440
sorption coefficient/retardation factor 243
source mass 249–255
sparge points 183
sparge wells 183
specific heat 396
spiral classifier 441, 446
stabilization, basic principles 461–462; important points 475–476
steady-state plume conditions 233–238
steam boilers 369
steam evaporator 94
steam strippers. See strippers, heated
step-wise regression analysis 278
stockpile locations 283
Streamlined Approach for Environmental Restoration. See SAFER
stripper emissions controls 122
strippers, heated 158–161, 162
strippers, packed air 182
strippers, recycled 157–158
stripping 283
stripping factor 143–144
strontium cations 82
strontium-90 82
submerged fixed-film reactors 223, 224
sulfate systems 230
sulfide precipitation 70, 71, 75

sulfur oxides 385
sumps 175
supercritical water oxidation 191, 192
Superfund sites. See CERCLA sites
supervisory control 61–62
supplies, costs 489–490
surface addition systems 230–231
surfactants 431
suspended growth bioreactor 218
suspended growth systems 217–222, 280
SVE. See soil vapor extraction
synthetic cation resins 82
synthetic polymetric adsorbents 161
synthetic resins 72, 81
system design, bioventing 289–300; groundwater stripping 169; metals removal 84; soil venting 346–354; soil washing 440–441; stabilization and solidification 466–467; suspended growth bioreactors 221–222; system design, thermal treatment 390; UV/peroxide systems 195, 197–198

taxes 485, 490
temperature 162, 284, 344; biological systems 211–212; control 54, 47; instrumentation 53–55; low 370; range 369
thermal desorbers 379–380, 392–395
thermal desorption 370–373, 413–414; costs 417–421; important points 422–426
thermal oxidizers 164–165, 181, 344–345, 346, 380–381
thermoplastic agents 465–466
thermosetting agents 465
thickener filter 447, 448
Thiessen polygon 235–236, 251
three-mode tuning 59
tilling frequency 267
total dissolved mass 234, 238
total petroleum hydrocarbons (TPH) 138–139, 277, 352–353
tower shell length 175
toxic metals, low-concentration 83
toxicants 212
TPH. See total petroleum hydrocarbons
tracer tests 290–291
tradeoff analysis 63–64
transmitters, signals from 49
transport models 241–242
tray designs 150–151
treatability studies 40–42, 305–306; aqueous oxidation 198–199; aqueous phase adsorption 128–131; bioremediation systems 310–311; bioslurping 295, 296; groundwater stripping 179; metals removal 94–99; soil venting 354–355; soil washing 453–456; stabilization and solidification 467, 470; thermal treatment 412–415
treated effluent concentration 94–96
treatment processes, selecting 1–2, 28–32
treatment technologies 6–9
trial burns 414–415
trickling filters 224
triethylamine (TEA) 440
trommel screen 441, 446
trough distributors 156–157
turndown 156–157

ultrafiltration 79, 80
underground storage tank 244, 257
unit costs. See relative costs
unsaturated soil treatment 22
utilities requirements 94, 127–128, 178–179, 361, 485–488
UV/hydrogen peroxide, ozone oxidation 197
UV light 189–190, 193–195
UV oxidation system vendors 196, 197

vacuum blowers 337–340, 348
vacuum-enhanced pumping 304
vadose zone 285; water addition 231
valve control, automatic 47
valves, sizing 47–48
vapor concentrations 171–172
vapor phase carbon 163–164
vapor pressure 392–395
vapor probe 309
vapor recompression 83
vapor treatment 181, 341–342
vapor-phase organic compounds 109
ventilation wells 354
Venturi scrubbers 411–412
VOCs. See volatile organic compounds
volatile organic compounds(VOCs) 122, 160–161, 163–164, 343, 344
volatilization 273
volitility rankings 182

waste application 267
waste characterization 391–392
waste components, effect on stabilization/solidification 468–469
waste disposal 261
waste hydrocarbon liquid 400–401
waste loading 262–264
Weldon Springs project 41
wells, 183, 325–326, 334–335, 336–337
wet air oxidation 191–192, 199
wood treatment facility 276
work plans 38

zeolites 72, 82